DC–DC CONVERTER TOPOLOGIES

IEEE Press
445 Hoes Lane
Piscataway, NJ 08854

DC–DC CONVERTER TOPOLOGIES

Basic to Advanced

GERRY MOSCHOPOULOS
University of Western Ontario
London, ON, Canada

IEEE PRESS

WILEY

Published by John Wiley & Sons, Inc., Hoboken, New Jersey.
Published simultaneously in Canada.

Library of Congress Cataloging-in-Publication Data applied for:

Hardback ISBN: 9781119612421

Cover Design: Wiley
Cover Images: © saginbay/Shutterstock; gremlin/Getty Images; artpartner-images/Getty Images

Set in 10/12pt TimesTen by Straive, Pondicherry, India

To My Parents

CONTENTS

ABOUT THE AUTHOR

Gerry Moschopoulos, PhD, is a professor at the University of Western Ontario, London, ON, Canada. He has published over 250 papers, mostly in various power electronic journals and conference proceedings and is an associate editor for the *IEEE Transactions on Power Electronics* and the *IEEE Journal on Emerging and Selected Topics in Power Electronics*. He has been involved in numerous technical program committees of major power electronic conferences such as the IEEE Applied Power Electronics Conference (APEC) and the Energy Conversion Congress & Expo (ECCE). He is a registered professional engineer in the province of Ontario.

PREFACE

Power electronics is the branch of electrical engineering that is concerned with the use of power semiconductor devices and passive electrical components to convert power from one form to another. Some form of power electronic conversion is needed for almost all electrical applications as it is very rare that the electrical power source used in any particular application and the load it must supply are inherently compatible. Some sort of electrical interface between sources and loads is needed in almost all applications and power electronic converters are the necessary interfaces for these applications.

Given that the power provided by a source and that fed to a load can be either AC (alternating current) or DC (direct current), power electronic converters can be classified as being one of the following four types: AC–DC converters, DC–DC converters, DC–AC converters, and AC–AC converters. The primary focus of this book is on DC–DC converters – power electronic converters that can convert the power from a DC power source into DC power that can be used to meet the requirements of a load. Such DC–DC converters can be found in renewable energy systems, electric vehicles, data servers and computer power supplies, telecommunications equipment, medical equipment, uninterruptable power supplies, electrical appliances, etc.

The design and implementation of any DC–DC converters involves the consideration of a number of aspects. Some of these include:

- The semiconductors used in the converter, such as transistors and diodes.
- The electrical circuit structure or circuit topology of the converter, which is the arrangement of the various electrical components to suit a particular application.

- The control of the converter to ensure that load requirements are satisfied.
- The design of electromagnetic elements such as inductors and transformers to ensure that they can process power and store energy, as required.
- The elimination of any electromagnetic interference (EMI) that may be caused by the high-frequency operation of any electrical switching devices.
- The application specific requirements that the converter must meet. For example, the requirements that a DC–DC converter for a solar energy system must satisfy are different from those for a DC–DC converter in an electric vehicle.

Emphasis is placed on DC–DC converter topologies in this book, with any discussion of other aspects occurring only when necessary.

Although numerous books on power electronics have been published and are readily available, almost all of these are for beginners who are new to the field or are on very specialized topics and dedicated to experts or to very senior graduate students. There are few, if any, books that are dedicated to presenting and explaining a variety of DC–DC converters, as is done in this book. This book was written to fill a gap in the power electronics literature. It can be considered to be a "second" book on power electronics by people who wish to learn about the topic of DC–DC converter topologies after learning fundamentals from a "first" book. Although some prior knowledge of power electronics is helpful, it is not absolutely necessary.

This book has 17 chapters and an Appendix and can be used as part of a single-semester university course or a two-semester course. It stems from a single-semester graduate-level course that the author has given at the University of Western Ontario for the past two decades. Several topics, such as single-switch pulse-width modulated (PWM) DC–DC converters, flyback and forward converters, active clamp and full-bridge converters, and resonant converters can be considered to be basic and are generally taught in most university courses that deal with DC–DC converters. Most of these topics can be found in the earlier chapters of this book. Other topics such as high-gain converters, three-level converters, and higher power three-phase DC–DC converters are more application specific and can be considered to be more advanced topics. Instructors wishing to use this book in their courses can use it to teach a cluster of basic topics, then supplement this teaching with topics selected from the rest of the book to suit the needs of their classes.

The following should be noted:

- The topic of DC–DC converter topologies is vast and seemingly limitless. Each year, hundreds, if not thousands, of new DC–DC topologies are presented in various power electronics publications or at various conferences and symposia. The main objective of this book is not to be a compendium of all DC–DC converter topologies but to serve as a guide for readers who are striving to learn more about the topic. Emphasis is placed on presenting and explaining fundamental principles in the field so that readers can then develop an ability to understand the operation and characteristics of a very wide variety of DC–DC converter topologies.

- The field of power electronics is undergoing a revolution at the time of this writing with the emergence of wide bandgap semiconductor devices made from materials such as silicon carbide (SiC) and gallium nitride (GaN). Semiconductor devices made with these materials have superior switching characteristics compared to those of traditional silicon (Si) devices and are expected to replace these devices sometime in the future. The transition away from Si devices to SiC and GaN devices, however, has been gradual due to several factors such as the cost and availability of the newer devices and the comfort that power electronic designers have with Si devices. They have used Si devices for decades and are much more comfortable with their reliability and characteristics than they are of those of the newer wide bandgap devices. Given that Si devices are still more widely used than wide bandgap devices at the time of this writing, the discussions presented in this book assume the use of Si devices in DC–DC converters unless specified otherwise.

The author would like to thank Dr. Mehdi Narimani for suggesting that he write this book. He would also like to express sincere appreciation to Jacqueline Le Feuvre and Adel Abosnina for assistance in the preparation of the figures of this book. He would like to thank the Chair of his department at the University of Western Ontario, Dr. Kenneth McIsaac, for his support during critical stages in the preparation of the manuscript. He would also like to thank the staff at Wiley Press for their help during the various stages of the preparation of this book: Brett Kurzman for commissioning the book, Sarah Lemore and Jayashree Saishankar for their help in managing the writing process, and Becky Cowan for her help in the design of the cover of the book. The author is especially grateful to Sundaramoorthy Balasubramani for his help in editing this book.

CHAPTER 1

BASIC CONCEPTS

This chapter will present a number of basic power electronic concepts that will be used throughout this book. It may serve as an introduction of sorts for novice readers who may be unfamiliar with power electronic concepts that are related to DC–DC switch-mode power converters with high switching frequency. It may serve as a review of fundamental concepts or be skipped altogether by more experienced readers and consulted if there is a need to clarify certain concepts that are pertinent to topics presented in later chapters.

1.1 LINEAR VOLTAGE REGULATORS

The most basic DC–DC converters are linear voltage regulators that are based on the principle of controlled voltage drops. Consider the circuit shown in Figure 1.1a, which consists of a DC source V_{in}, a variable resistor R_{adj}, and a load resistor R; the output voltage V_o is the voltage across R. Based on simple voltage division, the output voltage can be expressed as

$$V_o = V_s \frac{R}{R + R_{adj}} \tag{1.1}$$

DC–DC Converter Topologies: Basic to Advanced, First Edition. Gerry Moschopoulos.

Published 2024 by John Wiley & Sons, Inc.

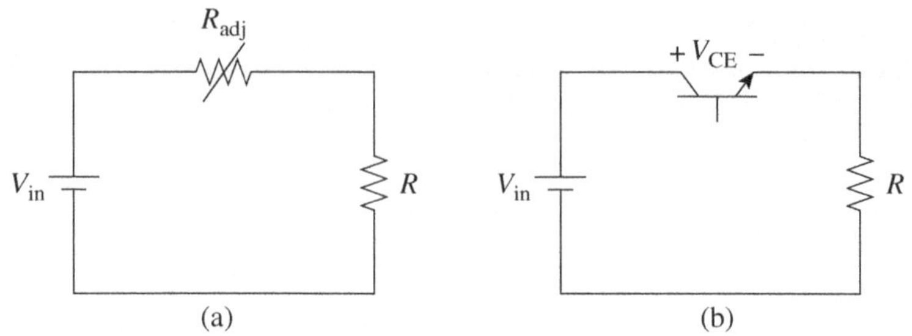

Figure 1.1 Linear voltage regulator. (a) With variable resistor. (b) With transistor.

V_o can be controlled simply by adjusting R. For example, if it is desired for V_o to be ½ V_{in}, then R_{adj} should be made equal to R. Similarly, if it is desired for V_o to be ¼ V_{in}, then R_{adj} should be equal to three times R. The output voltage can be regulated simply by adjusting the voltage drop across R_{adj} so that the output voltage becomes V_{in} minus the voltage drop across R_{adj}.

In a practical linear voltage regulator, R_{adj} is replaced with a transistor, as shown in Figure 1.1b. The output voltage is regulated by controlling the collector-emitter voltage across the transistor, V_{CE}, using some sort of feedback. In the circuit shown in Figure 1.1b, the output voltage is sensed, and then the base current of the transistor is adjusted by some feedback circuit that is not shown in the diagram so that an appropriate voltage drop that enables the output voltage to maintain the desired DC level is achieved.

Although linear voltage regulators are simple and rugged and they were widely used in the middle part of the previous century, their use in industrial applications is limited by the fact that they are inefficient. This is because a considerable amount of power must be dissipated in order to get the desired output voltage. For example, if it is desired that the output voltage is 25% of the input voltage, then the basic voltage divider circuit in Figure 1.1a requires that 75% of the source voltage be dropped across R_{adj}. This means that 75% of the power provided to the circuit by the source is dissipated in R_{adj} so that the overall converter efficiency is just 25%. Such poor efficiency is considered unacceptable in modern power converters and thus linear voltage regulators are rarely used today except in two popular applications: The first is in benchtop DC power supplies used in industrial and academic labs, as they are rugged and inexpensive. The second is for very low power, low current applications such as power the electronics of a printed circuit board (PCB). The power and current levels are so low for these applications, typically mAs of current and a couple of watts at most, that linear regulators dissipate little power.

1.2 SWITCH-MODE POWER SUPPLY FUNDAMENTALS

Most DC–DC converters today are switch-mode converters that operate with active power semiconductor devices that are used as on–off switches. Consider the circuit shown in Figure 1.2a, which consists of a voltage source, an ideal switch that is controllable, and a load resistor R. If the switch is turned on and off in a periodic manner with its on-time the same for all periods, then the output voltage across R will look like the square waveform shown in Figure 1.2b. This waveform has an average value that is dependent on the width of the on-time voltage pulse, DT, and this average value can be controlled by pulse-width modulation (PWM). D is referred to as the duty-cycle or duty-ratio in the literature and is defined as being the ratio of the on-time of the switch, t_{on}, over the switching period T as follows:

$$D = \frac{t_{on}}{t_{on} + t_{off}} = \frac{t_{on}}{T} \tag{1.2}$$

The average value of the output voltage can be expressed as

$$V_o = \frac{1}{T}\int_0^T v_o(t)\mathrm{d}t = \frac{1}{T}\int_0^{DT} V_{in}\mathrm{d}t = V_{in}D \tag{1.3}$$

1.2.1 Buck Converter

In order to achieve a DC output with the circuit shown in Figure 1.2a, the square voltage waveform shown in Figure 1.2b must be passed through a low-pass filter of some kind. This can be done with the circuit shown in Figure 1.3. This circuit is referred to as a buck converter or a step-down converter in the literature. It is essentially the same circuit as that shown in

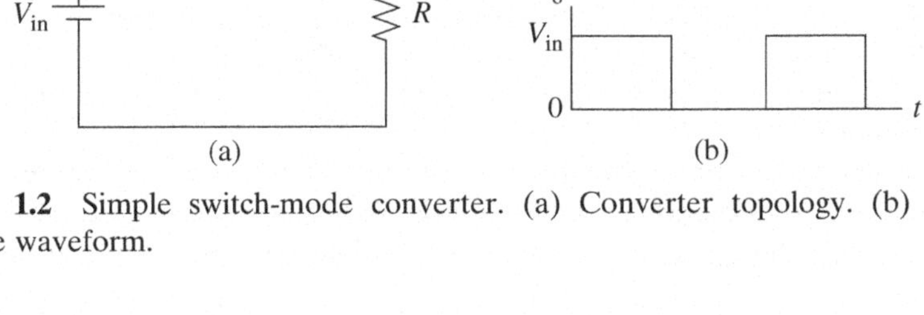

Figure 1.2 Simple switch-mode converter. (a) Converter topology. (b) Output voltage waveform.

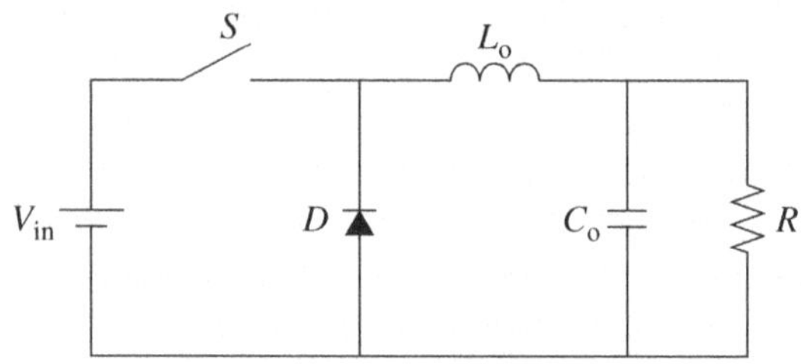

Figure 1.3 Buck converter.

Figure 1.2a, except that an output low-pass inductor/capacitor (LC) filter formed by inductor L_o and capacitor C_o has been added to smooth out the square waveform that is produced by the switch. Since an inductor has been added to the circuit, diode D must also be added to provide a path for the current in the inductor to flow through after the switch is turned off. Without this diode, turning the switch off without providing an alternate path for current to flow through will result in the catastrophic failure of the switch as the current will try to force its way through the switch. A buck converter is often called a chopper when it is used as the interface between a voltage source and the input terminals of DC motors. In this application, the input voltage V_{in} is "chopped" up into a train square waveform that is fed, unfiltered, into the terminals of the motor. As the motor has its own internal inductances, the voltage harmonics are filtered inside the machine so that only DC flows in the machine.

The buck converter operates as follows: When the switch is on, voltage V_{in} is impressed across the diode and the inductor current rises as V_{in} is greater than V_o and the voltage across the inductor is positive. When the switch is turned off afterwards, current in L flows through the diode and falls as the voltage across L_o is negative, as there is 0 V at one end and V_o at the other end. One of the two things can happen after the switch is turned off and before it is turned on again to start the next switching cycle: either the current falls to zero and remains at zero until the switch is turned on again to start the next switching cycle, or it falls but does not reach zero. The converter is said to be operating in discontinuous conduction mode or discontinuous current mode (DCM) of operation if the former occurs and in continuous conduction mode or continuous current mode (CCM) of operation if the latter occurs. CCM operation is preferred if the converter is operating with a high inductor current as this reduces peak current stresses. DCM operation is preferred if the converter is operating with a low inductor current as trying to make the inductor current continuous in this case would entail the use of a large, bulky, and heavy inductor.

The output voltage of the buck converter is the average value of the chopped voltage that appears across diode D, as expressed in Eq. (1.3). This is because only the DC component of this voltage remains after the low-pass filter. Equation (1.3), however, is only valid for the case when the converter is operating in CCM. When it is operating in DCM, the output DC voltage becomes

$$V_o = V_s \frac{2D}{D + \sqrt{D^2 + 8fL_o/R}} \tag{1.4}$$

where f is the switching frequency of the converter. It should be noted that when the converter operates in DCM, the parameters of the converter affect its voltage gain as they affect the shape of the converter waveforms and their average values. This is not the case when the converter operates in CCM.

1.2.2 Boost Converter

The boost converter, shown in Figure 1.4, is the simplest DC–DC converter that can step up voltage. It consists of the same elements as the buck converter – a switch S, a diode D, an inductor L_{in}, and a capacitor C_o, but arranged in a different manner. The converter works as follows: When the switch is on, input voltage V_{in} is impressed across L_{in} and the inductor current rises in a linear manner; when the switch is off, the current in L_{in} falls. The main physical reason why it is possible for the output voltage to be greater than the input voltage is the fact that when the switch is off, energy stored in the input inductor is added to the energy produced by the source so that it is the combined amount of energy that is delivered to the output. This is in contrast to the buck converter where only the energy that is stored in the inductor is delivered to the output when the switch is off, as the source is not connected to the inductor.

Like the buck converter, this current can fall to zero if the converter is operating in DCM or remain above zero if the converter is operating in CCM. Its output voltage can be expressed as

$$V_o = \frac{V_{in}}{1 - D} \tag{1.5}$$

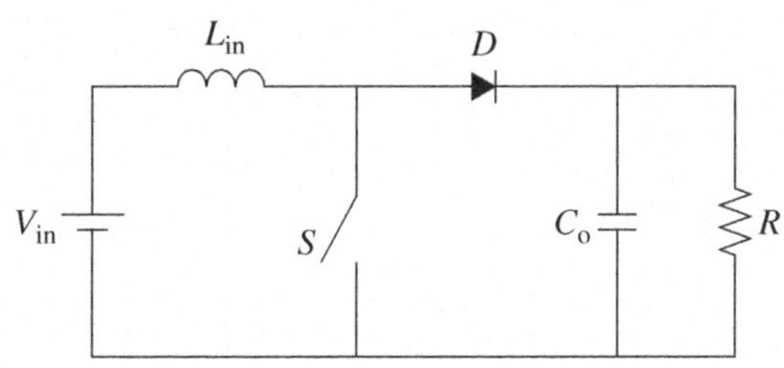

Figure 1.4 Boost converter.

if it is operating in CCM and

$$V_o = \frac{V_s}{2}\left(1 + \sqrt{1 + \frac{2D^2R}{L_{in}f}}\right) \tag{1.6}$$

if it is operating in DCM. As with the buck converter, whether the converter operates with CCM or DCM depends on parameters such as duty-cycle, switching frequency, and load. The boost converter has better input filtering than the buck converter as its input current has less ripple (AC fluctuation) on top of its DC component due to L_{in}, but worse output voltage ripple due to the presence of just a voltage-smoothing capacitor at the output and instead of both an inductor and a capacitor.

1.2.3 Buck–Boost Converter

The buck converter and the boost converter can only step up or step down voltage, respectively; therefore, they are not suitable for applications that require a converter that can operate over a very wide range of operating conditions where voltage must be stepped down under some conditions and stepped up under other conditions. The buck–boost converter, shown in Figure 1.5, is the simplest DC–DC converter that can step down and step up voltage. It has the same elements as the buck and boost converters but arranged differently.

The converter works as follows: When the switch is on, the full input voltage is placed across the inductor and the current through it rises, as is the case for a boost converter. When the switch is turned off, the inductor is disconnected from the source, as is the case for a buck converter, and the current in L falls as energy is transferred to the output, which is filtered only by a capacitor, like a boost converter. Since the converter shares some of the properties of a buck converter and some of the properties of a boost converter, the properties of the converter are a mix of the properties of the two, which is why it is called a buck–boost converter.

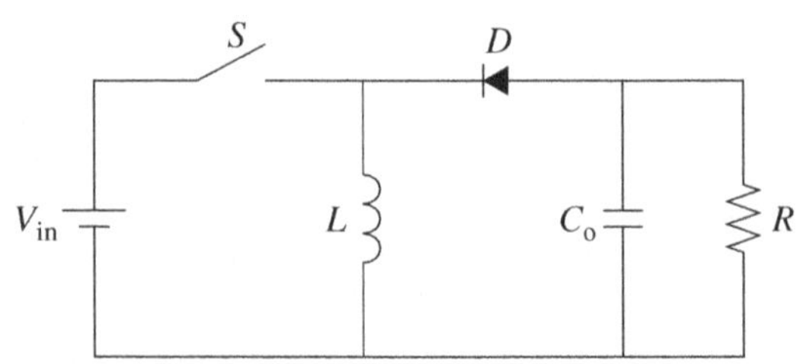

Figure 1.5 Buck–boost converter.

As with the buck and boost converters, the buck–boost converter can operate in either CCM or DCM. If the converter is operated in CCM, then its output voltage can be expressed as

$$V_o = -V_{in}\left(\frac{D}{1-D}\right) \tag{1.7}$$

If it is operated in DCM, then its output voltage can be expressed as

$$V_o = -V_{in}D\sqrt{\frac{R}{2Lf}} \tag{1.8}$$

The converter's ability to step voltage up or down can be clearly seen from the CCM voltage gain, Eq. (1.7). For example, when its duty-cycle is $D = 0.25$, then the input voltage is stepped down and the output voltage is 1/3 of the input voltage. When $D = 0.75$, then the input voltage is stepped up and the output voltage is three times the input voltage. The boundary value of D where the converter neither steps up voltage nor steps it down is $D = 0.5$. When $D = 0.5$, then the output voltage is equal to the input voltage, according to Eq. (1.7).

With the converter's ability to step voltage up or down, it would seem that there is no need for the buck and boost converters, but this would be mistaken for the following reasons:

- The buck–boost converter is less efficient than a buck converter when used in applications where only a step-down converter is needed and less efficient than a boost converter when used in applications where only a step-up converter is needed. This is because a buck converter or a boost converter can step the voltage down or up, respectively, over the full range of D while the buck–boost converter can only step voltage up or down over half the range of D. The narrowness of the duty-cycle range impacts the shape of the inductor current waveform, which in turn, impacts the losses and efficiency of the converter, compared to the buck and boost. The buck–boost converter should only be used in applications where voltage needs to be stepped down under certain operating conditions and stepped up under other operating conditions.
- The polarity of the output voltage of a buck–boost converter is different than that of the input voltage. Assuming that the polarity of the voltage fed to it is positive, the buck–boost converter will produce an output voltage that is negative, unlike the buck or the boost converters. This flipping of voltage polarity is considered to be disadvantageous as it complicates the implementation of the converter, mainly due to issues related to grounding and sensing and feedback.

- Both the input current and the output current are pulsating and not smooth, as opposed to the buck converter, which has a smooth output current, or the boost converter, which has a smooth input current. As a result, a buck–boost converter requires more filtering than a buck or a boost converter, which increases the size of the converter.

The derivation of key equations such as CCM voltage gain, CCM peak current, and the CCM/DCM boundary for the buck, boost, and buck–boost converters is presented in the Appendix of this book.

1.3 PWM CONVERTERS WITH VOLTAGE STEP-UP AND STEP-DOWN CAPABILITIES

The buck, boost, and buck–boost converters that have been presented thus far in this chapter are considered to be three of six fundamental six-switch non-isolated DC–DC converters in the power electronics literature. The other three fundamental converters are the Cuk, SEPIC, and Zeta converters. These three converters can step voltage up or down just like the buck–boost converter and have the same CCM operation voltage gain when the current in their main inductor is continuous,

$$V_o = V_{in} \frac{1}{(1 - D)} \tag{1.9}$$

In this section, the basic characteristics of the Cuk, SEPIC, and Zeta converters are presented and compared first to those of the buck–boost converter and then to those of each other.

1.3.1 Cuk Converter

The Cuk converter, shown in Figure 1.6, was introduced by Slobodan Cuk in [1]. Assuming that capacitor C_1 has been charged during the previous switching cycle so that the voltage across C_1 is positive, in accordance with

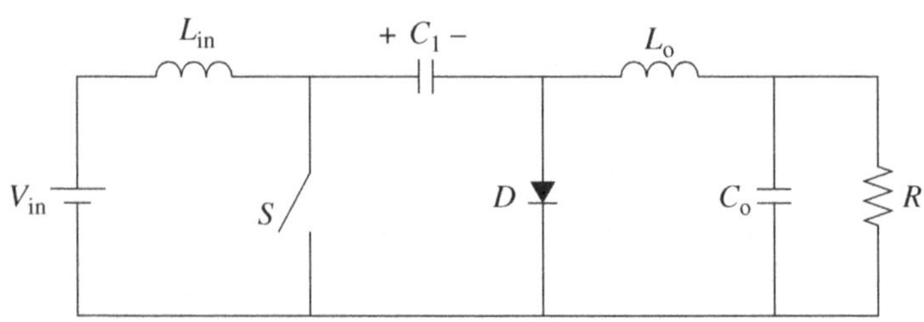

Figure 1.6 Cuk converter.

the polarity indicated in Figure 1.6, and that the current through L_o flows from right to left in Figure 1.6, the converter works as follows: When the switch is turned on, there are two currents that begin to flow through it. The first is the current that flows through L_{in}, which rises as the full input voltage is placed across it; the second is the current that flows through C_1, the C_o/R branch, and L_o, which also rises. When the switch is turned off, the current ceases to flow through the switch and begins to flow through diode D. As with the switch, there are two currents that flow through D when it conducts. The first is the current that flows through the source V_{in}, L_{in}, and C_1; the second is the current through L_o, which flows through the C_o/R branch. Both these currents fall when the switch is turned off and the current in each inductor may become discontinuous at some point during this interval.

1.3.2 Single-Ended Primary Inductance Converter (SEPIC)

A version of the SEPIC converter, shown in Figure 1.7, was first proposed in [2] by Bell AT&T Labs and was developed based on principles described in [3]. A variation of the SEPIC converter with transformer isolation was later proposed in the 1980s, as discussed in [4], but it was not until the converter shown in Figure 1.7, which is the most popular version of the converter, was proposed in [5] that power electronic researchers began to examine the characteristics of the converter closely. Assuming that capacitor C_1 has been charged so that its voltage has the polarity shown in Figure 1.7, the converter works as follows: When the switch is turned on, there are two currents that begins to flow through it. The first is the current that flows through the source V_{in} and the input inductor L_{in}, which rises; the second is the current that flows through C_1 as it discharges through the switch and through L_1. When the switch is turned off, there are two currents that flow through the diode. The first is the current that flows through V_{in}, L_{in}, C_1, which falls; the second is the current flowing through L_1, which must flow through the diode as it is the only path that is available to it.

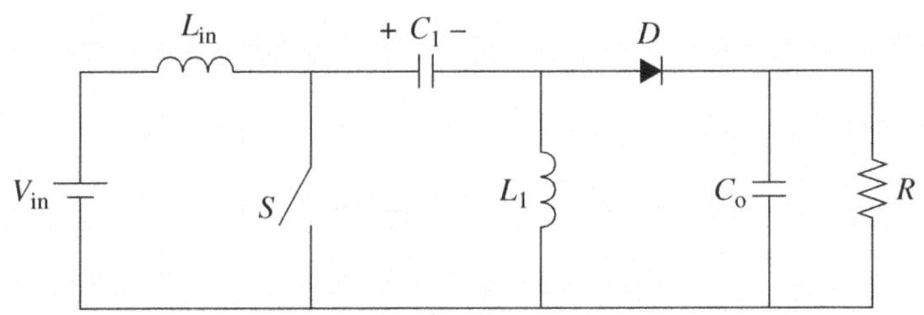

Figure 1.7 SEPIC converter.

1.3.3 Zeta Converter

The Zeta converter, shown in Figure 1.8, was first proposed in [6] and was initially called the dual SEPIC PWM converter because it is the dual of the SEPIC converter. If the SEPIC converter in Figure 1.7 has its output capacitor and load resistor replaced with a source V_{in}, its diode replaced with a switch, its switch replaced with a diode, its source replaced with a capacitor and resistor, and the resulting converter is operated with a power flow that goes from right to left instead of left to right on the diagram, then that converter is the converter shown in Figure 1.8. The converter was first referred to as the Zeta converter in [7]. It takes its name from the fact that it is the sixth fundamental single-switch DC–DC PWM converter and "Zeta" is the sixth letter of the Greek alphabet.

Assuming that capacitor C_1 is charged from a previous mode of operation, with a voltage that has a polarity that is the opposite of that shown in Figure 1.8, and that current is flowing through L_o from left to right on the diagram, then the converter works as follows: When the switch is turned on, there are two currents that flow through the switch. The first is through C_1, L_o, and the output; the second is the current flowing in L_1. When the switch is turned off, there are two currents that flow through the diode. The first is the current that flows through C_1 and L_1 as C_1 discharges through L_1; this current increases initially then decreases in a resonant manner. The second current is the current in L_o, which freewheels through diode D.

1.3.4 Comparison Between Converters with Voltage Step-Up and Step-Down Capabilities

Although the buck–boost converter is simpler and easier to implement than the Cuk, SEPIC, and Zeta converters, it lacks the filtering that these converters have. The Cuk converter has a low ripple input current and a low ripple output current, the SEPIC converter has a low ripple input current, and the Zeta converter has a low ripple output current. The buck–boost converter does not have such filtering either at the input or at the output. While its capacitor may provide some filtering at the output, this capacitor needs to be large if the output current is large, which slows down its dynamic response.

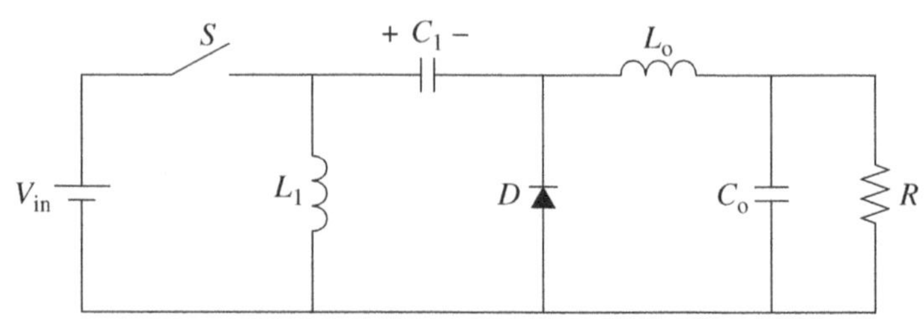

Figure 1.8 Zeta converter.

This makes the buck–boost converter unattractive for applications where the current coming out of a source or fed to a load must have a low ripple, such as battery charging, where a low current ripple is desired as it prolongs battery life.

When compared to the buck–boost converter, the Cuk, SEPIC, and Zeta converters share the following drawbacks:

- They are more expensive as they have more components. The intermediate capacitor in all these topologies is certainly a factor in this regard. It should be noted that since the buck–boost converter does not have the filtering that the other converters have, it may need to be implemented with additional filtering for a particular application, especially if input filtering is a concern. A buck–boost converter with such filtering will be more expensive, but will still be cheaper than the other converters in most cases.
- They are larger in size as they have more components. It is, however, possible to reduce their size if their two inductors are coupled together because the current waveforms in these inductors have the same shape. Coupling the inductors allows them to be implemented with a single core, instead of two separate cores, which reduces size. This, however, must be done with care as this can introduce ringing to the current waveforms in some cases, which can result in higher converter losses.
- Their switches and diodes may be subjected to more root-mean-square (RMS) current stress as they have currents from two circuit loops flowing through them instead of one. It may also cause conduction losses to increase as well, which will lower efficiency. These additional conduction losses may be offset by the reduction of turn-off switching losses if the current through the switch is smooth, as is the case for the Cuk and the SEPIC. In fact, their peak current stress may actually be less, especially for the Cuk, as they may be exposed to smoother currents. Conduction losses and switching losses in real-life switches will be discussed later in this chapter.
- The intermediate capacitor in the middle of each of these converters can cause their switch to be exposed to higher peak voltage stresses for the SEPIC and the Zeta converters. For all three converters, this capacitor can be large as the current flowing through it is pulsating.

The decision to use a Cuk, SEPIC, or Zeta will depend very much on the particular application under consideration and is not straightforward. As a result, there is very little literature that presents a direct comparison between these three converters and what few existing papers or application notes that do make such a comparison present results and conclusions that cannot be generalized or be extended to other applications. A paper design that considers factors such as stress, cost, and efficiency should be made when considering the Cuk, SEPIC, and Zeta converters for a particular application and the ultimate decision should be made based on the results of the design. The following is a list of considerations that can be taken into account for such as design:

- In applications where ripple considerations are paramount, the Cuk converter has low input and output current ripple, while the Zeta converter only has low output current ripple and the SEPIC converter only has low input ripple.
- The polarity of the output voltage of the Cuk converter is opposite that of its input voltage, as is the case with the buck–boost converter, but unlike the SEPIC and Zeta converters. The inverting output voltage polarity may complicate the implementation of the Cuk converter in terms of grounding and voltage sensing.
- The switch in the Zeta converter has no connection to the ground. This complicates the physical implementation of the converter as will be explained later in this chapter where gate drivers are discussed.
- It is much easier to design the feedback to regulate a Zeta converter than it is for the Cuk and SEPIC converters as it does not have a right-half-plane zero in its voltage gain transfer function. This allows the Zeta converter to operate with a wider loop bandwidth and better load-transient results with a smaller output capacitor. The Cuk and SEPIC converters are difficult to stabilize and are a key reason why they are not as popular as they could be as designing the feedback for these converters is a challenge to many power supply designers.
- The intermediate capacitor of the Zeta converter is larger than that of a SEPIC or a Cuk converter as the current flowing through it is more pulsating, with a higher ripple.

1.4 INTERLEAVED CONVERTERS

It is common in industry to parallel multiple DC–DC converters to reduce the size of the main inductor, especially for boost and buck converters. Such a practice, which is called interleaving, is also useful in reducing the current stress that individual converter components must handle. Figure 1.9 shows an interleaved boost converter that consists of two boost converter modules. The two boost diodes, D_1 and D_2, are connected to a common filter capacitor (C_o) and load. The switches are made to operate 180° out of phase. If more than two converters are interleaved, then the phase shift between switches is 360°/n, where n is the number of converters in the parallel configuration.

The gating signals for the two switches, the current in each inductor, and the overall input current of the two-module interleaved boost converter are

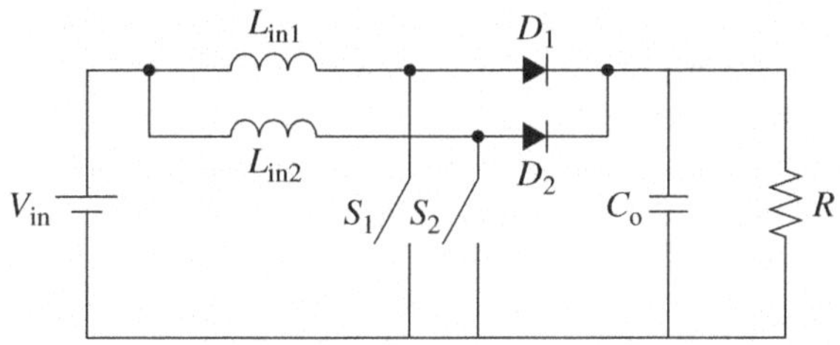

Figure 1.9 Topology of a two-module interleaved DC–DC boost converter.

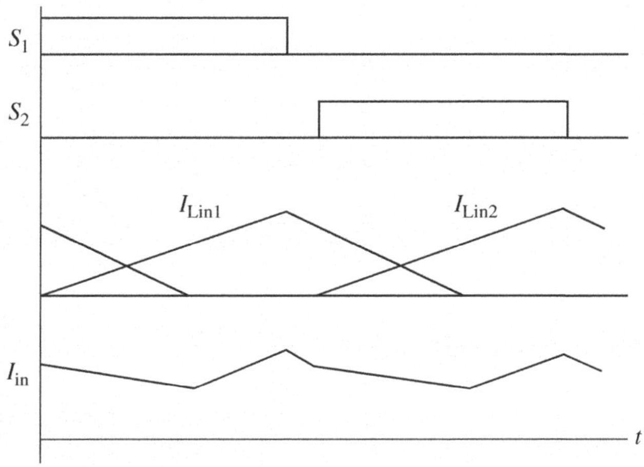

Figure 1.10 Typical gating signal and inductor current waveforms of a two-module interleaved DC–DC converter.

shown in Figure 1.10. It can be seen that the overall input current as seen by the source has a ripple that is significantly less than the ripple of each individual input inductor current and that the frequency of the overall input current waveform is doubled as well. The peak-to-peak variation in capacitor current is reduced as well, which leads to using a smaller filter capacitor in comparison with a single boost converter with the same output voltage ripple. If desired, the two input inductors, L_{in1} and L_{in2}, can be made smaller to achieve the same input current ripple as an individual boost module with a much larger inductor. The end result is that smaller inductors and a smaller output capacitor are needed so that the size and cost of these elements can be reduced significantly.

The following should be noted about interleaved converters:

- It is a common practice to reduce the number of modules that are active whenever the load is decreased. For example, an interleaved converter operating with four modules at full load may be operated with three modules at lighter loads, two modules at even lighter loads, and a single module under very light-load conditions. This practice is called phase-shedding in the literature and is done to minimize the power losses that may accumulate with the use of multiple modules. By shedding converter modules/phases as the load is decreased, switching losses, which are dominant under light-load conditions, decrease. Light-load efficiency has become increasingly important due to the proliferation of power converters in society at large. Although the light-load losses of individual converters may be insignificant, the combined light-load losses of millions of power converters are not.

- The implementation of interleaving can lead to greater converter reliability due to so-called $n+1$ redundancy. For example, a 1 kW four-module interleaved converter can be implemented with four individual modules that are each rated at 333 W. Under regular full-load operation, all four modules operate and process 250 W of power. Should one of the modules malfunction, then the other three converters can be made to operate at their full rated power until the defective converter is replaced. This increased reliability, however, leads to greater cost as such an interleaved converter with four modules is more expensive than a single 1 kW.

1.5 SEMICONDUCTOR DEVICES

The switches and diodes that are used in power electronic converters like the ones that have been described in this chapter are power semiconductor devices that are either made from silicon (Si) material or from wide bandgap material such as silicon carbide (SiC) or gallium nitride (GaN). Si devices are more widely used today, but this state of affairs is gradually changing and it is expected that they will be eventually replaced by SiC and GaN. In this section, the non-idealities of Silicon diodes and switches are discussed and a brief discussion of wide bandgap devices will follow.

1.5.1 Silicon Diodes

Silicon diodes have two types of losses: conduction losses due to their forward voltage drop and reverse recovery current losses. A small difference in voltage of around 0.7–2 V appears between the anode and cathode of a Si power diode whenever it conducts current. This forward voltage drop causes power losses that are proportional to the product of its value and the average value of the current that is flowing through it. Schottky diodes with low breakdown voltages have about half the forward voltage drop of regular Silicon diodes and thus about half of their conduction loss, but their use is limited to low voltage applications.

An ideal diode can turn off instantaneously and never allows current to flow from its cathode (K) to its anode (A), as defined in Figure 1.11a. A Si

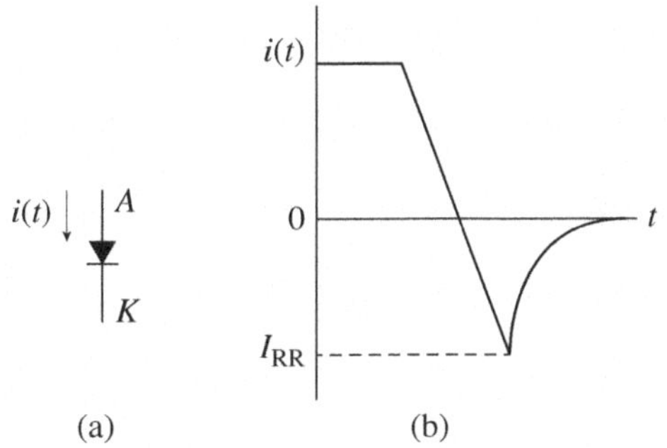

Figure 1.11 (a) Diode symbol. (b) Diode voltage and current waveforms.

diode, however, requires some time to turn off, as can be seen in Figure 1.11b where the current falls gradually. When the current in a diode becomes zero, the diode is still in the process of trying to turn off. Since the diode is not off and there is still charge in the device, current begins to flow from cathode to anode for a very short amount of time; the technical term for this is reverse-recovery current. After the diode current becomes negative, it increases in the negative direction until the diode begins to shut off so that the current is extinguished in the diode. The technical term for the length of time that a Si diode conducts current from its cathode to its anode is called reverse-recovery time. Reverse-recovery current in a Si diode can be eliminated if the transfer of current away from the diode is made to be gradual. What this does is that it allows the diode's turn-off to catch up with the decrease in current through the device so that there is no current in the device when it is turned off.

1.5.2 Silicon MOSFETs

The Si MOSFET (metal oxide field effect transistor) is the most common power semiconductor device that is used as a switch in lower-power electronic converters. It consists of three terminals as shown in its circuit symbol in Figure 1.12: gate (G), source (S), and drain (D). It can be considered to consist of an ideal switch in series with a resistance, an antiparallel diode, an output drain-source capacitance, and a gate-source capacitance as shown in the simplified model in Figure 1.12. A real MOSFET device has a gate-drain capacitance as well, but this capacitance is typically modeled as being distributed among the gate-source and drain-source capacitances.

The device is considered to be a field-effect transistor as the gate is insulated from the rest of the device and there is not a continuous current flow through it when it is on, as in the case of the bipolar junction transistor, which must be supplied with a continuous base current in order for the device to remain on. Instead, the gate-source capacitance is charged to a voltage that is greater than the threshold voltage of the device V_{TH} (typically about 4 V) and this induces a field that opens a channel between the

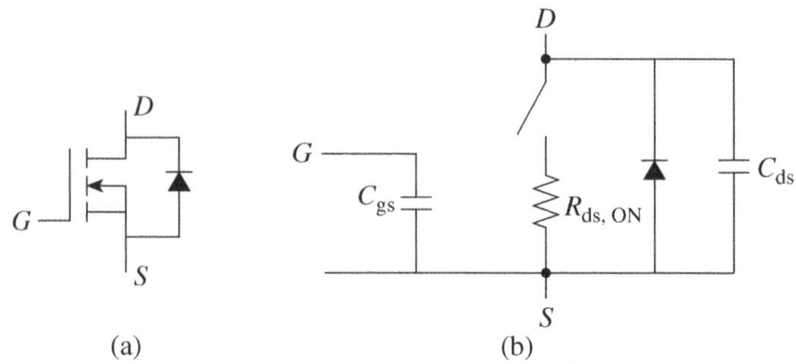

Figure 1.12 (a) MOSFET symbol (N-channel). (b) Simplified MOSFET model.

drain and source that allows current to flow. When a MOSFET is operated as an on/off switch, it is equivalent to a resistance, R_{DSon}, when it is on. Current flows from drain to source in the switch when it is on, but can flow from source to drain as well if the voltage across R_{DSon} is less than the voltage drop that would occur if it flowed through its antiparallel diode instead. This antiparallel diode is inherent to the body of the device and is often referred to as the body diode of a MOSFET in the literature. The diode can be blocked by adding another diode in series with a MOSFET, in the same device package, but this is not typically done. It is advantageous to use this diode, as will be seen in the next chapter. MOSFETs can be N-channel or P-channel devices, much like transistors, which can be NPN or PNP devices. P-channel MOSFETs are rarely used as their drain-source voltage must be low, due to properties that stem from their fabrication. They are used in low-voltage applications as this may simplify the gate drive for some of these applications.

MOSFETs are attractive in many applications where a high switching frequency converter is needed (>50 kHz) because they can be turned on and off faster than other power semiconductor devices. Although they have fast switching speeds, they are not ideal devices as there is an overlap between switch voltage and switch current during any switching transition, as shown in Figure 1.13. This overlap of voltage and current creates power losses that reduce converter efficiency – the higher the converter switching frequency, the higher the amount of power that is dissipated in the device. Of the two types of switching losses – turn-on losses and turn-off losses – turn-on losses are more dominant because the energy that is stored in the output capacitance is dissipated in the switch when it is turned on; there is an overlap between current and voltage until the capacitor is discharged completely. In the case of turn-off losses, the MOSFET device's output capacitance limits the rise in voltage across the switch after it is turned off so that the overlap between voltage and current during a turn-off transition is limited.

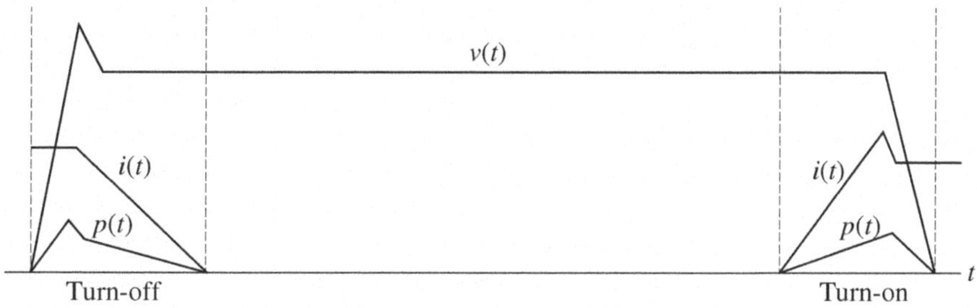

Figure 1.13 Typical non-ideal switch voltage and current waveforms.

1.5.3 Silicon IGBTs

The insulated gate bipolar transistor (IGBT) combines the best features of a MOSFET with the best features of a bipolar junction transistor (BJT). It has three terminals – a gate (G), an emitter (E), and a collector (C) – and combines the main structure of a BJT with the gate of a MOSFET. Its circuit symbol, shown in Figure 1.14a, is indicative of this combined BJT/MOSFET structure. Like a MOSFET, an IGBT has an antiparallel body diode whose anode is connected to the emitter of the device and cathode is connected to the collector. Unlike a MOSFET, IGBT devices that have a diode in series to block the body diode are commercially available. These reverse-blocking IGBTs tend to be used in higher power applications or in some zero-current switching circuits that are the subject of Chapter 3 of this book.

The main limitation of MOSFETs is that they have an R_{DSon} resistance when they conduct current. This means that the greater the current that flows in the device, the greater the voltage drop between drain and source will be and more conduction losses will be generated. The voltage drop of a BJT, however, is fixed to a saturation voltage $V_{CE,sat}$ that is almost fixed and varies little, regardless of the current that flows through its collector and emitter. When the current is low, a MOSFET will have fewer conduction losses than a BJT, but as the current is increased, there comes a point where the BJT with its fixed on-stage voltage drop has fewer conduction losses.

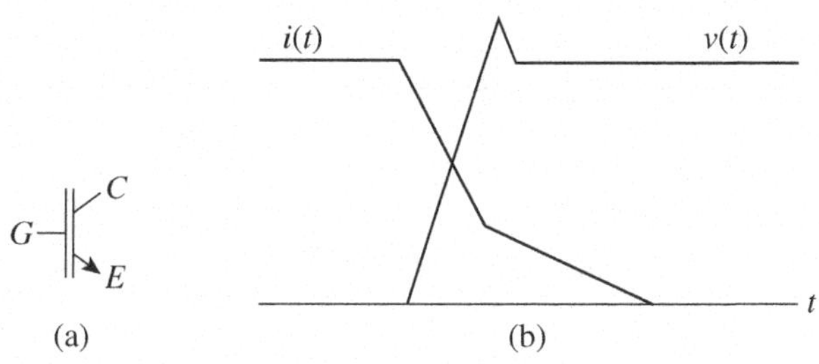

Figure 1.14 (a) IGBT symbol. (b) IGBT voltage and current waveforms.

Although IGBTs have the gate of a MOSFET, their switching speeds are lower than those of MOSFETs because they are minority carrier semiconductor devices. This means that electrons must be inserted and removed from an IGBT path after the device has been turned on or off and this slows down its switching transitions. This is especially true of the turn-off switching transition, which is slowed down considerably due to a current tail, as shown in Figure 1.14b. This current tail occurs at some point during the turn-off process and extends the overlap between switch voltage and switch current so that switching losses are increased. Although the turn-off switching losses of IGBTs are considerably higher than those of MOSFETs, they are still preferred over MOSFETs for higher-power, lower switching frequency applications (<100 kHz). Moreover, it is possible to eliminate the turn-off current tail if so-called zero-current switching methods are used, as discussed in Chapter 3.

1.5.4 Gate Drive Circuits

The signal that is used to turn MOSFETs and IGBTs on and off is generally derived from some sort of integrated circuit (IC). The output of such an IC needs to be connected across the gate and source of MOSFETs and IGBTs in order to turn them on, but this output cannot simply be fed directly to these devices. Although ICs can be used to produce voltage signals that can be fed to the gate of a MOSFET or an IGBT (gating signals) that will turn these devices on or off, the amount of current that is associated with these signals is very low. Given that these devices have a gate-source capacitance that needs to be charged to voltage levels that exceed their threshold level before they can be turned on, the amount of current that is typically outputted by an IC is too small to charge this capacitance. As a result, some sort of current amplifier circuit is needed to boost the current that is fed to the gate of these devices so that their gate-source capacitance can be quickly charged, which allows the devices to turn on quickly. Some sort of interface circuit called a gate drive circuit needs to be placed between the output of an IC and the gate of a MOSFET or an IGBT.

While gate current amplification is the main reason why gate drive circuits are needed, there is another consideration that needs to be taken into account. The output of an IC is generally referenced to ground. In converters such as the buck converter, the source of the switch is floating and not referenced to the ground. Any attempt to connect the output of an IC directly across the gate and source of a buck converter will result in a short circuit and a catastrophic failure of the converter. Some sort of gate drive circuit is needed to convert an IC output signal into a signal that can be fed across the gate and the source of a floating switch.

Numerous gate drive circuits have been proposed in the power electronic literature in various online applications notes, academic papers, or books such as [8, 9]. A review of such circuits, however, is beyond the scope of this book and for the purposes of this book, the following is all that needs to be known:

- Converters with switches whose source is connected to the ground have simpler gate drive circuit implementations than converters with floating switches. For example, when considering the use of a Cuk, SEPIC, or Zeta converter for a particular application, the fact that the Zeta converter requires a floating gate drive circuit while the other two converters do not is considered to be disadvantageous.
- Power MOSFETs and IGBTs are almost always N-channel devices rather than P-channel devices. This is because N-channel devices have higher voltage and current ratings than comparable P-channel devices. For low voltage and power applications, the floating switching of a converter such as a buck converter can be implemented with a P-channel MOSFET so that a non-floating gate drive circuit that is referenced to ground can be used. A gating signal that is referenced to ground can be fed to a P-channel device.

1.5.5 Wide Bandgap Devices

Wide bandgap devices share a common feature and that is that their physical structure has a wide bandgap. Although there are potentially numerous ways to implement wide bandgap devices, SiC and GaN devices are, by far, the most popular of these devices for power electronic applications. Given that semiconductor theory is beyond the scope of this book, there are only two characteristics that really need to be known about SiC and GaN devices for the purposes of this book.

The first is that diodes made from wide bandgap semiconductor material do not have reverse-recovery current as do Silicon diodes; thus, they have fewer losses and produce less electromagnetic interference (EMI)[1] noise than traditional Silicon diodes. This is because they are the majority carrier semiconductor devices. The second, and more important characteristic, is that the overlap between switch voltage and current that exists during a switching transition is much less than that of traditional Si devices. As a result, SiC and GaN devices have fewer turn-on and turn-off switching losses than Si devices. Power electronic designers take advantage of this characteristic by operating converters

[1] Very sharp transitions and spiking in waveforms can create very high frequency harmonics that are the source of electromagnetic interference (EMI) noise. This noise can be radiated or conducted and can affect the operation of surrounding electronic equipment if not attended to. Switching transitions by switches and diodes are a source of EMI noise. This noise can be mitigated by making the switching transitions in these devices more gradual.

with wide bandgap devices at switching frequencies that are higher than those typically used for converters with Si devices. Operating a power converter implemented with Si devices with high switching frequencies can reduce its size and weight as it reduces the size and weight of its magnetic and filtering elements, but it also reduces its efficiency as switching losses are increased. The penalty of reduced converter efficiency is mitigated with the use of SiC and GaN devices.

Although SiC and GaN devices are relatively new compared to Si devices, their use has been reported by power electronics experts almost since the start of this century. It is anticipated that SiC and GaN devices will become more widely used than traditional Si devices, whose use is expected for certain applications where very low cost is a must. At the time of this writing, Si devices are still more widely used for several reasons, including the following:

- *Availability*: SiC and GaN devices are not as widely available as Si devices. There are several reasons for this, mainly issues related to fabrication and patenting by semiconductor manufacturers and fabrication issues. Moreover, the voltage ranges for which SiC and GaN devices are available are not as wide as those for Si devices. As a result, it is much more difficult to find appropriate SiC and GaN devices for a particular application than it is to find Si devices.
- *Cost*: Since wide bandgap devices are not as widely available as Si devices, they are more expensive. Cost is expected to go down as wide bandgap devices become more available and most experts believe that this will only be a matter of time, but the cost of wide bandgap devices remains generally higher than that of Si devices at the time of this writing.
- *Reliability*: Si is a very well-established material for use in semiconductor devices. It has a long track record of success and is thus considered to be extremely reliable. As wide bandgap semiconductor technology is relatively new, it is not considered to be nearly as reliable as Si technology. Of the two wide bandgap devices, SiC devices are considered to be more reliable than GaN devices and extensive laboratory testing has shown that SiC devices may ultimately be as reliable as Si devices, but power converter designers continue to believe that Si devices are much less likely to fail than wide bandgap devices.
- *Si-based converter designs*: Si devices generally cannot be replaced by SiC and GaN devices on existing products. That is to say that one cannot take an existing commercially available power converter, pull out its Si semiconductor switches from a PCB, and then put in new SiC or GaN devices because SiC and GaN devices have different gate drive requirements and thus require different gate drive circuits. This means that wide bandgap devices can only be used in new converters and Si devices will still be required to make already existing converters work.

Most of the discussion in this book will assume the use of Si devices because of their enduring popularity. Mention will be made of wide bandgap devices only to differentiate their use in particular applications.

1.6 SNUBBERS

Power semiconductor devices are rarely installed in actual physical power converters without some sort of snubber attached to them. A snubber is a circuit that can be used to prevent overvoltage spiking and voltage ringing across an active switch or diode when the device is turned off or to prevent sudden current spikes from appearing when a switch is turned on or when a diode with a reverse-recovery current characteristic stops conducting. These phenomena are due to the presence of parasitic elements in the physical converter structure and the converter components. Snubbers can either be passive and thus have only passive elements or they can be active and contain at least one active switching device. Passive snubbers will be briefly reviewed in this section while active snubbers will be reviewed in Chapters 2 and 3.

The simplest passive snubber is the RC snubber shown in Figure 1.15a [10]. This snubber is typically used across diodes to help suppress any ringing that may be caused by the interaction of the diode's junction capacitance and any parasitic inductance that may be in the physical circuit due to copper traces of PCBs or wiring. The RC snubber has a capacitance that will decrease the amplitude and frequency of any voltage ringing and resistance that will dampen the ringing until it ceases. The energy that results from the diode/parasitic interaction is eventually dissipated in the resistor of the-snubber. The use of RC snubbers results in lower converter efficiency and the amount of this decrease depends on the stress that the component is subjected to, but converter components without a snubber may be subjected to catastrophic stress.

The RCD snubber shown in Figure 1.15b is widely used in parallel with active switches [10]. Like the RC snubber, it is used to suppress overvoltage spikes and ringing that may occur when the switch is turned off. When the switch is turned off, the energy that is trapped in parasitic inductance is

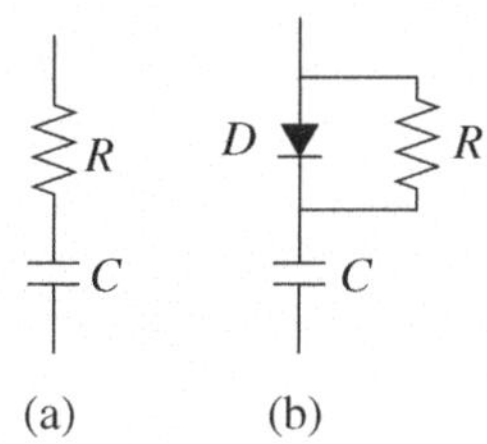

Figure 1.15 Basic passive snubbers. (a) RC snubber. (b) RCD snubber.

transferred to the snubber capacitor, which is sufficiently large to ensure that the voltage across the switch does not become excessive. The diode provides a more direct path to the snubber capacitor than a resistor would and thus it is used to bypass the snubber resistor. When the switch is turned on again in the next switching cycle, the snubber capacitor discharges through the snubber resistor, which prevents the discharge current through the switch from being excessive.

Many passive snubbers have been proposed in the literature, some simple, others much more sophisticated in their structure. Passive snubbers can be used to limit the rate of current rise in switches that have been turned on, but are more commonly used to suppress overvoltage spikes in switches after they have been turned off. The basic principle behind most of these snubbers is that energy that would otherwise contribute to switch overvoltage is transferred to a capacitor in the converter circuit that is large enough to absorb this energy. This energy is then removed from the capacitor later in the switching cycle. Most of these snubbers are non-dissipative as opposed to the RC and RCD snubbers shown in Figure 1.15 that dissipate energy that would otherwise contribute to component stress in a resistor.

An example of a more sophisticated snubber than the RC and RCD snubbers that is non-dissipative and that limits excessive voltage and excessive current in converter components is the one that was proposed in [11] and analyzed in [12]; it is shown in Figure 1.16. This snubber is implemented in a boost converter and consists of an inductor, L_r; two diodes, D_2 and D_3; and a capacitor C_r. The purpose of L_r is to make the transfer of current away from the main boost diode D_1 to the switch gradual so that reverse-recovery current is eliminated, as described in Section 1.5.1. When the switch is turned off, the current is forced to flow through D_2 and into C_r so that any overvoltage spike is clamped. Eventually, the current in L_r reverses direction and starts to discharge C_r through D_3. Current begins to flow through the main boost diode D_1 after C_r has been fully discharged.

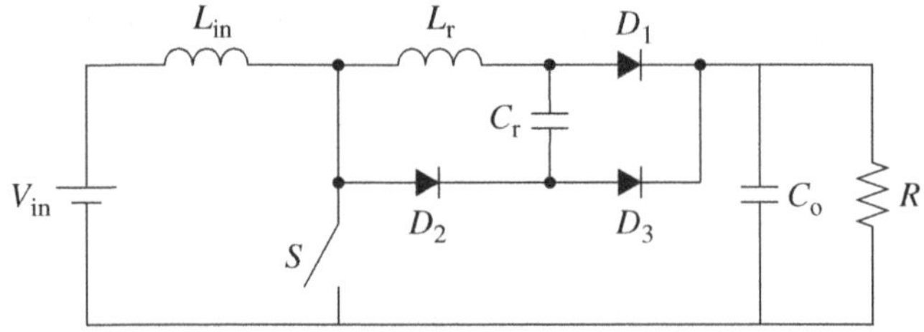

Figure 1.16 Example of a non-dissipative passive snubber for a boost converter switch.

It should be noted that diodes D_2 and D_3 are not the same devices as D_1. They conduct current for only a small portion of the switching cycle and have lower current ratings. They do not have the type of reverse-recovery characteristic that the main boost diode D_1 has.

1.7 CONCLUSION

This chapter presented several basic concepts that will be used throughout this book. These basic concepts include the structure and operation of the most basic DC–DC converter topologies such as the buck, boost, and buck–boost converters, the principles of interleaving converters, fundamentals about the semiconductors that are typically used in DC–DC converters; switching losses; and snubbers. Readers who encounter these concepts in later chapters of this book can return to this chapter should they need to be reminded of any basic concept that has been discussed in this chapter.

REFERENCES

1. S. Cuk and R. D. Middlebrook, "A new optimum topology switching DC-to-DC converter," *IEEE Power Electronics Specialist Conference (PESC)*, 1977, June 14–16, 1997, Palo Alto, USA, pp. 160–179.
2. R. P. Massey and E. C. Snyder, "High voltage single-ended DC-DC converter," *IEEE Power Electronics Specialist Conference (PESC)*, 1977, June 14–16, 1997, Palo Alto, USA, pp. 156–159.
3. P. W. Clarke, "Converter regulation by controlled conduction overlap," U. S. Patent 3,938,024, February 10, 1976.
4. C. D. Davidson, "Zero Voltage Switching isolated boost converter topology," *IEEE International Telecommunications Energy Conference (INTELEC)*, 2011, October 9–13, Amsterdam, Netherlands, pp. 1–8.
5. M. K. Kazimierczuk and J. J. Jozwik, "Optimal topologies of resonant DC/DC converters," *IEEE Transactions on Aerospace and Electronic Systems*, vol. 25, no. 3, pp. 363–372, May 1989.
6. J. J. Jozwik and M. K. Kazimierczuk, "Dual SEPIC PWM switching-mode DC/DC power converter," *IEEE Transactions on Industrial Electronics*, vol. 36, no. 1, pp. 64–70, Feb. 1989.
7. A. Peres, D. C. Martins, and I. Barbi, "ZETA converter applied in power factor correction," *IEEE Power Electronics Specialist Conference (PESC)*, 1994, June 20–25, 1994, Taipei, Taiwan, pp. 1152–1157.
8. D. A. Grant and J. Gowar, "Power MOSFETs: Theory and Applications," John Wiley & Sons, New York, 1989.

9. Z. Zhang and Y. F. Liu, "High Frequency MOSFET Gate Drivers: Technologies and Applications," The Institution of Engineering and Technology, London, 2017.
10. C. P. Todd, "Snubber circuits: theory, design and application," *Unitrode-Power Supply Design Seminar*, 1993
11. I. D. Jitaru, "Soft transitions power factor correction circuit," *Proceedings of High Frequency Power Conversion (HFPC) Conference*, 1993, May 23–27, 1993, Vienna, Virginia, pp. 202–208.
12. H. Levy, I. Zafrany, G. Ivensky, and S. Ben-Yaakov, "Analysis and evaluation of a lossless turn-on snubber," *IEEE Applied Power Electronics Conference (APEC)*, 1997, February 23–27, 1997, Atlanta, USA, pp. 757–763.

CHAPTER 2

NON-ISOLATED ZERO-VOLTAGE SWITCHING PWM CONVERTERS

Modern power electronics requires that power converters be small and compact in many applications. In order to reduce converter size, the power switches in a converter are operated with high switching frequencies (>50 kHz) to reduce the size of its magnetic and filtering components. Operating at such high switching frequencies limits the amount of energy that must be stored in these elements and it is this lessening of energy storage that allows their size to be reduced. The reduction of component size, however, comes at a price and that is the increase of switching losses and the accompanying drop in converter efficiency. As was discussed in Section 1.5, time is needed for a switch to make the transition from being fully on to fully off and vice versa and thus an overlap between the current through a switch and the voltage across it is created. Since power is related to the product of voltage and current, this overlap of voltage and current constitutes the main source of switching power losses of a switch device. The higher the converter switching frequency, the more common these voltage/current overlaps become and thus the greater the switching losses. As a result, converters operating with high switching frequencies are often implemented with some sort of soft-switching technique in order to reduce switching losses. This is especially true when silicon devices are used, which have slower switching transition times than wide bandgap devices made of SiC or GaN material.

The term "soft-switching" in power electronics refers to methods in which switching transitions are performed gradually (soft) as opposed to suddenly

DC–DC Converter Topologies: Basic to Advanced, First Edition. Gerry Moschopoulos.

Published 2024 by John Wiley & Sons, Inc.

(hard). A converter is said to operate with hard-switching if it operates naturally, with passive snubbers, as described in Section 1.6. With soft-switching, the voltage across or the current through a switching device can be forced to zero during a switching transition, thus achieving a significant reduction in switching losses as the overlap between voltage and current is diminished. Additional benefits of soft-switching include reduced electromagnetic interference (EMI) noise, as the gradual switching transitions result in lower levels of ultrahigh frequency harmonics and greater reliability of the switching devices themselves on account of their being less stressed.

Soft-switching methods involve making either the switch voltage or the switch current zero during a switching transition and thus they can be categorized as being either ZVS (zero voltage switching) or ZCS (zero current switching) methods. ZVS methods for MOSFET devices are the focus of this chapter and ZCS methods for IGBTs are the focus of the next one. ZVS methods are preferred when MOSFETs are used because they can take advantage of some of the inherent features of MOSFET devices.

ZVS methods for non-isolated, single-switch DC–DC converters that are operated with PWM control, such as the ones presented in Chapter 1, will be reviewed in this chapter. The chapter will begin with a brief discussion of ZVS and MOSFETs, then present several example converters of various types, and end with a brief discussion on several miscellaneous considerations when implementing ZVS-PWM in certain topologies and with wide bandgap devices. It should be noted that most of the discussion in this chapter will be related to ZVS with silicon MOSFET devices. The use of ZVS methods with wide bandgap devices will be briefly discussed toward the end of the chapter.

2.1 BASIC ZVS PRINCIPLES FOR MOSFETS

A real MOSFET device has an anti-parallel diode built into it as well as an output drain-source capacitance. This device can be approximated according to the model shown in Figure 2.1. If current is somehow injected through the anti-parallel diode by the operation of the rest of the converter before the switch is to be turned on, then the voltage across the switch will be clamped to a level that is equivalent to the forward voltage drop of the diode – a negligible level compared to the voltage typically impressed across a switching device in a power converter during its off-state. Consequently, there will be little if any overlap of voltage and current when the switch is turned on and thus there will be almost no turn-on switching losses.

The ZVS turn-off of a switch can be achieved with the use of the capacitor across the switch. When the switch is turned off, this capacitor limits the rate of voltage rise so that the voltage across the switch rises gradually instead of suddenly. As a result, the overlap between switch voltage and switch current during the turn-off switching transition time is reduced considerably and thus so too are turn-off switching losses. It should be noted that the anti-parallel

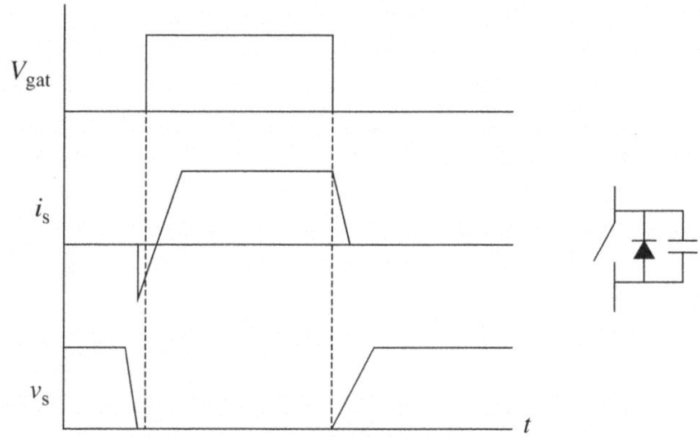

Figure 2.1 Typical switch waveforms for ZVS operation with simple switch model.

diode used for ZVS turn-on transitions and the capacitor used for ZVS turn-off transitions may be internal to the device – this is especially true of MOSFET devices. Depending on the specifications of the converter and the amount of voltage and current its switches must handle, a switching device can be operated with ZVS as is or external components such as an anti-parallel diode with better performance characteristics than the device's internal body diode or additional capacitance may be added across the device.

Figure 2.1 also shows example switch voltage, current, and gating signal waveforms of a converter that operates with ZVS. It can be seen that the switch current waveform becomes negative before the voltage waveform drops to zero. This is an indication that the converter's ZVS operation has forced the voltage across the switch to drop to zero by injecting current through the device's body diode (from source to drain) after the output capacitor of the switch has been discharged. It is while current is flowing through the body diode that the switch can be turned on with ZVS. Current eventually changes direction and flows from drain to source, some time after the switch has been turned on.

The turn-off of a switch in a ZVS converter is related to the size of its output capacitance. If this capacitance is large, then the rate of voltage rise across the switch is slowed down, which reduces the overlap between switch voltage and current during a turn-off switch transition, thus lowering turn-off switching losses. If this capacitance is small, then the rate of voltage rise across the switch is fast and significant switch voltage and current overlap exists, as do turn-off losses. The worst-case for turn-off losses is when there is no capacitance across a switching device.

A MOSFET's inherent switch output capacitance can act as a mechanism to slow down voltage rise whenever a switch is turned off. In the design of a ZVS converter, a key question is how large should this capacitance be. If the capacitance is large, turn-off losses may be decreased, but more energy may be needed to discharge this capacitance to ensure ZVS operation. This will result in either restricting the load range over which the converter can operate with ZVS or injecting additional circulating current through the body diode of the device, which will increase conduction losses and may act against any potential savings in turn-off losses as the switch will have to turn off more current when the current flowing through it eventually reverses direction and flows from drain to source. In the end, a compromise must be considered between turn-on losses and turn-off losses when designing a ZVS converter to maximize efficiency.

2.2 ZVS-PWM QUASI-SQUARE-WAVE DC–DC CONVERTERS

To summarize the discussion in this chapter thus far, a ZVS turn-on is ensured by injecting current through an anti-parallel diode before a switch is turned on, while a ZVS turn-off is ensured by a paralleled capacitor slowing down the rate of voltage rise across the switch. In terms of circuit implementation, what this means is that it is the ZVS turn-on mechanism that needs to be emphasized as the turn-off mechanism is inherent. In order for a ZVS turn-on to occur, the operation of the converter circuit must be such that current is injected in the body-diode of the device before it is to be turned on.

One of the first ZVS-PWM DC–DC converters to be proposed was the converter shown in Figure 2.2, which was proposed in [1]. This converter was referred to as a ZVS quasi-square-wave converter to differentiate it from ZVS quasi-resonant converters (discussed in Chapter 12) that were popular at the time. This converter is essentially a buck converter, but with a second active switch S_2 instead of a diode. The switches are shown with the switch model of Figure 2.1. The converter's modes of operation over a switching period are shown in Figure 2.3. The converter works as follows: When switch S_1 is on as shown in Figure 2.3a, current flows through the output inductor L_o in a left-to-right direction. When the switch is turned off, as shown in Figure 2.3b, current discharges the switch capacitance of S_2, C_{s2}, and charges the switch capacitance of S_1, C_{s1}. This is a ZVS turn-off as C_{s1} and C_{s2} slow down the rise of voltage across S_1, thus reducing the overlap between voltage

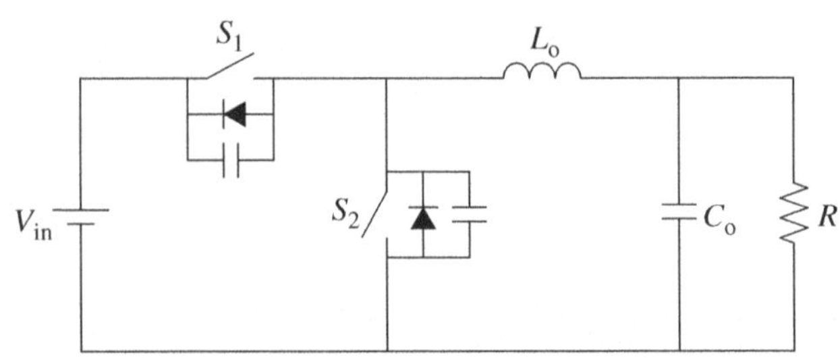

Figure 2.2 Quasi-square-wave ZVS-PWM converter.

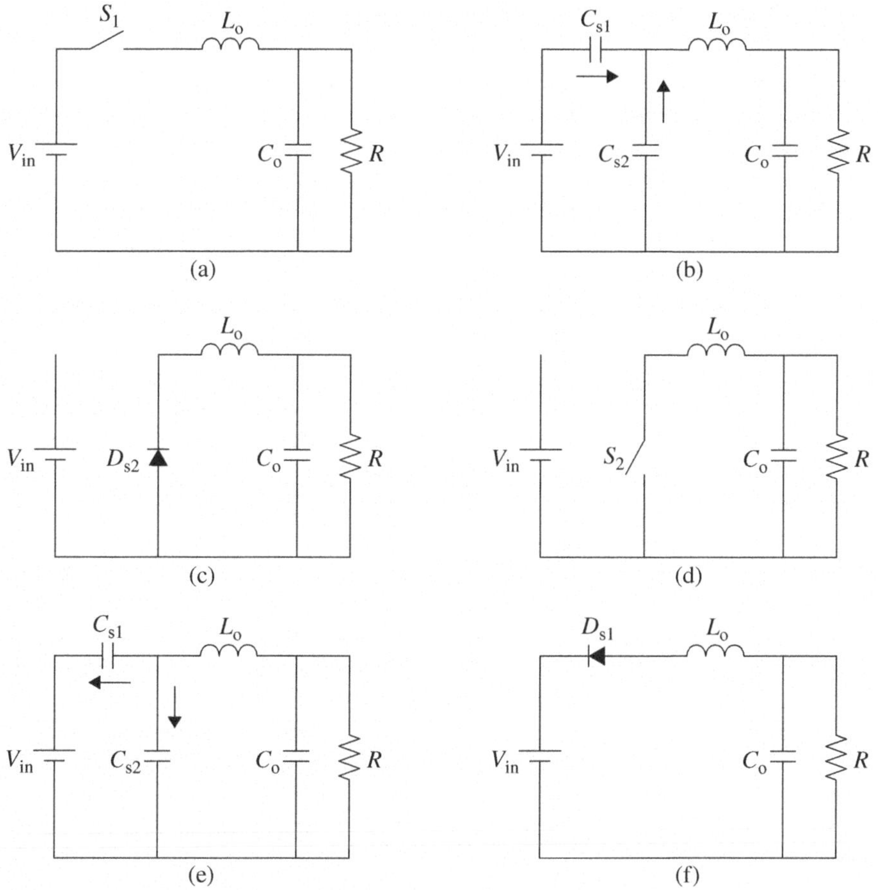

Figure 2.3 Modes of operation of the quasi-square-wave ZVS converter. (a) Mode 1. (b) Mode 2. (c) Mode 3. (d) Mode 4. (e) Mode 5. (f) Mode 6.

and current during this switching transition. Current then flows through the body-diode of S_2 as shown in Figure 2.3c. This clamps the voltage across this switch to zero, which allows the switch to be turned on with ZVS. The output inductor is designed so that its current would have been discontinuous if S_2 was a diode, but with S_2 on, the output inductor current reverses direction and flows from right-to-left, as shown in Figure 2.3d. Later in the switching cycle, S_2 is turned off and this action forces current to discharge C_{s1} and charge C_{s2}, as shown in Figure 2.3e, so that it then flows through the body-diode of S_1, as shown in Figure 2.3f, which allows the switch to turn on with ZVS. Eventually, the output inductor current reverses direction and flows from left-to-right in the output inductor and the converter operates as shown in Figure 2.3a.

The following about the operation of this converter should be noted:

- The on-time of S_1 must be greater than that of S_2 in order for a net amount of energy to be delivered to the load. If the load is replaced with a DC voltage source of some kind like a battery, then the on-time of S_2 can be greater than that of S_1, but energy will be delivered in the opposite direction to the input source and the converter will operate as a boost converter that steps up the voltage of the battery. This converter is commonly used in application where a bidirectional converter is required. Details of bidirectional converter operation are discussed in Chapter 16 of this book.
- The two converter switches will always turn on with ZVS unless the load is very light, which makes the on-time of the switches small. The switches must be on for a duration that is enough to store sufficient energy in L_o so that switch capacitors C_{s1} and C_{s2} can be discharged or charged; otherwise, the switches will not be able to turn on with ZVS.
- The main disadvantage with this converter is that the output inductor current flows in two directions instead of just one. This increases the peak current that the switches must handle, increases conduction losses and turn-off switching losses, and makes the ripple in the output inductor current large, which necessitates the use of a large output capacitor to filter out current harmonics. Although the converter can be made to operate with a higher efficiency than a hard-switching converter, it has been superseded by other approaches that will be discussed in the following sections.

2.3 ZVS-PWM DC–DC CONVERTERS WITH AUXILIARY CIRCUITS

The most commonly used approach to implementing ZVS in single-switch non-isolated DC–DC PWM converters over the past 30 years has been to use some sort of active auxiliary circuit placed in parallel to the main switch to help it turn on with ZVS. This auxiliary circuit consists of an active switch and passive components such as diodes, capacitors, inductors, transformers, etc., and is active for only a small part of the switching cycle. The auxiliary circuit is activated just before the main converter switch is to be turned on and discharges its output capacitance, which can be just the equivalent drain-source capacitance that is internal to the device or the sum of this capacitance and an additional external capacitance placed in parallel. Once the main switch has been turned on with ZVS, the auxiliary circuit can be deactivated and the converter continues to operate in the exact same manner as a conventional PWM DC–DC converter. Since the auxiliary circuit is active for only a short duration of time, a little bit longer than the time it takes to achieve a ZVS turn-on transition, the current ratings of the auxiliary circuit components are much less than those of the main converter components and thus can be smaller and cheaper than those components. ZVS-PWM converters with such auxiliary circuits are sometimes referred to as zero-voltage transition (ZVT) PWM converters as the auxiliary circuit is active only when the main converter switch is about to be turned on to ensure that it does so with ZVS.

Active auxiliary circuits for ZVS-PWM single-switch non-isolated DC–DC converters generally belong to one of the following categories:

- *Nonresonant auxiliary circuits:* These auxiliary circuits have some sort of external capacitance that can limit the rise of voltage across the auxiliary switch when it is turned off.
- *Resonant auxiliary circuits:* These auxiliary circuits have some sort of inductor–capacitor resonant network that shapes the current flowing through the auxiliary switch so that this current is forced to zero and the switch can be turned off softly.
- *Dual auxiliary circuits:* These auxiliary circuits combine the features of the other two types by having a nonresonant and a resonant branch. Like resonant auxiliary circuits, their switch can be turned off softly, but the amount of current circulating in the circuit is less.

In general, nonresonant auxiliary circuits are to be preferred in applications where the converter is operating with a very wide load range, resonant auxiliary circuits are to be preferred in applications where the converter is operating with a narrower load range, and dual auxiliary circuits are to be preferred as a compromise solution between the two other types. Each type of auxiliary circuit is reviewed in this section.

2.3.1 Nonresonant Auxiliary Circuits

An example of a ZVS-PWM DC–DC boost converter with a nonresonant auxiliary circuit is the converter that was proposed in [2] and shown in Figure 2.4. The converter is like a regular boost converter, but with an active auxiliary circuit that consists of an inductor L_r, a switch S_a, and a diode D_2. Equivalent circuit diagrams of the various modes of operation that the converter goes through during a switching cycle are shown in Figure 2.5. If it is assumed that the two converter switches are off and the input inductor current flows through the main boost diode before the start of Mode 1 at $t = t_0$, then the converter works as follows:

Mode 1 (t_0–t_1): S_a is turned on at $t = t_0$ and the process of transferring current away from D_1 to S_a begins. This process is gradual because of the presence of inductor L_r in series with S_a; and S_a is turned on with ZCS.

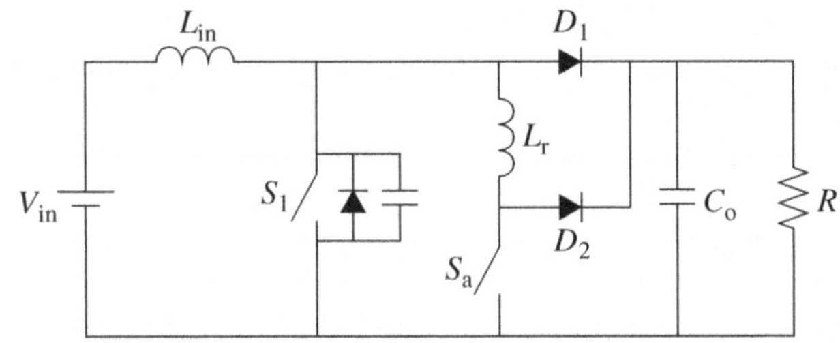

Figure 2.4 ZVS-PWM converter with simple auxiliary circuit.

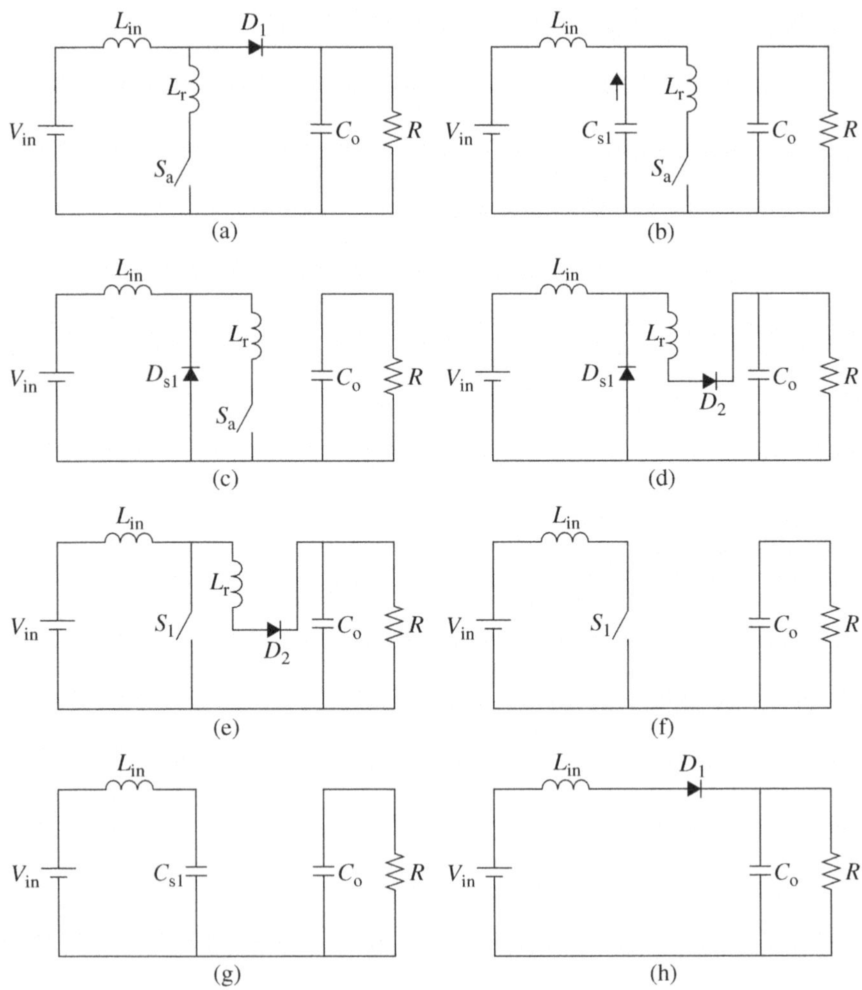

Figure 2.5 Modes of operation of the converter in Figure 2.4. (a) Mode 1. (b) Mode 2. (c) Mode 3. (d) Mode 4. (e) Mode 5. (f) Mode 6. (g) Mode 7. (h) Mode 8.

Mode 2 (t_1–t_2): The current transfer process that was started in Mode 1 is completed at $t = t_1$ and the full input inductor current flows through S_a. When this happens, the voltage across the main switch S_1 is no longer fixed at the output voltage V_o so that C_{s1} can begin to discharge into S_a. The current in S_a during this mode is the sum of the input current and the discharge current from C_{s1}.

Mode 3 (t_2–t_3): This mode begins at $t = t_2$ when C_{s1} has been fully discharged and current begins to flow through the body-diode of S_1. The current that flows through this diode is the difference in current between the current in L_r and S_a and the input current. S_1 can be turned on with ZVS while current is flowing through its body-diode. Current will continue to flow through the body-diode and not decline until S_a is turned off, as the voltage across L_r is zero.

Mode 4 (t_3–t_4): This mode begins when S_a is turned off at $t = t_3$. This forces current to flow through auxiliary circuit diode D_2 and to the output. Since a negative voltage of $-V_o$ is impressed across L_r, the current in this inductor begins to fall as does the current that is flowing through the body-diode of S_1. With the current in L_r falling, so too does the difference in current between the current in L_r and the input inductor current.

Mode 5 (t_4–t_5): At $t = t_4$, the current in S_1 reverse direction and no longer flows through its body-diode. Current continues to flow through L_r.

Mode 6 (t_5–t_6): At $t = t_5$, the current in L_r has been extinguished and the full input current flows through S_1. This mode is the same as the boost mode of a conventional PWM boost DC–DC converter.

Mode 7 (t_6–t_7): This mode begins at $t = t_6$ when S_1 is turned off. During this mode, the input current flows through the switch capacitance of S_1, C_{s1}. If C_{s1} is sufficiently large, it will delay the rise of voltage across the switch so that the overlap between voltage and current during this switching transition is reduced. Although this overlap of switch voltage and switch current is not fully eliminated, S_1 essentially is turned off with ZVS.

Mode 8 (t_7–t_8): At $t = t_7$, C_{s1} has been charged to the output voltage V_o so that the full input inductor current begins to flow through the main boost diode D_1. It continues to do so until S_a is turned on again to start the next switching cycle.

The following should be noted about the converter:

- Since current is gradually transferred away from the boost diode when the auxiliary switch is turned on at the start of Mode 1 as L_r limits the rate at which the current transfer takes place, it is possible to eliminate the reverse-recovery current in the boost diode if this transfer is sufficiently gradual. This helps reduce any reverse-recovery losses and electromagnetic interference (EMI) caused by the boost diode.

- The auxiliary switch should be a MOSFET device with a lower output capacitor than the device used for the main switch. Otherwise, there will be no savings in turn-on losses as the turn-on losses of the main switch would merely be transferred to the auxiliary switch. A MOSFET device with lower internal capacitance typically has a higher value of on-state resistance as the two are inversely proportional. Although such a device cannot be used as the main power switch as its high on-state resistance would result in considerable conduction losses, it can be used as the auxiliary switch because it is on for only a fraction of the switching cycle and thus conducts current for only a short amount of time.
- Since the auxiliary circuit is functional for only a small fraction of the switching cycle, smaller, lower-rated components can be used in the auxiliary circuit, thus justifying the use of the auxiliary circuit from the perspective of cost, size, and weight.
- The converter's auxiliary circuit can be implemented in other single-switch non-isolated converters, as shown in Figure 2.6. In these converters, the auxiliary circuit is attached in parallel to the main switch so that, regardless of how the main switch is oriented in the topology, the auxiliary circuit is oriented in a similar manner.
- In the actual implementation of the converter, it is standard practice to put a saturable inductor and a diode in series with L_r, before the $S_a - D_2$

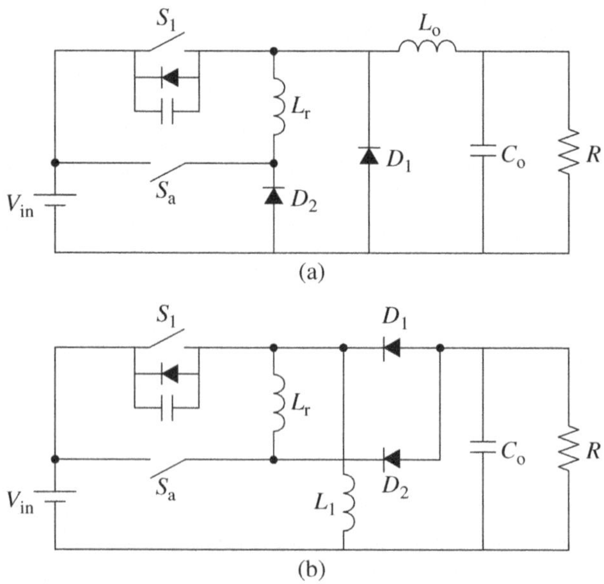

Figure 2.6 Other basic ZVS-PWM converter topologies with a simple auxiliary circuit. (a) buck, (b) buck-boost.

connection to suppress any voltage ringing that may occur due to the interaction of L_r and the output capacitance across S_a.

- The main switch can turn off with nearly ZVS as its output capacitance can limit the rate of voltage rise across the switch as its current falls to zero. This limits the overlap between switch voltage and switch current during the turn-off transition time, thus reducing turn-off switching losses.
- The auxiliary switch can turn on with ZCS because L_r limits the rate of current rise through this switch after it is turned on.

Although the ZVS method shown in Figure 2.4 increases converter efficiency in converters operating with high-switching frequencies, it is not as effective as it could be because its turn-off is hard and this creates turn-off losses. These losses are not insignificant as the full output voltage and the full input current can be present at some time during the switching transition. Numerous modifications to the basic ZVS-PWM auxiliary circuit shown in Figure 2.4 have been proposed to correct this deficiency, including the one that was proposed in [3] and shown in a PWM boost converter in Figure 2.7.

The main difference between this converter and the converter shown in Figure 2.4 is that this converter has a snubber capacitor C_r that is parallel to the auxiliary switch during the turn-off of this switch. This snubber capacitor helps suppress any overvoltage spikes that may appear across S_a when it turns off. Figure 2.8a shows a circuit diagram of this mode. It can be seen that the current flowing through the auxiliary circuit L_r flows through the output capacitance of the auxiliary switch and auxiliary circuit capacitor C_r instead of just the switch output capacitance. It is this combined capacitance that slows down the rate of voltage rise across the auxiliary circuit after it is turned off so that it can turn off softly and that also helps prevent overvoltage voltage spikes from appearing. The mechanism that discharges C_r is shown in Figure 2.8b. The discharge of C_r occurs after the main switch has been turned off and its switch capacitance is charged. During this charging process, a voltage level

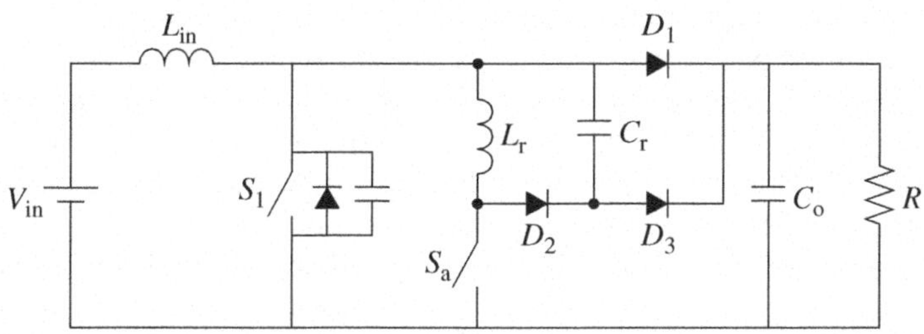

Figure 2.7 ZVS-PWM converter with nonresonant auxiliary circuit with flying capacitor snubber.

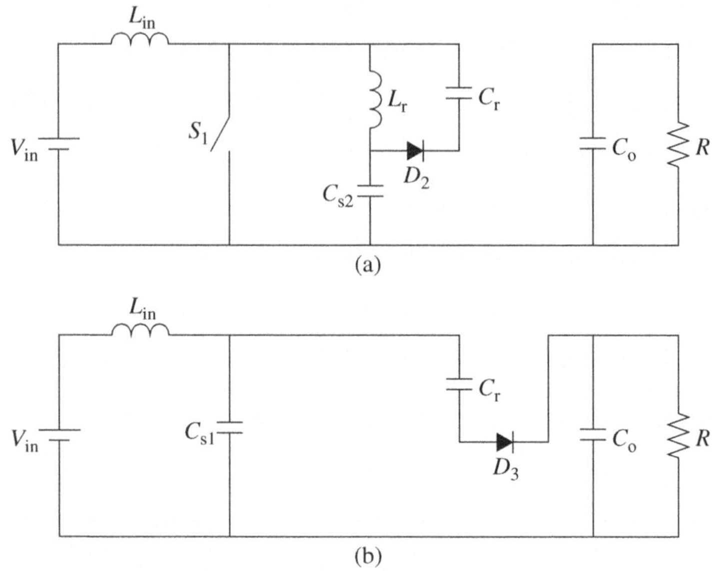

Figure 2.8 Key modes of operation of the converter in Figure 2.6: (a) Auxiliary switch turn-off. (b) Auxiliary circuit capacitor discharge.

is reached when the voltage across S_1 is equal to the difference between the output voltage and the voltage across C_r. When this happens, the output capacitance of S_1 continues to be charged and C_r begins to be discharged. It is only after C_r has been completely discharged that the input boost inductor current starts to flow through the boost diode, D_2.

The converter can be modified so that less current flows in the auxiliary circuit as was proposed in [4] and shown in Figure 2.9. The main modification to the converter is the addition of a center-tapped transformer to the auxiliary circuit and diode D_2. The concept of using a center-tapped transformer in this manner was proposed in [5, 6] for the buck and boost converter versions of the converter shown in Figure 2.7 as a way of reducing the turn-off losses produced by the hard turn-off of the auxiliary switch. The work that was done in [5, 6] was adapted to the converter shown in Figure 2.9.

The center-tapped transformer mechanism works as follows: When the auxiliary switch is turned on, current from the boost diode is transferred to the auxiliary circuit, but it is also transferred to diode D_2 as well. The current in D_2 flows directly to the output so that the amount of current that actually flows in the auxiliary switch can be reduced, thus reducing the peak and rms current stresses of this switch. The current flowing through D_2 mirrors that flowing through L_r and the converter with the center-tapped auxiliary circuit essentially operates with the same modes of operation as the converter shown in Figure 2.7.

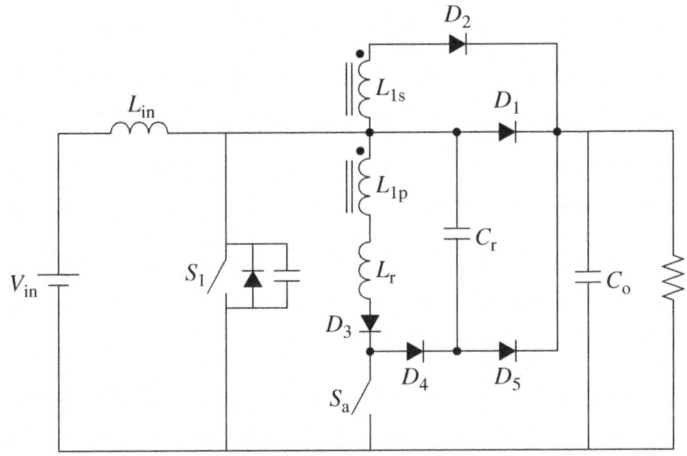

Figure 2.9 Modified ZVS-PWM converter with flying capacitor auxiliary circuit.

2.3.2 Resonant Auxiliary Circuits

Resonant auxiliary circuits were proposed as a way of ensuring that the switch in an active auxiliary circuit turns off softly. The general principle behind these circuits is that some sort of LC resonant circuit is used to shape the auxiliary switch current so that it falls to zero sometime after the auxiliary switch has been activated, which ensures a soft turn-off for the switch. One of the first resonant auxiliary circuits was the one that was proposed in [7] and shown in Figure 2.10 in a PWM boost converter. The auxiliary circuit consists of an inductor L_r, a capacitor C_r, a transformer T_1 with secondary diodes D_{a1} and D_{a2}, and an active switch S_a. When current flows through the transformer

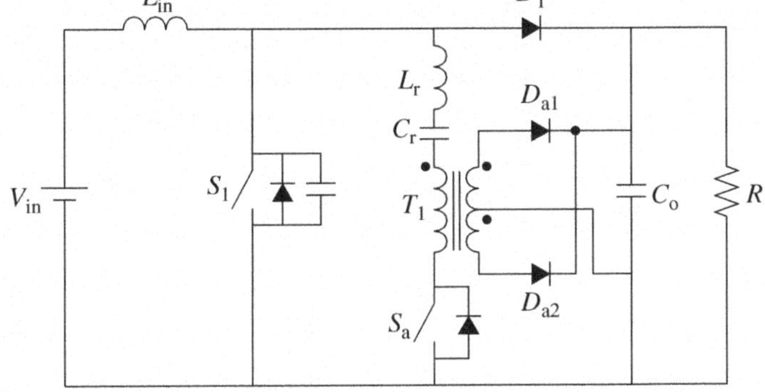

Figure 2.10 ZVS-PWM converter with resonant auxiliary circuit and auxiliary circuit transformer.

primary, one of the two secondary diodes conduct – if current flows from top to bottom in Figure 2.10, then diode D_{a1} conducts; if current flows from bottom to top, then diode D_{a2} conducts. This makes the transformer primary into a small DC voltage source that is a counter-voltage to the flow of current and that helps limit the amount of current in the circuit.

The modes of operation of this converter are very similar to those of the basic converter shown in Figure 2.4 – the auxiliary switch is turned on just before the main switch, current is diverted gradually away from the boost diode, the output capacitance of the main switch is discharged, and the main switch is turned on with ZVS sometime after the anti-parallel diode of the main switch begin to conduct. The main difference between the two converters is that the series resonant capacitor C_r is charged while current is flowing in the auxiliary circuit so that the voltage across the resonant inductor becomes negative, which results in the current in the auxiliary circuit falling to zero. By the time this happens, the main switch has been turned on so that current begins to flow through the anti-parallel diode across the auxiliary switch as shown in Figure 2.11. While this is happening, S_a can be turned off softly, thus avoiding the turn-off losses that exist in other non-resonant auxiliary circuits.

The following should be noted about the converter:

- In a real, practical implementation, the body-diode of S_a is often blocked with a diode that is placed in series with the device and an anti-parallel bypass diode is placed across the series diode – S_a combination. The negative resonant current then flows through the external bypass diode instead of the body-diode of S_a. The bypass diode is a device with a better reverse-recovery characteristic than the body-diode so that undesirable voltage ringing is avoided.
- The resonant cycle of the auxiliary circuit cannot be interrupted in the converter as there is no path for current to flow through if S_a is turned off while current is flowing through it. This removes the justification of the auxiliary circuit when the converter is operating under light load conditions as the conduction losses of the auxiliary circuit, along with the switching losses, become greater than the switching losses of the main switches that are to be saved. This issue can be addressed by adding

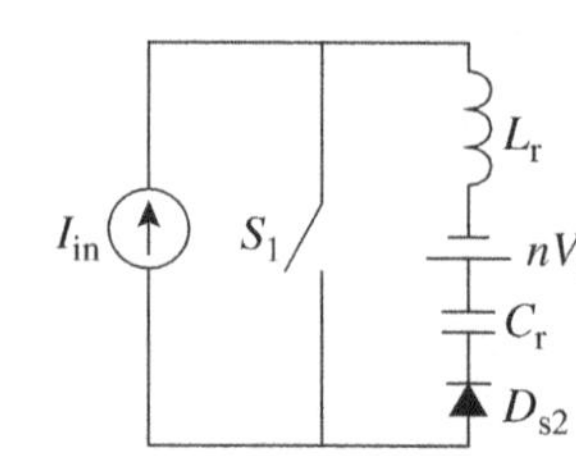

Figure 2.11 Auxiliary switch turn-off mode for the converter in Figure 2.10.

an additional diode to the auxiliary circuit, one with its anode connected to the drain of S_a and its cathode connected to the output.

- The voltage stress across S_a is slightly larger than the output voltage due to the negative voltage that appears across C_r when the auxiliary switch is off. This issue can also be resolved with the addition of a diode that has its anode connected to the drain of S_a and its cathode connected to the output.

Most resonant auxiliary circuits operate with the same basic principles as the converter shown in Figure 2.10. Each have their own strengths and weaknesses. The main difference between all of these circuits is how energy that would otherwise be trapped in the auxiliary circuit is transferred to the output. In the converter shown in Figure 2.10, this mechanism is the transformer, which allows energy to be transferred to the output through the secondary diodes. If the transformer was not in the auxiliary circuit, then energy would have to be dissipated in the auxiliary circuit. Although there could still be a net gain in efficiency compared to hard-switching operation where no auxiliary circuit is used due to the savings in boost diode reverse-recovery losses and reduced EMI, this gain would be very modest.

Another example of a resonant auxiliary circuit in a non-isolated single-switch PWM DC–DC converter is in the converter shown in Figure 2.12 that was first proposed in [8] then analyzed in [9]. It operates in a manner that is similar to the converter shown in Figure 2.10, but auxiliary circuit energy is transferred to the output by charging C_r then discharging it through D_2 after S_1 has been turned off, similar to the capacitor discharging mode shown in Figure 2.8b. Unlike the converter shown in Figure 2.10, the converter in Figure 2.12 is shown with a diode D_3, that allows current to be interrupted in the auxiliary circuit and clamps the voltage across S_a to the output voltage.

The main drawback with the converter shown in Figure 2.12 is that the resonant current that flows into the main switch from the auxiliary circuit is

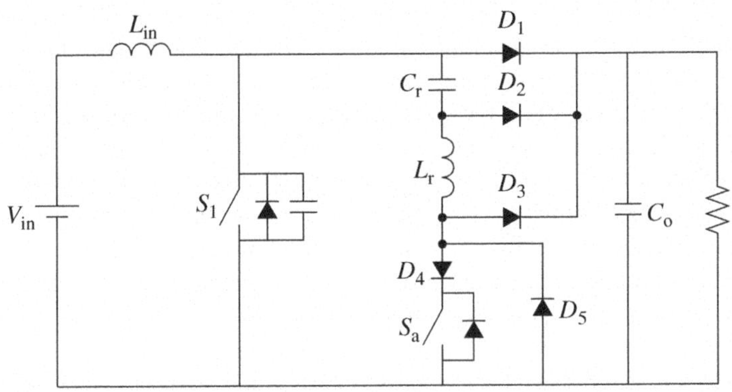

Figure 2.12 ZVS-PWM converter with clamped resonant auxiliary circuit.

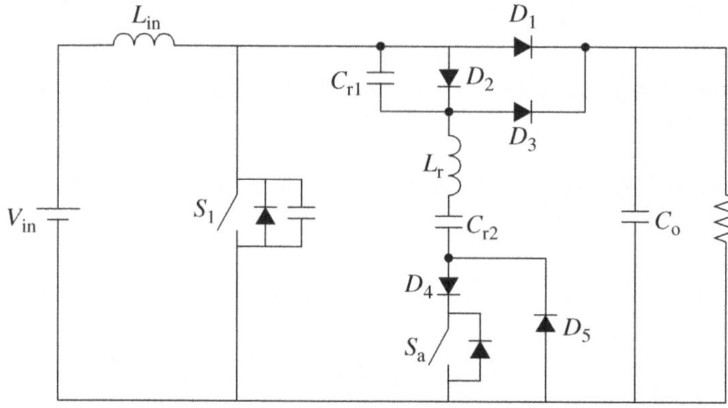

Figure 2.13 ZVS-PWM converter with modified clamped resonant auxiliary circuit.

higher than it is in the converter shown in Figure 2.10. This issue can be addressed by adjusting the impedance of L_r or C_r so that this resonant current is decreased. This is the approach proposed in [10] and shown in Figure 2.13. This auxiliary circuit is very similar to the resonant auxiliary circuit shown in Figure 2.12 except that an additional capacitor has been added to the auxiliary circuit, C_{r1}. The converter is shown without a clamping diode connected in the auxiliary circuit (D_3 in Figure 2.12), but this diode can be added to the circuit if desired. Capacitor C_{r1} increases the impedance in the auxiliary circuit and thus reduces the peak of the main switch current. It should be added that increasing the impedance of the resonant auxiliary circuit in this manner can result in the resonant cycle being extended significantly, which may limit the switching frequency of the converter or make the turn-off of the auxiliary switch more likely to be hard as the resonant cycle is interrupted.

2.3.3 Dual Auxiliary Circuits

The third general category of auxiliary circuits for non-isolated single-switch ZVS-PWM DC–DC converters that allow the main switch to turn-on with ZVS are dual auxiliary circuits that combine both types of auxiliary circuits in a single auxiliary circuit as a way of taking advantage of the strengths of each type of auxiliary circuits. Dual auxiliary circuits allow their active switch to turn off with ZVS just like resonant auxiliary circuits, but have less current flowing in auxiliary circuit, which is the result of the non-resonant part of the circuit. Dual auxiliary circuits can be synthesized according to the procedure presented in [11]. A couple of examples of dual auxiliary circuits are presented here.

Consider the ZVS-PWM boost converter that was proposed in [12] and shown in Figure 2.14. It can be seen that the auxiliary circuit in this converter

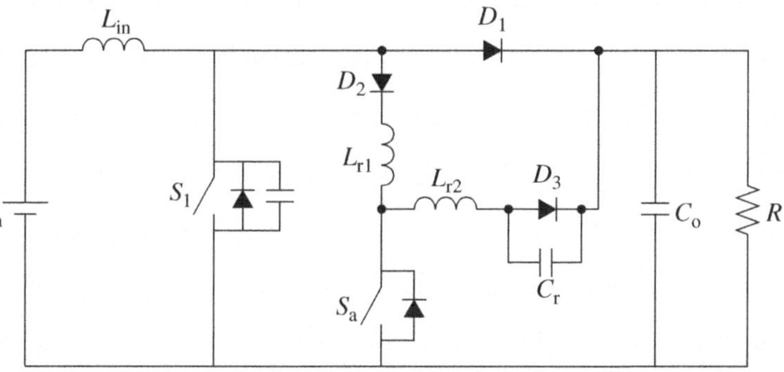

Figure 2.14 ZVS-PWM converter with dual auxiliary circuit with output branch connection.

has two branches that are connected to a common switch: a nonresonant branch consisting of diode D_2 and inductor L_{r1} and a resonant branch consisting of inductor L_{r2}, capacitor C_r, and diode D_3. The auxiliary circuit works as follows: When S_a is turned on, current begins to flow through both auxiliary circuit branches. The current in the resonant branch is sinusoidal with a current peak that is greater than the current peak of the nonresonant branch current. At some point in the resonant cycle of L_{r2} and C_r, the current in the resonant branch begins to reverse direction and starts to remove current from S_a. When the negative current of the resonant branch tries to exceed the current in the nonresonant branch, current is pulled away from S_a so that it can be turned off softly. Capacitor C_r in the resonant branch is eventually discharged and diode D_3, which is connected across it, begins to conduct. Current in the auxiliary circuit now flows through D_3 to the output and continues to do so until it has been fully extinguished.

Another example of a dual auxiliary circuit that was proposed in [11] and shown in Figure 2.15. In this circuit, the resonant branch is not connected to the cathode of the main boost diode, but is connected to its anode instead. The auxiliary circuit operates in a manner that is similar to that of the auxiliary circuit shown in Figure 2.14 as current flows through both non-resonant and resonant branches after the auxiliary switch S_a is turned on to initiate the process by which main switch S_1 is to be turned on with ZVS. When the current in auxiliary circuit capacitor C_r reverses direction, current begins to be removed from S_a. Like the auxiliary circuit shown in Figure 2.14, the peak of the resonant branch current must be such that it can divert the non-resonant branch current away from S_a. Sometime later in the switching cycle, the charge in C_r is removed in the same manner as what is shown in Figure 2.8b. The main difference between the two dual auxiliary circuit converters that are shown in Figures 2.14 and 2.15 is that the one in Figure 2.14 allows more current to flow in its auxiliary circuit, which increases the load range over which the main switch can be turned on with ZVS, but increases the conduction losses of the auxiliary circuit as well.

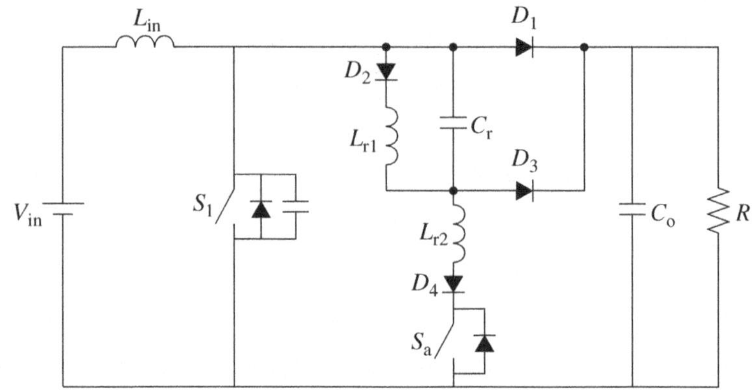

Figure 2.15 ZVS-PWM converter with dual auxiliary circuit with no output branch connection.

2.4 MISCELLANEOUS CONSIDERATIONS

Some miscellaneous considerations relating to nonisolated, single-switch, ZVS-PWM DC–DC converters are briefly covered in this section.

2.4.1 Application-Specific ZVS-PWM Converters

Each of the auxiliary circuits that have been presented in this chapter can be implemented in any of the six basic DC–DC converters, as shown in Figure 2.6. There are some auxiliary circuits, however, that are well suited for only one of these six converters due to their nature. An example of such a circuit is the auxiliary circuit that was proposed for buck converters in [13], then adapted for boost converters in [8] and shown in Figure 2.16. The auxiliary circuit is very simple as it consists of just a switch S_a, a blocking diode D_2 and an inductor L_r. The converter works as follows: The auxiliary switch is turned on just before the main switch is turned on. Since the input inductor current now has path of lower voltage that it can flow to it begins to flow through the auxiliary circuit and away from the main boost diode D_1. After

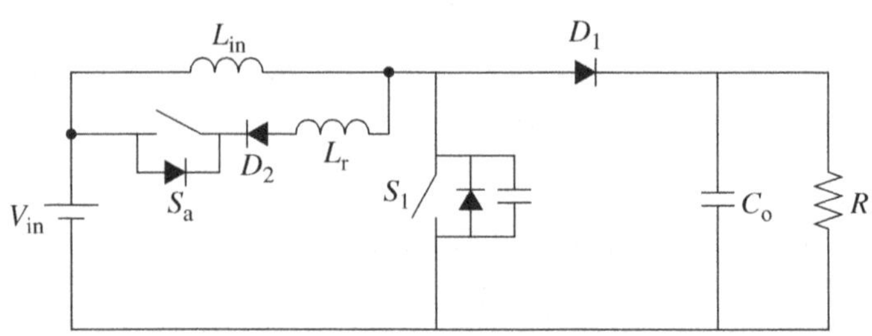

Figure 2.16 Application-specific ZVS-PWM boost converter with simple auxiliary circuit branch.

D_1 stops conducting current, switch capacitor C_{s1} begins to discharge through the auxiliary circuit. The switch can be turned on with ZVS after C_{s1} has been fully discharged. When the voltage at the end of the auxiliary circuit that is connected to the auxiliary circuit becomes less than the input voltage, current in the auxiliary circuit begins to fall. Eventually, current ceases to flow in the auxiliary circuit completely and S_a can be turned off softly.

The converter shown in Figure 2.16 is very simple and efficient. Its main drawbacks, however, are:

- The output voltage must be at least twice the input voltage so else the auxiliary switch will not turn off softly. This limits the applications that this converter can be used to applications where there is little variation in the input voltage.
- The auxiliary switch requires a floating gate drive circuit, which complicates the implementation.

Nonetheless, the converter is an attractive option for certain targeted applications.

2.4.2 ZVS-PWM Techniques in Converters with Wide Bandgap Devices

The ZVS-PWM converters that have been presented in this chapters were developed for use with silicon (Si) devices. As was discussed in the previous chapter, in Section 1.5.5, new, so-called, wide bandgap devices with superior switching characteristics have been developed in recent years and these devices are in the process of gradually replacing traditional Si devices. The turn-on and turn-off switching losses of these devices, the most popular type being made of either silicon carbide (SiC) or gallium nitride (GaN), are much lower than those of Si devices and it is believed that the difference in switching losses is such that converters with SiC or GaN devices need not be implemented with any type of ZVS-PWM converter, like the ones presented in this chapter.

ZVS-PWM converters in general introduce losses that are created by the mechanism needed to discharge the output switch capacitance of the main converter switch in single-switch PWM converters. These losses are smaller than the switching losses that are saved at nominal loads for Si converters and are only greater when the converter is operating with light loads. ZVS-PWM Si converters are typically operated with switching frequencies in the range of 50–200 kHz. In the case of SiC and GaN converters, the switching losses generated by the converter switches at these switching frequencies are generally too low to justify the use of any ZVS-PWM method. These losses only become significant at very high switching frequencies, but converters operating with such frequencies tend to be resonant-type converters operating with switching frequencies of 1 MHz and above as such converters can make use of parasitic inductances that become more important as such frequencies; resonant converters are discussed in Chapters 11 and 12 of this book.

The use of ZVS-PWM methods for converters with wide bandgap devices may be justified for a very limited range of applications where the switching frequency is between 400 and 800 kHz and only if the auxiliary circuit produces very few losses. This rules out most of the converters presented in this chapter except for the converter shown in Figure 2.16, which operates with very little rms current in the auxiliary circuit and a soft auxiliary switch turn off. At the time of this writing, however, only a few researchers have investigated the using wide bandgap devices with ZVS-PWM methods, and the results have generally been inconclusive.

2.5 CONCLUSION

The size of power electronic converters can be reduced if they are made to operate with high switching frequencies. Doing so helps reduce the size of energy storage components such as inductors and capacitors as they do not have to store as much energy as with lower frequency operation. High switching frequency operation, however, results in more switching losses as converter switches must be turned on and off more often and the switching process is not ideal. During any switching transitions, turn-on or turn-off, any overlap between switch voltage and switch current exists that results in power dissipating in the device and thus power losses that reduce converter efficiency.

Switching losses can be reduced if either the switch voltage or the switch current is made zero at the time of the switching transition. This can be done if soft-switching techniques are used. Soft-switching techniques make the switching transitions gradual instead of sudden and can be either zero-voltage-switching (ZVS) or zero-current-switching (ZCS); the focus of this chapter has been on ZVS methods for MOSFETS; the focus of the next chapter will be on ZCS methods for IGBTs. In this chapter, the basic fundamentals of ZVS were reviewed with emphasis on converters with silicon MOSFETS. A number of ZVS methods were presented with the operation of several ZVS-PWM converter discussed in detail. The converters presented in this chapter were developed for use with traditional silicon devices. They are rarely used with new wide bandgap devices and research into such use have been limited and inconclusive.

It should be noted that the converters presented in this chapter, as well as those that will be presented in the next chapter are often referred to as active-snubber converters. This is because the gradual switching transitions can often reduce voltage and/or current stresses in converter components.

REFERENCES

1. C. P. Henze, H. C. Martin and D. W. Parsley, "Zero-voltage switching in high frequency power converters using pulse width modulation," *IEEE Applied Power Electronics Conference and Exposition (APEC)*, 1988, Feb. 1–5, New Orleans, USA, pp. 33–40.
2. G. Hua, C.-S. Leu, Y. Jiang and F. C. Y. Lee, "Novel zero-voltage-transition PWM converters," *IEEE Transactions on Power Electronics*, vol. 9, no. 2, pp. 213–219, Mar. 1994.
3. R. Streit and D. Tollik, "High efficiency telecom rectifier using a novel soft-switched boost-based input current shaper," *IEEE International Telecommunications and Energy Conference (INTELEC)*, 1991, Nov. 5–8, 1991, Kyoto, Japan, pp. 720–726.
4. R. L. Lin, Y. Zhao, and F. C. Lee, "Improved soft-switching ZVT converters with active snubber," *IEEE Applied Power Electronics Conference and Exposition (APEC)*, 1998, Feb. 15–19, 1998, Anaheim, USA, pp. 1063–1069.
5. D. C. Martins, F. J. M. de Seixas, J. A. Brilhante and I. Barbi, "A family of DC-to-DC PWM converters using a new ZVS commutation cell," *IEEE Power Electronics Specialist Conference (PESC)*, 1993, June 20–24, 1993, Seattle, USA, pp. 524–530.
6. J. P. Gegner and C. Q. Lee, "Zero-voltage-transition converters using an inductor feedback technique," *IEEE Applied Power Electronics Conference and Exposition (APEC)*, 1994, Feb. 13–17, 1994, Orlando, USA, pp. 862–868.
7. G. Moschopoulos, P. Jain, Y.-F. Liu and G. Joos, "A zero-voltage switched PWM boost converter with an energy feedforward auxiliary circuit," in *IEEE Transactions on Power Electronics.*, vol. 14, no. 4, pp. 653–662, Jul. 1999.
8. K. M. Smith and K. M. Smedley, "A comparison of voltage-mode soft-switching methods for PWM converters," *IEEE Transactions on Power Electronics*, vol. 12, no. 2, pp. 376–386, Mar. 1997.
9. C.-J. Tseng and C.-L. Chen, "Novel ZVT-PWM converters with active snubbers," *IEEE Transactions on Power Electronics*, vol. 13, no. 5, pp. 861–869, Sept. 1998.
10. N. Jain, P. K. Jain, and G. Joos, "A zero voltage transition boost converter employing a soft switching auxiliary circuit with reduced conduction losses," *IEEE Transactions on Power Electronics*, vol. 19, no. 1, pp. 130–139, Jan. 2004.
11. W. Huang and G. Moschopoulos, "A new family of zero-voltage-transition PWM converters with dual active auxiliary circuits," *IEEE Transactions on Power Electronics*, vol. 21, no. 2, pp. 370–379, Mar. 2006.
12. M. L. Martins, H. Pinheiro, J. R. Pinheiro, H. A. Grundling, and H. L. Hey, "Family of improved ZVT-PWM converters using a self-commutated auxiliary network," *IEE Proceedings - Electric Power Applications*, vol. 150, no. 6, pp. 680–688, Nov. 2003.
13. L. Freitas and P. Gomes, "A high-power high-frequency ZCS-ZVS PWM buck converter using a feedback resonant circuit," *IEEE Power Electronics Specialist Conference (PESC)*, 1993, June 20–24, 1993, Seattle, USA, pp. 330–336.

CHAPTER 3

NON-ISOLATED ZERO-CURRENT SWITCHING PWM CONVERTERS

The focus of the previous chapter was on zero-voltage-switching (ZVS) DC–DC PWM converter topologies that allowed their switches to operate with reduced switch losses by ensuring that they are turned on and off with ZVS. Doing so eliminates the overlap that exists between the voltage and current in non-ideal switching devices during switching transitions, which makes the product of the two zero, thus eliminating switching losses, in theory. In practice, small amounts of switching losses remain as voltage and current may not actually be completely zero.

The focus of this chapter is on zero-current-switching (ZCS) DC–DC PWM converters that allow their switches to operate with ZCS. Such topologies are better suited for silicon IGBTs than ZVS methods because these devices are minority carrier devices. Charges in these devices cannot be quickly removed when they are about to be turned off; therefore, they turn off very slowly and with a considerable "current tail" that overlaps with voltage, thus creating significant switching losses. If the current can somehow be gradually removed from an IGBT device before it is turned off, then the device can do so without this current tail, thus reducing switching losses. Moreover, an IGBT without a current tail can be turned off quickly and can be made to operate with higher switching frequencies. The turn-on losses of silicon IGBTs are lower than those of silicon MOSFETs because they have less output capacitance.

DC–DC Converter Topologies: Basic to Advanced, First Edition. Gerry Moschopoulos.

Published 2024 by John Wiley & Sons, Inc.

ZCS in single-switch, non-isolated, DC–DC converters such as the ones reviewed in Chapter 1 can be implemented in resonant converters and in PWM converters; the focus of this chapter will be on the latter. The operation and properties of a number of ZCS-PWM boost converters are discussed and the strengths and weaknesses of each converter are stated. It should be noted that the topologies discussed in this chapter are best suited for converters that are implemented with silicon IGBT devices and are rarely considered when wide-bandgap devices such as silicon-carbide devices are used. This is because these devices do not have the turn-off current-tail that Si-IGBTs have and can be turned off very quickly so that there is little overlap between voltage and current during a turn-off switching transition. Although it is expected that wide bandgap devices will eventually supersede silicon devices in the future, Si-IGBTs remain widely used at the time of this writing due to their lower cost, greater availability, and perceived reliability.

3.1 ZCS-PWM CONVERTERS WITH SERIES-RESONANT AUXILIARY CIRCUITS

Most ZCS-PWM converters that have a single main switch are implemented with an auxiliary circuit that consists of an active switch and various passive elements. This active auxiliary circuit is placed parallel to the main power switch, similar to the ZVS-PWM converters presented in the previous chapter. The circuit is activated just before the main converter switch is to be turned off, instead of when it is about to be turned on, as is the case with ZVS-PWM converters. It diverts current away from the main switch and is deactivated shortly after the switch is turned off. ZCS-PWM converters with such auxiliary circuits are sometimes referred to as zero-current transition (ZCT)-PWM converters in the literature as the auxiliary circuit is active only during turn-off switching transitions.

Current can be directed away from the main switch after the auxiliary switch is turned on by using some sort of inductor–capacitor (LC) resonance in the auxiliary circuit to make the voltage across the auxiliary circuit resonant capacitor negative. An inductor is typically placed in series with the main power switch to slow down the rise in current when the switch is turned on to start the next switching cycle. This inductor allows the main power switch to have a ZCS turn on and it essentially eliminates the reverse recovery current of the main power diode, if it is a silicon device and turning on the main switch diverts current away from it. A Si-diode, like a Si-IGBT, is a minority carrier device and it benefits from the gradual transition of current away from it when the main power switch is turned on.

The first ZCS-PWM boost converters with active auxiliary circuits that were proposed generally had simple LC resonant circuits in their topologies. The operation and properties of several topologies of this type are discussed in this section.

3.1.1 ZCS-PWM Converter with Fully Resonant Auxiliary Circuit

One of the first ZCS-PWM converters to be proposed was the one proposed in [1–3] and shown in Figure 3.1. This converter has an auxiliary circuit that consists of an auxiliary switch S_a and a capacitor C_r. An inductor L_r is placed in series with the main switch S_1 so that it can turn on with ZCS. Equivalent circuit diagrams that show the various modes of operation that the converter goes through during a turn-off switching transition when it is operating with continuous input boost inductor current are shown in Figure 3.2. Assuming that current is flowing in the main switch and the auxiliary circuit capacitor C_r is charged to a negative voltage prior to the first mode of operation, the converter has the following modes of operation:

Mode 1 (t_0–t_1): Auxiliary switch S_a is turned on at time $t = t_0$. Current from the input boost inductor that was flowing into the main switch is gradually diverted into the auxiliary switch as the negative voltage across C_r provides a path of lower voltage for current to flow through. The current in L_r falls because there is a negative voltage across C_r.

Mode 2 (t_1–t_2): This mode begins when the current that was flowing in the main switch S_1 has been completely transferred to the auxiliary circuit. S_1 can now be turned off with ZCS. As mentioned earlier, the gradual transfer of current away from a Si-IGBT before it is turned off helps eliminate the current tail that would otherwise appear. C_r continues to be charged during this mode.

Mode 3 (t_2–t_3): This mode begins when the voltage across C_r is equal to the output voltage V_o. Current then begins to flow through the boost diode and continues to do so until the main switch is turned on. The auxiliary switch can be turned off with ZCS sometime during this mode.

Mode 4 (t_3–t_4): At $t = t_3$, the main switch S_1 is turned on and the current begins to be transferred away from D_1. Since L_r is in series with S_1, this current transfer is gradual, which eliminates the reverse-recovery current of a Si-diode that would appear if the converter was not operating with ZCS. Also during this mode, current begins to flow through C_r, through the antiparallel diode of S_2, into the main switch as it discharges.

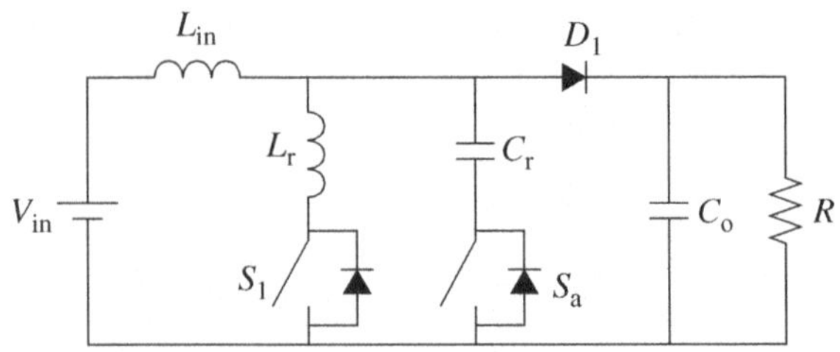

Figure 3.1 Converter with fully resonant auxiliary circuit proposed.

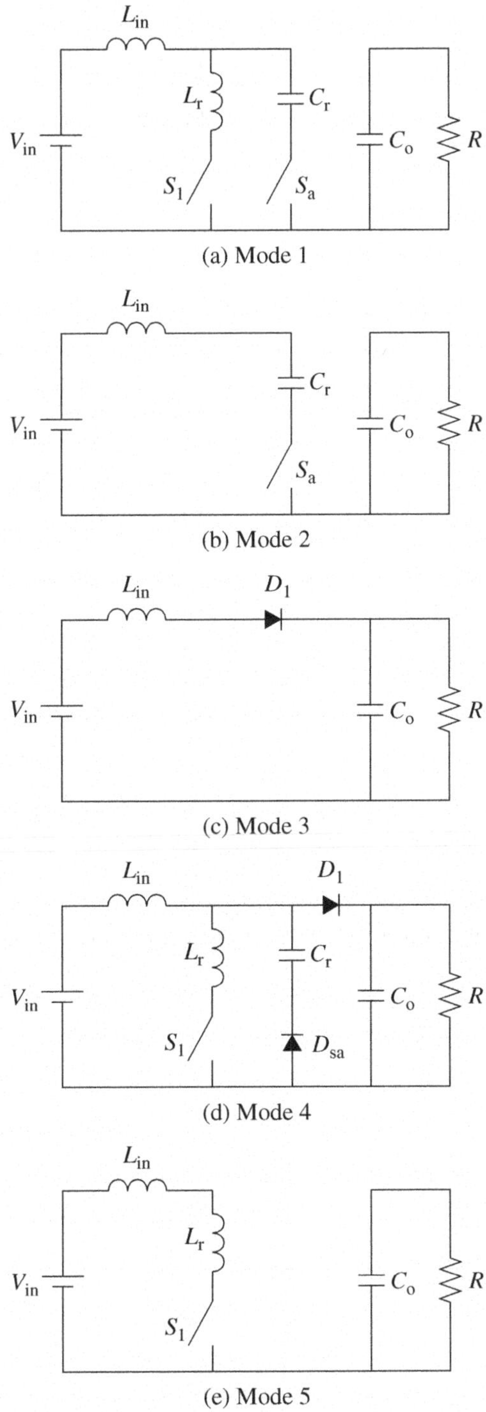

Figure 3.2 Modes of converter operation. (a) Mode 1. (b) Mode 2. (c) Mode 3. (d) Mode 4. (e) Mode 5.

C_r interacts with L_r in a resonant manner so that current flows out of C_r for a half resonant cycle and the polarity of C_r becomes negative. This sets up the auxiliary circuit for the next time that it is activated.

Mode 5 (t_4–t_5): This mode begins when there is no more current flowing out of C_r and the converter behaves just like a regular PWM boost converter. This mode continues until the auxiliary switch is activated to initiate the next turn-off of S_1.

The main advantages of this converter are that its auxiliary circuit is very simple and its main switch and auxiliary switch can turn on and off softly. The main drawbacks are:

- The auxiliary switch must be on for enough time to allow the auxiliary switch current to complete its resonant cycle. This increases conduction losses as more RMS current flows through the switch.
- The auxiliary circuit generates current in the converter that flows through the main switch. The current that flows out of the auxiliary circuit after S_1 is turned on is added to the boost inductor current. This additional current increases the RMS and peak current flowing through the main switch, thus increasing its current stress.
- The boost diode has a higher peak voltage stress as well, at least twice the output voltage V_o, as the voltage across C_r can swing from V_o to $-V_o$.

The converter can be modified by placing a diode in series with the auxiliary switch and then placing an antiparallel diode across the diode/switch combination if a diode with better characteristics than the diode of the auxiliary switch is desired (e.g. less reverse-recovery time). The converter can also be modified by adding two diodes to its topology, D_2 and D_3, as suggested in [4] and shown in Figure 3.3. The main purpose of these two diodes is to clamp any overvoltage ringing that may occur due to the interaction of the parasitic output switch capacitances of S_1 and S_a and inductances such as L_r or parasitic circuit inductances. The diodes also serve as protection in case

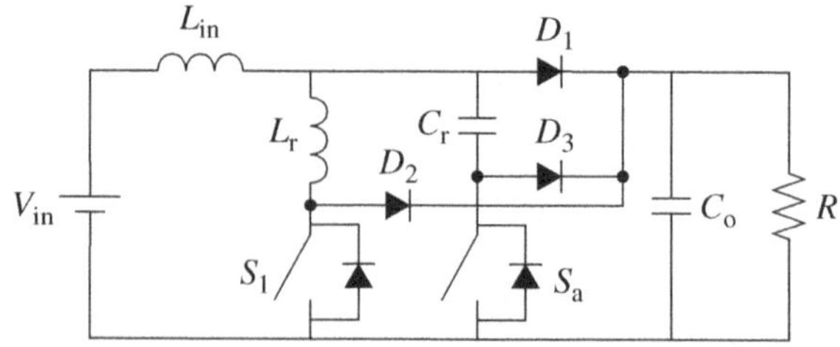

Figure 3.3 Converter with additional clamping diodes.

the switches are inadvertently turned off while they are conducting current as each diode provides a path for current to flow through if needed. This is especially true of D_2 as the current in L_r must be forced to zero before S_1 can be turned off. If the switch tries to turn off while the current is still flowing in L_r and D_2 is in the circuit, then the result may be a catastrophic failure.

3.1.2 ZCS-PWM Converter with Modified Resonant Auxiliary Circuit

A modified version of the converter shown in Figure 3.1 was proposed in [5] and is shown in Figure 3.4. This converter is like the converter shown in Figure 3.1 except that an extra inductor, L_{r2}, and two additional diodes, D_2 and D_3, have been added to the auxiliary circuit. The converter operates in the same way as the converter in Figure 3.1 except that C_r discharges through L_{r2} and D_3 instead of discharging through the antiparallel diode across S_2. Placing L_{r2} in the circuit reduces the peak of the current flowing in the main switch as this added inductance increases the impedance of the resonant circuit. As a result, the auxiliary switch turns softly on and off, but with a lower circulating current, thus reducing the main switch peak current.

The main weakness of the converter is that the negative portion of the auxiliary circuit current's resonant cycle is extended due to the larger inductance. This limits the operation of the converter to applications where the converter can operate under a narrow range of conditions with a large duty-cycle. Although the circulating current that flows from the auxiliary circuit to the main switch is reduced, there will still be an increase in the main switch RMS and peak current, and in conduction losses. The boost diode voltage stress can be as high as twice the output voltage as well.

3.1.3 Converter with Hard-Switching Auxiliary Circuit

A variation of the converters shown in Figures 3.1 and 3.4 is the converter that was proposed in [6] and shown in Figure 3.5. The main difference between this converter and the others is that a diode is connected between

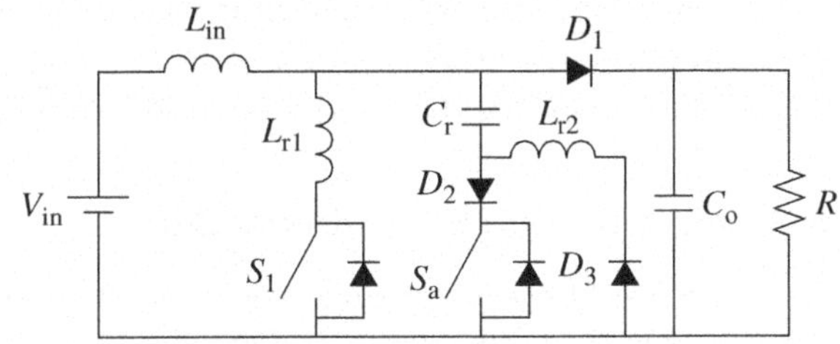

Figure 3.4 Converter with modified resonant auxiliary circuit proposed.

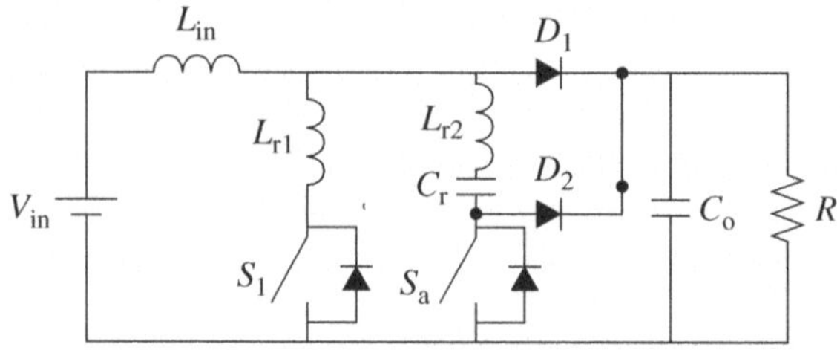

Figure 3.5 Converter with hard-switching auxiliary circuit proposed.

the collector of the auxiliary switch S_a and the output. This diode allows the current that circulates through the antiparallel diode of the auxiliary switch to do so for only a short duration of time after the main switch has been turned on, much less than that of the other ZCS-PWM converters shown in Figures 3.1 and 3.4. This is because the current flowing through the auxiliary switch can be interrupted during its resonant cycle shortly after S_a is turned on, which reduces the conduction losses caused by circulating current from the auxiliary circuit. It also results in lower boost diode peak voltage stress as the interrupted resonant cycle of the auxiliary switch current also results in the resonant cycle of the auxiliary circuit capacitor voltage being interrupted so that there can be less negative voltage across the capacitor and thus less voltage across the boost diode. The main weakness of this converter, however, is that the auxiliary switch has a hard turn-off, as it is turned off while current is flowing through it. This turn-off is dissipative and creates switching losses that somewhat offset the saving in turn-off losses in the main switch.

3.2 ZCS-PWM BOOST CONVERTERS WITH CONVENTIONAL PWM CONVERTER MAIN SWITCH CURRENT STRESS

The ZCS-PWM boost converters that have been discussed so far in this chapter share a common drawback. The resonant interaction in the auxiliary circuit that diverts current away from the main switch to the auxiliary circuit also causes current to flow through the auxiliary circuit back to the main switch, as discussed earlier. The next generation of ZCS-PWM converters that were proposed do not have this drawback and thus their main switch has the current stress of switches in conventional PWM converters. The operation and properties of some of these converters are discussed in this section.

3.2.1 ZCS-PWM Converter with Series Boost Diode

One of the first ZCS-PWM boost converters with an auxiliary circuit that did not cause an increase in the peak current of the main switch was proposed in [7] and is shown in Figure 3.6a. A variation of this converter with the

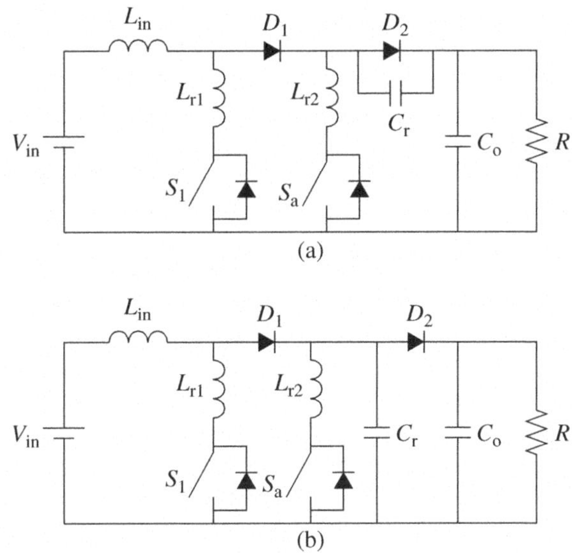

Figure 3.6 Converters with series boost diode: (a) Converter proposed. (b) Converter proposed.

resonant capacitor C_r placed across the auxiliary switch instead of the boost diode D_2 was later proposed by some of the same authors in [8] and is shown in Figure 3.6b. The two converters are very similar and operate in a similar manner, but the auxiliary circuit shown in Figure 3.6a can only be used in boost converters whereas the one shown in Figure 3.6b can be used in other single-switch PWM converters such as the buck and buck–boost converters. The converter in Figure 3.6a will be discussed in this section.

The auxiliary circuit consists of capacitor C_r, inductor L_{r2}, diode D_2, and switch S_a. Inductor L_{r1} is placed in series with S_1 so that the switch can turn on with ZCS and to eliminate the reverse recovery current of diodes D_1 and D_2. The circuit operates as follows: S_a is turned on just before S_1 is about to be turned off, and when it does, a resonance between C_r and L_{r2} is initiated as energy is fed from the output into the auxiliary circuit. At some point during the resonant cycle, the voltage across C_r is such that current begins to be diverted away from S_1 so that this capacitor can be charged. S_1 can be turned off with ZCS after all the input current has been transferred to C_r. Also, during the resonant cycle, the current flowing through L_{r2} reverses direction and S_a can be turned off softly. Eventually, the current ceases to flow in the auxiliary circuit and the converter operates like a conventional boost converter with input current flowing through diodes D_1 and D_2 to the output.

The auxiliary switch turns softly on and off and there is no circulating current from the auxiliary circuit to the main switch so that the main switch's peak current and conduction losses are not increased. The main weakness of the converter, however, is that the input current must flow through two boost diodes instead of one. This results in an increase in conduction losses and the need for an additional power diode instead of a device with a lower power-handling capability as this component is in the main power path. The auxiliary circuit also creates a current peak in the boost diode current, which results in the need for a device with a higher peak current rating. Current needs to flow from the output to the auxiliary circuit as part of the resonant process that diverts current away from the main switch, which creates additional conduction losses. The boost diode voltage stress is as high as twice the output voltage as well. The converter shown in Figure 3.6b does not allow current to flow from the output to the input. The voltage across resonant capacitor C_r swings to a negative level after the auxiliary switch is turned on and thus current can be diverted away from the main switch. The converter, however, shares many of the drawbacks of the converter in Figure 3.6a, most notably, the fact that the input inductor current must flow through two diodes went is flows to the output.

3.2.2 ZCS-PWM Converter with Output Resonance

An improved version of the converter shown in Figure 3.6 that does away with the additional series diode is the converter shown in Figure 3.7a, which was proposed in [9]. A variation of this converter that splits the output resonant inductor L_r into two inductors that are each placed in series with a switch

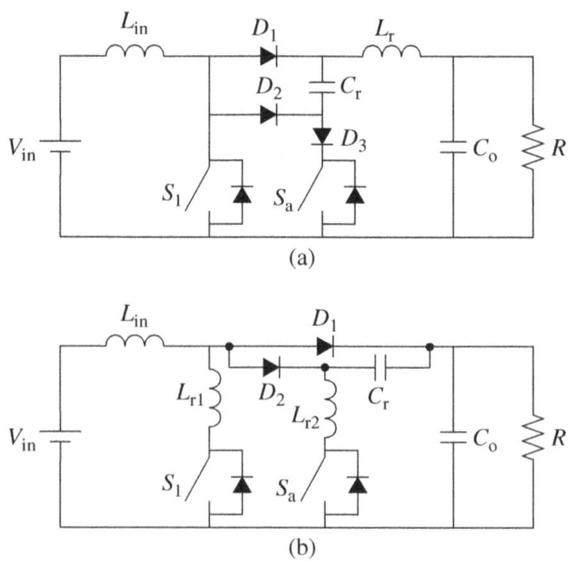

Figure 3.7 Converters with output resonance: (a) Proposed. (b) Proposed.

was proposed in [10] and is shown in Figure 3.7b. These two converters are essentially the same and thus only the converter shown in Figure 3.7a is discussed in this section.

The converter shown in Figure 3.7a is a boost converter with an auxiliary circuit consisting of switch S_a, diodes D_2 and D_3, and capacitor C_r. Inductor L_r is placed between the boost diode and the output capacitor to allow the main switch S_1 to be turned on with ZCS. The converter works as follows: Switch S_a is turned on just before switch S_1 is to be turned off; this initiates a resonance between L_r and C_r, and energy from the output is fed into the circuit. Since current cannot flow through S_a in the negative direction due to blocking diode D_3, the negative portion of the resonant cycle is made up of the current that flows through S_1 and its antiparallel diode. At first, the current through S_1 drops to zero, and then it flows its antiparallel diode, which allows the switch to be turned off softly. Eventually, C_r is fully discharged, and the full input current flows through the boost diode D_1.

Although an additional current flows through the antiparallel diode of S_1, this current is fairly small and is not added to the input current flowing through S_1 so that the peak current of this switch is the same as that that found in a conventional PWM boost converter. There is a direct path for energy to travel from the auxiliary circuit to the output so that energy is not trapped in the circuit where it creates losses, but is transferred to the load. Circulating current from the output to the auxiliary circuit creates losses, however, that reduce the overall efficiency at high power and the voltage stress on the boost diode is as high as twice the output voltage.

3.2.3 ZCT-PWM Converters with Parallel Auxiliary Circuit

Another converter whose auxiliary circuit does not cause an increase in the current stress of the main power switch was proposed in [11] and is shown in Figure 3.8. The converter is a boost converter with an auxiliary circuit that consists of a switch S_a, a diode D_2, a capacitor C_r, and an inductor L_{r2}. Inductor L_{r1} is placed in series with the main switch S_1 so that this switch

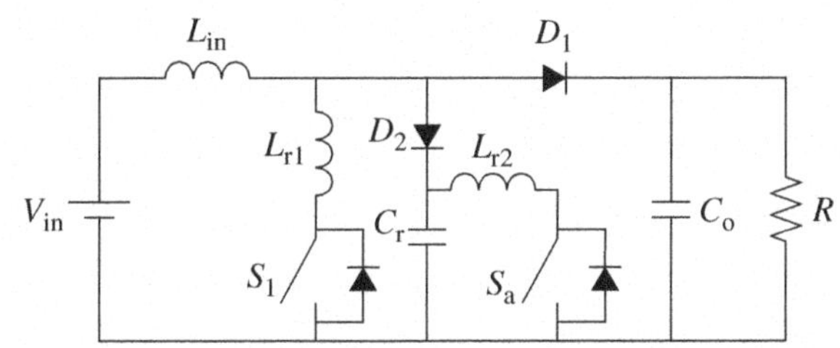

Figure 3.8 Converter with parallel auxiliary circuit proposed in [11].

can turn on with ZCS. The converter works as follows. Capacitor C_r is discharged sometime after S_a is turned on to activate the auxiliary circuit. The voltage across C_r drops to zero and then continues to swing to a negative voltage due to the resonant interaction with L_{r2}. This forward-biases diode D_2 and the diode starts to conduct. The current flowing through S_1 begins to be diverted away from this switch to the auxiliary circuit when the voltage across C_r is negative. S_1 can be turned off with ZCS when the current that was flowing through it is completely transferred to the auxiliary circuit. Current flowing through S_a eventually reverses direction due to the resonant interaction between L_{r2} and C_r, and flows through the antiparallel diode across S_a, which allows S_a to be turned off softly. The current in the auxiliary circuit is eventually extinguished as C_r is charged to the output voltage and the boost diode conducts.

The converter's auxiliary switch can turn on and off softly with no increase in peak current in the main switch and with few components in the main power path of the converter. There is, however, a significant amount of energy in the form of circulating current trapped in the auxiliary circuit and the boost diode peak voltage is close to twice the output voltage. Greater flexibility in the design of this converter can be achieved by adding an inductor between D_2 and C_r, as suggested in [12]. Doing so, however, increases the peak voltage stress of C_r and S_a as the voltage across C_r is no longer clamped to the output voltage.

A variation on this converter was proposed in [13] and is shown in Figure 3.9. This converter is similar to the converter shown in Figure 3.8 except that a transformer with two secondary diodes has been added to the auxiliary circuit. The addition of the transformer in the auxiliary circuit improves the performance of the ZCS-PWM converter and allows it to have advantages that the other converters of this type do not have. The transformer primary is clamped to $V_x = V_o(N_1/N_2)$ and the secondary diode D_{a1} is forward-biased when current is flowing through the auxiliary switch and the transformer primary is clamped to $V_x = -V_o(N_1/N_2)$ and diode D_{a2} is forward-biased when current is flowing through its body-diode.

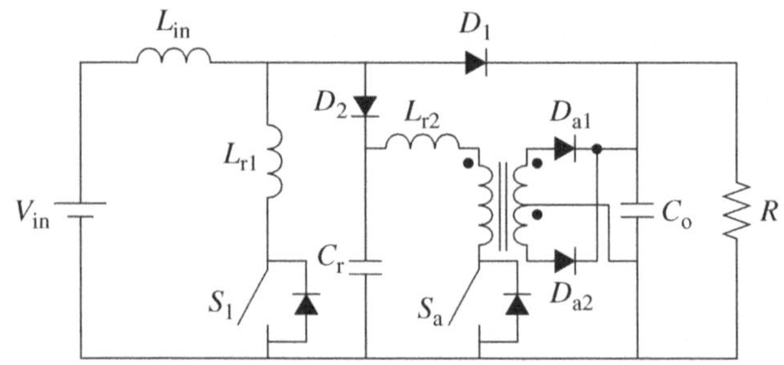

Figure 3.9 Converter with auxiliary circuit transformer proposed in [13].

The counter voltage created by the primary of the transformer reduces the negative peak voltage of the resonant capacitor C_r and thus reduces the voltage stress on the boost diode, which is typically around 2–2.5 times the output voltage in ZCS-PWM boost converters in general. Moreover, the circulating energy from the auxiliary circuit is transferred to the output whenever current is flowing through the transformer primary so that this energy is not trapped in the auxiliary circuit. This results in conduction losses that will be lower than those of other ZCS-PWM converters. The main drawbacks of this converter are that the additional transformer and secondary diodes increase the size and cost of the converter and decrease the converter's light-load efficiency as the transformer has losses that are only overcome when the converter is operating with heavier loads.

3.3 ZVS/ZCS-PWM BOOST CONVERTERS

Comparing the operation of the ZVS-PWM converters that were presented in the previous chapter with the ZCS-PWM converters that were presented in this chapter, it can be seen that there are similarities and differences in the operation of the two types of soft-switching converters. The main similarity is that an auxiliary circuit with an active auxiliary switch is used to help the main converter switch operate with soft-switching; the main difference is when the auxiliary switch is activated. The auxiliary switch is turned on just before the main switch is about to be turned on in ZVS-PWM converters, while it is turned on just before the main switch is about to be turned off in ZCS-PWM converters. Given that an auxiliary circuit is to be turned on before the turn-on of the main switch for one type of soft-switching and turned on before the turn-off of the main switch for the other type, soft-switching converters that use the same auxiliary circuit to ensure that the main switch can turn on with ZVS and turn off with ZCS have been proposed. The auxiliary circuit in such a ZVS/ZCS – PWM converter is activated twice during a switching cycle – once to assist the ZVS turn-on of the main switch, and a second time later in the same switching cycle to assist the ZCS turn-off of this switch.

An example of such a ZVS/ZCS – PWM converter was proposed in [14] and is shown in Figure 3.10. This converter is a boost converter with an auxiliary circuit that consists of a switch S_a with an antiparallel diode, a resonant inductor L_r, a resonant capacitor C_r, and a blocking diode D_2. The converter works as follows: Just before the main switch S_1 is to be turned on with ZVS, auxiliary switch S_a is turned on with ZCS as L_r slows down the rate of current rise in the switch, thus reducing the amount of overlap between switch voltage and switch current during this switching transition. L_r also helps ensure that the transition of current away from the main boost diode, D_1, is gradual so that reverse-recovery current is eliminated. C_r is charged to a voltage of V_{in}, the input voltage, and is clamped to this voltage by D_2. Since

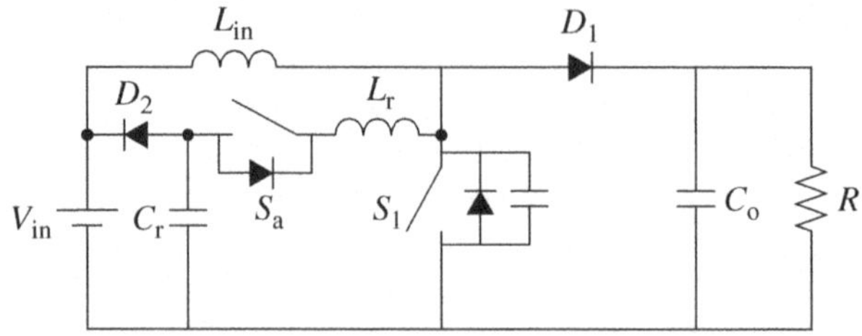

Figure 3.10 ZVS/ZCS-PWM converter proposed in [14].

the converter is a boost converter, the initial voltage across the main switch, which is equal to the output voltage, is greater than V_{in} so that current flows in S_a and the output capacitance of S_1 begins to discharge. Current continues to flow through S_a when the voltage across S_1 drops below V_{in} as L_r continues to draw current. Eventually, the voltage across S_1 drops to zero and S_1 can be turned on with ZVS.

After S_1 has been turned on with ZVS, the current in L_r begins to reverse direction and flow through the body-diode of the auxiliary switch and S_1 so that the current flowing through S_1 is the summation of the input inductor current and the auxiliary circuit current. This auxiliary circuit current is a resonant hump as the current is initially zero, then rises to its peak value, and then falls to zero as C_r is discharged. It should be noted that the voltage across C_r becomes negative and remains negative after the current has ceased to flow through L_r. This prepares the auxiliary circuit for the ZCS turn-off of the main switch, later in the switching cycle.

Just before the main switch is to be turned off, the auxiliary switch is turned on a second time during the switching cycle with ZCS. Even though the voltage across the main switch S_1 is zero as it is on, the current is diverted away from S_1 as the voltage across C_r is negative, so that the current flows toward a path of less potential difference. S_1 can be turned off with ZCS once the current through it is zero. The current in L_r is eventually extinguished and the voltage across C_r is positive, in preparation for the next switching cycle. Diode D_2 ensures that this voltage does not exceed V_{in}.

An alternative ZVS/ZCS – PWM boost converter is the converter that was proposed in [15] and shown in Figure 3.11. This converter has an auxiliary circuit that consists of a switch S_a with an antiparallel diode D_a, a resonant capacitor C_r, and a resonant inductor L_r. The converter works as follows: Just before S_1 is to be turned on, S_a is turned on with ZCS as L_r slows down the rate of current rise in the switch after it is turned on. As with the previous converter, reverse-recovery current does not appear in the main boost diode. The voltage across C_r is positive, but less than the output voltage so that current is gradually transferred away from the boost diode and flows in the auxiliary circuit. After the current ceases to flow in D_1, the output capacitance

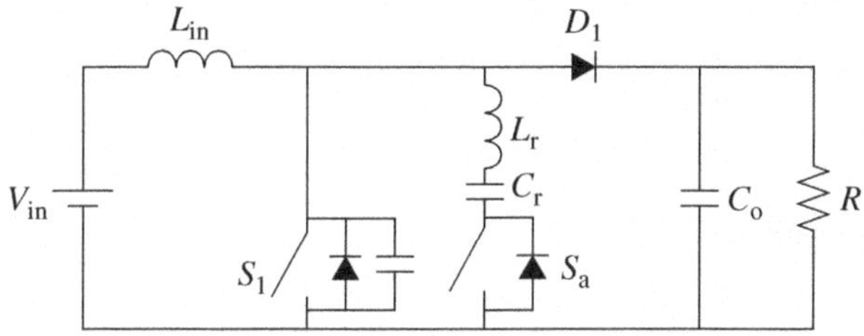

Figure 3.11 ZVS/ZCS-PWM converter proposed in [15].

of S_1 begins to discharge and eventually drops to zero. When this happens, S_1 can then be turned on with ZVS.

With the voltage across S_1 clamped to zero, the current in L_r ceases to flow then reverses direction and flows through the antiparallel diode across S_a, thus allowing S_a to be turned off softly. Eventually, current in L_r ceases to flow altogether and the voltage across C_r is negative, in preparation for when the auxiliary circuit is to be activated to assist in the ZCS turn-off of S_1, later in the switching cycle.

When S_1 is about to be turned off, S_a is turned on with ZCS and the current through S_1 begins to be diverted into the auxiliary circuit. After the full input inductor current has been transferred to S_a, current begins to flow through the antiparallel diode across S_1 and it is while this is happening that S_1 can be turned off with ZCS. It should be noted that the auxiliary resonant circuit must be designed properly to ensure that the current drawn by the auxiliary circuit is greater than the input inductor current so that current does flow in the anti-parallel diode across S_1. Moreover, the design should be such that after S_1 is turned off and the voltage across this switch rises, the voltage across C_r remains higher than the voltage across C_{s1} so that current can be transferred into C_{s1} and S_a can be turned off with ZCS. With the full input current flowing through C_{s1}, the voltage across S_1 continues to rise and is clamped to the output voltage by the boost diode, D_1. The voltage across C_r remains positive, but less than the output voltage in preparation for the next switching cycle, while current is flowing through D_1.

The main drawback that ZVS/ZCS-PWM converters have is that the duty-cycle of the converter must be limited so that time is given to the auxiliary circuit to operate twice during a switching cycle. Since ZVS/ZCS-PWM converters must have resonant-type auxiliary circuits with a resonant capacitor that can present a negative voltage path for the main switch current to be diverted to when it is time to turn this switch off, the time that the auxiliary switch must be on is part of a resonant cycle with a duration that is not insignificant. In addition to duty-cycle loss, the fact that the auxiliary switch must be active twice during a switching cycle means that the current stress of this switch is increased so that it must be implemented with a more expensive device. The loss of duty-cycle and the increased current stress of the auxiliary switch have made the use of ZVS/ZCS-PWM converters less popular than purely ZVS-PWM or ZCS-PWM converters, but these converters can be effective under the right circumstances.

3.4 CONCLUSION

ZCS-PWM converters are converters that use an active auxiliary circuit to allow their main switch to turn off with ZCS. ZCS methods are better suited for silicon IGBTs than ZVS methods because these devices are minority carrier devices and as such, charge cannot be quickly removed from them when they are about to be turned off. IGBTs, therefore, turn off very slowly and with a considerable current tail that creates switching losses. If the current can somehow be gradually removed from an IGBT device before it is turned off, then the device can do so without this current tail, thus turning off with fewer switching losses. Moreover, without the current tail, an IGBT can be turned off quickly and thus can be made to operate with higher switching frequencies. The turn-on losses of silicon IGBTs are lower than those of silicon MOSFETs because they have less output capacitance. ZCS methods are not suited for converters with wide bandgap devices because these devices do not have a turn-off current-tail that slows down their turn-off time and generates turn-off switching losses. Nonetheless, ZCS-PWM converters continue to be used at the time of this writing as Si-IGBTs continue to be used due to their lower cost, greater availability, and perceived reliability.

A number of ZCS-PWM converters were reviewed in this chapter, each with its own advantages and disadvantages. The ZCS-PWM converters discussed in Section 3.1 have been mostly superseded by the converters discussed in Section 3.2 and are only used when the cost is the overriding issue in a design. This is mainly because the converters discussed in Section 3.2 do not have any auxiliary circuit current flowing in the main converter switch, which reduces conduction losses in the converter. Of the converters discussed in Section 3.2, the most efficient converters are the ones shown in Figures 3.8 and 3.9. As demonstrated in [16], the converter shown in Figure 3.9 is more efficient than the converter in Figure 3.8 for heavy loads as auxiliary circuit energy can be transferred to the output through the transformer, but is less efficient for light loads because of the losses that this transformer has. Regardless of which ZCS-PWM converter is used, it will never be as efficient as a hard-switching converter when operating under light-load conditions due to the losses that the auxiliary circuit generates. It is only under heavy-load conditions that the savings in switching losses that would otherwise be generated by the main switch exceed the auxiliary circuit losses.

ZVS/ZCS-PWM converters are converters that use a single active auxiliary circuit to help the main converter switch turn on with ZVS and turn off with ZCS. The auxiliary switch is turned on just before the main converter switch is to be turned on so that this switch can do so with ZVS. It is turned on before the main switch is to be turned off so that it can do so with ZCS. ZVS/ZCS-PWM converters can be more efficient than purely ZVS or purely ZCS converters, but only under certain operating conditions. What limits the effectiveness of ZVS/ZVS-PWM converters is the fact that the auxiliary circuit must be a resonant-type circuit that is activated twice during a switching cycle. This results in duty-cycle loss to allow the auxiliary circuit resonant cycle to be completed, which limits the gain of the converter. It also increases the current stress on the auxiliary switch as it must be on for a longer duration over the switching cycle.

REFERENCES

1. I. Barbi, J.C.O. Bolacell, D.C. Martins, and F.B. Libano, "Buck quasi-resonant converter operating at constant frequency: analysis, design, and experimentation," *IEEE Transactions on Power Electronics*, vol. 5, no. 3, pp. 276–283, Jul. 1990.
2. K. Wang, G. Hua, and F.C. Lee, "Analysis, design and experimental results of ZCS-PWM boost converters," *IEEE International Power Electronics Converter (IPEC)*, 1995, Apr. 3–7, 1995, Yokohama, Japan, pp. 1197–1202.
3. G. Ivensky, D. Sidi, and S. B. Yaakov, "A soft-switcher optimized for IGBTs in PWM topologies," *IEEE Applied Power Electronics Conference (APEC)*, 1995, Mar. 5–9, 1995, Dallas, USA, pp. 900–906.
4. T. Mishima and M. Nakaoka, "A practical ZCS-PWM boost DC-DC converter with clamping diode-assisted active edge-resonant cell and its extended topologies," *IEEE Transactions on Industrial Electronics*, vol. 60, no. 6, pp. 2225–2236, Jun. 2013.
5. R. C. Fuentes and H. L. Hey, "An improved ZCS-PWM commutation cell for IGBT's application," *IEEE Transactions on Power Electronics*, vol. 14, no. 5, pp. 939–948, Sep. 1999.
6. G. Hua, E. X. Yang, Y. Jiang, and F. C. Lee, "Novel zero-current-transition PWM converters," *IEEE Transactions on Power Electronics*, vol. 9, no. 6, pp. 601–606, Nov. 1994.
7. C. A. Canesin and I. Barbi, "Novel zero-current-switching PWM converters," *IEEE Transactions on Industrial Electronics*, vol. 44, no. 3, pp. 372–381, Jun. 1997.
8. F. T. Wakabayashi, M. J. Bonato, and C. A. Canesin, "Novel high-power factor ZCS-PWM pre-regulators," *IEEE Transactions on Industrial Electronics*, vol. 48, no. 2, pp. 322–333, Apr. 2001.
9. D.-Y. Lee, M.-K. Lee, D.-S. Hyun, and I. Choy, "New zero-current transition PWM DC/DC converters without current stress," *IEEE Transactions on Power Electronics*, vol. 18, no. 1, pp. 95–104, Jan. 2003.
10. C.-M. Wang, "A novel ZCS-PWM power-factor preregulator with reduced conduction losses," *IEEE Transactions on Industrial Electronics*, vol. 52, no. 3, pp. 689–700, Jun. 2004.
11. H.-S. Choi and B. H. Cho, "Novel zero-current-switching (ZCS) PWM switch cell minimizing additional conduction loss," *IEEE Transactions on Industrial Electronics*, vol. 49, no. 1, pp. 165–172, Feb. 2002.

12. X. Wu, G. Wu, J. Zhang, and Z. Qian, "New ZCT-PWM cell for dc-dc converters with reduced current stress and conduction loss," *IEEE International Telecommunications Energy Conference*, 2005, Sept. 18–22, 2005, Berlin, Germany, pp. 539–544.
13. P. Das and G. Moschopoulos, "A zero-current-transition converter with reduced auxiliary circuit losses," *IEEE Transactions on Power Electronics*, vol. 22, no. 4, pp. 1464–1471, Jul. 2007.
14. C. M. de Olivera Stein and H. L. Hey, "A true ZCZVT commutation cell for PWM converters," *IEEE Transactions on Power Electronics*, vol. 15, no. 1, pp. 185–193, Jan. 2000.
15. N. Altintas, A. F. Bakan, and I. Aksoy, "A novel ZVT-ZCT-PWM boost converter," *IEEE Transactions on Power Electronics*, vol. 29, no.1, pp. 256–265, Jan. 2014.
16. P. Das and G. Moschopoulos, "A comparative study of zero-current-transition PWM converters," *IEEE Transactions on Industrial Electronics*, vol. 54, no. 3, pp. 1319–1328, Jun. 2007.

CHAPTER 4

BASIC ISOLATED CONVERTERS

The basic converters that were presented in Chapter 1 can step voltage up or down, but they are limited in their ability to do so. For example, if a DC voltage is to be stepped down to 10% of its original level with a conventional pulse-width modulation (PWM) buck converter, then the converter's duty-cycle must be very small, approximately 10%. The small duty-cycle is impractical as this implies that energy transfer from the input to the output occurs for only 10% of the switching cycle so that a tremendous amount of energy must be transferred over a very short amount of time. Such a scenario would result in very high peak converter component stresses and would be very inefficient. Similar issues would arise if a very high voltage gain using a boost converter is desired as this would require the boost converter switch to be on for at least 90% of the switching cycle.

Extreme duty-cycles can be avoided in DC–DC converters if they are implemented with transformers. Although transformers cannot step DC voltage up or down directly by impressing a DC voltage across their primary winding, they can be used in DC–DC converters if care is taken to ensure that they are not saturated. This entails storing energy in a transformer during the initial stages of a switching cycle and then removing this energy during the latter stages so that flux in the transformer is not built up to the extent that it leads to transformer saturation.

Basic isolated DC–DC converters are presented in this chapter. This chapter will begin by reviewing basic transformer models and reviewing the two most fundamental single-switch isolated DC–DC converters: the flyback

DC–DC Converter Topologies: Basic to Advanced, First Edition. Gerry Moschopoulos.

Published 2024 by John Wiley & Sons, Inc.

converter and the forward converter. Several variations of the basic forward converter will then be presented and the chapter will conclude with a discussion of several multiswitch converter topologies that can be used in higher power applications where single-switch isolated converters are unsuitable.

4.1 TRANSFORMER MODELS

An ideal transformer is typically depicted in circuit diagrams as an element with two or more windings on a core that can step voltage up or down without any losses of any kind. In reality, a transformer has several non-idealities that must be taken into account when designing a power converter. A commonly used transformer model that considers certain key non-ideal effects is shown in Figure 4.1a. This model consists of primary winding resistance R_1, primary self-inductance L_1, core resistance R_c, core magnetizing inductance L_m, an ideal transformer that can step down or step up voltage, secondary winding resistance R_2, and secondary resistance L_2. Inductances L_1 and L_2 model flux that is not contained in the core, but leaks out of the core in air; thus, they are often referred to as leakage inductances. L_m models the flux that is contained in the core, which is generated at the primary and is coupled to the secondary winding; thus it is often referred to as the mutual inductance. R_c models transformer core losses.

The model shown in Figure 4.1a is often simplified by referring the secondary impedances, R_2 and L_2, to the primary, as shown in Figure 4.1b, then lumping them with R_1 and L_1 so that the approximate model shown in Figure 4.1c is obtained. The lumping of the reflected R_2 with R_1 and the reflected L_2 with L_1 is an approximation that can be justified by the fact that impedances of R_c and L_m are much larger than those of the reflected impedances and thus considerably less current flows in R_c and L_m. The model can be further simplified by removing all resistances, as shown in Figure 4.1d, as well as the leakage inductance so that only the magnetizing inductance and the ideal transformer remain, as shown in Figure 4.1e. This simplified model is sufficient for the discussion of isolated converters in this chapter.

4.2 FLYBACK CONVERTER

The most fundamentally isolated DC–DC converter is the flyback converter as shown in Figure 4.2. Figure 4.2a shows the converter with a conventional transformer symbol while Figure 4.2b shows the converter with the simplified transformer model shown in Figure 4.1e. The flyback converter can be considered to be the isolated version of the buck–boost converter that was discussed in Section 1.2.3. It can be synthesized from the buck–boost converter by replacing the inductor with a transformer and rearranging the primary and secondary components.

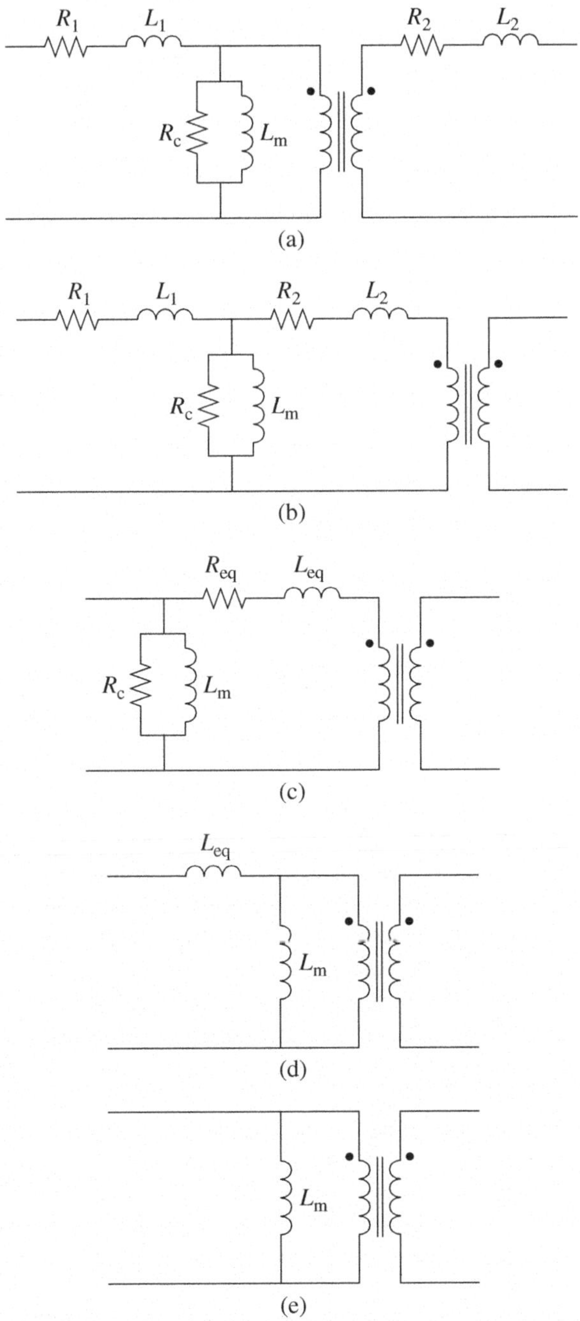

Figure 4.1 Transformer models. (a) Full transformer model. (b) Full transformer model with secondary impedances referred to the primary. (c) Approximate transformer model. (d) Simplified approximate transformer model with resistive losses removed. (e) Ideal transformer with magnetizing inductance.

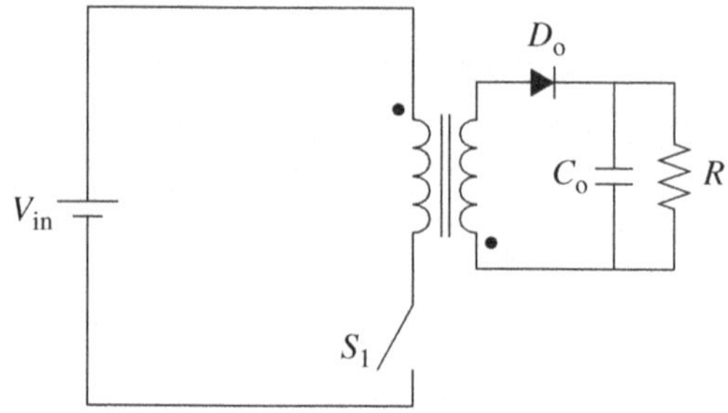

(a)

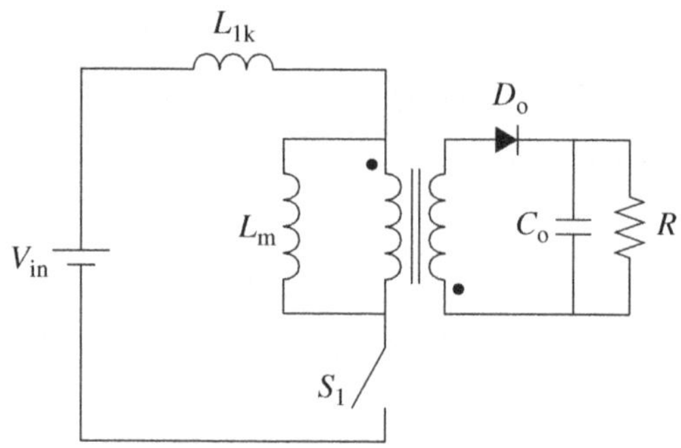

(b)

Figure 4.2 Flyback converter (a) with conventional transformer symbol (b) with simplified model.

The operation of the flyback converter is very similar to that of the buck–boost converter as energy is placed into the magnetic element when the switch is turned on, and then transferred to the output through the output diode when the switch is turned off. Moreover, just as the inductor in the buck–boost converter can operate with continuous or discontinuous current, so too can the flyback converter operate with continuous or discontinuous flux in its core, which is modeled as continuous or discontinuous current in the core magnetizing current flowing in L_m in the circuit shown in Figure 4.2b. A more common way of stating this type of operation is to say that the magnetic core of the flyback converter can be magnetized throughout a switching cycle, which corresponds to continuous current in L_m, or it can become demagnetized with all its flux removed sometime later in the cycle after the switch has been turned off, which corresponds to discontinuous current flowing through L_m.

The following should be noted about the flyback converter:

- The converter has a natural reset because the voltage of the opposite polarity can be impressed across the transformer after the switch has been turned off. The volts-seconds of the transformer can be naturally balanced.

- Since the flyback converter is essentially an isolated buck–boost converter, its voltage gain is that of a buck–boost converter, but with an additional N_2/N_1 factor that is the transformer's turns ratio. The converter's gain when the core is never fully demagnetized is

$$V_o = V_s\left(\frac{D}{1-D}\right)\left(\frac{N_2}{N_1}\right) \tag{4.1}$$

- The peak voltage stress across the switch is

$$V_{s1,pk} = V_s + V_o\left(\frac{N_1}{N_2}\right) \tag{4.2}$$

which is the sum of the input voltage and the reset voltage across the primary winding of the transformer.

- It is possible to operate a flyback converter with a duty-cycle D that is greater than 0.5, but doing so results in increased peak voltage stress as the voltage needed to reset the transformer must be increased to make up for the shorter reset time. As a result, such an operation tends to be avoided or restricted to $D < 0.8$.

4.3 FORWARD CONVERTER

Although the flyback converter is simple, it has several fundamental drawbacks, including the following:

- All the energy that must be transferred from the input to the output must be stored in the transformer first as energy is transferred to the transformer when the switch is on and is only transferred to the output when it is off. As a result, the converter is limited to low-power applications, typically no more than 150 W, because the transformer becomes bulky otherwise.
- The output of the converter is just a C filter instead of an LC filter. As a result, the converter should not be used in applications where the output voltage is low and the output current is high, even when the maximum load that the converter must supply is less than 150 W. For example, a 100 W converter with 5 Vdc output requires 20 A of output current. Its output current has a large ripple as current only flows through the output diode for a fraction of the switching cycle when the switch is off. This necessitates the use of a very large capacitor in a flyback converter to filter the current.

The forward converter, shown in Figure 4.3 addresses these drawbacks, but it is not as simple as the flyback converter and it is more expensive as it has more components. This converter is essentially an isolated buck converter as the left hand of the converter is primarily a mechanism that places a chopped voltage at the front of the LC output filter, much like a buck converter's switch and diode, which act as a voltage chopper circuit. The converter works as follows: When the switch is turned on, the input voltage is impressed across the primary of the transformer and is either stepped up or down, depending on the transformer's turns ratio. Since diode D_{o1} at the secondary is forward-biased, this voltage, V_x, appears at the left-hand side of the output inductor. When the switch is turned off, D_{o1} is reverse biased and current freewheels through D_{o2}, much like the output inductor of a buck converter freewheels through the buck diode.

Energy can be transferred directly from the output to the input when the switch is on so that the transformer in a forward converter can be smaller than that of a flyback converter, which has no such energy transfer. Some energy, however, is stored in the transformer during this time and it must be removed later in the switching cycle to avoid transformer saturation. In the conventional forward converter topology, this can be done using a third or tertiary winding that is taken from the main transformer core and, with a series diode, is connected across the input voltage source. This winding is very similar to the secondary winding of a flyback converter and this is indicated in Figure 4.3 by an opposing dot notation. Although this winding can be connected to the output, this is rarely done because doing so would require that the winding be thicker to handle the current it would deliver to the output. More importantly, however, it would offset a key advantage that the forward converter has over the flyback converter and that is its LC output filter. Connecting the tertiary winding to the output would increase the output current ripple and would thus defeat the purpose of the filter.

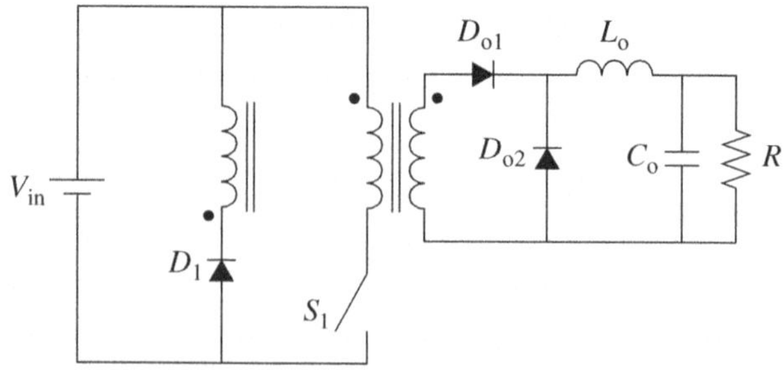

Figure 4.3 Forward converter.

4.4 VARIATIONS ON THE FORWARD CONVERTER

A number of variations to the conventional forward converter have been proposed to reset the transformer without using a tertiary winding, as this increases the size, cost, and design complexity of the transformer. Three such variations are reviewed in this section: the forward converter with RCD snubber, the forward converter with LCDD snubber, and the forward converter with regenerative energy snubber. It should be noted that the RCD snubber, the LCDD snubber, and the regenerative energy snubber can all be used in the flyback converter. The operation of these snubbers in flyback converters will not be discussed here as it is almost identical to that when used in forward converters. When they are used in flyback converters, however, they do not need to be designed to reset the transformer since these converters can reset naturally. Instead, they can be designed just to suppress potential overvoltage spikes that might appear when the switch is turned off. When these snubbers are used in forward converters, they must be able to reset the transformer and suppress overvoltage spikes across the switch.

4.4.1 Forward Converter with RCD Snubber

Figure 4.4 shows a forward converter with a RCD snubber that consists of diode D_1, capacitor C_1, and resistor R_1. The converter works as follows: When the switch is turned on, the input DC voltage is impressed across the primary of the transformer. Secondary output diode D_{o1} is forward-biased and voltage is impressed on the left-hand terminal of output inductor L_o. When the switch is turned off, current flows through snubber diode D_1 so that the voltage across C_1 is impressed across the transformer primary, thus resetting the transformer. While this is happening, diode D_{o1} is reverse-biased and D_{o2} is forward-biased so that current freewheels in the output. Depending on the size of L_o, the current may freewheel until the switch is turned on again to start

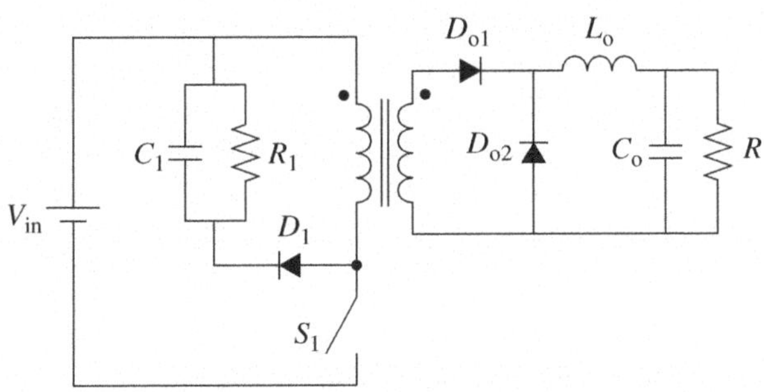

Figure 4.4 Forward converter with RCD snubber.

the next switching cycle or it may fall to zero and remain at zero until the start of the next switching cycle. The resistor across capacitor C_1 allows the capacitor to discharge so that the voltage across it does not become excessive.

The main advantage that the RCD snubber forward converter has over the conventional forward converter is its simplicity. Transformer reset can be achieved with a simple RCD snubber and a two-winding transformer rather than with a three-winding transformer that is bulkier and more expensive. Another advantage is that the voltage across C is adaptive. When the converter's duty-cycle is low, the amount of energy placed in the transformer when the switch is on is relatively low so that the amount of available reset voltage is low as well. With a low-duty-cycle, the time during which the switch is off and the transformer can be reset is larger, so that it is possible to reset the transformer with a low reset voltage. Likewise, when the converter's duty-cycle is high, the amount of energy placed in the transformer when the switch is on is relatively high so that the amount of available reset voltage is high as well. With a high-duty-cycle, the time during which the switch is off and the transformer can be reset is lower, but it is possible to reset the transformer in the shorter reset time available since the reset voltage is higher. This is unlike a conventional forward converter with tertiary reset winding where the reset voltage is fixed, regardless of load or duty-cycle. Given that forward converters with tertiary windings are generally designed so that the primary and the reset winding have the same number of turns, the maximum converter duty-cycle is limited to $D = 0.5$ so that the peak voltage stress across the switch is $2\,V_s$. In the case of the forward converter with RCD snubber, the adaptive nature of the voltage across snubber capacitor C allows the converter to be operated with duty-cycles that are greater than 0.5.

The main drawback of the forward converter with RCD snubber is that the energy in the snubber capacitor is discharged through R, which results in losses. This limits the converter to low-power applications (<50 W) to limit power dissipation. It should be noted that the low power limit of the converter makes the converter attractive as a back-end converter of a two-stage AC–DC converter with a passive diode-bridge front-end. This is especially true since at low power, active AC–DC power factor correction is not a requirement and the converter can operate with low-line and high-line voltage over the universal input voltage range (85–265 Vrms). Since the output of the passive input diode-bridge rectifier is low under low-line conditions and high under high-line conditions, the converter's ability to operate with duty-cycles greater than 0.5 is very attractive compared to the conventional forward converter's inability to do so.

4.4.2 Forward Converter with LCDD Snubber

In order to improve the efficiency of the forward converter with an RCD snubber, the converter snubber is replaced with an LCDD snubber, as shown in Figure 4.5. The converter works in the following manner: The input voltage

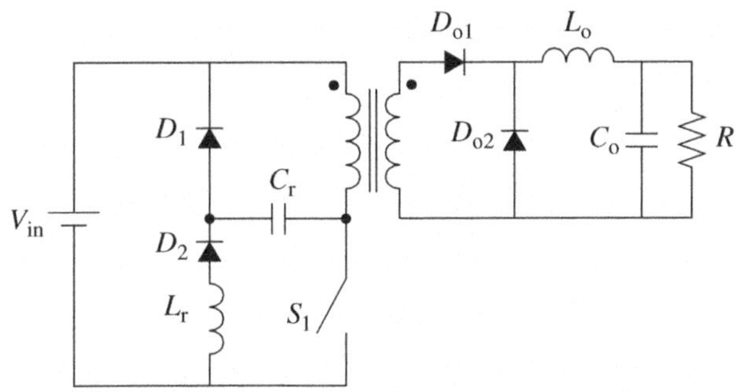

Figure 4.5 Forward converter with LCDD snubber. (Note: The windings share the same core).

is impressed across the transformer primary when the switch is turned on. After it is turned off, energy from the transformer is transferred to the snubber capacitor, C_r, through diode D. When this happens, the voltage across C_r is impressed across the transformer and resets it. When the switch is turned on again to begin the next cycle, voltage is again impressed across the transformer primary, but the capacitor is discharged through the switch as well, through D_2 and snubber inductor L_r. The polarity of the voltage across C_r is changed during this resonant discharge process. The converter operates in the same way as a convention forward converter at its secondary side. It has the same adaptive snubber capacitor voltage ability as a forward converter with an RCD snubber but is more efficient because energy is not dissipated in a resistor. It is also bulkier and more expensive as it has more components.

4.4.3 Forward Converter with Regenerative Energy Snubber

Researchers have determined that the efficiency of the LCDD snubber can be improved if the inductor is replaced with a transformer winding. One way of doing so was proposed in [1] and shown in Figure 4.6. In the diagram shown in Figure 4.6, it is assumed that the snubber inductor L_r is embedded in the transformer.

The converter operates in the same manner as the forward converter with LCDD snubber except that energy that would otherwise be trapped in the primary side of the converter can be transferred to the output, thus improving converter efficiency. It is the fact that energy can be recycled and transferred to the output with the additional winding that the converter is referred to as a forward converter with a regenerative snubber in the literature. The additional winding can also reduce the peak current stress of the switch as it creates a counter voltage in the C_r–D_2, switch loop that reduces the amplitude of the current hump in the switch that is caused by discharging of C_r after it is turned on. The additional inductance may be added in series with the winding if it is desired to further reduce the peak of the current hump. The forward

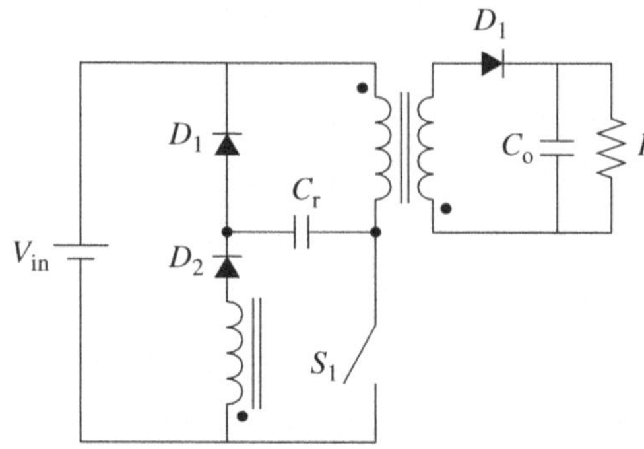

Figure 4.6 Flyback converter with energy regenerative snubber. The windings share the same core.

converter with regenerative snubber performs better than the forward converter with RCD or LCDD snubbers, but it must be implemented with a transformer that has a tertiary winding. This partially offsets the advantages of the converter over the RCD snubber forward converter and the LCDD snubber forward converter.

4.5 BASIC TWO-SWITCH ISOLATED CONVERTERS

Two-switch isolated converters have two main switches in their topology and can be operated at higher power levels than flyback or forward converters. They are mainly isolated buck converters with a different reset mechanism than the forward converter that addresses some of the drawbacks found in conventional forward converters. Three of the most basic converters, two-switch converters, are reviewed in this section: the two-switch forward converter, the push-pull converter, and the half-bridge converter.

4.5.1 Two-Switch Forward Converter

Figure 4.7 shows a two-switch forward converter, which is also known as the double-ended forward converter in the literature. The converter has two switches and two diodes in the primary but has the same secondary-side circuit as a conventional forward converter so that its secondary operates in the same manner. Although two-switch flyback implementations are possible, they are rarely used as they detract from the simplicity of the basic flyback topology.

The converter shown in Figure 4.7 operates as follows: When both switches are turned on simultaneously, voltage is impressed across the primary winding of the transformer. When the switches are turned off, also simultaneously, primary current flows from the transformer through diode D_2, the input source, and D_1 so that a current loop is formed. Given how D_1 and D_2 are connected

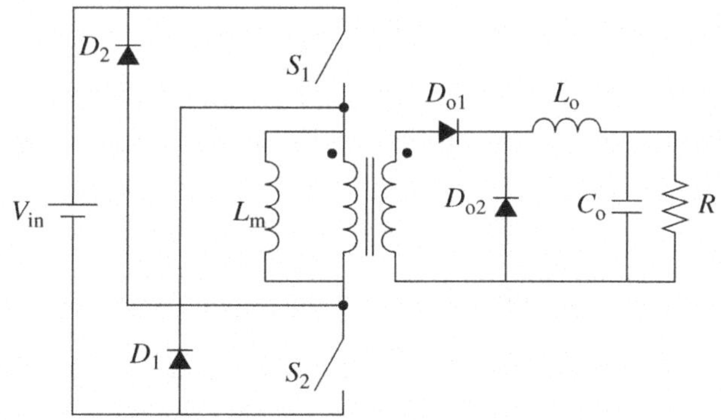

Figure 4.7 Two-switch forward converter.

to the transformer, a negative voltage is impressed across the transformer when these diodes conduct, thus resetting the transformer. The transformer may or may not become fully demagnetized after the switches have been turned off.

The main advantage that the two-switch forward converter shown in Figure 4.7 over the conventional forward converter topology is that the peak voltage stress of the two converter switches is the input voltage. This is unlike the peak voltage stress of the switch in a conventional flyback converter, which is at least twice the input voltage and maybe even greater, depending on the design. Another advantage is that a smaller and less bulky two-winding transformer can be used instead of a transformer with a tertiary winding. The main drawback of the two-switch forward converter is that it requires more switches and diodes than a conventional forward converter, but for high-input voltage applications, it is not uncommon to find that using two switches with conventional voltage ratings (e.g. 600 V) is less expensive than using a single switch with a very high voltage rating (e.g. 1200 V). Another drawback is that the duty-cycle of the converter cannot be greater than 0.5 as the only possible reset voltage is the input voltage and positive and negative volt-seconds must match.

The advantages of the two-switch forward converter are:

- The peak voltage stress of the switches is equal to the input voltage, unlike the push-pull converter.
- It is immune to transformer saturation as it has a natural reset mechanism in the form of the two primary diodes, unlike the push-pull converter.
- It has fewer bulk capacitors than the half-bridge converter.
- The full input voltage is impressed across the transformer, unlike the half-bridge converter.

The disadvantages of the two-switch forward converter are:

- The frequency of the voltage fed to the output LC filter is the same as that of the transformer primary and secondary voltage as power is transferred to the output only once during a switching cycle.
- The duty-cycle of the converter cannot be greater than 0.5.
- The converter requires a floating gate-drive circuit for S_1, which is more difficult to implement than a gate-drive circuit that can be referenced to the ground. Some sort of isolation or additional circuitry is needed to ensure that the source of the S_1 switch is not somehow connected to the ground.

4.5.2 Push–Pull Converter

Figure 4.8 shows the push–pull converter. This converter has two primary switches, a transformer with a primary set and a secondary set of center-tapped windings, two output diodes and an LC output filter. All the windings shown in Figure 4.8 are wrapped on the same transformer core. Unlike the two-switch forward converter, the switches in the push–pull converter are never on at the same time and their gating signals are shifted by 180° in phase. Switch S_1 is turned on at the start of the switching cycle, then turned off before the switching cycle is halfway through while S_2 is turned on at the halfway point of the switching cycle, and then turned off before the switching cycle ends. The gating signals for the two switches have the same pulse width.

The converter operates as follows: When switch S_1 is turned on, positive V_{in} voltage is impressed across primary winding N_{p1}. This results in negative V_{in} being impressed across primary winding N_{p2} and positive voltage impressed across both secondary windings. The voltage across the secondary windings is either lower or higher than V_{in}, depending on whether V_{in} has

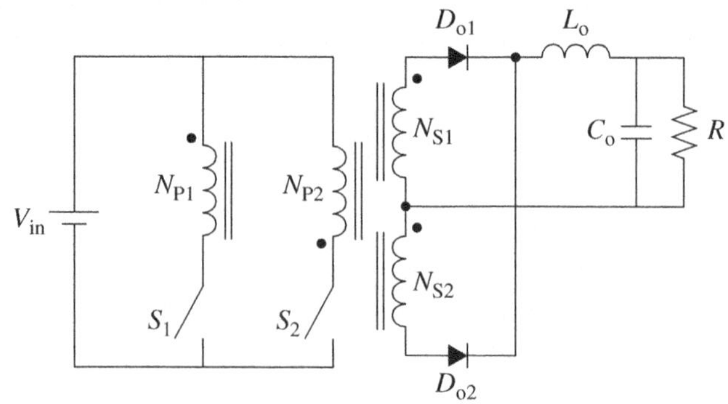

Figure 4.8 Push–pull converter.

been stepped down or up. Of the two secondary-side diodes, only D_{o1} is forward-biased and the voltage across secondary winding N_{s1} appears at the front end of the output LC filter. When switch S_2 is turned on, positive V_{in} voltage is impressed across primary winding N_{p2}. This results in negative V_{in} being impressed across primary winding N_{p1} and negative voltage impressed across both secondary windings. Of the two secondary-side diodes, only D_{o2} is forward-biased and the voltage across secondary winding N_{s2} is rectified and appears at the front end of the output LC filter. When neither switch is on, then the output current is ideally split between the two secondary diodes equally and freewheels through them. This current distribution is generally uneven in real-life push–pull converters.

The advantages of the push-pull converter are:

- The frequency of the voltage fed to the output LC filter is twice that of the transformer primary and secondary voltage as power is transferred to the output twice during a switching cycle, unlike a two-switch forward converter.
- The converter does not require any floating gate-drive circuits for its switches, unlike a two-switch forward converter and a half-bridge converter.
- The full input voltage can be impressed across a transformer primary winding, unlike a half-bridge converter.
- It has fewer bulk capacitors than the half-bridge converter.

The disadvantages of the push–pull converter are:

- The converter switches are exposed to a higher peak voltage stress than the other two-switch converters as it is twice the input voltage. This is because when one of the two switches is on, the other will have the input voltage plus the voltage across its primary winding, which is also the input voltage, across it. The two primary windings have the same turns ratio.
- The push–pull converter is prone to transformer saturation as it is difficult to ensure a volt-second balance for the transformer. Voltage is placed across the transformer when S_1 is on and the transformer is reset when S_2 is on. Given that the voltage placed across the transformer is the input voltage, this means that the amount of time that one is on must be identical to the on-time of the others or else saturation will result. This drawback can be corrected by monitoring the transformer primary waveforms and then adjusting the width of the gating signal pulses appropriately, but this adds to the complexity and cost of the converter.

4.5.3 Half-Bridge Converter

Figure 4.9 shows the half-bridge converter. This converter consists of two switches in one leg; two bulk capacitors in the second leg; a transformer, two rectifying diodes, and an LC output filter. Each end of the transformer primary winding is connected to the midpoint of one of the legs. The converter gets its name from the fact that it has two switches instead of the four that a full-bridge converter has.

The converter works as follows: Voltage is impressed across the transformer primary whenever one of the converter switches is on. If S_1 is on, current flows through the transformer and through C_2; if S_2 is on, then current flows through the transformer in the opposite direction and through C_1. No voltage is impressed across the transformer when neither switch is on. A typical gating signal sequence during a switching cycle is: S_1 on, then all switches off, then S_2 on, then all switches off. S_1 and S_2 should be on for the same amount of time, but can never be on at the same time or else the converter will be short circuited.

The advantages of the half-bridge converter are:

- The frequency of the voltage fed to the output LC filter is twice that of the transformer primary and secondary voltage, unlike a two-switch forward converter.
- The peak voltage stress of the switches is equal to the input voltage, unlike a push–pull converter.
- The converter is immune to transformer saturation, unlike a push–pull converter. For example, if S_1 is on for a longer time than S_2 or the S_1–C_2 current path impresses more voltage across the transformer primary than the S_2–C_1 current path due to less voltage drop, then C_2 will be charged to a higher voltage than C_1 so that less voltage is impressed across the transformer when S_1 is on than when S_2 is on, thus balancing the volt-seconds of the transformer.

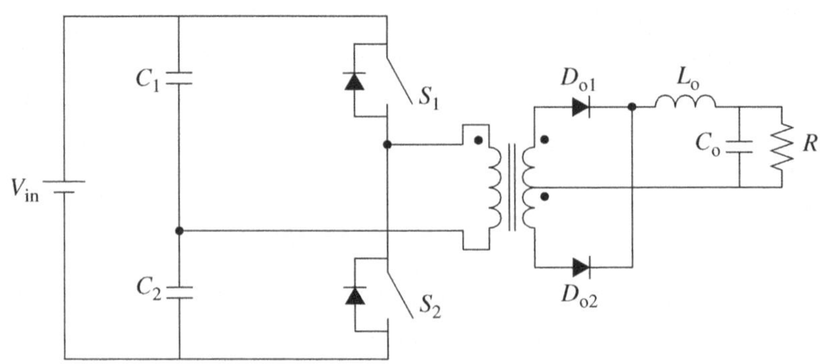

Figure 4.9 Half-bridge converter.

The disadvantages of the half-bridge converter are:

- Only half the input voltage is impressed across the transformer, which means that the current flowing in the primary side of the converter is twice that of the two-switch forward and push–pull converters. This results in more conduction losses and turn-off switching losses that limit the power level of the converter.
- The converter requires a floating gate-drive circuit for S_1.
- The converter has two DC bus capacitors, which can make the converter bulkier, depending on the amount of capacitance needed to reduce primary voltage ripple.

4.6 FULL-BRIDGE CONVERTER

For applications where a DC–DC converter with a power rating of 500 W or higher is needed, the full bridge DC–DC converter is the most popular option. The basic topology is shown in Figure 4.10a and variations on this basic topology are shown in Chapter 8. It consists of four switches, S_1–S_4, a high-frequency transformer, two rectifying diodes, D_{o1} and D_{o2}, and an LC output filter. The basic operating principle of the converter is that the four-switch bridge generates an AC square-wave voltage across the primary of the transformer that is stepped down or up, rectified, and then filtered to produce a DC output voltage.

A four-diode bridge can be used instead of a center-tapped rectifier at the converter's secondary side, as shown in Figure 4.10b. The choice of which secondary arrangement is used is dependent on the output voltage and current. If the output voltage is low and the output current is high, then a center-tapped secondary is used as the current will only flow through one diode and thus there will be fewer conduction losses. If the output voltage is high and the output current is low, then the full-bridge arrangement is preferred as the peak voltage stress of each diode is equal to the output voltage as opposed to a voltage stress of twice the output voltage that each diode of the center-tapped arrangement has. Although the number of diodes is greater, it may be cheaper in some cases to use four diodes with lower voltage ratings than two diodes with high voltage ratings. If the output current is low, then conduction losses are less of an issue so that the full-bridge diode option becomes more attractive. The decision as to whether to use a center-tapped arrangement of a full-bridge diode arrangement ultimately depends on cost and efficiency and thus there is no standard rule as to when to use which one for a particular application.

There are several ways to turn the switches of a full-bridge converter on and off, basic PWM and phase-shift PWM. Of these two methods, basic PWM is simpler, but phase-shift PWM is more efficient and thus more

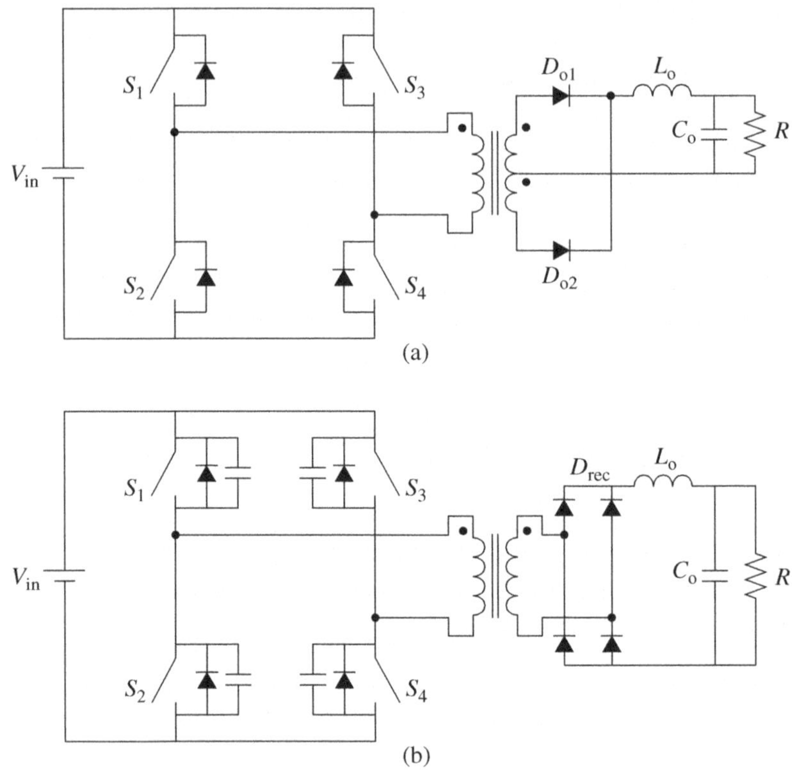

Figure 4.10 Full-bridge converter (a) with two-diode secondary (b) with four-diode bridge secondary.

popular. Basic PWM operation is discussed in this section and a discussion of phase-shift PWM operation is deferred to Chapter 7 where it will be the main focus of the chapter. When operating the full-bridge converter with basic PWM, the input DC voltage is impressed across the transformer primary whenever a pair of diagonally opposed switches is on. If S_1 and S_4 are on, then a positive voltage is impressed across the winding; if S_2 and S_3 are on, then a negative voltage is impressed across the winding. No voltage is impressed across the transformer primary if none of the converter switches is off.

The converter's gating signals and its transformer primary voltage are shown in Figure 4.11. Switches S_1 and S_4 are turned on simultaneously at the start of the switching cycle, and then turned off sometime before the start of the next half-cycle. As can be seen from the primary voltage waveform in Figure 4.11, the primary voltage is high until S_1 and S_4 are turned off and then becomes zero afterward. When the converter is halfway through its switching cycle, switches S_2 and S_3 are turned on simultaneously and a negative voltage is impressed across the transformer primary. The primary voltage turns to

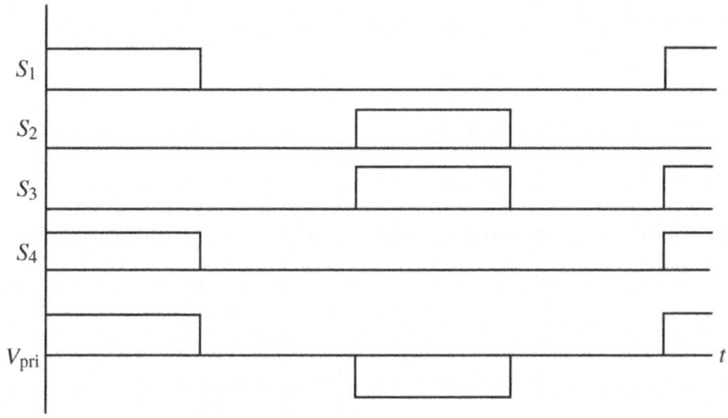

Figure 4.11 Full-bridge converter gating signal and transformer primary voltage waveforms.

zero after these two students are turned off and remains zero until the start of the next switching cycle. It should be noted that the width of the gating signals pulses of S_2 and S_3 must be the same as those of S_1 and S_4 to ensure a voltage-seconds balance of the primary voltage so that the transformer does not get saturated.

When the voltage across each of the center-tapped secondary winding is positive, then diode D_{o1} becomes forward-biased and conducts current and D_{o2} becomes reverse-biased and does not conduct current. The opposite happens when the secondary winding voltage is negative – D_{o1} is reverse-biased and D_{o2} is forward-biased. When there is no voltage across the transformer, then, ideally, the output inductor current splits into two with half the current flowing through D_1 and the other half flowing through D_2. This current distribution is uneven in real-life converters.

The rectified and stepped primary voltage has the same shape as the voltage that is at the input of the LC output filter of a buck converter; in essence, a full-bridge converter is a type of isolated buck converter. The width of the voltage pulses of this waveform can be varied by changing the on-time of the gating signal pulses. Since the rectifying process doubles the frequency of the primary and secondary voltage waveforms, the duty-cycle of the converter is defined as the ratio of the switches' on-time to half the length of the original switching cycle.

The full-bridge converter is superior to the two-switch topologies discussed in Section 4.5 for higher power applications as it combines many of the best features of these converters:

- The voltage ratings of its switches do not exceed the input voltage.
- The frequency of the voltage that is fed to the output filter is doubled so that the output filter size can be reduced.

- The full input voltage can be impressed across the transformer so that high conduction losses are not a limiting factor as they are in half-bridge converters.

The main drawbacks of the converter are:

- The converter must be implemented with four switches and floating gate drive circuits, which increases the cost and size of the converter.
- It is easy to saturate the transformer as it is difficult to ensure that the gating signals of all the switches are exactly the same. This, however, can be addressed by adding a DC-blocking capacitor in series with the transformer or by sensing the transformer primary waveforms and then adjusting the converter gating signals accordingly.

4.7 CONCLUSION

Non-isolated DC–DC converters are limited in their ability to step voltage up or down because of limitations in their duty-cycle. Extremely small or large duty-cycles are to be avoided because they increase the peak stresses of the components and result in poor converter efficiency. Although transformers cannot step DC voltage up or down directly, they can be used in converter topologies to make the gain less dependent on the duty-cycle and thus avoid the problems that can be created by extreme duty-cycles. The key to such implementations is the reset of the transformer. A DC–DC converter with a transformer in its topology must have some mechanism by which the volt-seconds that have been impressed across its primary winding during the initial stages of switching are somehow counteracted by impressing equivalent volt-seconds of opposite polarity across the winding, sometime later in the same switching cycle.

After a brief review of fundamental transformer modeling, the two basic isolated DC–DC converters were presented: the flyback converter and the forward converter. The flyback converter, which can be considered to be an isolated buck–boost converter, is the simplest isolated converter as it consists of just a switch, a transformer, a diode, and an output filter capacitor. Moreover, it has an inherent transformer reset mechanism so that it does not need to be implemented with an additional reset circuit, unlike other isolated converters. The flyback converter, however, is limited in power because of the size of its transformer as it must first store all the energy that is to be transferred from the input to the output before releasing it to the output. Another drawback that the flyback converter has is that it only has a capacitor as an output filter, which must be large if a large amount of ripple current needs to be filtered out. Flyback converters are best suited for applications where the maximum load is 100 W and the minimum voltage is 12 V.

The forward converter can be considered to be an isolated buck converter and it does not have the drawbacks that the flyback converter has such as transformer size and a capacitive output filter. It can be used in higher power applications (<200 W – 300 W) and lower output voltage applications ($\leq$5 V). The forward converter, however, does not have a natural reset mechanism like the flyback converter and is dependent on some circuit modification to do so. Examples of such modifications include a tertiary winding that is taken from the main transformer and used to implement a flyback-type reset mechanism, an RCD snubber, an LCDD snubber, and a modified LCDD snubber with a transformer winding in place of an inductor.

Two-switch isolated converters have two main switches in their topology and can be operated at higher power levels than flyback or forward converters. They are mainly isolated buck converters with a different reset mechanism than the forward converter that addresses some of the drawbacks found in conventional forward converters. Three of the most basic converters two-switch converters were reviewed in this section – the two-switch forward converter, the push–pull converter, and the half-bridge converter – and their relative strengths and weaknesses were compared. For higher power applications, greater than 500 W, the most popular option is the full-bridge converter, which has four switches. This converter can operate with either basic PWM or with phase-shift PWM. Basic PWM operation was discussed in this chapter and phase-shift PWM will be discussed in Chapter 7.

REFERENCE

1. C. Vartak, A. Abramovitz, and K. M. Smedley, "Analysis and design of energy regenerative snubber for transformer isolated converters," *IEEE Transactions on Power Electronics*, vol. 29, no. 11, pp. 6030–6040, Nov. 2014.

CHAPTER 5

SECONDARY-SIDE IMPLEMENTATIONS IN ISOLATED DC–DC CONVERTERS

The secondary of most isolated DC–DC converters is typically implemented with some sort of diode rectifier to rectify the voltage from the secondary winding so that it can be filtered and made to be DC. This diode rectifier is either a four-diode bridge or a center-tapped rectifier, with the latter configuration being much more popular as it uses only two diodes and results in fewer conduction losses. A four-diode bridge rectifier is used in applications where the output voltage is very high as its diodes have a peak voltage stress that matches the output voltage whereas the diodes in the center-tapped rectifier have a peak voltage stress that is double. There are, however, applications where the secondary needs to be implemented using some other means for various reasons and this is the focus of this chapter. In this chapter, three different alternatives to the conventional secondary-side diode rectifiers are reviewed: synchronous rectifiers (SRs), current doubler rectifiers, and implementations for multiple output rectifiers. The basic principles of each of these alternatives are discussed and implementation examples are given. The main points of the chapter are summarized at the end.

5.1 SYNCHRONOUS RECTIFIERS

SRs are MOSFETs with very low values of on-state resistance that are used as diodes at the secondary of power converters with low output voltage and high output current [1–6]. They are meant to reduce the conduction losses that

DC–DC Converter Topologies: Basic to Advanced, First Edition. Gerry Moschopoulos.

Published 2024 by John Wiley & Sons, Inc.

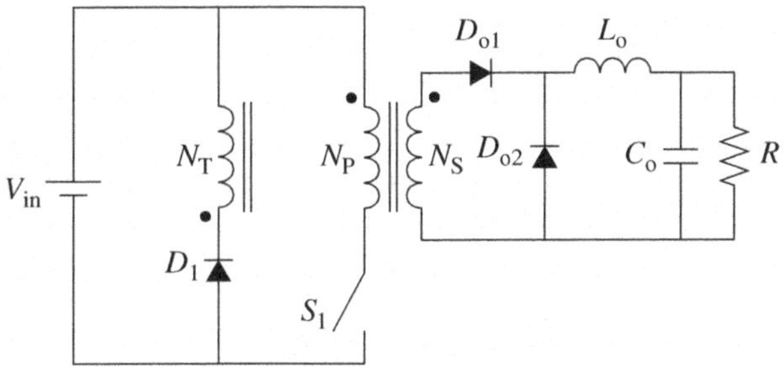

Figure 5.1 Forward converter.

are caused by having power diodes in the converter. Although low-voltage Schottky diodes have a voltage drop that is half that of conventional power diodes (0.25–0.5 V), their conduction losses are still significant in low-output voltage and high-output current applications.

Consider a forward converter with a center-tapped transformer secondary, two output diodes, and an LC filter, as shown in Figure 5.1. In steady-state, the full output inductor current will flow through only one of the diodes at any given time in real-life converters, with the exception of certain brief transition modes of operation. If each diode has a voltage drop of 1 V, and the converter has a 100 W, 5 VDC output, then the output inductor will have an average value of 20 A so that the power lost to conduction losses will be 20 A. This means that it will require at least 120 W to deliver 100 W power to the load if all other losses are ignored, which will result in a converter efficiency of about 83%. If Schottky diodes with 0.5 V drop are used instead, then the conduction losses due to diode forward voltage drop are reduced to 10 V so that the converter efficiency becomes approximately 91%. Although this figure is an improvement, the conduction losses of Schottky diodes are still significant.

A MOSFET may be considered to be a device that consists of a switch with an on-state resistance of $R_{DS,on}$, a body diode connected in parallel to the switch with its anode connected to the source of the device and its cathode connected to the drain, and an output capacitance connected in parallel to the switch and diode. Unlike a bipolar junction transistor, the switch part of the MOSFET is bidirectional and can conduct current from the drain to the source and vice versa. When current flows from source to drain in a MOSFET device, it can flow through the switch part of the device or the diode part, depending on the voltage drop that the current encounters. If current flows through the antiparallel diode, then the forward voltage drop is a fixed value, V_f. If it flows through the switch part of the device, then the forward voltage drop is $I \times R_{DS,on}$ where I is the current. If $I \times R_{DS,on}$ is less than V_f, then the current will flow through the switch part; if it is greater, the current will flow through the antiparallel body diode.

SRs get their name from the fact that they are MOSFETs used as diodes (rectifiers) and their operation is synchronized with the operation of the rest of the converter. SRs can be self-driven or they can be control-driven. The main difference between the two is that self-driven SRs are turned on using gating signals that are derived from the converter circuit itself whereas control-driven SRs are turned on using gating signals that are generated by an external controller of some sort. In general, self-driven SRs are used in topologies with LC output filters whereas control-driven SRs are used in topologies with C output filters or in topologies with LC output filters where efficiency is of particular concern over cost.

Control-driven SRs rely on sensing to determine whether current is flowing through the body diode of an SR device and then have an external controller generate a gating signal that will turn on the device. This voltage sensing is typically done using a current transformer to sense the current through a SR device or with a shunt resistor and sensing the voltage across it, then processing the resulting signal and feeding it to the controller to generate the gating signal. More sophisticated control-driven methods are predictive in the sense that they manage to turn SRs on just as current is needed to flow through them so that the amount of time that current flows through the body diode of an SR is negligible. This makes these methods more efficient than any self-driven SR method. Since control-driven SRs are mainly synthesized from sensing circuitry rather than from the topology itself, which is the case for self-driven SRs, control-driven SR implementations will not be covered in this chapter and the focus of the rest of this section will be on self-driven SRs. Many self-driven SR implementations have been proposed in the power electronics literature and the ones discussed here represent just a small sample.

It should be noted that in addition to being used in isolated DC–DC converters, SRs are often used in non-isolated topologies for low-voltage/high-current applications to reduce conduction losses, especially in buck converters used in distributed point-of-load power architectures. A standard approach is to replace the diode of a conventional buck PWM converter with an SR MOSFET and synchronize its operation to the rest of the converter so that it is on when current is supposed to flow through the buck diode and off when current is not supposed to flow. The control of the SR can be done with a dedicated controller and such controllers are commercially available from manufacturers of power converter controller ICs. These controllers have the control circuitry in a single IC. The current through the SR is sensed and a signal is then fed to the controller, which then turns the SR on or off, depending on whether current flows through the device.

Figure 5.2 shows a conventional forward converter with self-driven SRs, SR_1 and SR_2. It can be seen from this figure that the two output diodes of the forward converter have been replaced with MOSFETs. The two MOSFETs are oriented based on the position of the diodes. The source of SR_1 is connected to one of the transformer windings as the anode of its antiparallel diode must be connected to that point. Similarly, the source of SR_2 is connected to the secondary-side ground as the anode of its antiparallel diode must be connected to that point. In order to turn on SR_1 and SR_2 in a timely

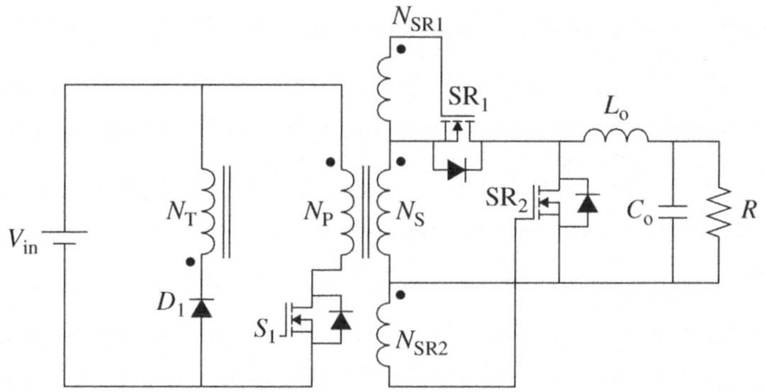

Figure 5.2 Forward converter with self-driven synchronous rectifiers.

manner, the gating signals of these switches must be synchronized in an appropriate manner as the secondary side will be short circuited otherwise. For example, if SR_2 is on while there is a positive voltage across the main secondary winding, then current can flow through SR_1, either through its switch or its body diode, then through SR_2 until a catastrophic failure occurs.

Given that the converter is a forward converter, SR_1 should be on when the voltage across the secondary winding is positive and off when this voltage is negative. The opposite is true for SR_2 as it should be off when this voltage is positive and on when it is negative. The mechanism by which SR_1 and SR_2 can be turned on is the set of gating signal windings, N_{SR1} and N_{S2}, as shown in Figure 5.2. When the voltage across the main secondary winding is positive, the voltage across each gating signal winding is positive. When the voltage across the main secondary winding is negative, the voltage across each gating signal voltage is negative.

Based on how the SRs are implemented in the forward converter in Figure 5.2, one end of the winding end N_{SR1} is connected to the gate of SR_1 and the other is connected to its source. When the voltage across the main secondary winding is positive, SR_1 is turned on and remains on. The operation of SR_1 corresponds to what a diode would do if one was used instead of a switch – just as a diode in this position conducts when the voltage across the secondary winding is positive, so too does SR_1. In other words, just like a forward converter in an energy-transfer mode where power is transferred from the converter input to the output load when the primary-side switch is on and current flows through a secondary-side diode, the converter in Figure 5.2 acts in the same manner except that current flows through a MOSFET instead of a diode. Similarly, when the primary-side switch S_1 is turned off and the converter enters a freewheeling mode of operation, the voltage across the secondary winding and winding N_{SR2} is negative as the transformer is being reset. A negative voltage across N_{SR2} means that the gate-source voltage of SR_2 is positive, according to the dot notation in Figure 5.2, so that the switch is turned on and the output inductor current flows through it.

The N_{SR1} and N_{SR2} windings are needed as the SR MOSFETs cannot be turned on if their gate-source voltage is less than threshold voltage. Since these two windings are coupled to the same core as the main N_s winding, they must have a sufficient number of turns to ensure that an appropriate voltage can be generated across them. The N_{SR1} and N_{SR2} windings are only necessary if the voltage across the N_s winding is below or close to the threshold voltage of the SR MOSFETs. If the voltage across the N_s winding is greater than the threshold voltage of the MOSFETs, then the implementation shown in Figure 5.3 can be used.

Figure 5.4 shows some typical waveforms of the forward converter shown in Figure 5.2. The following should be noted:

- SR_1 will always be on when S_1 is on, but SR_2 will always be off as its gate-source voltage is negative.
- Switch SR_2 is on when S_1 is off and the transformer is in the process of being reset. The voltage across the primary winding is reflected from the tertiary winding and voltage will appear across the primary winding as long as the transformer is being reset.
- When S_1 is off, the voltage across it is the sum of the input voltage and the voltage impressed across the transformer.
- If the transformer is fully demagnetized, as shown in the waveforms in Figure 5.4, then the voltage across S_1 becomes the input voltage. More importantly, however, is that there is no voltage across N_{SR1} and N_{SR2} so that SR_2 is turned off as its gating signal has been removed. When this happens, there is a certain amount of so-called "dead time" when none of the converter switches is on and the output current flows through the body diode of SR_2 instead of the switch part. This leads to an increase in conduction losses as the switch part of the SR_2 MOSFET is not operational.

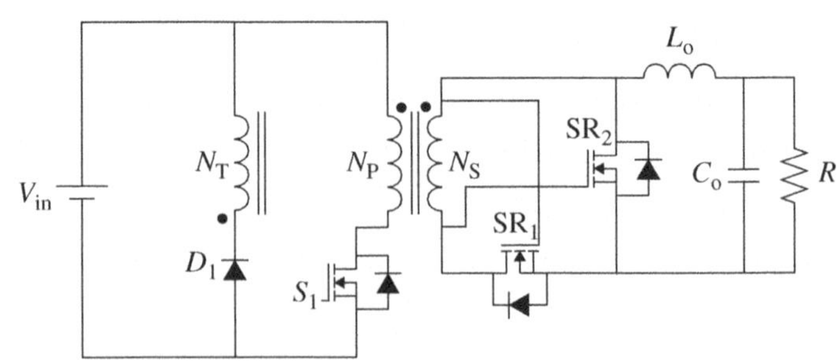

Figure 5.3 Forward converter with self-driven synchronous rectifiers with higher secondary output voltage.

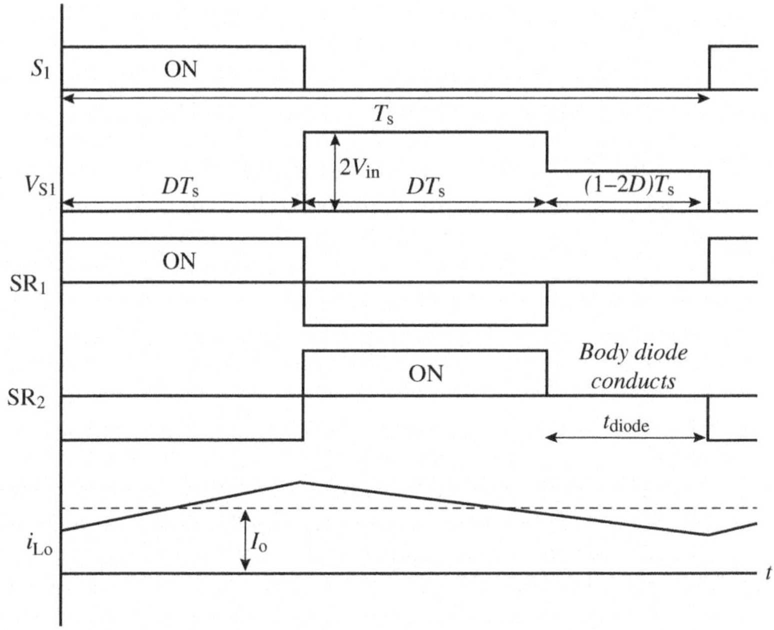

Figure 5.4 Key waveforms of the converter in Figure 5.2.

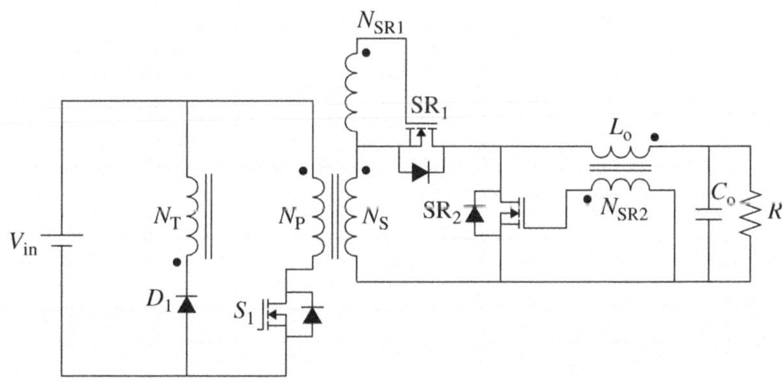

Figure 5.5 Modified forward converters with output inductor/transformer.

Figure 5.5 shows another SR implementation scheme that was proposed in [2], that tries to correct the dead-time issue that the converter in Figure 5.2 has. This converter is very similar to the converter shown in Figure 5.2 except that the gating signal for SR_2 is obtained using the voltage across the output inductor, which has been replaced with a transformer. When the primary-side switch S_1 is on, the converter is in an energy-transfer mode, SR_1 conducts current and the voltage across the output inductor is positive. Given the dot notation of this element that is shown in Figure 5.5, the voltage at the gate of SR_2 is negative so that the gate-source voltage of the device is less than its threshold voltage and it does not turn on. When S_1 is off and the converter enters a freewheeling mode, the voltage across the output inductor becomes negative, and the voltage at the gate of SR_2 becomes positive so that the gate-source voltage of SR_2 is positive and thus turns on. SR_2 will remain on regardless of whether the main transformer of the converter is fully demagnetized by the reset process, as long as there is current flowing in the output inductor. If there is no current in the output inductor, it is indicative of light-load operation, which means that the output current is low and SRs are not needed as the conduction losses are low.

Figure 5.6 shows the key waveforms of the converter shown in Figure 5.5. It can be seen that SR_2 has a gating signal (SR_2 in Figure 5.5) during the whole time that S_1 is off, even though the transformer has been fully demagnetized and the voltage across S_1 becomes equal to the input voltage. As a result, the body diode of SR_2 never conducts during a freewheeling mode of operation and only does so when the load is light and the output inductor current

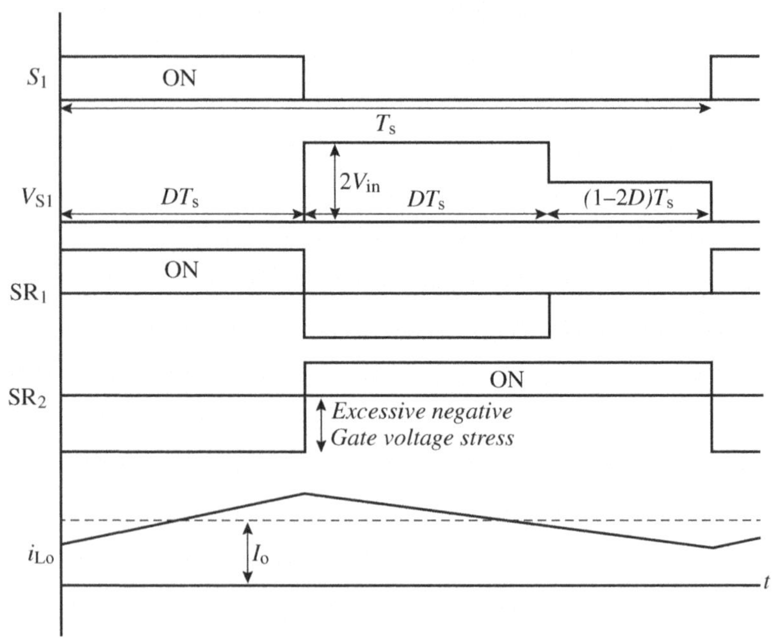

Figure 5.6 Key waveforms of the converter in Figure 5.5.

becomes discontinuous. There is, however, a drawback that limits the effectiveness of the SR gate-driving implementation shown in Figure 5.5 and that is the fact that $(1 - D)$ is large when the converter operates with a small duty-cycle D. Since the volt-seconds of the output inductor must be balanced and the voltage at the gate of SR_2 must be greater than the threshold voltage, then the negative voltage at the gate of SR_2 when the converter is in a freewheeling mode will be high and may exceed the ratings of the device. For example, if D is 0.2 and V_{g3} is made to be 10 V to turn on S_3, then the gate-source voltage of SR_2 must be 40 V to ensure a volt-seconds balance, which is excessive for most MOSFETs.

A modification to the converter shown in Figure 5.5 that corrects this excessive negative voltage was proposed in [5] and is shown in Figure 5.7; key waveforms of the converter are shown in Figure 5.8. Like the converter shown in Figure 5.5, this converter uses voltage across its output inductor to generate a gating signal for SR_2 that allows it to stay on while the converter is in a freewheeling mode of operation. The main difference between the two converters is that the converter in Figure 5.7 has a blocking diode D_a connected to the gate of SR_2 and a small, low-power signal MOSFET, S_a, connected between the gate of SR_2 and the secondary ground. The blocking diode protects the gate of SR_2 from the excessive negative voltage that may otherwise arise when operating the converter with a low duty-cycle. The presence of D_a in the converter, however, does not allow voltage to be removed from the gate of SR_2 when the converter switches from a freewheeling mode to an energy-transfer mode and thus S_a is needed. S_a is turned on when the converter enters an energy-transfer mode after S_1 is turned on and clamps the gate of SR_2 to zero while the voltage across the secondary winding of the output inductor transformer is negative and D_a is reverse-biased.

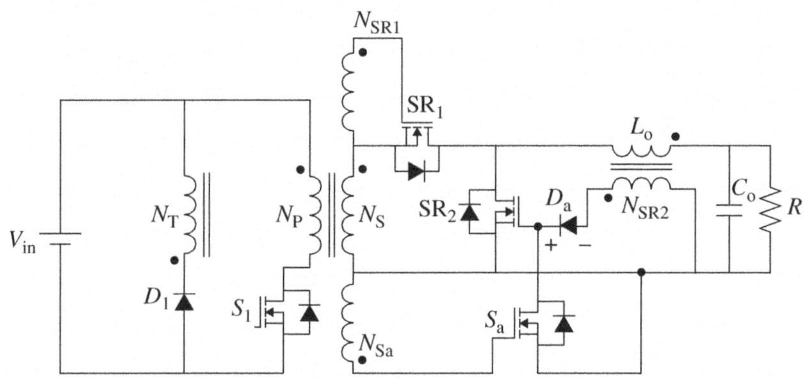

Figure 5.7 Forward converter with synchronous rectifier dead-time elimination.

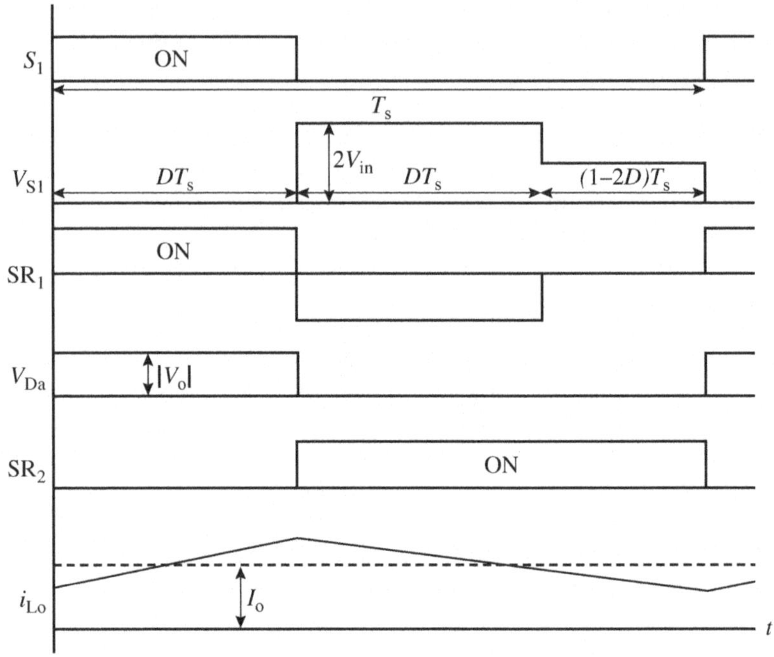

Figure 5.8 Key waveforms of the converter in Figure 5.7.

5.2 CURRENT DOUBLERS

A current doubler can be used as an alternative to the conventional center-tapped diode rectifier configuration at the secondary of an isolated DC–DC converter in certain applications [6–12]. Figure 5.9 shows two current doubler circuits. The two circuits shown in Figure 5.9 are electrically equivalent. The current doubler gets its name from the fact that the current flowing

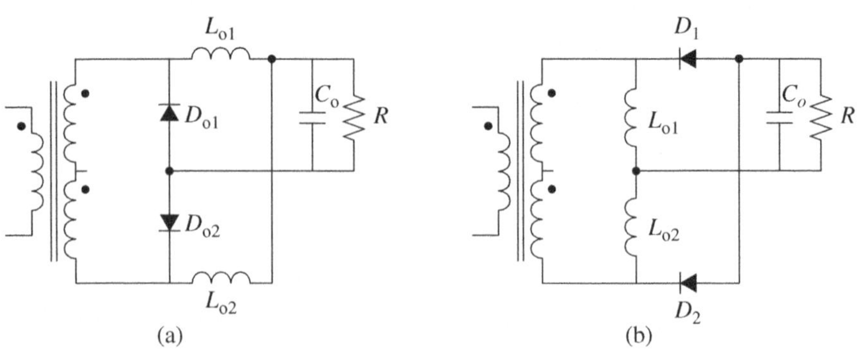

Figure 5.9 Current doubler implementations (a) Current doubler #1 (b) Current doubler #2.

out of its secondary winding is doubled due to the two inductors in its circuit. It has the following three modes of operation, as shown in Figure 5.10 for the implementation shown in Figure 5.9a:

- Positive secondary winding voltage (Figure 5.10a,b): There are two current paths in this mode: the first path is the loop that consists of the secondary winding, output inductor L_{o1}, the output capacitor C_o and the load, and device D_{o2} (Figure 5.10a). The second path is the loop that consists of output inductor L_{o2}, C_o and the load, and D_{o2}. During this mode, the current in L_{o1} rises and the current in L_{o2} falls, while D_{o2} conducts the sum of the currents in both inductors; this combined current flows to the output as well.
- Negative secondary winding voltage (Figure 5.10c,d): There are two current paths in this mode: The first path is the loop that consists of output inductor L_{o1}, C_o and the load, and D_{o1} (Figure 5.10c). The second path is the loop that consists of the secondary winding, output inductor L_{o2}, the output capacitor C_o and the load, and device D_{o1}. During this mode, the current in L_{o1} falls and the current in L_2 rises, while D_{o1} conducts the sum of the currents in both inductors; this combined current flows to the output as well.
- No secondary winding voltage (Figure 5.10e,f): There are two current paths in this mode, which is a freewheeling mode of operation that takes place when no voltage is impressed across the primary winding: The first path is the loop that consists of output inductor L_{o1}, C_o and the load, and D_{o1}. The second path is the loop that consists of output inductor L_{o2}, C_o and the load, and D_{o2}. The current in both inductors falls during this mode.

The implementation shown in Figure 5.9a is attractive for applications where D_{o1} and D_{o2} are replaced with SRs, as the two SRs are referenced to ground. The implementation shown in Figure 5.9b is attractive if integrated magnetics are used. A feature that the current doubler has that the conventional center-tapped rectifier does not is that the frequency of the current in its two inductors is the same as the frequency of the voltage and current at the secondary winding. As a result, it is possible to integrate the two inductors and the transformer in one core as shown in Figure 5.11 so that the size and weight of the overall converter where the current doubler is used are decreased [12]. It should be noted, however, that since the frequency of the current in each inductor is half that of the inductor in a center-tapped rectifier and since there are two inductors instead of one, the amount of inductance needed to ensure that the output ripple produced by the current doubler matches that produced by the center-tapped rectifier may be as much as four times.

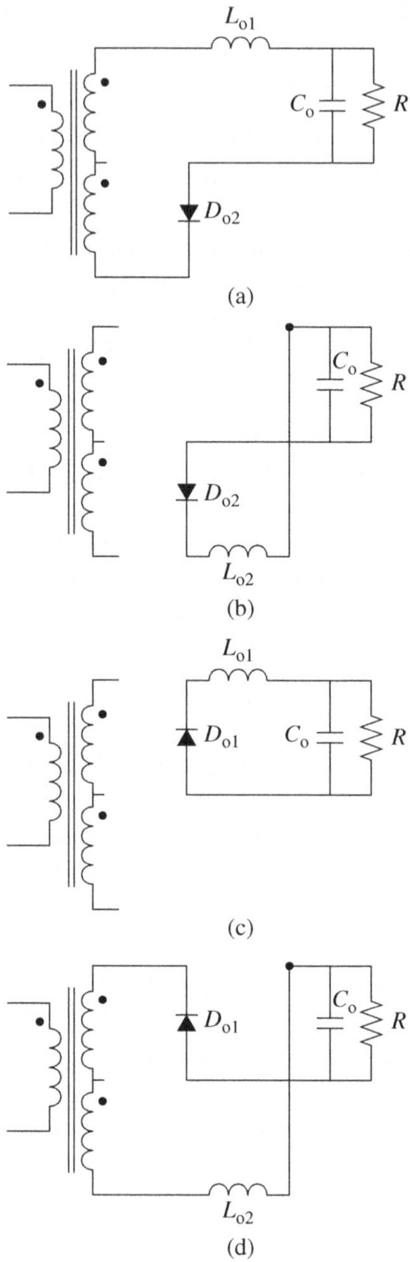

Figure 5.10 Current doubler modes of operation. (a) Positive secondary winding voltage, L_{o1} current flow. (b) Positive secondary winding voltage, L_{o2} current flow. (c) Negative secondary winding voltage, L_{o1} current flow. (d) Negative secondary winding voltage, L_{o2} current flow. (e) Zero secondary winding voltage, L_{o1} current flow. (f) Zero secondary winding voltage, L_{o2} current flow.

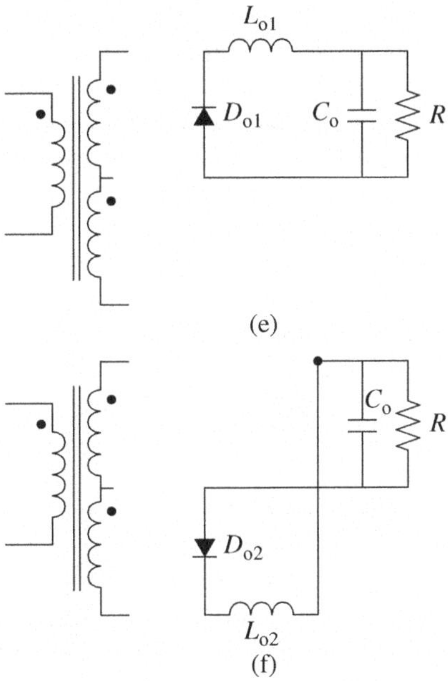

Figure 5.10 (*Continued*)

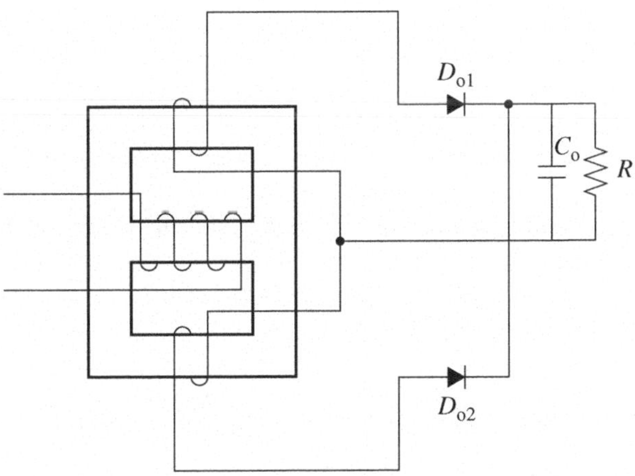

Figure 5.11 Current doubler with integrated magnetic implementation.

In addition to simpler SR implementation and the possible use of integrated magnetics, another advantage that the current doubler has over the conventional center-tapped rectifier is that its transformer is not center-tapped. This makes the transformer structure simpler and finer steps in turns ratio possible. The number of secondary turns of the current doubler secondary winding is double that of each of the center-tapped secondary windings. This makes the voltage that appears across the secondary winding of the current doubler twice that of each winding of the center-tapped secondary and the current through this winding is half. The secondary current that is reflected to the primary and the peak voltage stress of the secondary are thus the same for both secondary implementations.

Ultimately, the decision on whether or not to use a conventional center-tapped rectifier or a current doubler depends on the application and the specifications of the converter. There are no hard and fast rules as to which is the better implementation. This can only be determined for a specific application by designing with both implementations in mind and then selecting the design that best meets the desired criteria.

5.3 MULTI-OUTPUT CONVERTERS

It is often the case that a power converter is required to produce multiple outputs in order to avoid using multiple single-output converters and thus lower cost and size. In such cases, one of the outputs is regulated and the rest of the outputs fall in line naturally or are made to fall in line through some means [13]. Consider the multi-output flyback converter shown in Figure 5.12, which produces two output voltages, V_{o1} and V_{o2}. Given that the converter has just one switch, only one of the output voltages can be regulated by adjusting the duty-cycle of the switch. The other output must be regulated by some other method such as adding a post-regulator circuit in series that can actively adjust the voltage (e.g. a secondary converter with an active switch) or by the output can be left by itself with the assumption that the voltage will fall within an acceptable tolerance range. When a selection needs to be made of which output of a multi-output converter is to be actively regulated, the general guiding principle is that it should be the output with the greatest load variation. For example, if the flyback converter shown in Figure 5.12 has an output of V_{o1} = 12 Vdc and 2.5 A and an output of

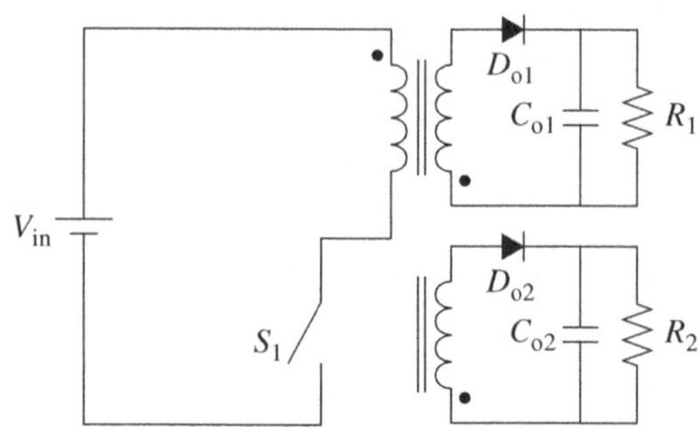

Figure 5.12 Flyback converter with multiple outputs.

$V_{o2} = 5$ Vdc and 10 A, then V_{o2} is the output that should be regulated by the converter. If the transformer is designed with little secondary leakage inductance, then V_{o1} can be restricted to 12 Vdc plus/minus some small tolerance.

This method of regulating multiple outputs is suitable for flyback converters because they are used in low-power applications with low output currents. Under such conditions, the voltage drop that can be caused by winding resistance and leakage inductance is low because the output current is low. This method, however, is rarely used in forward converters for the following reasons:

- Forward converters are used in applications where the output current is high. This makes it difficult to maintain the unregulated output voltages outside their tolerance band.
- Each forward converter output would have to have its own, separate, filter inductor. The presence of a filter inductor between each secondary winding and output would increase the impedance seen by each secondary current, which would result in additional voltage drops.

For forward converters, the unregulated output voltages must be regulated by some means.

A simple way of doing so is to couple all the output inductors together, as shown in Figure 5.13 [14–17]. What this coupling does is link all the output inductors to the inductor of the output that is regulated, which makes the

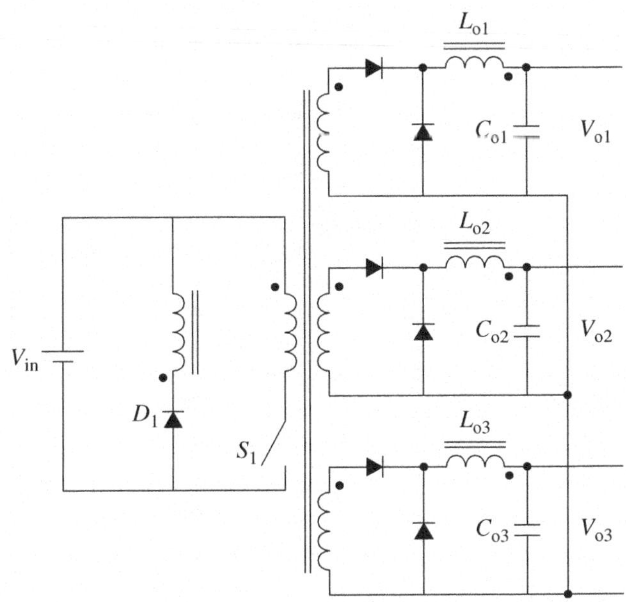

Figure 5.13 Forward converter with coupled inductor outputs.

voltage drop across each inductor less susceptible to change due to load variations, It also reduces the winding resistance and leakage inductance that would otherwise exist with individual, non-coupled, inductors, which further reduces voltage drop. This method, however, does have its limitations in applications where the required voltage regulation for each output must be tight. For such applications, some sort of post-regulation for each secondary winding is needed, where each output is regulated by some sort of active mechanism between the winding and the output.

The simplest and cheapest method of secondary post-regulation is to use small, commercially available linear voltage regulators for each output. This approach can be used for low power outputs of several watts, but not for higher power due to the inefficiency of linear voltage regulators. Another way of performing post-regulation is to use so-called "magnetic amplifiers" or "mag-amps" for short, as shown in Figure 5.14 [18–20]. Mag-amp circuits for post-regulation were especially popular in the1980s, and while they are still used, they are not as popular as they once were. They are more expensive than linear regulators, but not as expensive as an independent converter.

The basic principle behind a mag-amp is the use of a saturable reactor as a magnetic switch that can be turned on and off. The typical mag-amp has two similar transformer magnetic cores that are separate. Each of these cores typically has a control winding and an AC winding. The two control windings are generally connected in series and are fed by a low-impedance source while the two AC windings may be connected either in series or in parallel. The amount

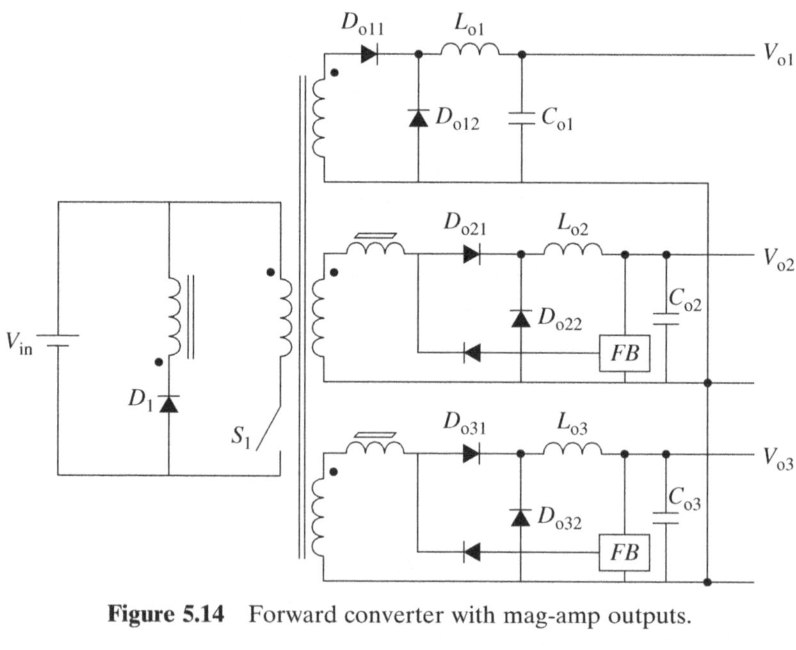

Figure 5.14 Forward converter with mag-amp outputs.

of control current fed into the control winding of each core sets the point in the AC winding waveform at which either core will saturate. The impedance of this saturable reactor is large if the core is not saturated so that no power is transferred to the output. The impedance of the core drops when the core is saturated and this is analogous to the turning on of a switch. The control current controls the voltage level at which the mag-amp switch turns on and a small DC on the control winding is able to switch larger ACs on the AC windings. Two magnetic cores are generally used in mag-amps as AC in the AC windings will generate a high voltage in the control windings. If the mag-amp cores are connected in opposite phases, the voltages that are induced in the control circuit cancel each other so that no current is induced in the control circuit.

Although mag-amps can be effective in voltage regulation for multi-output converters, they are difficult to design, they require numerous inductive components, and their operating frequency is limited, as it takes time to turn a mag-amp switch on and off. The slow speed of a mag-amp switch also limits the transient response of a converter output to any load variation. An approach that is widely used today is to use a MOSFET with a low on-state $R_{DS,on}$ value for each output, in between the secondary output and the rest of the converter [21–23]. Doing so is like placing an additional converter stage at the secondary with the MOSFET switch controlled by secondary-side feedback. An example of such an implementation is shown in Figure 5.15. This

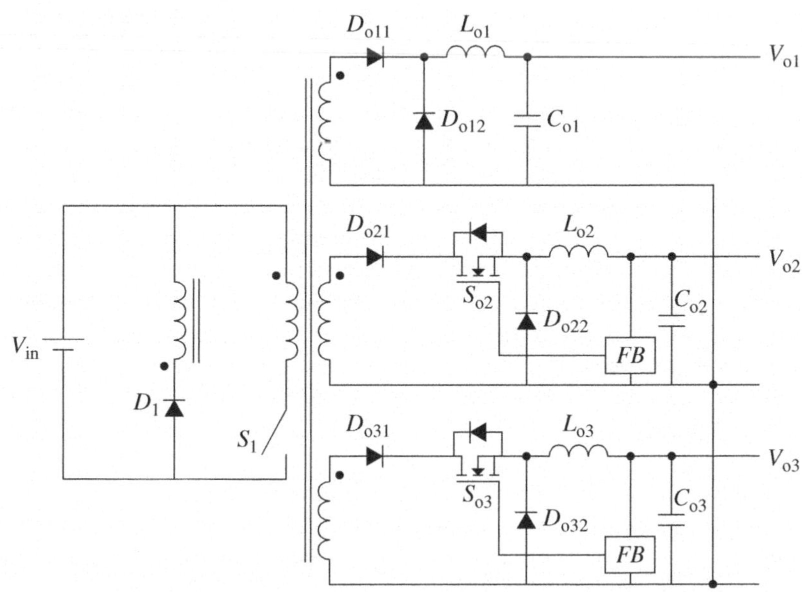

Figure 5.15 Forward converter with post-regulation converters.

places a small buck converter in series with each secondary winding and this buck converter can be regulated by its own controller. Any switching losses generated by the buck converter MOSFET can be limited if the output voltage is very low, less than 12 V. The main drawback to this approach is cost, as multiple switches, controllers, and gate driver circuits are needed. Consequently, it should be avoided for low power applications where cost is typically the overriding concern and where the output load variation is such that a coupled inductor approach is suitable.

5.4 CONCLUSION

This chapter reviewed alternatives to conventional secondary-side implementations in isolated DC–DC converters and focused on three topics: SRs, current doubler rectifiers, and multiple output converters. SRs are MOSFETs with low on-stage resistance values that are used in place of diodes to reduce conduction losses. They get their name from the fact that they are used like secondary output diodes (rectifiers) and their operation is synchronized with that of the converter. SRs conduct current from source to drain instead of the usual drain to source path. When implemented, the main objective is to have current flow through the switch part of a MOSFET device instead of its body diode. Care must be taken that they are not turned on at inappropriate instants that will result in the secondary side of the converter being short circuited.

SRs can be control-driven or self-driven. Control-driven SRs get their gating signal from some externally powered controller that uses sensing to generate a gating signal that is synchronized with the operation of the converter. Self-driven SRs get their gating signals from secondary windings and thus do not need an external controller. Self-driven SRs are less expensive to implement but are prone to the appearance of dead time. When the converter is in a free-wheeling mode of operation and the transformer is fully demagnetized, that the gating signal of the SR that is conducting current is lost and current flows through the body diode of the device, which increases conduction losses. Dead time can be eliminated by adding more components to the converter, but this increases cost. Control-driven SRs do not have dead time issues, but are more expensive. They are generally used in converters with capacitor-only output filters whereas self-driven are more popular in converters with LC output filters.

Current doubler rectifiers are secondary-side rectifying circuits that have two output inductors instead of the one that conventional center-tapped secondary rectifiers have. They get their name from the fact that the current at the output is double that flowing in the secondary winding of the converter's transformer. Current doubler rectifiers can simplify SR's implementations and allow the use of integrated magnetics that integrate the two output inductors with the main transformer in a single core. Such an implementation is possible as the frequency of the current flowing in the output inductors matches that of the current flowing out of the main secondary winding, which is not the case for the conventional center-tapped rectifier. The fact that current doublers can be implemented with a two-winding instead of a three-winding transformer is advantageous, but bridge for high output voltage/low output current applications and large inductance values are needed to achieve the same output current ripple. It is not obvious whether a current doubler is more efficient than a center-tapped rectifier for a particular application as this very much depends on the application.

Multi-output converters, single converters that can produce multiple output voltages, are attractive as they can be used to avoid the cost and size that are associated with using multiple single-output converters. The general principle behind multi-output converters is to regulate the output with the largest variation in load and leave the other output unregulated. This approach can work for lower power flyback converters with low output current or lower power forward converters with coupled output inductors. For higher power applications, some sort of secondary post-regulation approach such as the use of mag-amps or a switch post-regulator approach is preferred. Of the two approaches, the latter offers better performance, but at a higher cost.

REFERENCES

1. E. Oxher, "Using power MOSFETs as high-efficiency synchronous and bridge rectifiers in switch-mode power supplies," MOSPOWER Applications Handbook, Siliconix Technical Article, Siliconix Inc. 1984, pp. 5-87–5-94
2. G. Chryssis, "High Frequency Switching Supplies: Theory and Design," 2nd ed., McGraw-Hill, 1989, pp. 144–152.
3. H. K. Ji and H. J. Kim, "Active clamp forward converter with MOSFET synchronous rectification," *IEEE Power Electronic Specialists Conference (PESC)*, 1994, June 20–25, 1994, Taipei, Taiwan, pp. 895–901.
4. J. A. Cobos, "Optimized synchronous rectification stage for low power output voltage (3.3 V) DC/DC conversion," *IEEE Power Electronic Specialists Conference (PESC)*, 1994, June 20–25, 1994, Taipei, Taiwan, pp. 902–908.
5. Y. Xi, P. K. Jain, and G. Joos, "An improved gating technique for the synchronous rectifier MOSFETs in the forward converter topology," *IEEE Canadian Conference on Electrical and Computer Engineering (CCECE)*, 1997, May 25–28, 1997, St. John's, Canada, pp. 552–555.
6. F. D. Tan, "The forward converter: extension to include synchronous rectifiers and current doublers," *IEEE International Telecommunications Energy Conference (INTELEC)*, 2002, Sept. 28–Oct. 2, 2002, Montreal, Canada, pp. 40–47.

7. C. Peng, M. Hannigan, and O. Seiersen, "A new efficient high frequency rectifier circuit," *High Frequency Power Conversion (HFPC)*, 1991, June 9–14, 1991, Toronto, Canada, pp. 236–243.
8. K. O'Meara, "A new output rectifier configuration optimized for high frequency operation," *High Frequency Power Conversion (HFPC)*, 1991, June 9–14, 1991, Toronto, Canada, pp. 219–225.
9. L. Balogh, "The performance of the current doubler rectifier with synchronous rectification," *High Frequency Power Conversion (HFPC)*, 1995, May 6–12, 1995, San Jose, USA, pp. 216–225.
10. L. Balogh, "Design review: 100 W, 400 kHz, DC/DC converter with current doubler synchronous rectification achieves 92%," *Unitride Power Supply Design Seminar*, Sem-1100, 1995, pp. 216–225.
11. L. Huber and M. M. Jovanovic, "Forward-flyback converter with current-doubler rectifier: analysis, design, and evaluation results," *IEEE Transactions on Power Electronics*, vol. 14, no. 1, pp. 184–192, Jan. 1999.
12. J. Sun, K.F. Webb, and V. Mehrotra, "Integrated magnetics for current-doubler rectifiers," *IEEE Transactions on Power Electronics*, vol. 19, no. 3, pp. 582–590, May 2004.
13. T. G. Wilson, Jr., "Cross regulation in an energy storage DC-to-DC converter with two regulated outputs," *IEEE Power Electronics Specialists Conference (PESC)*, 1977, June 14–16, 1997, Palo Alto, USA, pp. 190–199.
14. H. Matsuo and K. Harada, "New energy-storage DC–DC converter with multiple outputs," *IEEE Transactions on Magnetics*, vol. 14, no. 5, pp. 1005–1007, Sep. 1978.
15. H. Matsuo, "Comparison of multiple-output DC–DC converters using cross regulation," *IEEE Transactions on Industrial Electronics and Control Instrumentation*, vol. 27, no. 3, pp. 176–189, Aug. 1980.
16. A.F. Witulski, "Introduction to modeling of transformers and coupled inductors," *IEEE Transactions on Power Electronics*, vol. 10, no. 3, pp. 349–357, May 1995.
17. D. Maksimovic, R.W. Erickson, and C. Griesbach, "Modeling of cross-regulation in converters containing coupled inductors," *IEEE Transactions on Power Electronics*, vol. 15, no. 4, pp. 607–615, Jul. 2000.
18. R. Hiramatsu, K. Harada, and T. Ninomiya, "Switch mode converter using high-frequency magnetic amplifier," *IEEE International Telecommunications Energy Conference (INTELEC)*, 1979, Nov. 26–29, 1979, Washington DC, USA, pp. 282–288.
19. D. Y. Chen, J. Lee, and C. Jamerson, "A simple model predicts small-signal control loop behavior of magamp post-regulator," *IEEE Transactions on Power Electronics*, vol. 4, no. 4, pp. 402–408, Oct. 1989.
20. Y.-T. Chen and J.-M. Liang, "Paralleling magamp-postregulator modules with sliding-mode-control method," *IEEE Transactions on Industrial Electronics*, vol. 53, no. 3, pp. 974–983, Jun. 2006.
21. Y.-L. Lin and K. H. Liu, "A new synchronous switch post regulator for multi-output forward," *IEEE Applied Power Electronics Conference (APEC)*, 1990, Mar. 11–16, 1990, Los Angles, USA, pp. 693–696.

22. Y. T. Chen, "The overall small-signal model of the synchronous switch postregulator," *IEEE Transactions on Power Electronics*, vol. 13, no. 5, pp. 852–860, Sep. 1998.
23. Y. Xi and P. K. Jain, "A forward converter topology with independently and precisely regulated multiple outputs," *IEEE Transactions on Power Electronics*, vol. 18, no. 2, pp. 648–658, Mar. 2003.

CHAPTER 6

SOFT-SWITCHING FORWARD AND FLYBACK CONVERTERS

Just like non-isolated DC–DC converters, forward and flyback converters can be implemented with soft-switching to reduce their switching losses. Although many of the soft-switching methods that are used in non-isolated converters, such as the ones discussed in Chapters 2 and 3, can be used in isolated converters, there is one fundamental difference that must be taken into account when doing so – the transformer reset. Regardless of the soft-switching mechanism used, it must be integrated with the reset mechanism, which flips the polarity of the primary transformer voltage. The reset mechanism is less of a concern in two-switch converters such as the half-bridge and thus these converters can be implemented with soft-switching using more conventional methods.

In this chapter, the two most popular ZVS forward converters, the resonant reset and the active clamp are presented, with the active clamp forward converter in particular discussed in detail, given its widespread use in industry. The flyback versions of these converters are discussed and the chapter concludes with a discussion of alternatives to resonant reset and active clamp converters. It should be noted that the converters that are discussed in this chapter are almost always implemented with traditional silicon devices and are rarely implemented with newer wide bandgap devices. Forward and flyback converter designs are cost sensitive and the cost of wide bandgap devices at the time of this writing are still not such as to justify the use of these devices in these converters.

DC–DC Converter Topologies: Basic to Advanced, First Edition. Gerry Moschopoulos.

Published 2024 by John Wiley & Sons, Inc.

6.1 FORWARD CONVERTERS WITH RESONANT RESET

Forward converters with resonant reset mechanisms are attractive as they can operate with ZVS in a cost-effective manner [1–3]. One such converter is the ZVS forward converter that was proposed in [3] and shown in Figure 6.1. This converter is very simple as the reset mechanism is just a capacitor that is connected across its switch. In many cases, this capacitor is simply the output capacitance of a switch and no external capacitance is added, as shown in Figure 6.1. The fundamental principle behind this converter is that this capacitor interacts in a resonant manner with the transformer's magnetizing inductance so that it can be charged to a voltage that is greater than the input voltage. This allows a negative voltage to appear across the transformer primary for a sufficient length of time to reset the transformer. Since this converter has this resonant reset mechanism, there is no need for the main transformer to have a tertiary reset winding and thus its size can be reduced. The secondary side of the converter is identical to that of a conventional forward converter.

The converter works as follows: Voltage is impressed across the primary winding when the switch is turned on. Secondary-side diode D_{o1} then becomes forward-biased and the output inductor current flows through it. When the switch is turned off, the transformer's magnetizing current flows through the output switch capacitance, C_{s1} and it begins to charge. When the voltage across the transformer is no longer positive and the reset process begins, current at the secondary starts to be transferred away from D_{o1} and into D_{o2}. This current transfer, which is not instantaneous due to leakage inductance in the secondary winding, is eventually completed and the full output inductor current freewheels through D_2. At the primary side, capacitor C_{s1} continues to be charged unless there is no more energy stored in the transformer. When this happens, the primary current will begin to reverse direction as the voltage across C_{s1} is greater than the input voltage. Eventually, C_{s1} is fully discharged and the voltage across it is zero, which allows the switch to be turned on with ZVS. With the voltage across C_{s1} clamped to zero, the primary current reverses direction and flows through the switch. At the secondary side, current

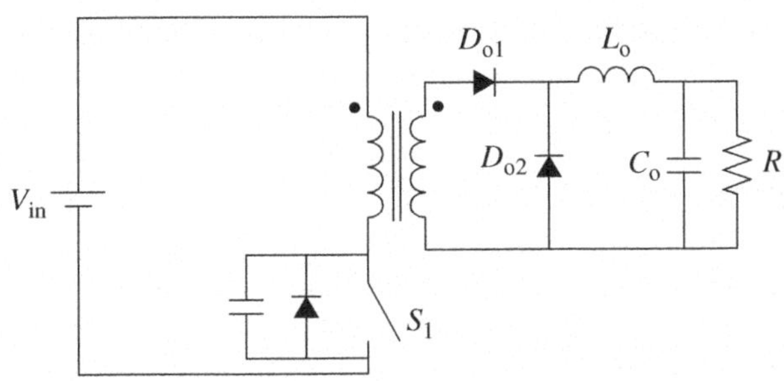

Figure 6.1 Forward converter with resonant reset.

begins to be transferred away from D_{o2} to D_{o1} when the voltage across the transformer primary is positive so that the full output inductor current flows through D_{o1} by the time primary current begins to flow through the switch.

If the converter is operated with a high switching frequency, close to 1 MHz, then an external capacitor across the switch is not needed and the MOSFET's output capacitance is sufficient for the resetting of the transformer. The converter shown in Figure 6.1 is very attractive as it can be implemented with just the main switch and a regular two-winding transformer instead of a transformer with a tertiary winding or a passive snubber as is the case with the forward converters discussed in Chapter 4. Moreover, unlike these converters, it can operate with ZVS so that its switching losses are lower. The converter, however, does have one key drawback and that is that the instant when the switch is to be turned on with ZVS is not controllable and is dependent on the converter parameters, the input voltage, and the load. The switch must be turned on when the resonant interaction between the output switch capacitance, C_{s1}, and the transformer's magnetizing inductance forces the voltage across the switch to zero or else the primary current will reverse direction and begin to charge the capacitor up again. If the switch is turned on while this charging occurs, it will do so without ZVS, but the opportunity to do so varies with operating conditions.

From a practical point of view, this means that the converter cannot be operated with a fixed switching frequency PWM and must be operated with a variable switching frequency instead. This is disadvantageous because it complicates the design of the transformer and the controller as they cannot be optimized for one specific frequency. Given that the window of opportunity for the switch to be turned on varies with load and the difficulties of designing the converter's transformer and controller to operate over a wide range of switching frequencies, the converter's operation is typically limited to light loads and is rarely used above 100 W.

It should be noted that a flyback converter can be implemented with a resonant capacitor across its switch to convert it into a ZVS converter. The modes of operation of such a converter compared to a resonant reset forward are very similar with the main difference being that the selection of the resonant capacitor is less dependent on the converter's reset mechanism, given that the flyback converter can be reset naturally, without any additional means. The ZVS mechanism is caused by the interaction of the transformer's leakage inductance and the resonant capacitor.

6.2 ACTIVE CLAMP CONVERTER

The converter shown in Figure 6.1 can be modified in order to operate with ZVS and with a fixed switching frequency if another switch is added in series to the external capacitor as shown in Figure 6.2a. What the switch does is that it makes it possible to control the mechanism by which the main switch, S_1, can be turned on with ZVS, so that it can be activated when it is needed, just

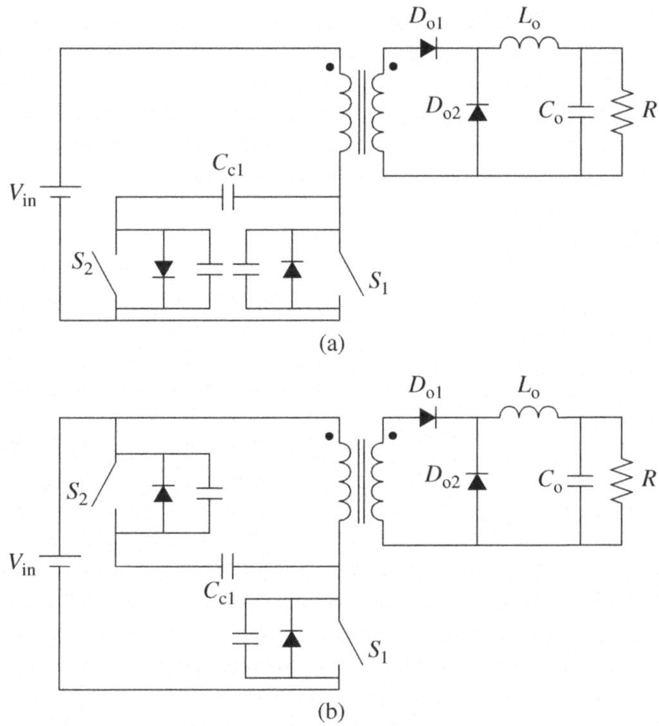

Figure 6.2 Active clamp forward converters: (a) Low-side active clamp converter (b) High-side active clamp converter.

before the switch is to be turned on. This is in contrast to the converter in Figure 6.1 where the switch must be turned on at an uncontrollable instant when the voltage across it is zero. Since there is a switch in series with a capacitor that can be considered to be a clamping capacitor to suppress possible overvoltage spikes, the converter is referred to as the active clamp converter in the literature. This converter, which was first proposed in the late 1970s–early 1980s [4–5] is a very popular and commonly used converter topology – arguably the most popular DC–DC lower power topology – and thus its operation and design will be explained in detail in this chapter.

The implementation shown in Figure 6.2a is common for applications with low input voltage, but not for applications with higher input voltages. For example, this implementation is never used for applications where the converter is supplied by a 400 V DC bus. When the input voltage is low, the active clamp switch can be implemented with a P-channel MOSFET device instead of an N-channel device. The use of a P-channel MOSFET as the active clamp switch allows a simple non-floating gate drive circuit to be used to turn the device on and off instead of a more sophisticated floating gate driver circuit. The use of P-channel devices, however, is limited, to low voltage applications as they are far inferior to N-channel devices at higher voltage. Consequently, the implementation shown in Figure 6.2b, which is generally referred to as a high-side active clamp converter, is more commonly used and the implementation shown in Figure 6.2a, which is generally referred to as a low-side active clamp converter, is not as common. Consequently, this section will focus on the operation of the high-side active clamp converter shown in Figure 6.2b.

6.2.1 Modes of Operation

Circuit diagrams that illustrate the modes of operation that the high-side active clamp converter goes through during a switching cycle are shown in Figure 6.3. For the explanation of the modes of operation, it is assumed that the converter is already operating under steady-state conditions with steady-state output inductor current and output voltage, that clamp capacitor C_{cl} is already charged, that the transformer has negligible parasitic capacitance, and that the main switch S_1 has already been turned on. The switching cycle actually begins sometime before the start of the first mode with the turning on of S_1, as the current is going through its body-diode. Since this action does not actually change the operation of the converter when it occurs, the description of the converter's operation that follows will start at the time when the converter enters an energy-transfer mode at Mode 1. The operation of the converter during a switching cycle is as follows:

Mode 1 ($t_0 - t_1$): At time $t = t_0$, both the main switch S_1 and secondary-side diode D_{o1} conduct current. S_1 was turned on with ZVS in a previous mode of operation. During this mode, the input voltage is impressed across the transformer primary and a stepped-down or stepped-up voltage appears across the secondary winding, typically a stepped-down voltage. Diode D_{o1} is forward-biased and the output inductor current flows through it. The converter is in an energy-transfer mode as energy is transferred from the input to the output.

Mode 2 ($t_1 - t_2$): At time $t = t_1$, the main switch S_1 is turned off and the primary current begins to charge the output capacitance of S_1 and discharge the output capacitance of S_2. During this mode, current flows through the clamp capacitor C_{cl} and the secondary side of the converter operates as it did during Mode 1, as positive voltage continues to be impressed across the primary winding.

Mode 3 ($t_2 - t_3$): At time $t = t_2$, the output capacitance of switch S_2 has been fully discharged and current begins to flow through its body-diode. S_2 can be turned on with ZVS as soon as this occurs. During this mode, the current is transferred away from the secondary-side diode D_{o1} to D_{o2}. This current transfer does not take place instantaneously due to the leakage inductance of the transformer. At some point during this mode, the voltage across C_{cl} is greater than the input voltage and a negative voltage begins to be impressed across the transformer. When this happens, the current in the transformer primary begins to decrease

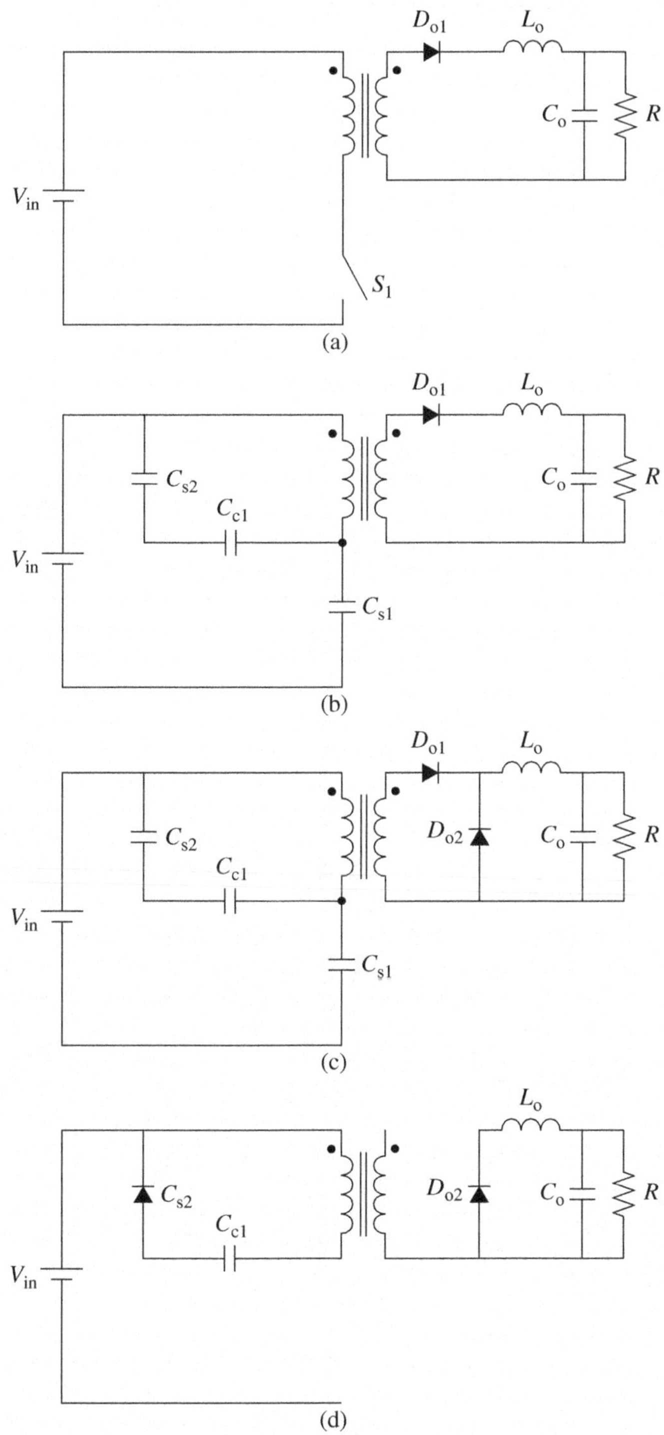

Figure 6.3 Modes of operation of the high-side active clamp forward converter. (a) Mode 1. (b) Mode 2. (c) Mode 3. (d) Mode 4. (e) Mode 5. (f) Mode 6. (g) Mode 7. (h) Mode 8.

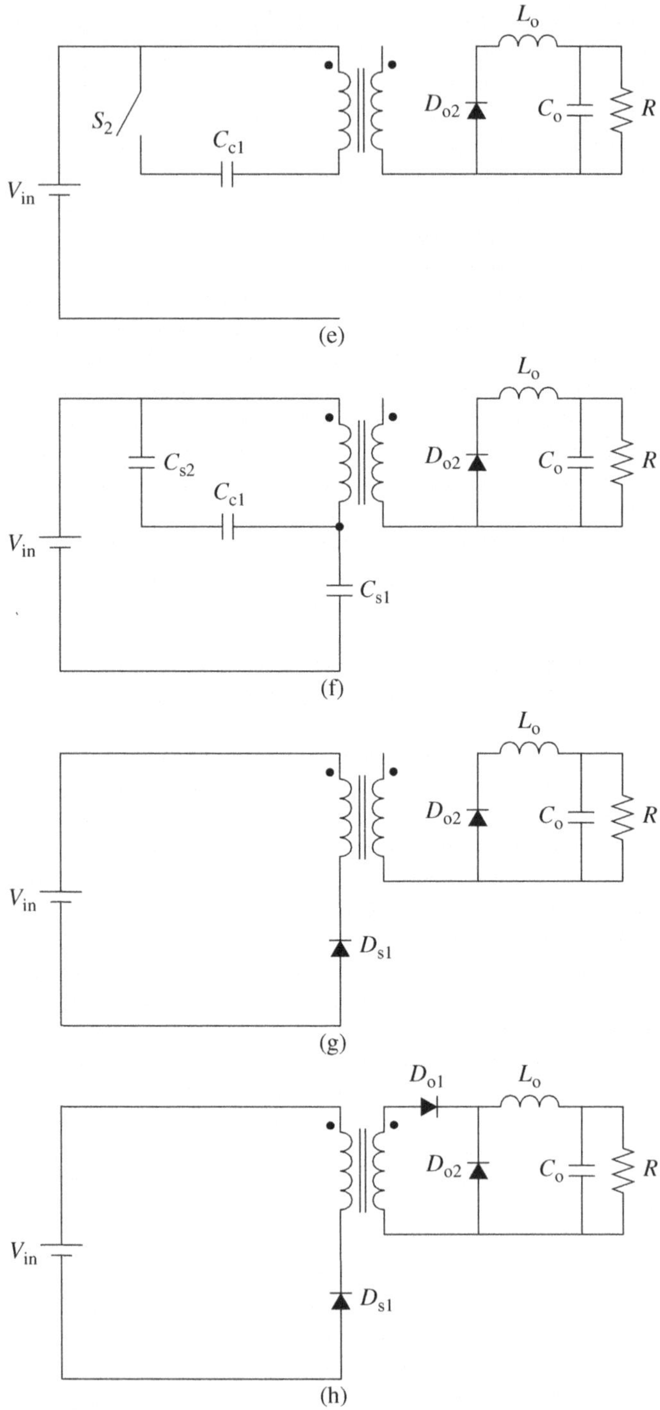

Figure 6.3 (*Continued*)

as does the current flowing out of the secondary to the point where this current is less than the output inductor current. The difference between the output inductor current and the transformer secondary current starts to flow through D_{o2}. The sum of the current flowing through D_{o1} and the current flowing through D_{o2} must be equal to the output inductor current. The transfer of current from D_{o1} to D_{o2} is not instantaneous due to the presence of leakage inductance in the transformer secondary, which slows down this current transfer.

Mode 4 ($t_3 - t_4$): At time $t = t_3$, the full output inductor current is flowing in D_{o2}. At the primary side, the current is flowing through the clamp capacitor C_{cl} and the voltage across C_{cl} is impressed across the primary winding, thus starting the transformer reset process. Since the current is flowing into C_{cl} and C_{cl} is the only voltage element with voltage in a loop with the transformer and S_2, this current is decreasing throughout this mode. The mode ends when the current flowing through C_{cl} is zero. It should be noted that current can flow through the switch in the direction from source to drain after it has been turned on if the voltage drop across the switch is such that it is less than the forward voltage drop of the body-diode of the device. This is similar to what happens with synchronous rectifiers, as was discussed in the previous chapter.

Mode 5 ($t_4 - t_5$): At time $t = t_4$, the primary current flowing in capacitor C_{cl} reverses direction and begins to flow through switch S_2. The secondary side operates in the same manner as it did in Mode 4. The resetting of the transformer continues during this mode.

Mode 6 ($t_5 - t_6$): At time $t = t_5$, switch S_2 is turned off and the process for eventually turning on the main switch S_1 is started. Energy stored in the transformer's magnetizing inductance is used to discharge the output capacitance of S_1 and charge the output capacitance of S_2. The secondary side operates in the same manner as it did in Mode 5. The resetting of the transformer continues during this mode.

Mode 7 ($t_6 - t_7$): At time $t = t_6$, sometime during the discharge of the output capacitance of S_1, the voltage impressed across the transformer's primary winding becomes positive and thus positive voltage is impressed across the transformer's secondary winding. This initiates the process by which current is transferred away from D_{o2} to diode D_{o1}. As with the process that transferred current from D_{o1} to D_{o2}, this process is not instantaneous due to the presence of leakage inductance at the secondary. Although the transformer primary voltage is zero, no power is transferred to the output from the input as D_{o2} continues to conduct and the left end of the output inductor has zero voltage until the current transfer to D_{o1} has been completed.

Mode 8 ($t_7 - t_8$): At time $t = t_7$, the output capacitance of S_1 has been completely discharged and current begins to flow through the body-diode of this switch and continues to do so during this mode. S_1 can be turned on with ZVS while the current is flowing through its body-diode. At the secondary side of the converter, the current continues to be transferred away from D_{o2} to D_{o1}. Eventually, the converter enters Mode 1 to start another switching cycle either with the completion of the current transfer to D_{o1} at the secondary after the current has changed direction and begun to flow through S_1 or with the change of direction in the flow of the current through S_1 after the transfer of output inductor current to D_1 has been completed, with the latter possibility more likely to take place. This mode ends when both these events have taken place.

6.2.2 Design Considerations

The popularity of the active clamp forward converter is such that a number of manufacturers of controller integrated chips (ICs) for power supplies have developed application notes for the design of this converter. Some of these, along with some more academic papers, are listed at the end of this chapter, in the References section [6–12]. Designing an active clamp converter is generally an iterative process where a number of factors are considered and a number of iterations are made before a satisfactory design is achieved. Some general considerations for the design of an active clamp converter are presented in this section. Readers are referred to the active clamp design publications listed in the References section for more detailed examples of how an active clamp converter can be designed. It should be noted that the design considerations that are listed here are not listed in terms of importance as they are all interrelated. One consideration is dependent on one or more other considerations, thus the need for some sort of iterative procedure. Most design examples that are presented in the literature tend to fix some considerations beforehand to simplify the presentation of the design procedure.

6.2.2.1 *Transformer Reset*

The converter must be designed in a manner that ensures that its transformer can be reset. Although it is possible for the converter to operate with a duty cycle of $D > 0.5$, extreme duty cycles should be avoided as this will increase the peak voltage stress of the converter switches. It may also result in the transformer saturating under transient conditions as the duty cycle may approach $D = 1$, which means that there will be little opportunity for the transformer to be reset. As a result, the duty cycle of the converter is typically restricted to a maximum of $D = 0.7$.

6.2.2.2 *ZVS Operation*

A key feature of the active clamp converter is that both its switches can operate with ZVS, but this is possible only if the converter is properly designed. In order for a switch to turn on with ZVS, its output capacitance must first be fully discharged and then its body-diode must have current flowing through it. The requirements for the ZVS turn-on of S_1 are different than those for the ZVS turn-on of S_2. In this section, the ZVS turn-on is discussed first, and then the ZVS turn-on of S_2 is discussed next.

Looking at Mode 6 in Figure 6.3f, when switch S_1 is about to be turned on, it can be seen that when the output capacitance of one switch is being discharged, the output capacitance of the other switch is being charged. The swing of voltage is from $V_{in} + V_{cl}$ to zero volts for S_1 and zero volts to $V_{in} + V_{cl}$ for S_2. In addition to this swing of voltage, there is another swing of voltage to consider and that is the swing of voltage from $-V_{cl}$ to V_{in} that occurs at the transformer primary during Modes 3–5 of operation, as shown in Figure 6.3. Given that the transformer primary has a winding capacitance of C_{TR} that is caused by the fact that the winding that is wrapped around the transformer core can be considered to be a set of capacitances consisting of conductors with non-conducting dielectric layers, there must be sufficient energy to swing the voltage across the net winding capacitance C_{TR}. If the output capacitance of S_1, C_{S1}, the output capacitance of S_2, C_{S2}, and C_{TR} are each considered to be parallel to the other capacitances and each capacitance experiences a voltage switching of $V_{in} + V_{cl}$, then the output of energy that is needed to discharge these capacitances is

$$E_c \geq \frac{1}{2} C_r (V_{in} + V_{cl})^2 \tag{6.1}$$

where

$$C_r = C_{S1} + C_{S2} + C_{TR} \tag{6.2}$$

Some manufacturers of silicon MOSFETs recommend that a 4/3 factor be multiplied by the output capacitance of their devices in order to ensure a more accurate determination of the actual, effective output capacitance of a device that is being charged and discharged during the operation of the converter that it is used in. If this 4/3 factor is considered, then Eq. (6.2) becomes

$$C_r = \frac{4}{3}(C_{S1} + C_{S2}) + C_{TR} \tag{6.3}$$

The energy that is required to swing the voltages across these capacitances from $V_{in} + V_{cl}$ to zero or zero to $V_{in} + V_{cl}$ comes from the transformer. Recalling the simplified transformer model that was presented in Section 4.1 in Chapter 4 and that is included in the converter as shown in Figure 6.4, a transformer can be considered to have a magnetizing inductance L_m and a leakage inductance L_{lk}. It is the energy that is stored in these inductances that are used to swing the voltages of C_{S1}, C_{S2}, and C_{TR}. Given that the current that flows in the magnetizing inductance, I_{Lm} is the magnetizing current (the current through L_m in the transformer model) and the current that flows in the leakage inductance, I_{Llk}, is the sum of this current and the reflected secondary current (the current through the primary winding of the ideal transformer in the transformer model) $I_{sec}N$, where N is the transformer turns ratio. The sum of

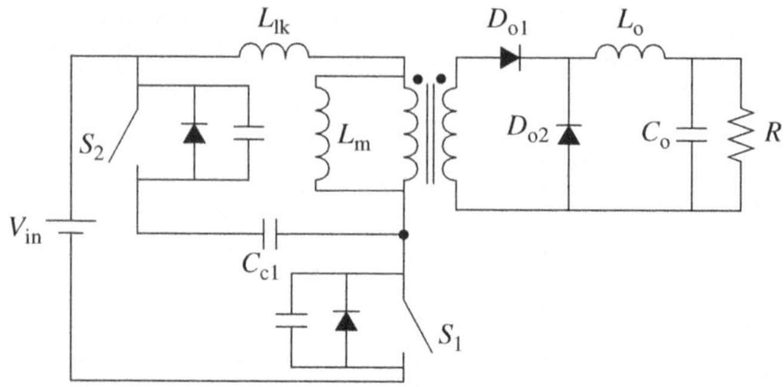

Figure 6.4 High-side active clamp forward converter shown with simplified transformer model (L_{lk}).

these currents is actually the transformer primary current I_{pri} so that I_{Llk} is equal to I_{pri}. It should be noted that the reflected secondary current is much larger than the magnetizing current so that I_{pri} can be approximated by $I_{sec}N$. This I_{pri} is the current that was placed in the leakage inductance during previous modes of operation.

Since the energy that is stored in an inductor is ½ LI^2, the energy that is stored in the transformer and that is available to swing the voltages across C_{S1}, C_{S2}, and C_{TR} is

$$E_{TR} = \frac{1}{2} L_m I_m^2 + \frac{1}{2} L_{lk} I_{pri}^2 \tag{6.4}$$

If Eqs. (6.1) and (6.4) are equated together, then the expression that must be satisfied to ensure that switch S_1 is turned on with ZVS is

$$\frac{1}{2} L_m I_m^2 + \frac{1}{2} L_{lk} I_{pri}^2 \geq \frac{1}{2} C_r (V_{in} + V_{cl})^2 \tag{6.5}$$

There is less need to be concerned about the ZVS operation of S_2 as energy can be accessed from the secondary through the current that is reflected to the primary and from the input source as well. From the circuit diagrams of Modes 2–3 of operation shown in Figure 6.3, it can be seen that current flows out of the input source and there is still a magnetic link between primary and secondary as C_{S2} is being discharged. Given the amount of energy that is available during Modes 2–3, S_2 will always be able to turn on with ZVS except under extremely light load conditions.

6.2.2.3 *Core Considerations*

Although the subject of magnetic design is outside the scope of this book, a few considerations about the transformer of an active clamp forward converter are discussed in this section. One characteristic that makes the transformer of an active clamp forward converter different from a standard forward converter with a tertiary winding or an RCD snubber is that the transformer is forced to operate in the so-called third quadrant of its *B–H* curve. If the first quadrant of a *B–H* curve is defined as being the quadrant where both the ampere-turns, *H*, and the flux density, *B*, are positive, then the third quadrant is the quadrant where both *H* and *B* are negative. With respect to the transformer model in Figure 6.4, the third quadrant operation translates to the negative voltage across the transformer primary and a negative magnetizing current. Although a regular forward converter can be designed so that its transformer is fully demagnetized by the end of a switching cycle (discontinuous magnetizing current) or with a DC bias current (continuous magnetizing current), the transformer of an active clamp forward converter will always have a DC bias current, as explained in [7].

The fact that the active clamp forward converter transformer can swing between the first and the third quadrant of its *B–H* gives the impression that the swing of flux in the transformer is much greater than that of a transformer in a conventional forward converter, but this is not so, as explained in [6]. The flux swing is almost the same except that it is not confined to the first quadrant as it is in a transformer in a conventional forward converter; since the flux swing is the same, so too are the core losses. According to [6], at lower frequencies, or with better core materials where core loss is not the dominant factor, the active reset technique offers the benefit of higher flux density swings. This can translate into reducing the number of turns in the transformer primary or secondary by half or operating with lower frequency or higher voltages, taking volt-seconds into account.

6.2.2.4 Magnetizing Current and Clamp Capacitor Voltage

Two important parameters that need to be considered in any design of an active clamp forward converter are the values of the magnetizing inductance of the transformer and the active clamp capacitor. These two are usually selected in conjunction with each other as the selection of a value for one of these parameters affects the selection of that of the other. Considerations that need to be taken into account when selecting these parameters include the following:

Magnetizing current: Contrary to a conventional forward transformer, the transformer of an active clamp forward converter should have low magnetizing inductance in order to generate a higher magnetizing current in the converter. This should be done for two reasons: The first is that current is needed to charge the clamp capacitor so that the voltage across it is greater than the input voltage, which allows the transformer to be reset by a negative voltage. Also, the amount of energy that is available to swing the voltages across C_{S1}, C_{S2}, and C_{TR} is dependent on the energy that is stored in the magnetizing and leakage inductances of the transformer. Of the two inductances, the magnetizing inductance is more dominant when it comes to energy storage and it is typically at least 100 times that of the leakage inductance, which is considered to be a parasitic element that should be minimized. Given that the energy stored in the magnetizing inductance is ½ $L_m I_m^2$, the magnetizing inductance should be low so that sufficient is stored in it to swing the voltages across C_{S1}, C_{S2}, and C_{TR}. It should not, however, be too high as this would increase primary current-related power losses and component stresses. The magnetizing current should be between 10% and 25% of the reflected load current in a typical design.

Clamp capacitor voltage: A key consideration in the selection of a clamp capacitor value is the voltage across it. It must be sufficiently high to reset the transformer when the main switch is off, but cannot be too high as this will increase component stress. For example, the peak voltage of the main switch S_1 is $V_{in} + V_{cl}$ so that a high V_{cl} results in S_1 having a high peak voltage stress when it is off.

Resonant frequency: The magnetizing inductance of the transformer and the clamp capacitor interact in a resonant manner during the switching cycle. The amount of time it takes to charge and discharge the clamp capacitor depends on the resonant cycle that is established by the magnetizing inductance of the transformer, L_m, and the clamping capacitor, C_{cl}. If the resonant cycle is too long, then the duty cycle of the converter is impacted, as is the maximum switching frequency that the converter can operate with. If the values are too low, then the magnetizing current and the clamp capacitor voltage can be high, which increases component stress and losses.

To summarize these design considerations: L_m and C_{cl} should be small so that sufficient magnetizing and clamp voltage is generated to ensure ZVS operation and appropriate transformer reset. Moreover, low values of L_m and C_{cl} allow the converter to operate at higher switching frequencies or with less duty-cycle loss. If L_m and C_{cl} are too small, however, then the voltage and current stresses of the primary side components rise as do power losses. A proper design of an active clamp converter is thus an iterative process that must consider a variety of factors before the end design is achieved.

6.2.3 Active Clamp Flyback Converter

The active clamp technique for ZVS operation can also be implemented in flyback converters, as was proposed in [13] and shown in Figure 6.5. The mechanism of ZVS operation in this converter is the same as that of the active clamp forward converter, with one very important difference and that is that an additional inductor, L_1 must be added to the primary. This inductor is needed to prevent a short-circuit that can occur after the switch is turned off and the primary voltage reverses direction to start the reset process.

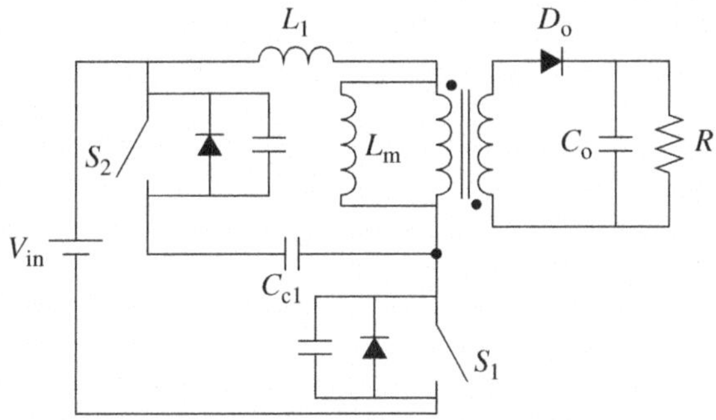

Figure 6.5 Active clamp flyback converter.

Without L_1, there would be very little impedance in the circuit loop formed by the transformer, the body-diode of S_2 and C_{cl}. Moreover, unlike a forward converter, the energy stored in the transformer of a flyback converter is transferred to the output when the switch is turned off. This means that there will be no energy to transfer to the active clamp capacitor and thus no energy that can be used to discharge the output capacitance of the switch.

Other than the addition of L_1, the active clamp mechanism is identical to that of the active clamp forward converter; thus the operation of the active clamp flyback converter will not be discussed here.

6.3 ALTERNATIVES TO THE ACTIVE CLAMP CONVERTER

Although active clamp forward and flyback converters are popular and widely used, they do have one key drawback that has prompted researchers to investigate alternative approaches to soft-switching in forward and flyback converters and that is the fact they lose their ZVS capability when operating under light-load conditions. When these converters operate with light loads, the primary current is very low and there is not enough stored in the converter's magnetics that can be used to discharge the output capacitance of the main switch. Most of the alternatives that have been proposed involve the use of some sort of active auxiliary circuit such as the ones presented in Chapter 2 for ZVS operation and Chapter 3 for ZCS operation. This section looks at how some of these auxiliary circuits can be implemented in forward and flyback converters.

6.3.1 Forward Converters

Examples of forward converters with active auxiliary circuits that allow their main switches to turn on with ZVS are shown in Figure 6.6. The converter shown in Figure 6.6a, which was proposed in [14], is the same as the resonant reset converter shown in Figure 6.1 except that a simple auxiliary

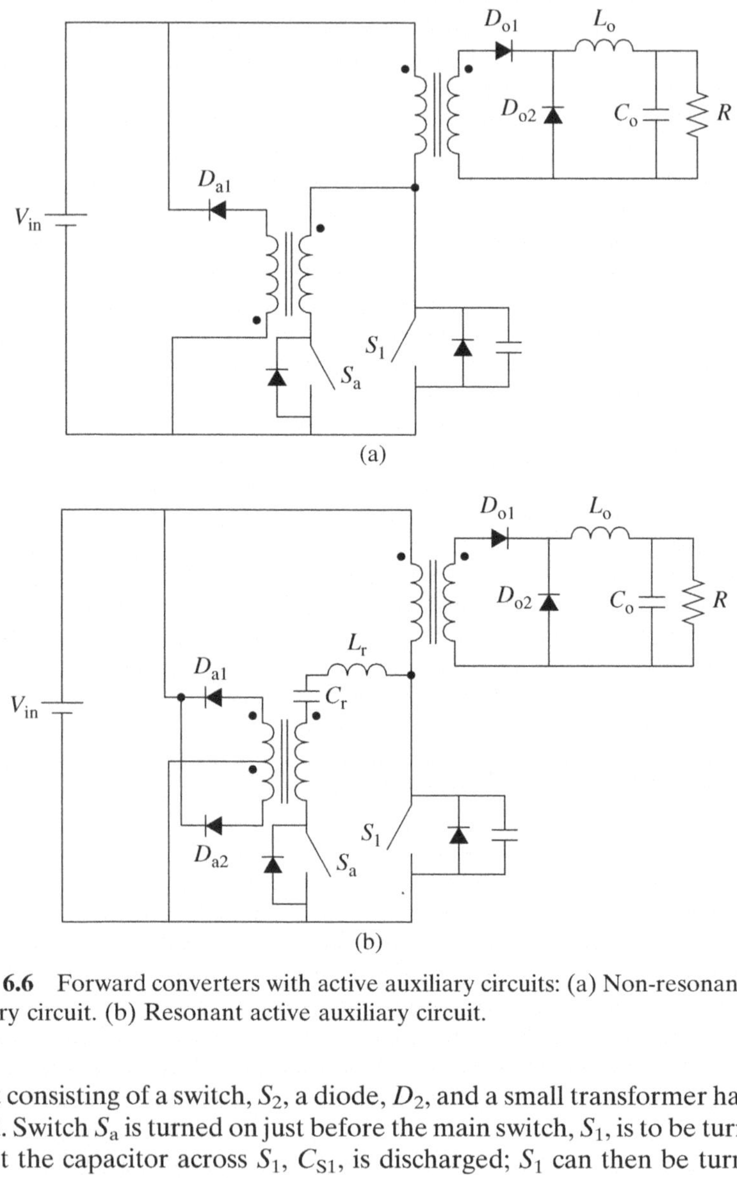

Figure 6.6 Forward converters with active auxiliary circuits: (a) Non-resonant active auxiliary circuit. (b) Resonant active auxiliary circuit.

circuit consisting of a switch, S_2, a diode, D_2, and a small transformer has been added. Switch S_a is turned on just before the main switch, S_1, is to be turned on so that the capacitor across S_1, C_{S1}, is discharged; S_1 can then be turned on with ZVS after the voltage across it falls to zero. Regardless of the load that the converter is supplying, S_1 will always be able to turn on with ZVS simply by turning the auxiliary switch on and discharging C_{S1}. S_a must be a device with a lower output capacitance than the capacitance of C_{S1}, which is typically the combined capacitance of the output capacitance of S_1 and some external capacitance.

The turn-off of the auxiliary switch of the converter shown in Figure 6.6a is hard, however. When this switch is turned off, energy in the auxiliary circuit transformer is fed to the input just as it would be in a flyback converter and, in fact, the auxiliary circuit can be considered to be a mini-flyback converter. This hard turn-off creates turn-off losses that somewhat offset the gain in converter efficiency due to the reduction of S_1 turn-on losses and thus the converter shown in Figure 6.6b was proposed. This converter, which was proposed in [15], has the same resonant type auxiliary circuit that was discussed in Section 2.3.2. This circuit allows S_a to turn off softly as the resonant interaction between L_r and C_r forces current to flow through the body-diode of the device so that when S_a is turned off, it is turned off with ZVS and ZCS.

In general, any active auxiliary circuit that can be used in non-isolated converters such as the ones presented in Chapters 2 and 3, can be used in a forward converter as long as the converter is implemented with some sort of reset mechanism. Unlike resonant reset and active clamp converters, the active auxiliary circuit in converters such as the one shown in Figure 6.6 do not contribute to the reset of the transformer and must be implanted with a tertiary winding or a resonant reset capacitor. If a resonant capacitor is used then care must be taken to design the converter properly to allow time for the resonant reset to occur. An example of how to design the resonant reset of a ZVS forward converter with an auxiliary circuit is presented in [14] and the reader is referred to this paper for details.

It should be noted for the converter shown in Figure 6.6 that although the main converter switch S_1 can be turned on with ZVS, the auxiliary circuit switch S_a cannot be, as energy that is stored in its output capacitance is dissipated in this switch when it is turned on. This switch, however, does turn on with ZCS so that its switching losses are reduced. Moreover, the amount of current that it has to turn off is less than that of the active clamp switch so that a device with a lower output capacitance can be used. What this means is that although S_a may have some switching losses, these are not as significant as they may appear to be.

It should also be noted that the auxiliary circuit switch S_a does not conduct current for as long a time as the active clamp switch in an active clamp converter. S_a is on for only a small fraction of the switching cycle while the active clamp switch is on whenever the main switch is off, with the exception of deadtimes. Consequently, a smaller, less expensive device can be used for S_a.

6.3.2 Flyback Converters

As with forward converters, if it is desired to implement soft-switching in a flyback converter with a method other than with a purely resonant reset or an active clamp, then the reset mechanism of the converter must be considered. There is no need to be concerned about the actual reset as this can be done naturally; it is the possible interrelation between the reset mechanism and the soft-switching mechanism that needs to be taken into account. This can be done with ZVS and with ZCS and an example of a ZCS-PWM flyback converter is presented here.

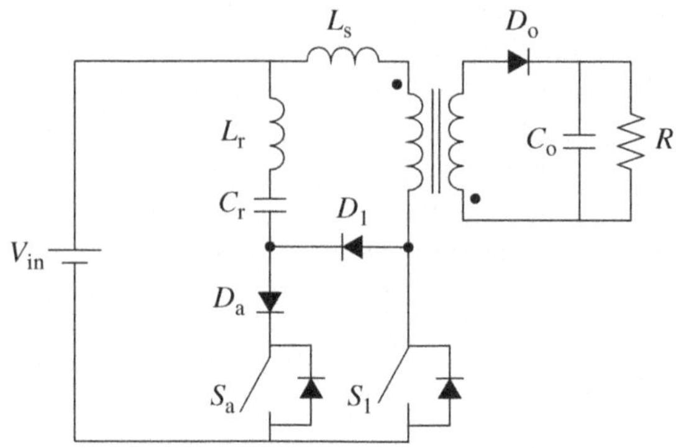

Figure 6.7 ZCS flyback converter.

Several ZCS-PWM flyback converters that take advantage of the flyback converter's natural reset mechanism have been proposed. Although ZVS is the preferred method of soft-switching for flyback converters, which are implemented with MOSFETs in most applications, the use of ZCS can be justified in applications where the input voltage is low (e.g. 12 V) and the input current is high as the output capacitance of the switch device is charged to little voltage and current-related losses are more significant. Figure 6.7 shows a ZCS-PWM converter that was proposed in [16]. This converter is a conventional flyback converter with an auxiliary circuit consisting of switch S_a, diodes D_1 and D_a, and resonant elements L_r and C_r. An additional inductance L_s is added to the converter to assist in the ZCS turn-on of the main switch S_1 and to simplify the design of L_r, as explained here.

The converter works as follows: When the transformer is in the process of being reset and has a negative voltage across it due to the inherent reset mechanism of a flyback converter, C_r is charged to a negative voltage by the transformer current as it is in a circuit loop with L_r, L_s, D_1, and the transformer. This prepares C_r for the next time the auxiliary circuit needs to be activated to turn off the main switch S_1 with ZCS. At some time during the reset process, the transformer should be fully demagnetized so that the core no longer has a magnetizing current. This allows S_1 to be turned on with ZCS to begin the next switching cycle as the current starts from zero when it begins to flow in the switch. L_s slows down the rate of the current rise in S_1 after it has been turned on, but whether it is actually needed will depend on the specifications of the converter. When it is time to turn off S_1, the auxiliary switch is turned on and this starts a resonant interaction between L_r and C_r. The voltage across C_r changes polarity and becomes greater than the input voltage. When the current in C_r changes direction and starts to flow upwards, it draws current away from S_1, which allows this switch to be turned off with ZCS. After the resonant cycle of the current in the L_r–C_r branch is completed, C_r is set up for the next time the auxiliary circuit is activated.

The auxiliary switch can turn on with ZCS because it is in series with L_r, which has zero current when the switch is turned on and which makes the rise of current through the switch gradual. Current from the transformer will not flow through S_a when it is turned on as a path of lower voltage potential does not exist. Current will only be diverted later on in the resonant cycle of the L_r–C_r branch current. S_a can be turned off with ZCS as the L_r–C_r resonant circuit shapes its current so that it is forced to zero.

6.3.3 Converters with Regenerative Energy Snubber

It is believed that active clamp converters are always more efficient than conventional flyback or forward converters because they are soft-switching converters while the conventional converters are hard-switching converters whose switches do not operate with ZVS. While this is generally true, it is worth considering the use of flyback converters with regenerative energy snubbers for certain applications, such as the one that was presented in Section 4.4.3 and shown in Figure 6.8. It was demonstrated in [17] that the efficiency of a flyback converter with a regenerative energy snubber can be comparable with that of an active clamp flyback converter when the input voltage is high (>200 V) and when the load is light (<50 W). By comparison, the difference in power losses is slight.

An active clamp converter ensures that its switches turn off with ZVS by using inductive energy to discharge the output capacitances of its switches. If, however, the converter is operating with high input voltage and light load, then the energy available in the converter will not be sufficient to discharge the output capacitances of the switches so that they will not turn on with ZVS, but will turn on with switching losses instead. Given the following:

- An active clamp converter operating under high input voltage/light-load conditions will have switching losses that a hard-switching flyback converter will not have as it has two switches in its topology instead of one.

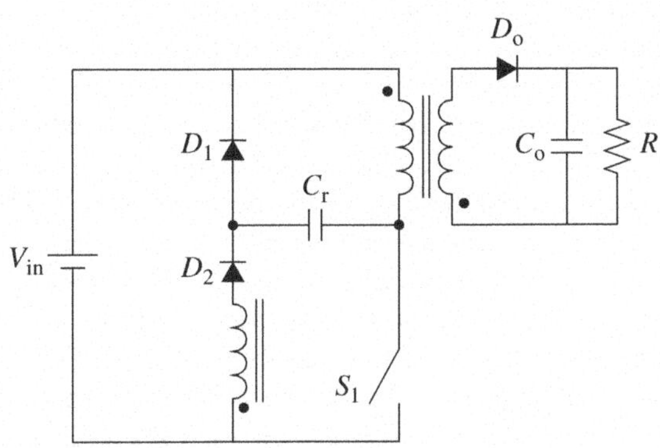

Figure 6.8 Forward converter with energy-regenerative snubber. (Note: The windings share the same core).

- A flyback converter with a regenerative energy snubber will have energy transferred to the load that would otherwise be dissipated in a lossy snubber.
- A flyback converter with a regenerative energy snubber that requires only one switch will be smaller and cheaper than an active clamp flyback converter that requires two switches.
- A flyback converter with a regenerative energy snubber can be considered instead of an active clamp converter when operating under high input voltage/light-load conditions.

6.4 CONCLUSION

Forward and flyback converters can be implemented with soft-switching to reduce their switching losses, just like non-isolated converters, but any method that is used to do so must take into account the transformer reset mechanism of these converters. In this chapter, two popular methods of implementing soft-switching in forward converters were presented: the resonant reset and the active clamp. The resonant reset forward converter uses a resonant capacitor that is connected across the switch to shape the switch voltage so that it can fall to zero sometime after the switch is turned off, which allows the switch to be turned on with ZVS. This resonant capacitor, which interacts with the magnetizing inductance of the transformer, can be charged to a voltage that is higher than the input voltage sometime during its resonant cycle and this allows negative voltage to be impressed across the primary of the transformer. The resonant reset forward converter is simple as it does not require a transformer with a tertiary winding, and it can be made even simpler if it is operated with high switching frequencies so that the output capacitance of the switching device is sufficient to act as the resonant capacitor in the converter. The main drawback of this converter is that it must be operated with variable switching frequency control, which complicates the design of the transformer and limits its load to a narrow range.

The active clamp forward converter can operate with ZVS and with fixed switching frequency as it has an active switch in series with the resonant capacitor that allows the mechanism for ZVS to be activated when desired. It too does not need to be implemented with a transformer with a tertiary winding as the resonant/clamping capacitor can reset the transformer. The active clamp switch works in a way that is complementary to the main switch and is on whenever the main switch is off and vice versa. The ZVS of the main switch is initiated by placing energy in the transformer when the active clamp switch is on, then turning off this switch to force current to flow through the body-diode of the main switch so that its voltage drops to zero before it is turned on. The active clamp switch itself operates with ZVS. The main drawback of active clamp converters is that they lose their ZVS capability at light loads as the amount of energy that can be stored in the transformer and used to discharge the capacitance of the main switching device is insufficient. The flyback version of the resonant reset and active clamp converters were briefly discussed in this chapter. Active clamp implementations for two-switch forward and flyback converters are also possible but were not discussed in this chapter as their operation is essentially the same as that of active clamp forward and flyback converters, with only a few minor differences.

Researchers have proposed alternatives to resonant reset and active clamp converters. Most of these involve the use of some active auxiliary circuit whose operation is independent of the reset process of the converter. In the case of forward converters, the active auxiliary circuit cannot reset the transformer and thus soft-switching forward converters must be implemented with some reset scheme such as with a tertiary winding off of the main converter transformer or a resonant reset. The same is true for flyback converters, but the natural reset of their transformer can be used advantageously in ZCS implementations for low input voltage/high current applications. Examples of ZVS-PWM forward and ZCS-PWM flyback converters were presented in this chapter.

Active flyback clamp converters may not be an attractive option for high input voltage/light-load applications as the amount of voltage swing that the switch and the transformer go through is high and the amount of energy that is available to enact this switch is insufficient. Flyback converters with regenerative energy snubbers may be a more attractive option as they do not have auxiliary switch switching losses and they can transfer energy from the primary to the output through the regenerative energy mechanism. As with any design option, the decision of whether to use an active clamp flyback converter or a flyback converter with a regenerative energy snubber is very much dependent on the application and on the operating conditions.

REFERENCES

1. P. Vinciarelli, "Forward converter switching at zero current," U. S. Patent 4,415,959, Nov. 15, 1983.
2. K. Wallace, "Resonant reset," *IBM Technical Disclosure Bulletin*, vol. 26, no. 11, pp. 6179–6180, Apr. 1984.
3. N Murakami and M. Yamasaki, "Analysis of a resonant reset condition for a single-ended forward converter," *IEEE Power Electronics Specialists Conference (PESC)*, 1988, Apr. 11–14, 1988, Kyoto, Japan, pp. 1018–1023.
4. B. Carsten, "Epic data multi-output switched mode power supply," *Epic Data internal memo*, Dec. 1978. expanded in B. Carsten, "Design techniques for transformer active reset circuits at high frequencies and power levels," *Interect Communications Inc. High Frequency Power Conversion Conference (HFPCC)*, 1990, May 6–11, 1990, Santa Clara, USA, pp. 235–246.

5. P. Vinciarelli, "Optimal resetting of the transformer's core in single ended forward converters," U. S. Patent, 4,441,146, Apr. 3 1984.
6. B. Andreycak, "Active clamp and reset technique enhances forward converter performance," *Unitrode Power Supply Design Seminar*, SEM-1000, 1994.
7. Q. Li, F. C. Lee, and M. M. Jovanovic, "Design considerations of transformer DC bias of forward converter with active-clamp reset," *IEEE Applied Power Electronics Conference (APEC)* 1999, Mar. 14–18, 1999, Dallas, USA, pp. 553–559.
8. D. Dhaval, "Design considerations for active clamp and reset technique," *Unitrode Power Supply Design Seminar*, SEM-1100, 2000.
9. Q. M. Li and F.C. Lee, "Design consideration of the active-clamp forward converter with current mode control during large-signal transient," *IEEE Transactions on Power Electronics*, vol. 18, no. 4, pp. 958–965, Jul. 2003.
10. H. Huang, "Design guidelines on the effect of resonant transitions of forward converter on efficiency with active clamp," *IEEE Applied Power Electronics Conference (APEC)*, 2008, Feb. 24–28, 2008, Austin, USA, pp. 600–606.
11. K. S. Bhanuprasad, "Designing active-clamp forward converters using peak-current-mode controllers," *Maxim Application Note*, 5780, 2014.
12. "Design of a 100 W active clamp forward DC–DC converter for telecom systems using the NCP1562", *ON Semiconductor Application Note*, AND8273/D, 2020.
13. R. Watson, F. C. Lee, and G. C. Hua, "Utilization of an active-clamp circuit to achieve soft switching in flyback converters," *IEEE Transactions on Power Electronics*, vol. 11, no. 1, pp. 162–169, Jan. 1996.
14. Y. Xi., P. Jain, Y.-F. Liu, and R. Orr, "A self core reset and zero voltage switching forward converter topology," *IEEE Transactions on Power Electronics*, vol. 15, no. 6, pp. 1192–1203, Nov. 2000.
15. Y. Xi., P. Jain, "A forward converter topology employing a resonant auxiliary circuit to achieve soft switching and power transformer resetting," *IEEE Transactions on Industrial Electronics*, vol. 50, no. 1, pp. 132–140, Feb. 2003.
16. C.-M. Wang, "A novel ZCS-PWM flyback converter with a simple ZCS-PWM commutation cell," *IEEE Transactions on Industrial Electronics*, vol. 55, no. 2, pp. 749–757, Feb. 2008.
17. A. Alganidi, A. Abosnina, and G. Moschopoulos, "A comparative study of DC–DC flyback converters for telecom applications," *IEEE International Telecommunications Energy Conference (INTELEC)*, 2017, Oct. 22–26, 2017, Gold Coast, Australia, pp. 425–431.

CHAPTER 7

THE ZVS-PWM FULL-BRIDGE CONVERTER

The basic DC–DC PWM full-bridge converter that was presented near the end of Chapter 4 is the most popular type of DC–DC converter for applications that require DC–DC conversion of at least 500 W power with transformer isolation. Isolated DC–DC converters that have been discussed in previous chapters such as the flyback and forward converters and the higher power, two-switch converters are rarely used in higher power applications due to the drawbacks that they have, such as switch stress, transformer resetting, and output ripple. Although the basic DC–DC PWM full-bridge converter is often modified when used in various applications, as will be discussed in the next chapter, these modifications are based on the basic converter topology that is the focus of this chapter.

The full-bridge converter topology can be operated with basic PWM and phase-shift PWM. The operation of the converter with basic PWM was discussed in Section 4.6; this discussion is briefly reviewed in this chapter. The focus of this chapter is on the operation of the converter with phase-shift PWM. Phase-shift PWM is much more widely used due to its ability to ensure that the converter switches can operate with ZVS. A full-bridge converter that operates with phase-shift PWM is referred to as a ZVS-PWM full-bridge converter, although there is no actual modification to the actual topology. This chapter will focus on the operation of a DC–DC full-bridge PWM converter with phase-shift PWM and on some of the issues that can arise when operating the converter in such a manner. A discussion as to which of the two PWM

DC–DC Converter Topologies: Basic to Advanced, First Edition. Gerry Moschopoulos.

Published 2024 by John Wiley & Sons, Inc.

methods should be used for particular loads is presented in this chapter, as well as a discussion of the impact that using wide bandgap devices instead of traditional silicon devices has on the design of the converter.

7.1 DC–DC PWM FULL-BRIDGE CONVERTER WITH BASIC PWM CONTROL

As was explained in Section 4.6, the DC–DC PWM full-bridge converter topology can be implemented with either a two-diode secondary, as shown in Figure 7.1a, or with a four-diode bridge secondary, as shown in Figure 7.1b. The first implementation is more popular as full-bridge converters are popular in applications where a DC voltage is stepped down to a much lower level and must supply considerable current to a load. Conduction losses are decreased if current flows through one secondary diode instead of two diodes in series, as is the case with the four-diode bridge implementation.

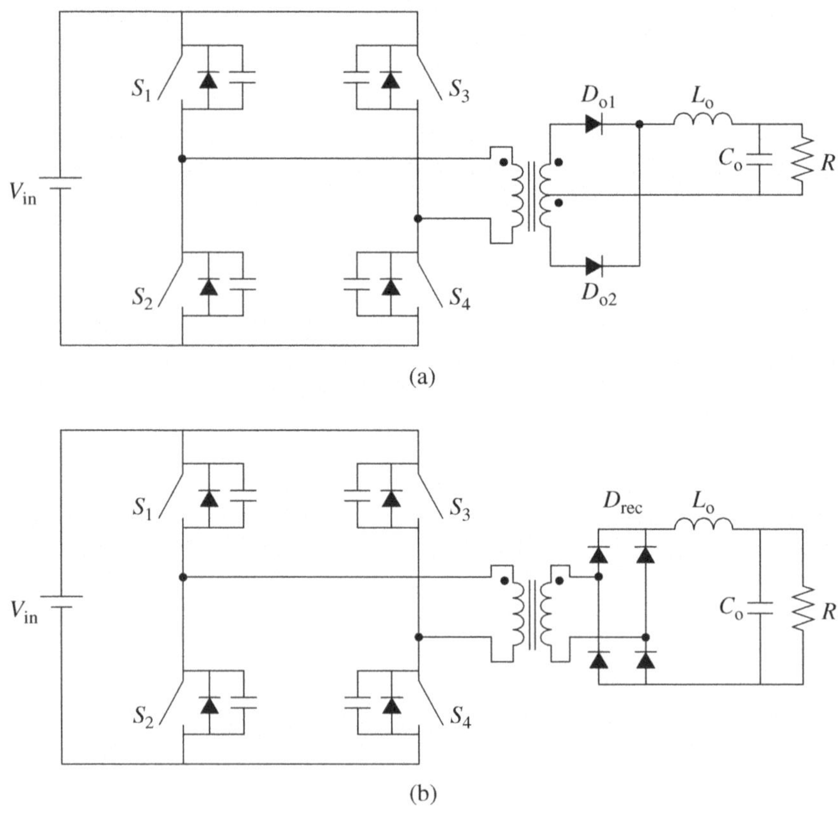

Figure 7.1 DC–DC full-bridge converter. (a) Implementation with center-tapped transformer. (b) Implementation with non-tapped transformer.

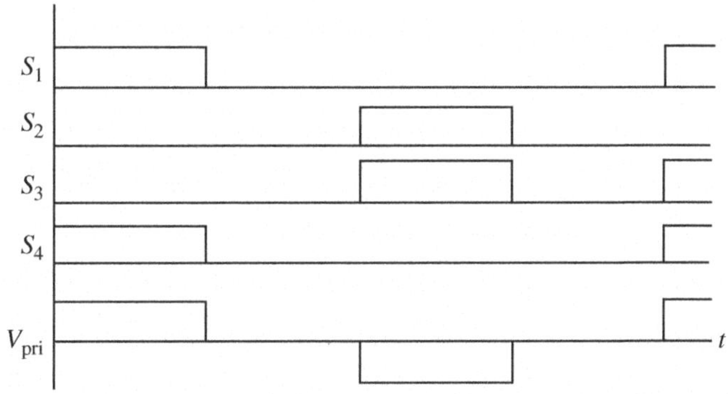

Figure 7.2 Gating signals and transformer primary voltage of a DC–DC full-bridge converter operating with basic PWM control.

The four-diode bridge implementation is used in applications where the output voltage is high since the voltage stress of a diode in the bridge is the output voltage instead of twice the output voltage, as explained in Section 4.6.

The operation of a full-bridge converter with basic PWM control was explained in Section 4.6 and is reviewed here to contrast it to the operation of the converter with phase-shift PWM, which is discussed in the next section. When operating with basic PWM control, the input DC voltage is impressed across the transformer primary whenever a pair of diagonally opposed switches is on – if S_1 and S_4 are on, then a positive voltage is impressed across the winding; if S_2 and S_3 are on, then a negative voltage is impressed across the winding. No voltage is impressed across the transformer primary if none of the converter switches is off.

The gating signals of the converter switches along with the transformer primary voltage when the converter is operating with basic PWM control are shown in Figure 7.2. It can be seen from Figure 7.2 that switches S_1 and S_4 are turned on simultaneously at the start of the switching cycle, and then turned off simultaneously sometime before the start of the next half-cycle. As can be seen from the primary voltage waveform, the primary voltage is high until S_1 and S_4 are turned off and then becomes zero afterwards. When the converter is halfway through its switching cycle, switches S_2 and S_3 are turned on simultaneously and a negative voltage is impressed across the transformer primary. The primary voltage turns to zero after these two students are turned off simultaneously and remain zero until the start of the next switching cycle.

7.2 ZVS-PWM FULL-BRIDGE CONVERTER WITH PHASE-SHIFT PWM

Although operating the DC–DC full-bridge converter with basic PWM control is the simplest way to operate it, it is not the most common way of doing so. By far, the most common way of operating a basic DC–DC full-bridge

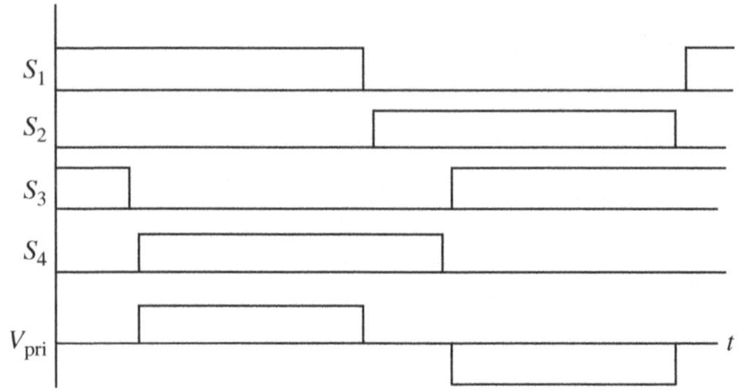

Figure 7.3 Gating signals and transformer primary voltage of a DC–DC full-bridge converter operating with phase-shift PWM.

converter is to operate it with phase-shift PWM [1, 2] because the converter can operate with ZVS when this is done [3–6]. The gating signals of the four switches of the DC–DC full-bridge converter and the transformer primary voltage are shown in Figure 7.3. Unlike basic PWM control, the gating signal of each converter switch is almost 50% of the switching cycle and is not varied when phase-shift PWM is used. The switches of a converter leg are complimentary so that when one switch of a leg is on the other is off except for small time intervals when both switches of a leg are off, which are referred to as dead-times in the literature. These dead-times ensure that the short-circuiting of a converter leg is avoided by never having the gating signals of the switches of a leg overlap and that a switch has sufficient time to turn on with ZVS, as will be explained later in this chapter.

The width of the voltage pulses is varied by varying the phase shift between the gating signals of the two legs. If S_1 and S_4 overlap completely, then S_2 and S_3 overlap completely as well so that the input DC bus voltage is applied to the transformer primary, with one polarity when both S_1 and S_4 are on and with the opposite polarity with S_2 and S_3 are on. This situation represents a phase shift of 0° between the gating signals of the two converter legs. The other extreme is a phase shift of 180°, which occurs when the gating signals of S_1 and S_3 overlap completely and those S_2 and S_4 overlap completely as well. A 180° phase shift in the gating signals of the two converter legs ensures that no voltage is ever impressed across the transformer primary as both ends of the transformer will either be connected to the positive rail of the bus and each has a voltage of V_{in}, or be connected to the negative rail of the bus and each has a voltage of 0.

A phase shift φ between 0° and 180° means that the converter will have states when a voltage of one polarity or the other is impressed across the transformer and states when the converter will have states when there is no voltage impressed across the transformer. The general switching sequence when the converter is operating with phase-shift PWM is S_1 and S_4 on, S_1 and S_3 on, S_2 and S_3 on, and S_2 and S_4 on. Voltage is impressed across the transformer whenever a pair of diagonally opposed switches is on, S_1 and S_4 or S_2 and

S_3, and is not impressed across the transformer whenever two top switches, S_1 and S_3, or two bottom switches are on, S_2 and S_3. These two states can also be described as being energy-transfer states or freewheeling states. In the former case, energy can be transferred from the primary to the secondary side of the converter and thus the output through the secondary diodes when voltage is impressed across the transformer primary as the secondary diodes are forward-biased and a path exists from input to output for energy to be transferred. In the latter case, no such energy transfer can occur as the primary and secondary are not magnetically coupled so that the current in both the primary and secondary just circulates or freewheels.

When the voltage across each center-tapped secondary winding is positive, then diode D_{o1} becomes forward-biased and conducts current and D_{o2} becomes reverse-biased and does not conduct current. The opposite happens when the secondary winding voltage is negative – D_{o1} is reverse-biased and D_{o2} is forward-biased. When there is no voltage across the transformer, then, ideally, the output inductor current splits into two with half the current flowing through D_{o1} and the other half flowing through D_{o2}. In real-life, non-ideal converters, this current distribution is usually uneven. The rectified secondary voltage has the same shape as the voltage that is at the input of the LC output filter of a buck converter so that in essence, a full-bridge converter is a type of isolated buck converter. The width of the voltage pulses of this waveform can be varied by phase-shifting the gating signal pulses and since the rectifying process doubles the frequency of the primary and secondary voltage waveforms, the duty-cycle of the converter is defined as the ratio of the switches' on-time to half the length of the original switching cycle.

The modes of operation that a DC–DC full-bridge converter goes through during a half-switching cycle are shown in Figure 7.4. Only the modes for a half-switching cycle are shown as the modes for the other half are identical except with different switches – the modes when the converter is trying to transition from an energy-transfer mode with S_1 and S_4 on to one when S_2 and S_3 are on are identical to those when the converter is trying to transition from S_2 and S_3 on to an energy-transfer mode when S_1 and S_4 are on. The inductor current is shown as a current source. Assuming that switches S_1 and S_4 are on, the converter operates with the following modes over a half-switching cycle:

Mode 1 $(t_0 - t_1)$: Switches S_1 and S_4 are on and the converter is in an energy-transfer mode as a voltage is impressed across the transformer primary and energy is transferred from the input to the output. At the secondary, current is flowing through diode D_{o1} as this diode is forward-biased while D_{o2} is reverse-biased.

Mode 2 $(t_1 - t_2)$: Switch S_4 is turned off while S_1 remains on. The output capacitance of S_4 begins to be charged while that of S_3 begins to be discharged. Since the primary and the secondary windings of the transformer are magnetically coupled throughout this mode, the reflected secondary current ensures that there is sufficient energy to charge and discharge the output capacitances of S_4 and S_3, respectively.

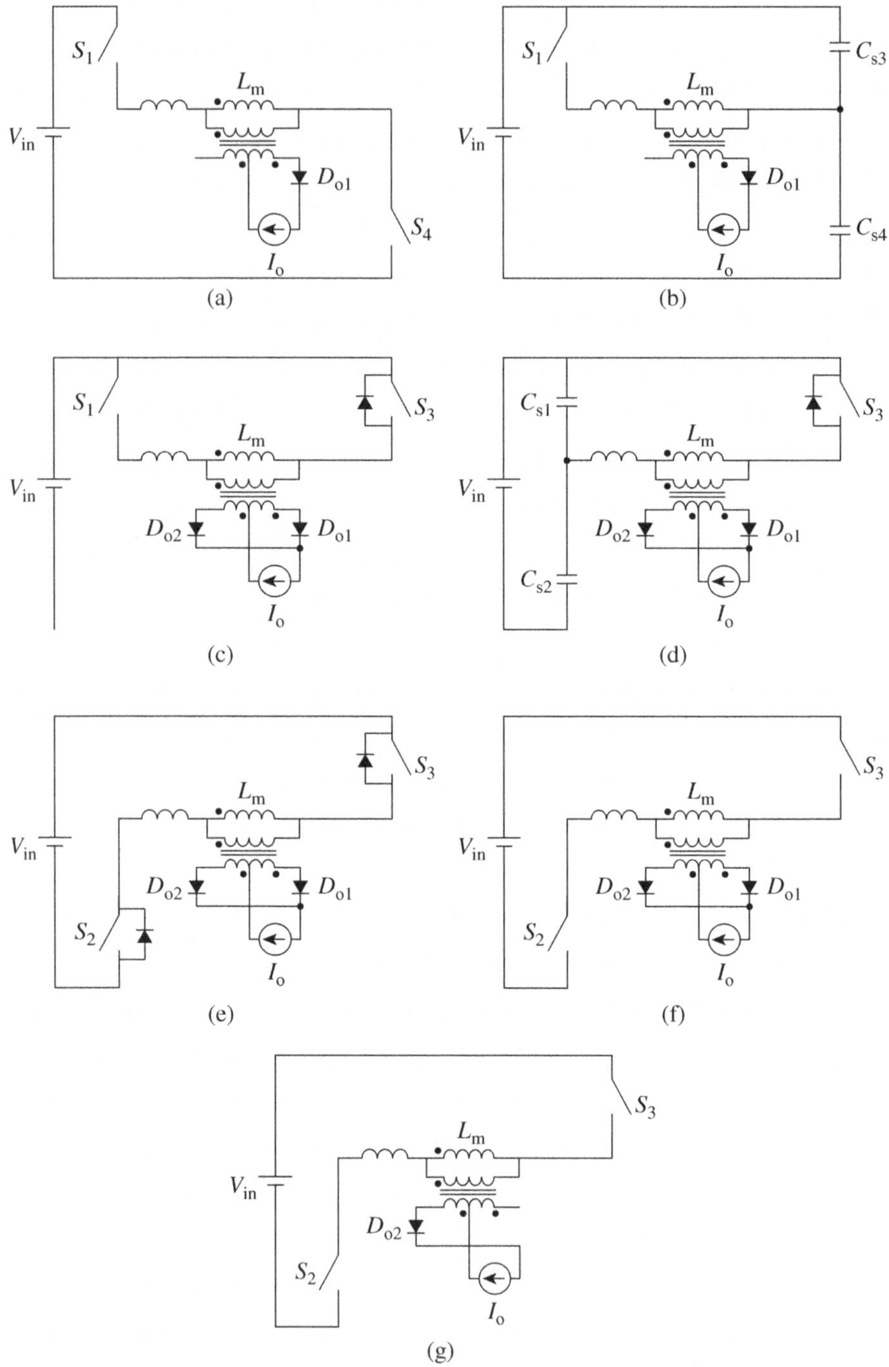

Figure 7.4 Modes of operation of a DC–DC full-bridge converter operating with phase-shift PWM. (a) Mode 1. (b) Mode 2. (c) Mode 3. (d) Mode 4. (e) Mode 5. (f) Mode 6. (g) Mode 7.

Mode 3 ($t_2 - t_3$): This mode begins after the output capacitance of S_3 has been fully discharged and current begins to flow through the anti-parallel body-diode of the device. When this happens, the switch can be turned on with ZVS as the voltage across it is just one diode drop, almost zero. Whether current continues to flow through the body-diode or it begins to flow through the switch in a negative direction will depend on which path results in a lower voltage drop. Regardless of the path of current in the device, the converter is in a freewheeling mode of operation as current circulates in the primary and the primary and secondary windings are no longer magnetically coupled. At the secondary, since the voltage across both primary and secondary windings is zero, the current is split between D_{o1} and D_{o2}. It is assumed here that this split is even, but it may be the case that the split is uneven if non-ideal diodes are used. The transition of current from D_{o2} to D_{o1} begins when the voltage across the secondary winding drops to zero. So, the conduction of D_{o1} no longer forces D_{o2} to become reverse-biased. This current transition is not instantaneous because of the presence of secondary leakage inductance and parasitic inductances but is sudden as these parameters are small.

Mode 4 ($t_3 - t_4$): This mode begins when S_1 is turned off and the converter begins to exit the freewheeling mode. The output capacitance of S_1 begins to be charged while that of S_2 begins to be discharged. Unlike the transition from S_4 to S_3, there is no reflected secondary current available to charge and discharge S_1 and S_2, respectively, so that only leakage and magnetizing inductance energy are available to do so. Current in the secondary continues to freewheel through both diodes while the switching transition is taking place, but since the voltage across the transformer primary becomes non-zero, the process of current transferring away from D_{o1} to D_{o2} begins.

Mode 5 ($t_4 - t_5$): This mode begins after the output capacitance of S_2 has been fully discharged and current begins to flow through the body-diode of the device. When this happens, the switch can be turned on with ZVS as the voltage across it is just one diode drop. Similar to S_3 in Mode 3, whether current continues to flow through the diode or it begins to flow through the switch in a negative direction will depend on which path results in a lower voltage drop. Regardless of the path of current in the device, primary current flows upwards in the circuit diagram through the S_2 and S_3 devices and voltage is impressed across the transformer primary with the opposite polarity of what it was in Mode 1. The current in the secondary diodes continues to flow through both secondary diodes as the current flowing through D_{o1} is in the process of being fully transferred to D_{o2}.

Mode 6 ($t_5 - t_6$): This mode begins when the primary current reverses direction and flows from the positive terminal of the input source to the negative terminal. Secondary diode D_{o1} continues to conduct current during this mode.

Mode 7 ($t_6 - t_7$): The transfer of current away from D_{o1} that began in previous modes is completed at the start of this mode and current begins to flow through just one secondary diode instead of both diodes. The converter is in an energy-transfer mode and remains so until switch S_3 is turned off at time $t = t_7$, which corresponds to the end of the first half switching cycle.

The following should be noted when the DC–DC full-bridge converter is operated with phase-shift PWM:

- A DC–DC full-bridge converter operated with phase-shift PWM is commonly referred to as the ZVS-PWM full-bridge converter, even though the actual topology is the same as that when it is operated with regular PWM control.
- The benefits of the converter are the same as those when basic PWM control is used: the voltage rating of its switches does not exceed the input voltage, and the frequency of the voltage that is fed to the output filter is doubled so that the output filter size can be reduced, it does not have the conduction losses that limit half-bridge converters.
- The drawbacks of the converter are the same as those when basic PWM control is used: the number of switches, floating gate drives, and transformer saturation.
- The anti-parallel diode across each switch is typically the body-diode of a MOSFET or an IGBT device. Sometimes, an anti-parallel diode with a lower forward voltage drop than that of the body-diode and with better reverse-recovery characteristics is placed parallel to the device to improve the performance of the converter, especially if the converter is to be operated with high switching frequency and with silicon devices. This is not common practice, however, due to the cost of the additional diodes, but it is an option that can be considered when silicon devices are used.
- The capacitance across each switch can be the output capacitance of the device or it can be the combination of the output capacitance and an additional external capacitance. Whether an external capacitance is added depends on the amount of current that flows in the converter primary – if this current is low, then additional capacitance is not needed, if it is high, it is. It should be noted that more capacitance across the device means more energy is needed to discharge it before the device can be turned on with ZVS. As a result, the converter can lose ZVS at higher loads and thus there is a trade-off to be considered between turn-on losses and turn-off losses.

- Dead-times, which are short time intervals when neither switch of a converter leg is on (thus rendering the converter leg inactive or "dead") are needed to ensure that switch output capacitances can be discharged and charged before the switches can be turned on and off. If a switch is turned on before its output capacitance is fully discharged, then it will not do so with ZVS.

7.3 ISSUES RELATED TO THE OPERATION OF ZVS-PWM PWM FULL-BRIDGE CONVERTER

Several issues need to be considered when operating a DC–DC PWM full-bridge converter with phase-shift PWM to make it into a ZVS-PWM full-bridge (FB) converter. These include

- ZVS operation
- Duty-cycle loss
- Voltage ringing

Each of these issues is reviewed in this section.

7.3.1 ZVS Operation

The leading leg switches of the converter – the switches that are turned on or off to transition the converter from an energy-transfer mode to a freewheeling mode – have different ZVS requirements than the lagging leg switches – the switches that are turned on and off to transition the converter from a freewheeling mode to an energy-transfer mode. The leading leg switches will discharge and charge output capacitances due to the availability of output inductor energy. Consider Mode 2 of operation as explained above and illustrated in Figure 7.4b. When the converter is operating in Mode 2, which is the mode when a leading leg switch is turned off to initiate the transfer into a freewheeling mode, current reflected from the secondary is still available in the primary to help discharge the capacitance of the switch that is to be turned on with ZVS in the same leg. The amount of output inductor energy available to discharge and charge output capacitances is half the product of the value of the output inductor and the amount of current that flows through it during the transition of modes. Since the output inductor is a relatively large value, typically at least in the tens of micro-henrys, there will always be sufficient inductive energy to ensure ZVS operation during the transition of modes.

In order to ensure ZVS operation of the leading leg switches, time must be given for the switch capacitance of a switch to discharge completely after its complementary switch in the same leg has been turned off. The dead time can be determined from the basic relation between current and voltage for a capacitor, which can be expressed as

$$I_C = C\frac{dV_c}{dt} \tag{7.1}$$

Note that current is needed not only to discharge the switch capacitance of a switch but to charge the switch capacitance of the complimentary switch in the same leg as well, in order to satisfy Kirchhoff's Voltage Law. If the magnetizing current is ignored, this current can be considered to be the reflected load current at the primary, which can be expressed as

$$I_{Mode2} = \frac{N_S}{N_P}\left(I_o + \frac{\Delta I}{2}\right) \tag{7.2}$$

where ΔI is the peak-to-peak output inductor current ripple. Since the current can be considered to be constant during this interval, the change in switch capacitance-voltage will be linear so that

$$I_C = C\frac{\Delta V_c}{\Delta t} \tag{7.3}$$

which can be rewritten as

$$\Delta t = C\frac{\Delta V_c}{I_C} \tag{7.4}$$

Δt is the dead time t_{dead}, $\Delta V_S = V_S$, $I_C = I_{Mode2}$ from Eq. (7.2) and

$$C = 2C_S + C_{TR} \tag{7.5}$$

where C_S is the value of a switch capacitance and C_{TR} is the parasitic capacitance of the transformer, which can be modeled as a capacitor in parallel with an ideal transformer. This capacitance is typically very small and is often neglected in Eq. (7.5). The dead time for a leading leg switching transition needed to ensure the ZVS operation of its switches can be expressed as

$$t_{dead} \geq (2C_S + C_{TR})\frac{V_S}{I_p} \tag{7.6}$$

The output inductor current is not available when the converter is in the process of exiting a freewheeling mode of operation as the transformer primary and secondary windings are decoupled and current circulates in these windings independently of each other. Assuming that the energy that is available from the magnetizing inductance is negligible, as is generally the case unless the magnetizing inductance has been purposely designed to be low (a case that will be discussed in Chapter 8) or the load is very light, then the only inductive energy that is available to charge and discharge capacitances is the energy stored in the leakage inductance of the power transformer. The key relation that needs to be satisfied to ensure the ZVS operation of the lagging leg switch is

$$\frac{1}{2}L_{\text{lk}}I_{\text{pri}}^2 \geq C_{\text{S}}V_{\text{S}}^2 + \frac{1}{2}C_{\text{TR}}V_{\text{S}}^2 \tag{7.7}$$

where L_{lk} is leakage inductance, C_{S} is the output capacitance of a switch, V_{S} is the switch voltage, which is the DC source voltage or bus voltage, C_{TR} is the parasitic capacitance of the transformer. I_{pri} is the transformer primary current and can be assumed to be the reflected current from the secondary if a rough calculation is acceptable. Equation (7.7) accounts for the total energy of the output capacitances of the two switches of the lagging leg, twice $\frac{1}{2}CV^2$, and the parasitic capacitance of the transformer. C_{TR} can be ignored in Eq. (7.7) for low switching frequencies (<50 kHz).

The minimum amount of current that needs to be circulating in the primary to ensure that the lagging leg switches operate with ZVS, the critical current I_{crit}, can be determined by rearranging Eq. (7.7) to give

$$I_{\text{crit}} = \sqrt{\frac{2}{L_{\text{lk}}}\left(C_{\text{S}}V_{\text{S}}^2 + \frac{1}{2}C_{\text{TR}}V_{\text{S}}^2\right)} \tag{7.8}$$

The available current through L_{lk} at the start of Mode 4 when the converter begins to exist in a freewheeling mode of operation can be calculated as

$$I_{\text{Mode4}} = \frac{N_{\text{S}}}{N_{\text{P}}}\left(I_{\text{o}} + \frac{\Delta I}{2} - \frac{V_{\text{o}}}{L_{\text{f}}}(1-D)\frac{T}{2}\right) \tag{7.9}$$

ZVS is achieved when $I_{\text{Mode4}} > I_{\text{crit}}$.

The dead time needed to ensure the ZVS operation of the converter's lagging leg switches can be expressed as

$$t_{\text{dead}} \geq \frac{\pi}{2}\sqrt{L_{\text{lk}}(2C_{\text{S}} + C_{\text{TR}})} \tag{7.10}$$

This represents a quarter of the resonant cycle that is the result of the interaction between the transformer leakage inductance and the switch and transformer capacitances. Unlike the leading leg switches, it is imperative that the lagging leg switches are turned on soon after the minimum allowable dead time is reached. This is because the dead time is dependent on the transformer's leakage inductance, which is much smaller than the output inductor. If the dead time is too long, then the switch capacitance of the switch that is about to be turned on will begin to recharge again due to the resonant interaction with the small leakage inductance. As a result, the opportunity for ZVS operation will be lost and the switch will not turn on with ZVS. In other words, there is a time window for the lagging leg switches during which a lagging leg switch must be turned on after its complementary switch has been turned off if it is to be turned on with ZVS. If the switch is turned on before the time specified in Eq. (7.10) or after the time when its switch capacitance begins to recharge, then it will not be turned on with ZVS and turn-on switching losses will result.

It should be noted that some manufacturers of MOSFETs suggest that the $C_S V_S^2$ term in Eq. (7.7) be multiplied by a factor of 4/3 to give a more accurate estimation of the amount of energy needed to ensure the ZVS operation of the lagging leg [5]. This corresponds to twice the energy that is stored in a nonlinear drain-to-source capacitor whose capacitance is inversely proportional to the square root of the voltage and that mainly defines the capacitance of a MOSFET switch. If this is done, then the equations derived in this section can be adjusted accordingly. For example, Eq. (7.7) becomes

$$\frac{1}{2} L_{\mathrm{lk}} I_{\mathrm{pri}}^2 \geq \frac{4}{3} C_{\mathrm{S}} V_{\mathrm{S}}^2 + \frac{1}{2} C_{\mathrm{TR}} V_{\mathrm{S}}^2 \tag{7.11}$$

and Eq. (7.8) becomes

$$I_{\mathrm{crit}} = \sqrt{\frac{2}{L_{\mathrm{lk}}} \left(\frac{4}{3} C_{\mathrm{S}} V_{\mathrm{S}}^2 + \frac{1}{2} C_{\mathrm{TR}} V_{\mathrm{S}}^2 \right)} \tag{7.12}$$

7.3.2 Duty-Cycle Loss

Figure 7.5 shows three ZVS-PWM full-bridge converter waveforms: the voltage across the transformer primary, V_{pri}, the voltage across the transformer secondary, V_{sec}, and the primary current, I_{pri}. It can be seen from these waveforms that the square pulses of the secondary voltage are narrower than those of the primary voltage so that there is less voltage applied to the input of the output LC filter than expected. This difference in pulse width is referred to as duty-cycle loss and can be explained as follows: When a lagging leg switch is turned off and its complimentary switch in the same leg is turned on, this action starts a process that takes the converter out of a freewheeling mode of operation and into an energy-transfer mode, as shown in Figure 7.4. The transition into an energy-transfer mode cannot be instantaneous due to the presence of leakage inductance in the transformer. Starting from a mode such as Mode 5 in Figure 7.4e, where the primary current flows upwards from the

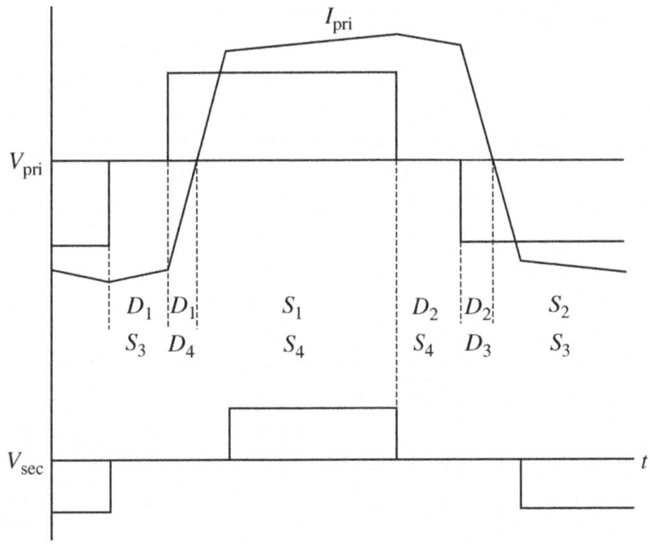

Figure 7.5 Primary voltage and current waveforms and secondary voltage waveforms of a DC–DC full-bridge converter operating with phase-shift PWM.

0 V rail to the DC voltage rail and the output inductor current is split among the two secondary diodes, the current begins to decrease in one of these secondary diodes and increase in the other so that all the output inductor current eventually flows through just one secondary diode. Although voltage may be applied to the primary, voltage will not appear across the secondary until this transition of current has been completed because the voltage across the secondary winding is clamped to zero as long as the current is flowing in both secondary diodes.

Given that the full input DC voltage is placed across the primary winding while this current transition is happening and the voltage across the secondary is zero, the determining factor as to how quickly or slowly this transition occurs is the transformer leakage inductance. Since

$$V_{Llk} = L_{lk} \frac{dI_{Llk}}{dt} \tag{7.13}$$

where L_{lk} is the primary leakage inductance, V_{Llk} is the voltage across this inductance, which is the switch voltage $V_S = V_{in}$, and I_{Llk} is the current flowing through this inductance, which is the transformer primary current Ipri, then this equation can be rearranged to give

$$\frac{dI_{pri}}{dt} = \frac{V_S}{L_{lk}} \tag{7.14}$$

It can be seen in Figure 7.5 that voltage across the secondary appears when the primary current is equal to the reflected secondary current (I_{pri} with a gradual positive slope), assuming that the magnetizing current in the transformer is negligible. This corresponds to the full output inductor current flowing in only one of the secondary windings and one of the secondary diodes so that this secondary winding is magnetically coupled to the primary winding.

Based on Eq. (7.14), it would seem that the transformer leakage inductance should be as small as possible to make the rate of current rise faster and thus reduce duty-cycle loss. Duty-cycle loss should be minimized as it limits the load range that the converter can operate with. To correct for duty-cycle loss in the case of a step-down converter, the turns ratio may need to be adjusted so that the transformer steps the voltage down less, but this means that the current reflected from the secondary to the primary side of the transformer is increased so that conduction losses and turn-off losses are increased.

7.3.3 Voltage Ringing

The secondary diodes of a ZVS-PWM-FB converter, regardless of whether a center-tapped or a full-bridge diode implementation is used, are prone to voltage ringing. That is to say that when the current in these diodes ceases to flow in these diodes, the voltage across the diodes will rise to the off-state voltage level ($2V_o$ for a center-tapped implementation, output voltage V_o for a full-bridge implementation) and then overshoot this level and begin to oscillate or ring at a very high-frequency. This diode voltage ringing is dependent on the interrelation between the leakage inductance of the transformer, its winding capacitance, and the junction capacitance of the diodes. In order words, a diode can be considered to be like a small capacitor when it is off. After current stops flowing in a secondary diode, the leakage inductance of the transformer resonates and rings with winding capacitance and diode capacitance. The voltage ringing that is produced can result in secondary diodes experiencing peak voltages that are several times greater than the expected off-state voltage level and in electromagnetic interference (EMI) noise being produced.

There are several ways to suppress this voltage ringing. One way is to place a simple RC snubber, like the one described in Section 1.6. The snubber capacitance helps to slow down the rate of voltage rise across the diode while the snubber resistance helps to dissipate the resonant energy that is producing the ringing. Although this is the simplest way to deal with voltage ringing, it is the least effective and the least efficient. Another way to suppress voltage ringing across secondary diodes is to add some sort of voltage clamp to the secondary. One such clamp was proposed for the full-bridge secondary diode implementation in [5] and is shown in Figure 7.6. The voltage clamp shown in Figure 7.6 is essentially an RCD snubber, as was described in Section 1.6, with a diode in series with a clamping capacitor and a resistor that provides a path for the snubber capacitor to discharge. Although voltage

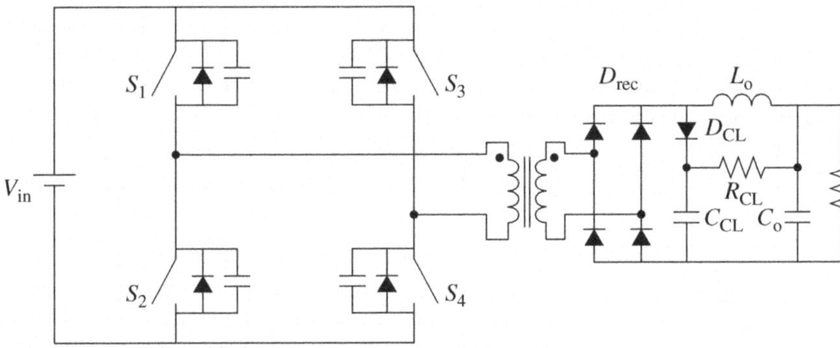

Figure 7.6 Voltage clamping circuit attached to the secondary side of a DC–DC full-bridge converter.

clamps such as the one shown in Figure 7.6 are more efficient than RC snubbers across diodes, they still dissipate energy. More sophisticated and more efficient approaches involve more significant modifications to the ZVS-PWM-FB converter and some of these approaches are discussed in Chapter 9.

7.4 ZVS-PWM PWM FULL-BRIDGE CONVERTER DESIGN CONSIDERATIONS

There are several considerations that need to be taken into account when designing a ZVS-PWM FB converter. The design process is, by its very nature, iterative, as the various factors that need to be considered are interrelated with one affecting another and vice versa. There is no particular order as to how one should approach a design, but some factor or parameter needs to be fixed first so that the rest of the design can follow. If the maximum converter duty-cycle is set first, then the design can proceed by taking into account the following design considerations:

Duty-cycle: The converter should be designed so that it operates with a maximum secondary duty-cycle that is as large as possible, but no greater than $D = 0.7$ or 0.8. Output inductor current ripple is at its minimum when $D = 0.5$ and thus the converter should be designed to operate with a maximum duty-cycle of around $D = 0.5$ at full-load. Some margin should be added to allow for transient operation as duty-cycle loss should be kept in mind, given that the converter's actual duty-cycle is greater than its effective duty-cycle. A 10% reduction of duty-cycle loss is generally considered to be acceptable.

Transformer turns ratio: The turns ratio of the transformer should be such that it is the highest possible value that allows the converter to operate with its maximum duty-cycle. The lower the turns ratio is (ratio of the number of primary winding turns to secondary winding turns), the greater the voltage that will appear across the transformer's secondary windings so that the converter can operate with a narrower effective duty-cycle, but a low turns ratio means that more current is reflected to the primary, which results in more conduction losses and turn-off switching losses. Making the transformer turns ratio higher helps reduce these losses.

ZVS range and leakage inductance: The ZVS range of the converter should be considered after the transformer turns ratio has been set. As was discussed earlier, the ZVS operation of the leading leg switches is ensured by the availability of output filter inductor energy and it is the ZVS operation of the leading leg switches that is problematic. With the transformer turns ratio set, the amount of current that is flowing in the primary when the converter is in a freewheeling mode can be approximated and can be compared to the ZVS criterion, as shown in Eqs. (7.8) and (7.9). It can be seen from Eq. (7.8) that the amount of current needed to ensure the ZVS operation of the lagging leg switches can be reduced with greater transformer leakage inductance, but increased transformer leakage inductance results in duty-cycle loss and thus a trade-off between ZVS range and duty-cycle loss needs to be considered. A typical design is to make the leakage inductance such that the converter begins to lose ZVS when it is operating with half-load. As the load is decreased, there will be less energy available to discharge the switch output capacitances before the switches are turned on and efficiency will decrease, as shown in the generic efficiency vs load graph in Figure 7.7. Although the switch capacitances may not be fully discharged before the switches are turned on, they will be partially discharged so that some switching loss reduction may be achieved. For

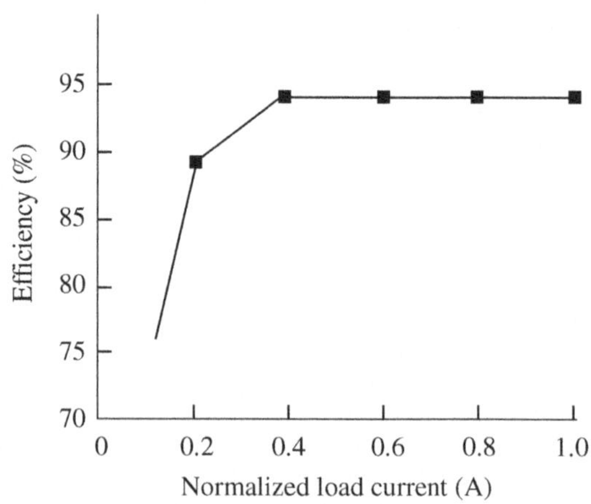

Figure 7.7 Generic efficiency vs load curve of a DC–DC full-bridge converter.

example, if a converter is designed so that its lagging leg switches can turn on with ZVS when the converter is operating with 50% of its full load, then when the converter is operating with, say, 45% of its load, the output capacitances of its lagging leg switches will be almost fully discharged before these switches are on, so that the turn-on switching losses of these switches will still be significantly reduced, but not quite eliminated.

Switching frequency: The switching converter of the converter should be considered after the leakage inductance of the transformer has been set. The maximum switching frequency that the converter can be operated is the inverse of the minimum switching cycle that the converter can operate with the duty-cycle, transformer turns ratio, and leakage inductance that has been set. The selection of a converter switching frequency will vary from application to application. In general, higher switching frequency results in smaller and lighter converters, but also results in less efficiency as switching losses are not completely eliminated, potential EMI issues, and more challenging designs as parasitic impedances and their effect on converter operation become more prominent, and greater cost. ZVS-PWM FB converters with silicon devices are typically designed to operate with switching frequencies in the range of 50–200 kHz, with 100 kHz being a common selection. Some designs can go as low as 25 kHz switching frequency, but these are limited to applications where size and weight are of less concern. It is rare for ZVS-PWM FB converters with silicon devices to operate with switching frequencies that are greater than 200 kHz.

If it turns out that the previous steps in the design process have resulted in a design that is unsatisfactory, that the converter cannot operate with the desired switching frequency as the switching cycle too short to account for the desired maximum duty-cycle and the maximum duty-cycle losses, then adjustments need to be made. There are several options that can be considered:

- The switching frequency can be reduced, but doing so will increase the converter size and weight.
- The duty-cycle loss can be reduced by reducing the leakage inductance of the transformer, but doing so will result in the converter losing ZVS at heavy loads.
- The transformer turns ratio can be decreased, but doing so increases the amount of current that is reflected from the secondary to the primary, thus increasing primary-side conduction losses and turn-off switching losses.

Ultimately, it is up to the designer to determine rank the significance of these factors and perform several iterations before settling on a final design.

7.5 LIGHT LOAD OPERATION AND HYBRID PWM

The lagging leg switches of a ZVS-PWM FB converter operating with phase-shift PWM are not able to turn on with ZVS if the converter is operating under light-load conditions because there will not be sufficient inductive energy to discharge their output capacitances. This causes the converter to have lower efficiency than its typical heavier-load efficiency. In addition to losing ZVS, another reason why the efficiency of the converter drops as lighter loads is that it is operating with a lower duty-cycle. Ideally, the duty-cycle of a converter should be independent of whether a converter is operating with a light load or a heavy load if the input and output voltage are kept the same. In reality, however, the amount of current flowing in the converter does affect the amount of voltage that is dropped as it flows through parasitic impedances such as on-state resistances, winding resistance, parasitic inductances, etc. As load is decreased, duty-cycle needs to be decreased as well as less voltage is needed to make up for less voltage being dropped, which means that it spends more time in freewheeling modes of operation when current is just circulating in the primary and less time in energy-transfer modes when energy is actually transferred from primary to output.

The issue of freewheeling mode circulating current creating conduction losses is an issue that exists when a full-bridge converter is operating with phase-shift PWM as this current is needed to discharge and charge the output capacitances of the lagging leg switches to ensure a ZVS switching transition to take the converter out of a freewheeling mode. If the converter is operated with basic PWM control, as described in Section 7.1, then any circulating current in the converter is quickly extinguished so that it has little opportunity to create conduction losses. Since the converter switches do not turn on with ZVS when basic PWM control is used, there is no need to generate any freewheeling mode circulating current.

Given that phase-shift PWM is better for the heavy-load operation of a PWM full-bridge converter and basic PWM control is better for the light-load operation of the converter, several researchers have proposed various hybrid PWM schemes that make use of each of these PWM methods [7–9]. A converter operating with hybrid PWM operates with phase-shift PWM when its load is heavy and with basic PWM control when its load is light so that efficiency can be optimized throughout the entire load range. Research on hybrid PWM scheme is focused on the boundary between the load region where phase-shift PWM is best, the load region where basic PWM control is best, and on how to make the switch from one scheme to another.

7.6 ZVS PWM FULL-BRIDGE CONVERTERS WITH WIDE BANDGAP DEVICES

If a ZVS-PWM-FB converter is implemented with wide bandgap devices such as SiC or GaN devices, then designing for ZVS operation becomes less of a concern that it is when Si devices are used due to their superior switching characteristics. There are two options for switching frequency range to consider when designing a ZVS-PWM-FB converter with wide bandgap devices:

- The first is to design the converter to operate with switching frequencies in the range of 50–200 kHz and take advantage of their superior switching characteristics to ensure an efficiency that is greater than what can be achieved with silicon devices.
- The second option is to design the converter to operate with switching frequencies that are greater than 200 kHz in order to reduce the size of the converter. Such designs, however, are typically restricted to 400 kHz, mainly due to the limitations of magnetics and the significance of parasitic inductances and capacitances at higher frequencies. With high switching frequencies, resonant-type converters like the ones that are discussed in Chapters 11 and 12 are preferred as the converter parasitics can be used in an advantageous manner in a way that is not possible with PWM converters. Another limiting factor to very high switching frequency operation is EMI noise. The turning on and off of wide bandgap devices can generate more EMI noise than that of silicon devices due to their faster switching transitions. EMI noise can be reduced with more gradual switching transitions, but this would defeat the purpose of using wide bandgap devices.

Although ZVS is less of a concern than not a concern when wide bandgap devices are used, using secondary diodes made of wide bandgap material will result in voltage ringing that is greater than that generated when Si devices are used and greater in frequency as well, which contributes to EMI noise. This is due to the lower junction capacitance of wide bandgap devices, which, although contributes to their superior switching characteristics, also creates higher frequency diode voltage ringing across a secondary diode when current stops flowing through it. As a result, the voltage ringing considerations that are described in Section 7.3.3 become more imperative when designing a ZVS-PWM-FB converter with wide bandgap devices.

7.7 CONCLUSION

This chapter reviewed the basic principles behind the PWM full-bridge converter, which is the most popular type of DC–DC converter for applications that require DC–DC conversion of at least 500 W power with transformer isolation. In this chapter, the basic topology was described and its operation with two methods of PWM, basic PWM control and phase-shift PWM, was explained. Basic PWM control involves turning pairs of diagonally opposed switches on and off simultaneously and changing the width of the gating signals of the converter switches to adjust the width of the voltage waveform pulses across the transformer primary. With phase-shift PWM, the gating signal pulses are kept fixed with a duty-cycle of about $D = 0.5$, with the gating signals of the two switches of a leg complimentary to each other. The width of the voltage pulses across the transformer primary is adjusted by shifting the gating signals of the switches of one leg relative to those of the switches of the other leg. PWM full-bridge converters are generally operated with phase-shift PWM as this ensures that the converter switches operate with ZVS and basic PWM control is only used for light-load operation as part of hybrid PWM schemes where the converter is operated with phase-shift PWM for heavy loads and with basic PWM control for light-loads.

Emphasis was placed on the phase-shift PWM operation of PWM full-bridge converters as this is the standard method of operation in these converters. Certain issues such as duty-cycle loss, ZVS operation range, and voltage ringing were highlighted and the design of the converter was discussed in detail. It was explained that there are several trade-offs that need to be considered when designing a ZVS-PWM full-bridge converter operating with phase-shift control - there is no such thing as a perfect design. The weighting of these various trade-offs is very much dependent on the application and required specifications of the converter that is to be designed. This weighing changes when wide bandgap devices are used instead of traditional silicon devices.

REFERENCES

1. R. A. Fisher, K. D. T. Ngo, and M. H. Kuo. "A 500 kHz 250 W DC–DC converter with multiple outputs controlled by phase-shifted PWM and magnetic amplifiers," *Interect Communications Inc. High Frequency Power Conversion Conference (HFPCC)*, 1988, May 1–5, 1988, San Diego, USA, pp. 100–110.
2. O. D. Patterson and D. M. Divan, "Pseudo-resonant full bridge DC/DC converter," *IEEE Transactions on Power Electronics*, vol. 6, no. 4, pp. 671–678, Oct. 1991.
3. L. H. Mweene, C. A. Wright, and M. F. Schlecht, "A 1 kW 500 kHz front-end converter for a distributed power supply system," *IEEE Transactions on Power Electronics*, vol. 6, no. 3, pp. 398–407, July 1991.
4. D. B. Dalal, "A 500 kHz multi-output converter with zero voltage switching," *IEEE Applied Power Electronics Conference (APEC)*, 1990, Mar. 11–16, 1990, Los Angeles, USA, pp. 265–274.
5. J. A. Sabate, V. Vlatkovic, R. B. Ridley, F. C. Lee, and B. H. Cho, "Design considerations for high-voltage high-power full-bridge zero-voltage-switched PWM converter," *IEEE Applied Power Electronics Conference (APEC)*, 1990, Mar. 11–16, 1990, Los Angeles, USA, pp. 275–284.
6. "Phase-shifted full-bridge zero-voltage transition design considerations," Texas Instruments Application Report SLUA107A, Aug. 2011.
7. B. Y. Chen and Y. S. Lai, "Switching control technique of phase-shift-controlled full-bridge converter to improve efficiency under light-load and standby conditions without additional auxiliary components," *IEEE Transactions on Power Electronics*, vol. 25, no. 4, pp. 1001–1012, Apr. 2010.

8. Y. S. Lai and Z. J. Su, "New integrated control technique for two-stage server power to improve efficiency under the light-load condition," *IEEE Transactions on Industrial Electronics*, vol. 62, no. 11, pp. 6944–6954, Nov. 2015.
9. M. Pahlevani, S. Pan, and P. Jain, "A hybrid phase-shift modulation technique for DC/DC converters with a wide range of operating conditions," *IEEE Transactions on Industrial Electronics*, vol. 63, no. 12, pp. 7498–7510, Dec. 2016.

CHAPTER 8

VARIATIONS ON THE CONVENTIONAL ZERO-VOLTAGE-SWITCHING DC–DC PWM FULL-BRIDGE CONVERTER

A number of modifications to the conventional zero-voltage-switching (ZVS) pulse-width-modulated (PWM) full-bridge (FB) DC–DC converter that was the subject of Chapter 7 have been proposed to improve its performance. Although this converter is very popular and widely used in applications where DC–DC power conversion that is greater or equal to 500 W and primary/secondary isolation are required, it does have several drawbacks. Among the most important that were discussed in Chapter 7 are:

- Limited load range for the ZVS operation of the switches caused by the lack of available inductive energy to discharge switch output capacitances at lighter loads.
- Duty-cycle loss caused by the presence of a transformer leakage inductance that must be sufficiently large to contain the energy needed to discharge the output capacitances of the switches before they are turned on.
- Voltage oscillations across the output secondary diodes that are caused by the interaction between their junction capacitance and the transformer leakage inductance.

Researchers thus have proposed variations to the conventional ZVS-PWM-FB converter to remove or mitigate these drawbacks and improve

DC–DC Converter Topologies: Basic to Advanced, First Edition. Gerry Moschopoulos.

Published 2024 by John Wiley & Sons, Inc.

performance. Many of these variations can be classified into certain types and several of these types are reviewed in this chapter. First, the general principles behind these types are introduced, then examples of each type are given. The chapter concludes with a general comparison of the strength and weakness of these types.

8.1 MODIFIED ZVS-PWM DC–DC FULL-BRIDGE CONVERTER WITH SATURABLE REACTORS

Saturable rectors can be used to increase the load range for ZVS in ZVS-PWM-FB converters. A saturable reactor or saturable inductor is a small magnetic core that can be easily saturated when a small amount to volt-seconds is applied to it. The core can be reset by applying an equivalent amount of volt-seconds of opposite polarity across it. When the core is saturated, it is as if the inductance has been removed from the circuit it is in.

A saturable reactor is useful in applications where inductance in a circuit is needed for a brief duration of time and then must be removed afterward. In the case of ZVS-PWM-FB converters, saturable reactors can be used to:

- Store energy that can be used to discharge the output capacitance of converter switches that supplements the energy that is stored in the leakage inductance. After a switch has been turned on with ZVS, the saturable reactor can then be made to saturate, thus effectively removing it from the converter.
- Impact the flow of current at the converter's secondary side by adding inductance during critical modes of operation that can then be removed afterward. This affects the nature of the current that can be reflected to the converter's primary side and the converter's ability to operate with ZVS at lighter loads.

Both approaches are examined in this section.

8.1.1 Modified ZVS-PWM-FB Converter with Primary-Side Saturable Reactor

Figure 8.1 shows a ZVS-PWM-FB converter that has been modified with the addition of a small saturable reactor place in series with the primary of the transformer, as proposed in [1]. The saturable reactor helps the converter's lagging leg switches turn on with ZVS in the following manner: Assume that the converter is in a freewheeling mode with current flowing from switch S_1 through S_3, either through its body-diode or its source and drain, and that the saturable reactor has been saturated as a result of the volt-seconds previously placed across the core. When switch S_1 is turned off to exit the freewheeling mode, the voltage across its output capacitance begins to rise while that of S_2

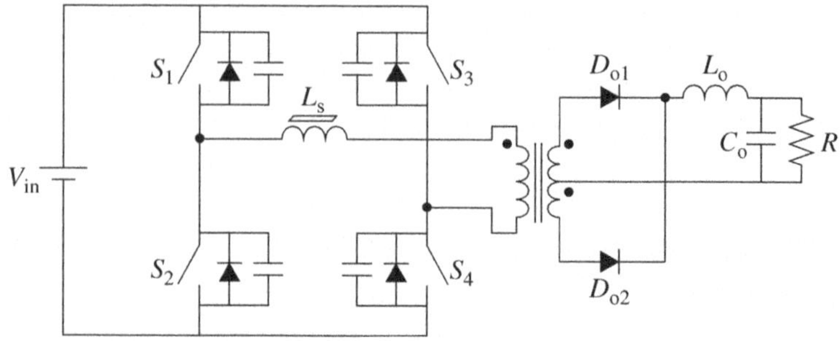

Figure 8.1 ZVS-PWM-FB converter with primary-side saturable reactor.

begins to fall. While this is happening, volt-seconds begin to be applied to the saturable reactor so that the saturable reactor becomes unsaturated and becomes an inductance that can be added to the leakage inductance of the transformer. As a result, the amount of energy that is available to fully discharge the output capacitance of S_2 so that it can be turned on with ZVS is greater than what it would have been without the saturable inductor in the circuit. Since there is more inductance available to store energy, there is less dependence on the amount of primary current, which is related to the amount of load current, so that the range of output load that the converter switches can turn on with ZVS can be expanded. Soon after the switch is turned on with ZVS, the primary current reverses direction and the saturable reactor saturates so that it is effectively removed from the circuit.

As the saturable reactor acts like an inductance that is in the circuit for only a fraction of the switching transition then effectively removed, there is little loss in converter duty-cycle. If desired, the converter can be designed so that the saturable reactor is used to compensate for lower leakage inductance so that voltage ringing across the secondary diodes can be reduced. Although the addition of a simple saturable reactor results in better light-load efficiency with less duty-cycle loss compared to the conventional ZVS-PWM-FB converter, the saturable core placed on the primary side has a thermal problem at high frequency because the flux swings between positive and negative saturation levels. As a result, a full-bridge converter with a saturable reactor core cannot be operated with as heavy a load as can a full-bridge converter that does not have this core.

8.1.2 Modified ZVS-PWM-FB Converters with Secondary-Side Saturable Reactors

An alternate approach to using saturable reactors to improve the performance of the conventional ZVS-PWM-FB converter is to place two saturable reactors in series with the secondary diodes and an optional freewheeling diode. The modified converter, which was proposed in [1, 2], is shown in

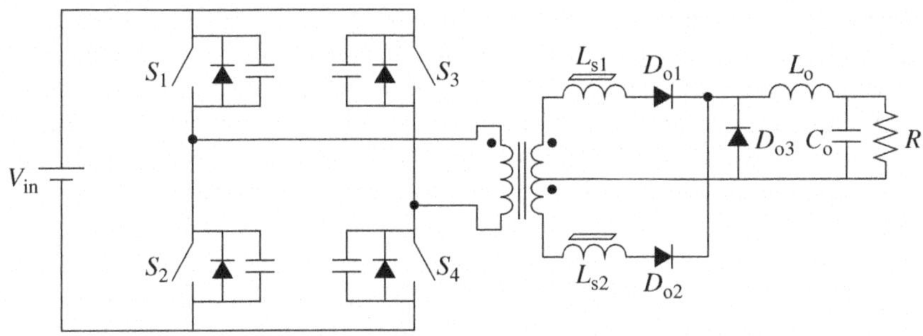

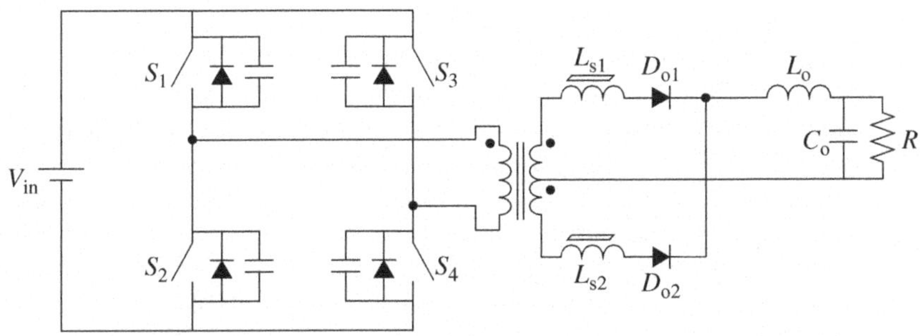

Figure 8.2 ZVS-PWM-FB converters with secondary-side saturable reactors: (a) With additional freewheeling diode. (b) Without additional freewheeling diode.

Figure 8.2 with the freewheeling diode and without the freewheeling diode in Figure 8.2a,b, respectively. For both converters, the saturable reactors act as magnetic ON/OFF switches during freewheeling mode to energy-transfer mode switching transitions so that more energy is available to discharge the output capacitances of the lagging leg switches. For the converter shown in Figure 8.2a, the saturable reactors temporarily block current from flowing through the secondary diodes, as if the diodes have been turned off, so that the transformer's magnetizing current can be used to help the lagging leg switches turn on with ZVS. The saturable reactors force the secondary current to flow though just one of the secondary diodes instead of being distributed among the two diodes – as if one diode has been switched on and the other off – so that output current can be reflected to the converter's primary side and used to help the lagging leg switches turn on with ZVS. The saturable reactor cores can be reset by the natural operation of the converter so that the converter operates as if it had no saturable reactors for most of the switching cycle.

For example, if the converter is in an energy-transfer mode of operation with diagonally opposed switches S_1 and S_4 on, then current flows through secondary diode D_{o1} and does not flow through D_{o2}. The saturable reactor in series with D_{o1} is saturated, whereas the one in series with D_{o2} is not so that D_{o2} has an inductor in series with it. When S_4 is turned off, the voltage across the two secondary diodes drops to zero, but the status of the two saturable reactors has not changed – the one in series with D_{o1} has not been reset as no negative volt-seconds have been applied to it and the one in series with D_{o2} has not been saturated. If there is a freewheeling diode in the converter, as in the case of the converter shown in Figure 8.2a, then this diode becomes forward-biased and conducts the full output current so that there is no current flowing in the secondary windings. If there is no freewheeling diode, as in the case of the converter shown in Figure 8.2b, then current continues to flow through diode D_{o1} and no current flows through D_{o2} due to the inductance of the saturable reactor in series with it.

When S_4 is turned off so that current can be transferred to S_3, then the magnetizing current can be used to help discharge the output capacitance of S_3 if there is no current flowing in the secondary winding, so that this switch can be turned on with ZVS. In a conventional ZVS-PWM full-bridge converter without secondary-side saturable reactors, magnetizing current can be used to charge and discharge the output capacitances of the converter's MOSFETs only when it is larger than the current reflected from the output. In this case, the difference between these two currents becomes available to perform this task. This current difference can be generated by reducing the magnetizing inductance in the transformer, but doing so would increase the magnetizing current considerably and add a significant amount of current to the amount already flowing in the primary, thus increasing primary-side conduction and turn-off losses. With the saturable reactors and the freewheeling diode in the converter, the full magnetizing current becomes available to discharge the output capacitance of S_3 so that the load range of ZVS can be increased without having to increase the magnetizing current so that any potential increase in primary-side conduction losses and turn-off losses can be avoided.

When the freewheeling diode is not added to the converter, as is the case with the modified converter shown in Figure 8.2b, the full reflected output current becomes available to discharge the output capacitance of S_3, just like it is available to discharge the output capacitance of a leading leg switch. After S_3 has been turned on with ZVS, current begins to be transferred from D_{o1} to diode D_{o2} as voltage appears across the transformer secondary. Eventually, the saturable reactor in series with D_{o2} becomes saturated while that in series with D_{o1} becomes unsaturated.

The main drawback of using the magnetizing inductance for ZVS operation is that more magnetizing energy circulates in the primary, resulting in increased conduction losses, but not as much as there would be if the magnetizing current was increased by reducing the size of the magnetizing inductance. The main drawback of using the output current for ZVS operation is that the duty cycle loss is slightly larger than that of the approach using the magnetizing inductance. Its total duty cycle loss, however, is still small compared to that of the conventional ZVS-PWM-FB converter. In both converters, the converter can be implemented with lower transformer leakage inductance so that there is less voltage ringing across the secondary diodes.

Both converters, as well as the modified converter with primary-side saturable reactor discussed above, have been shown to operate with greater efficiency than the standard ZVS-PWM-FB converter in [2]. Of these three converters, the modified converter with secondary-side saturable reactors and no freewheeling diode shown in Figure 8.2b is slightly more efficient than the other converters, as demonstrated in [2]. This is because it uses the reflected output current for ZVS and does not have the drawbacks of the other converters such as saturable core losses due to high voltage swings or higher conduction and turn-off losses due to increased magnetizing current. It should be noted that the two converters with secondary-side saturable reactors can be implemented with mag-amps, as described in Section 5.3. This is especially advantageous if a modified ZVS-PWM-FB converter needs to be implemented with multiple outputs.

8.2 MODIFIED ZVS-PWM-FB CONVERTERS WITH PASSIVE SERIES AUXILIARY CIRCUITS

The load range for the ZVS operation of the switches in a ZVS-PWM-FB converter can be also increased if passive auxiliary circuits are added to the converter. One of the first of this type of modified ZVS-PWM-FB converter was the converter proposed in [3] and shown in Figure 8.3a. This converter has the same topology as the conventional ZVS-PWM-FB converter except that a passive circuit consisting of inductor L_a and diodes D_{a1} and D_{a2} have been added to the converter. The auxiliary circuit has two functions: The first is to increase the amount of inductance that can store energy that can be used to help discharge the output capacitances of the lagging leg switches before they can be turned on with ZVS. The second is to clamp any voltage ringing that may occur as a result of the interaction between the leakage inductance of the transformer and the junction capacitance of the output diodes.

The presence of L_a in the auxiliary circuit means that there is another energy storage element in addition to the transformer leakage inductance that can be used to store energy that can be used to discharge the output capacitance of the lagging leg switches before they are turned on. This inductance can be used to reduce the amount of leakage inductance in the transformer to reduce the amount of voltage ringing that can occur. As the transformer still has some leakage inductance, there will still be voltage ringing, but this can be clamped by diodes D_{a1} and D_{a2}. If the voltage across the transformer secondary tries to overshoot, this overshoot voltage will try to appear across the transformer primary where it will forward-bias one of the clamp diodes. There is no need to put any clamping diodes on the other side of the transformer primary as the transformer secondary has a center-tapped ground connection and thus there is no voltage ringing at that end.

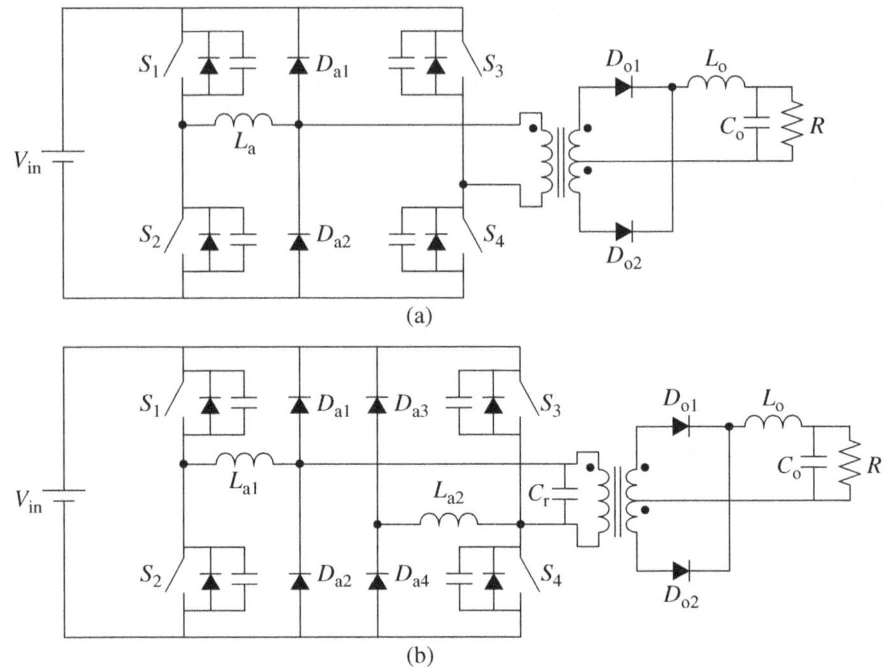

Figure 8.3 ZVS-PWM-FB converters with passive series auxiliary circuits: (a) Single auxiliary circuit implementation. (b) Dual auxiliary circuit implementation.

The main drawback of the converter shown in Figure 8.3a is that the auxiliary circuit inductor is in the main path of current in the transformer primary. This means that the inductor must be able to handle the full primary current and that there is some duty-cycle loss. This duty-cycle loss, however, is not as severe as it is in the conventional ZVS-PWM-FB converter as one of the clamp diodes will quickly conduct once a switching transition is made so that the voltage at the lagging end of the transformer primary can change relatively quickly. Since there is little leakage in the transformer, the secondary voltage is able to change quickly as well. It should also be noted that there will be some increase in conduction losses as well as current flows through the two clamp diodes during a switching cycle. This increase is the difference between the auxiliary circuit inductor and the transformer primary current and need not be severe if a sufficient large auxiliary circuit inductor is implemented, but care should be taken that this is not at the expense of duty-cycle loss.

A variation of this converter was proposed in [4] and is shown in Figure 8.3b. The main differences between this converter and the converter shown in Figure 8.3a are that an additional passive auxiliary circuit has been added to the leading leg switches of the converter and a capacitor has been added across the transformer primary. The purpose of the additional auxiliary circuit is to help the leading leg switches operate with ZVS even when the converter is operating under no-load conditions. This is done for applications where the lifetime of the converter, especially the converter devices, is especially important and in applications where the environment is noise-sensitive and the EMI noise that can be caused by ringing in the switches when they are off must be suppressed. When the output load is very light, there is very little current flowing in the transformer primary. When this current ceases to flow, the output capacitances of the switches may interact with the leakage inductance of the transformer when they are off. The presence of the auxiliary circuits in the converter ensures that there is always some current in the circuit so that this ringing can be suppressed.

The capacitor across the transformer primary helps clamp voltage ringing that may appear during switching transitions; it also provides an additional path for current to flow through in the primary. As a result, the primary current is the sum of the reflected load current and the current through the capacitor, which comes as a result of the resonant interaction between C_r and the auxiliary circuit inductors L_{a1} and L_{a2}. This means that there is more current and thus more energy available to discharge switch capacitances, but this comes at the cost of increased conduction losses and turn-off losses.

8.3 ZVS-PWM-FB CONVERTERS WITH PASSIVE PARALLEL AUXILIARY CIRCUITS

A different approach to implementing passive auxiliary circuits in ZVS-PWM-FB converters is to place these circuits away from the main power path and to the side of the converter as was done in the modified ZVS-PWM-FB converter that was proposed in [5] and shown in Figure 8.4. In this converter, a simple passive auxiliary circuit consisting of an inductor that is connected to the mid-point of two capacitors is added to each side of the circuit. Although the auxiliary circuit is needed only for the lagging converter leg, another passive auxiliary circuit is added to the leading leg of this converter to ensure that its switches can operate with ZVS, under no-load conditions, when there is no reflected load current that can discharge the output capacitances of the switches before they are turned on. This is for applications where the converter may need to operate under no-load conditions for some time and such operation may reduce the lifetime of the switching devices unless they are operated with ZVS [5].

The converter operates as follows: A voltage of $V_{in}–V_{Ca2}$ is impressed across an auxiliary circuit inductor whenever a top switch is on, and a voltage of $-V_{Ca2}$ is impressed whenever a bottom switch is on. Whenever a switch in a converter leg is turned off, the current in the auxiliary circuit inductor attached to that leg must flow through another path. It does so by discharging the switch capacitor of the other switch in the leg and flowing through its body-diode. This allows the switch to be turned on with ZVS. The auxiliary circuit supplements the energy that is available from the transformer leakage inductance.

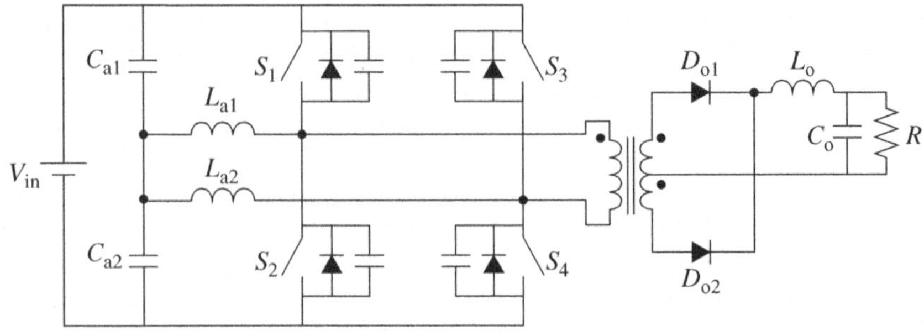

Figure 8.4 ZVS-PWM converter with passive parallel auxiliary circuits.

It should be noted that the modified ZVS-PWM-FB converter shown in Figure 8.4 is especially attractive for applications where the input DC voltage varies, as may be the case in applications where this voltage is obtained from the output of a front-end AC–DC converter in a two-stage converter system. A voltage of V_{in}–V_{Ca2} is impressed across the auxiliary inductors for half the switching cycle and a voltage of $-V_{Ca2}$ is impressed for the other half. Since each switch in a converter leg is on for about 50% of a switching cycle with respect to the complementary switch in the same leg regardless of what the converter duty-cycle is, the mid-point auxiliary circuit capacitor voltage is always half the input voltage. The auxiliary circuit current decreases when the input DC bus voltage decreases as less voltage is placed across the auxiliary circuit inductor. As a result, there is less auxiliary circuit current available to discharge the output capacitances of the switches at lower DC input voltages, when there is less need for energy to do so as there is less voltage across the switches before they are turned on.

Instead of connecting auxiliary circuit inductors to a capacitor mid-point, the auxiliary circuit can be connected directly across the transformer primary. Examples of such an implementation were proposed in [6, 7] and are shown in Figure 8.5. For both these converters, the passive auxiliary circuit is not connected in series with the main path of current in the primary side, but is connected in parallel across the transformer instead so that the auxiliary circuit components are not required to handle the full primary current. The purpose of these auxiliary circuits is to supplement the transformer primary current so that more energy is available to discharge the output capacitances of the switches so that the load range over which they can be turned on with ZVS can be extended to no-load operating conditions.

The main advantage of the auxiliary circuits shown in Figure 8.5 is that the additional current that the auxiliary circuit produces has a lower RMS value that the additional current produced by the auxiliary circuit of the converter shown in Figure 8.4 so that it has lower conduction and turn-off losses. This is because the auxiliary circuit does not produce a triangular current with a high peak, but a more rounded current that is shaped by the auxiliary circuit.

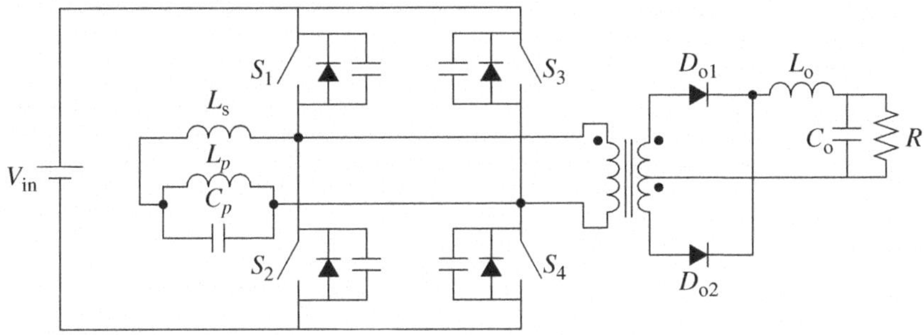

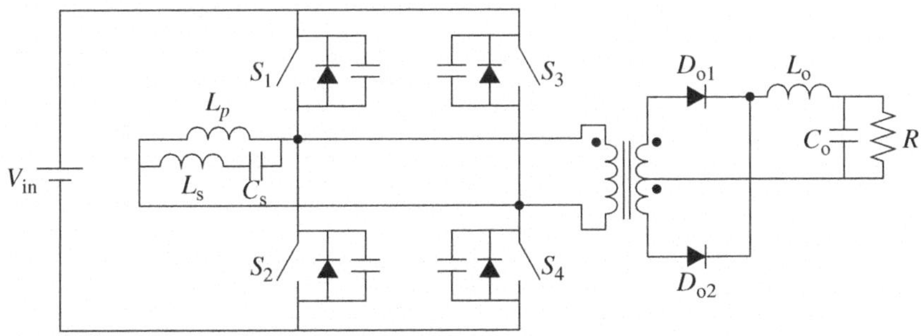

Figure 8.5 ZVS-PWM converters with passive parallel auxiliary circuits and reduced RMS current. (a) Implementation proposed in [6]. (b) Implementation proposed in [7].

The main drawback of these paralleled auxiliary circuits is that they are not as simple to design.

8.4 ZVS-PWM-FB CONVERTERS WITH PASSIVE PARALLEL AUXILIARY CIRCUITS WITH A TRANSFORMER

The performance of the passive auxiliary circuits that have been discussed in this chapter this far can be enhanced if they are implemented with transformers. In this section, examples of two such circuits are given: one based on a series auxiliary circuit, the other based on a parallel auxiliary circuit.

8.4.1 ZVS-PWM-FB Converter with a Passive Auxiliary Series Auxiliary Circuit with a Transformer

The passive auxiliary circuit shown in Figure 8.3a can be modified to include a transformer as was proposed in [8] and shown in Figure 8.6a. In this converter, a transformer T_2 is added to the auxiliary circuit. The primary of the transformer is connected to auxiliary circuit inductor L_a while the

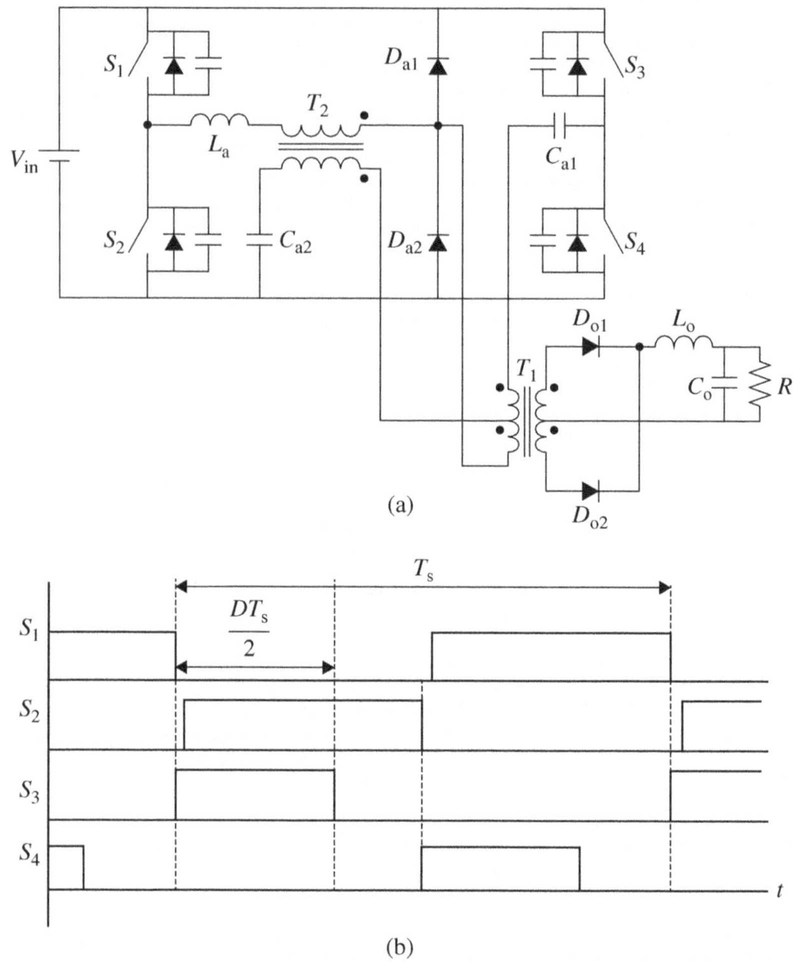

Figure 8.6 ZVS-PWM-FB converter with passive coupled inductor series auxiliary circuit: (a) Converter topology. (b) Gating signals for variable duty-cycle PWM control.

secondary is connected to a blocking capacitor at one end and to the mid-point of the primary of the main power transformer at the other end. The converter has two diodes that are connected to the positive and ground rails to provide a path for the difference in current between the auxiliary circuit inductor L_a and the transformer primary current to flow through, like the modified converter shown in Figure 8.3a. The diodes are also used to clamp any voltage ringing that may occur due to the interaction of the junction capacitance of the output rectifier diodes and the leakage inductance of the main power transformer. Like the modified ZVS-PWM-FB converter in Figure 8.3a, however, this ringing will be less than that found in the conventional ZVS-PWM-FB converter because the leakage inductance of the main transformer does not have to be as large to ensure the ZVS operation of the switches.

The purpose of T_2 is to generate current in the primary that will supplement the current that is available to discharge switch output capacitances. Current is made to flow through the secondary of T_2 and this current is reflected to its primary so that it supplements the current through L_a. This is on top of the current that is flowing in the secondary winding, which is also added to the primary current. With more current in the primary, the auxiliary circuit inductance L_a in Figure 8.6a can be reduced as it becomes less important in providing ½ LI^2 energy to discharge switch capacitances, thus reducing duty-cycle loss. Moreover, T_2 can be small in size as it is only used to store a relatively small amount of energy.

Ideally, T_2 should have more current through it when the converter is operating under light-load conditions than when it is operating under heavy-load conditions, when there is less need to generate additional energy to discharge the output capacitances of the lagging leg switches. In other words, the energy that is made available for ZVS operation by T_2 should be adaptive so that the energy stored in L_a increases as the load current or input voltage decreases. The reflected current that is generated can be made adaptive if the PWM method used to operate the converter is changed so that it is not standard phase-shift PWM, but the variable duty-cycle method shown in Figure 8.6b. For this PWM method, the gating signals of the leading leg switches, S_1 and S_2, are complimentary with a 50% duty-cycle like those for the leading leg switches when phase-shift-PWM is used, but the duty-cycle of the gating signals of the lagging leg switches are varied so that it ranges from 0 to about 50%, with appropriate dead-time added.

The adaptive nature of the converter stems from the voltage that appears at the mid-point of the transformer primary. Voltage is stored in the auxiliary circuit capacitor C_{a2} whenever the converter is in an energy-transfer mode of operation and diagonally opposed converter switches are on. During such a mode, voltage that is greater than the voltage across C_{a2} appears at the mid-point of the transformer primary. Current flows out of C_{a2} when the converter is in a freewheeling mode of operation as the voltage at the mid-point of the transformer primary is zero. It is this current, reflected to the primary of T_2 that helps discharge the output capacitances of the lagging leg switches S_3 and S_4. When the converter is operating under light-load conditions, the width of the gating signals of S_3 and S_4 is reduced and the converter is more likely to operate in freewheeling modes of operation. This results in more current flowing out of C_{a2} and thus more reflected current in the primary of T_2. As the load is increased, the duty-cycle of the lagging leg switches is increased, which reduces the duration of the freewheeling modes of operation so that less reflected current is generated, which is beneficial as less is needed. It is in this manner that the converter has an adaptive ZVS mechanism that can reduce the current that circulates in the primary side of the converter as the reflected current in T_2 can be adjusted to meet the ZVS needs of the converter.

The fact that this converter cannot be operated with conventional phase-shift PWM is the main drawback of this converter. Phase-shift PWM is by far the most popular way of operating a ZVS-PWM-FB converter and thus commercially available integrated circuit (IC) controllers are readily available for this purpose. These controllers, however, cannot be used with this converter, which complicates the generation of its gating signals and the implementation of its control.

8.4.2 ZVS-PWM-FB Converters with Passive Parallel Auxiliary Circuits and Reduced Output Current Ripple

If the modified ZVS-PWM-FB converter shown in Figure 8.4 is implemented with only one auxiliary circuit inductor and that inductor is replaced with a transformer winding as was proposed in [9], then the additional benefit of output ripple reduction can be achieved. This converter, shown in Figure 8.7a, has

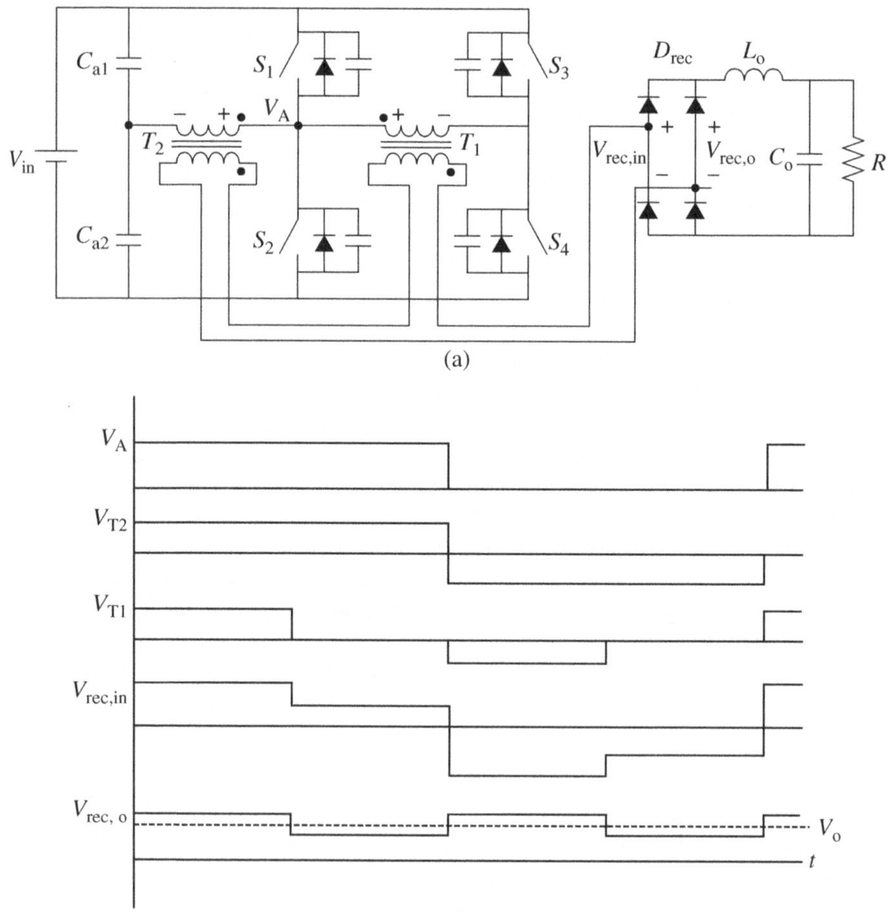

Figure 8.7 ZVS-PWM converter with passive parallel auxiliary circuit and reduced output ripple: (a) Converter topology. (b) Key voltage waveforms.

a winding from the main power transformer connected to the mid-point of the auxiliary circuit capacitors and the mid-point of the lagging leg switches. This voltage is a square-wave voltage that is either $+V_{in}/2$ or $-V_{in}/2$ at any given time. The voltage at the input of the diode bridge, $V_{rec,in}$, is the sum of the voltage across the main primary winding, V_{T1}, a square-wave voltage with zero-voltage states and the voltage across this auxiliary winding, V_{T2}, as shown in Figure 8.7b. This voltage, which is rectified, appears at the output of the diode bridge and is fed to the input of the output LC filter, which filters out the AC harmonic components.

This voltage has significantly less ripple than the voltage that is fed to the output LC filter of a conventional ZVS-PWM-FB converter, which results in the current flowing through the output inductor L_o having less ripple as well. The allows the size of the output inductor and capacitor to reduced significantly, which makes the overall size of the converter smaller as well. If the size of the two converters is made the same, then the reduced-ripple converter can be operated with lower switching and thus higher efficiency as lower frequency increases converter size. Whether size or efficiency is the dominant design factor, or whether some compromise between the two is desired, is dependent on the application that the converter is being used for and its specifications.

The following should be noted about the converter:

- The auxiliary circuit transformer can be used to store energy that can be used to discharge the output capacitance of leading leg switches, as if it was an inductor. The converter can therefore has the extended ZVS range of the original modified converter in Figure 8.4, but with less output ripple.
- It is difficult to implement the output ripple reduction mechanism if the secondary of the main power transformer is center-tapped because of the multiple windings and taps that would be needed. This would increase the size and cost of the main transformer in a way that would offset the size reduction that can be achieved with a single-output secondary. Since a single-output secondary requires that the output diode bridge be a four-diode bridge rectifier, then it means that secondary current must flow through at least two secondary diodes at all times, which increases conduction losses. The reduced-ripple converter is therefore less advantageous to use in applications where the output voltage is low and the output current is high.

8.5 ZVS-PWM-FB CONVERTERS WITH ACTIVE AUXILIARY CIRCUITS

Although the use of passive auxiliary circuits in PWM-FB converters is effective in extending the load range over which the converter switches can operate with ZVS, the main drawback with their use is the additional current that must circulate in the primary side of the converter. This additional current increases RMS switch current stresses and conditional losses. It also increases turn-off switching losses as well, as the converter switches must turn off with more current. Ideally, additional current should only be made available to help discharge the output capacitance of a switch that is about to be turned on during the switch's turn-on switching transition time and not exist when it is not needed. This can be ensured if active auxiliary circuit that contain at least one active switch are used instead of passive auxiliary circuits.

One of the first converters to use active auxiliary circuits to extend the load range of ZVS of a ZVS-PWM-FB converter is the full-bridge converter that was proposed in [10] and is shown in Figure 8.8. In this converter, an auxiliary circuit consisting of two active switches, S_{a1} and S_{a2}, and an inductor L_a have been added to the converter. The auxiliary circuit is added to help the lagging leg switches turn on with ZVS. In Figure 8.8, S_1 and S_2 are the lagging leg switches as the auxiliary circuit is connected between the two switches, while S_3 and S_4 are the leading leg switches. S_3 and S_4 do not need the help of an active auxiliary circuit to help them turn on with ZVS. As the leading leg switches, there will always be sufficient energy available to discharge their output capacitance before they are turned on, as discussed in Section 7.3.1.

The converter shown in Figure 8.8 operates as follows: Assuming that S_2 and S_4 are on and the converter is in a freewheeling mode, switch S_{a1} is turned on to allow current from the input to flow through the auxiliary circuit inductor. Just before S_1 is to be turned on, both S_2 and S_{a1} are turned off. Current flowing through L_a is forced to flow through the body-diode of S_{a2} and discharges and charges the output capacitances of S_1 and S_2 respectively, along with the energy in the leakage inductance of the transformer. Current then begins to flow through the body-diode of S_1 and this switch can be turned on with ZVS. The current flowing through L_a flows back to the input after S_1 is turned on. The converter then continues to operate as a standard ZVS-PWM-FB converter until S_{a2} is turned on to help S_2 turn on with ZVS.

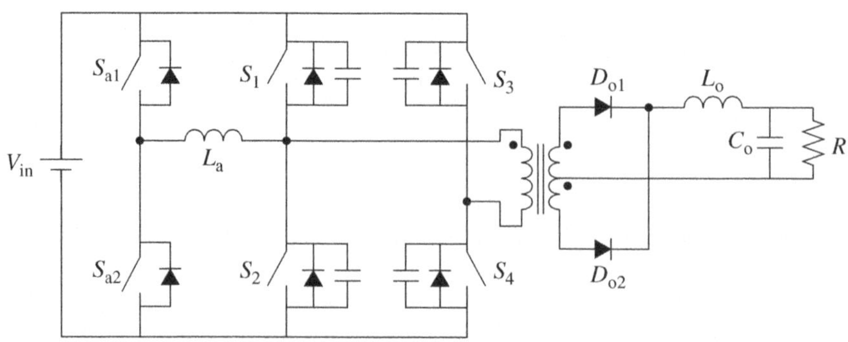

Figure 8.8 ZVS-PWM converter with single two-switch active auxiliary circuit.

Since the auxiliary switches are on for only a small fraction of the converter's switching cycle, only when a lagging leg switch is about to be turned on, the converter operates as a standard ZVS-PWM-FB converter most of the time. If desired, the auxiliary circuit can be made to be active only during light-load converter operation as it is not needed when the converter is operating with heavy loads, given that sufficient energy to discharge the output capacitances of lagging leg switches already exists. The auxiliary circuit, therefore, helps extend the ZVS range of the converter while having minimal impact on its operation. Moreover, the components in the auxiliary circuit handle very little current as the auxiliary circuit is active for only a small fraction of the switching cycle.

If desired, the ZVS operation of the converter can be made to be even less dependent of the transformer leakage inductance by reducing its value so that the effects of duty-cycle loss can be eliminated and the converter can be made to operate at higher switching frequencies. The lower leakage inductance would also reduce the amount of voltage ringing across the output rectifier diodes. If this is done, however, then more energy would have to be placed in L_a so that more current would have to flow in the auxiliary circuit, but the RMS current stress of the auxiliary circuit components would still be small. If the use of a larger leakage inductance is desired, saturable reactors or passive snubbers at the secondary would be needed to dampen and/or clamp secondary parasitic ringing. The circuit shown in Figure 8.8 does have several drawbacks. The first is that the addition of an active auxiliary circuit increases the cost and complexity of the converter. The increased cost and complexity comes from the need for two extra switches plus the circuitry needed to drive these switches, including an extra high-side gate driver. Other drawbacks include the fact that the top auxiliary switch requires an additional floating gate drive and the auxiliary switches have turn-off losses since they turn off with hard-switching.

A greater variety of possible auxiliary circuit implementations can be used by considering the research that has been done on the use of active auxiliary circuits to achieve ZVS operation in boost converters. It was shown in [11] that a full-bridge converter leg can be considered to be topologically equivalent to a boost converter, with the bottom switch of the leg acting as the main boost power switch. Many active auxiliary circuits that have been proposed to assist the ZVS operation of a boost converter switch can be used in the ZVS-PWM-FB converter, as long as the circuit is attached to a bottom switch of a converter leg and as long as they are not the type of circuit that was discussed in Section 2.3.1. This type of auxiliary circuit has a direct connection of the output of a non-isolated converter, which is not possible in an isolated converter.

A ZVS-PWM-FB converter with such an auxiliary circuit placement was proposed in [11] and is shown in Figure 8.9a. It should be noted that if auxiliary circuits are placed to assist the ZVS operation of the two bottom switches, then the turn-on of these two switches should occur when the converter is

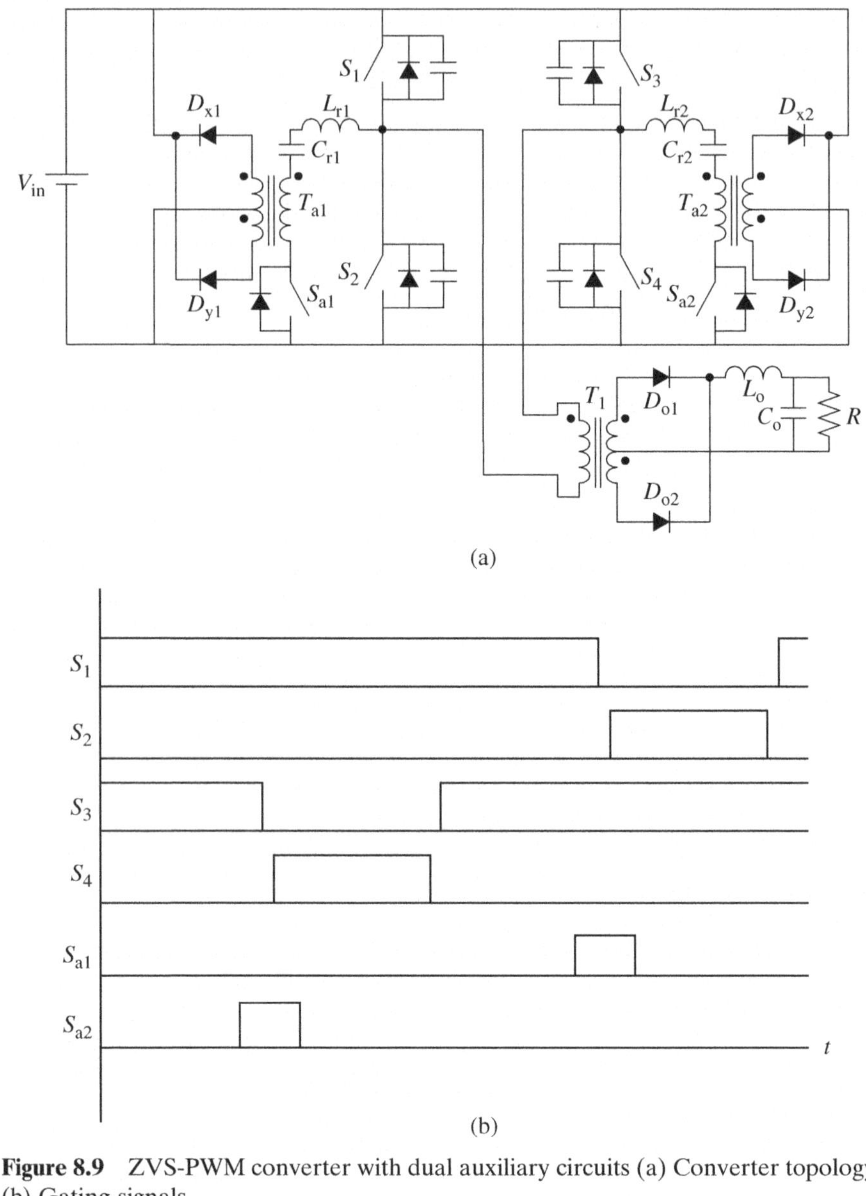

Figure 8.9 ZVS-PWM converter with dual auxiliary circuits (a) Converter topology. (b) Gating signals.

going from a freewheeling to an energy-transfer mode. This means that the standard phase-shift PWM technique that is normally used for ZVS-PWM-FB converters cannot be used since this transition occurs for both switches in one of the two converter legs. One possible alternative PWM technique is shown in Figure 8.9b. The gating signals are made asymmetrical so that the bottom switches are on for a duration of approximately DT while the top switches are on for the remainder of the period, a duration of approximately $T/2+(1-D)T/2$, where T is the period and D is the duty-cycle. The PWM of the transformer primary voltage is done by adjusting the width of the bottom switch gating signals instead of phase-shifting the gating signals of the switches in one converter leg with respect to those of the switches in the other leg. As can be seen from Figure 8.9b, the transition from a freewheeling mode to an energy-transfer mode when a diagonally opposed pair of switches is on now occurs whenever a transition is made from a top switch to a bottom switch – from S_1 to S_2 or from S_3 to S_4.

The modified ZVS-PWM-FB converter shown in Figure 8.9a has the same advantages as those of the converter shown in Figure 8.8, but with several additional advantages, such as the possibility of more auxiliary circuit implementations, ones with potentially better performance. This may be the case given that any auxiliary circuit can be used for a ZVS-PWM boost converter can be used and the auxiliary circuit switch gate drives can be simpler as both auxiliary switches can be referenced to ground. Moreover, the energy required to circulate in certain types of auxiliary circuits (most notably non-resonant circuits) can be reduced thus reducing the auxiliary switch turn-off losses normally found in these circuits. This is because all that has to be done to discharge the equivalent output capacitance of a bottom switch is to turn on its auxiliary switch. No additional energy has to be placed in the auxiliary circuit beforehand. The main drawback of the modified ZVS-PWM-FB converter shown in Figure 8.9a is that it needs two auxiliary circuits so that the cost and complexity of the converter are increased. An additional drawback is that the converter switch stresses are more uneven as the top converter switches conduct more current than the bottom ones, due to the asymmetrical nature of the PWM technique that is used. This results in slightly more conduction losses than the modified ZVS-PWM-FB converter shown in Figure 8.8.

8.6 ZVS-PWM-FB CONVERTER WITH A SINGLE ACTIVE AUXILIARY CIRCUIT

It is possible to implement a single-switch auxiliary circuit in a ZVS-PWM full-bridge converter if the asymmetrical PWM technique shown in Figure 8.9b is used and the bottom two switches of the converter are considered to be similar to a two-module interleaved boost converter with its two switches operating with the same gating signal, but with a 180° phase difference. An example of such an implementation is shown in Figure 8.10. For such a converter, according to the PWM method in Figure 8.9b, only the top two switches should be allowed to conduct current whenever the converter is in a freewheeling mode of operation and the converter must enter an energy-transfer mode by turning on a bottom switch. The auxiliary circuit would have to be activated just after a top switch is turned off, in order to discharge the

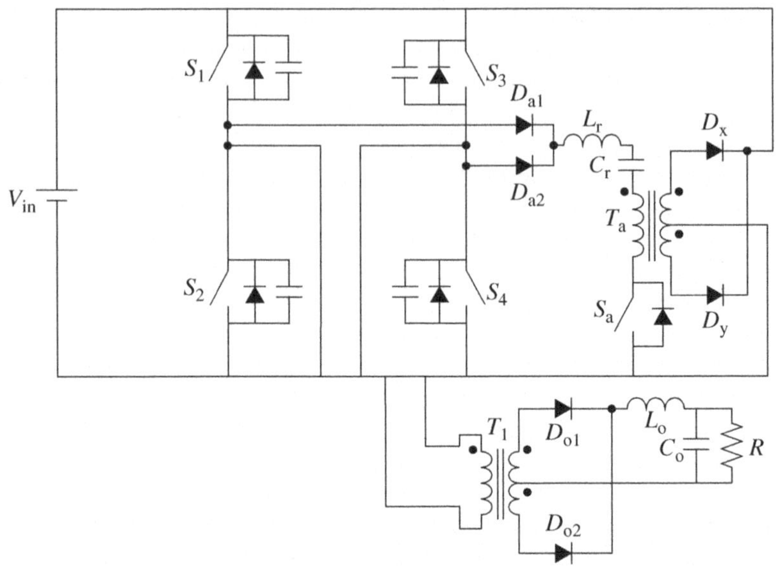

Figure 8.10 ZVS-PWM converter with single active auxiliary circuit based on interleaved boost converters.

output capacitance of the bottom switch whose top switch has just been turned off. For example, if the converter is in a freewheeling mode of operation where current flows down through the drain and source of S_1 and up through the source and drain of S_3, then S_1 would have to be turned off first before the auxiliary circuit is activated to discharge the output capacitance of S_2. The auxiliary circuit can be deactivated with the turn-off of the auxiliary switch after S_2 has been turned on with ZVS. The converter can then operate in an energy-transfer mode with switches S_2 and S_3 on.

Although operating the converter in this manner saves the cost of an auxiliary circuit, the asymmetrical operation of the converter can increase conduction losses. The main drawback to this approach, however, is that the auxiliary circuit must be operated twice during a switching cycle, which can increase the RMS current stress of the auxiliary circuit components to an unacceptable level. For example, it may be fine to have an auxiliary circuit on for a quarter of the switching cycle to help the main power switch of a boost converter operate with ZVS, but if this auxiliary circuit is activated twice during each cycle, then the auxiliary circuit is active for half the switching cycle and the auxiliary circuit components would have the same RMS current stresses as the main circuit components, which would increase cost and size. As a result, the application of this type of converter is limited unless the auxiliary circuit, especially the switch, is active for only a very short duration.

Another single active auxiliary circuit implementation for a ZVS-PWM-FB converter was proposed in [12]. In this converter, an auxiliary circuit is

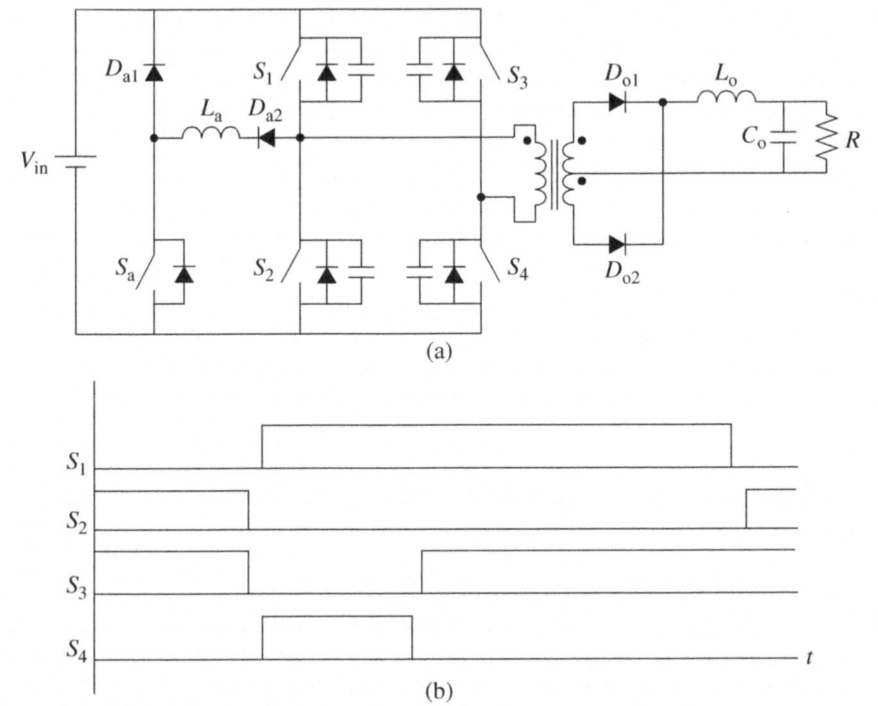

Figure 8.11 ZVS-PWM-FB converter with single active auxiliary circuit: (a) Converter topology. (b) Gating signals.

attached to only one of the bottom switches as shown in Figure 8.11a. Any auxiliary circuit that can be used in a boost converter, can be implemented in this converter. The PWM technique shown in Figure 8.11b is used to control the output voltage. This PWM technique is considerably different from standard phase-shift PWM as voltage control is achieved by varying the duty-cycle of the bottom switches. Switches S_2 and S_3 are made to turn-off simultaneously, and switches S_1 and S_4 are made to turn on simultaneously. This changes the sequence of converter modes so that instead having the energy-transfer and the freewheeling modes alternate (as is the case with phase-shift PWM), the converter first enters a power-transfer mode and then enters another power-transfer mode before going into the freewheeling mode. This is done so that the energy built up in inductor L_a during an energy-transfer mode can be used to discharge the output capacitance of switches S_1, S_3, and S_4 at appropriate times. Since S_2 is the only switch that is turned on when the converter is going from a freewheeling to an energy-transfer mode with this modified PWM technique, therefore, it is the only switch that needs an auxiliary circuit to help it operate with ZVS.

The converter, however, does have a number of drawbacks such as uneven converter switch current stresses and increased conduction losses since an asymmetrical PWM technique is used. Another drawback is that the converter does not have an output filter inductor as is normally found on a standard ZVS-PWM-FB converter. The output current is made to have a high ripple on purpose to build up the energy in L_a needed for ZVS operation to occur. Since the output current is discontinuous, the converter therefore has all the drawbacks caused by such a current. These drawbacks include higher turn-off losses of the main power circuit switches, higher peak current stress of the main power circuit components (both primary and secondary), and the need for a very large output capacitor. The converter also has considerable duty-cycle loss since L_a must be fairly large to make up for the absence of an output filter inductor. In general, implementing ZVS-PWM full-bridge converter with just a single active auxiliary circuit is an approach with a number of drawbacks that should only be considered for lower power applications.

8.7 ZVS-PWM-FB CONVERTERS BASED ON DUAL HALF-BRIDGE CONVERTERS

By definition, a full-bridge converter has four switches, as opposed to the other isolated DC–DC converters that were presented in Chapter 4, which have one or two switches. As such, a full-bridge converter can be considered as consisting of two half-bridge converters with two switches each. If a half-bridge converter is operated with a duty-cycle of $D = 0.5$, then it will have no zero-voltage states as the voltage between the mid-point of the leg with the two switches and that of the leg with the two capacitors will be either positive or negative, but never zero. As a result, a half-bridge converter operated in such a manner will always be in an energy-transfer mode of operation so that it will not have any freewheeling mode circulating current that contributes to conduction losses.

If a half-bridge converter is operated with $D < 0.5$, then its switches will have switching losses, but these can be made to be less than those found in a ZVS-PWM-FB converter. This is especially true if the half-bridge is designed so that there is no circulating current in the primary after a switch is turned off. If the primary current in a half-bridge converter is quickly extinguished after a switch is turned off, the voltage across the switches in the switch-leg will be divided among the two switches so that there will be a voltage of $V_{bus}/2$ across each switch, where V_{bus} is the DC bus voltage. This means that the amount of energy that is stored in the output capacitance of each switch will be related to $(½\, V_{bus})^2$ instead of the expected $V_{bus}{}^2$ so that CV^2 losses are significantly lower.

Figure 8.12 shows a dual half-bridge converter that was proposed in [13]. This converter consists of two half-bridge converters: one consisting of switches S_1 and S_2, which can be considered to be the leading converter, and one consisting of S_3 and S_4, which can be considered to be the lagging converter. The leading converter has clamping diodes to help suppress voltage

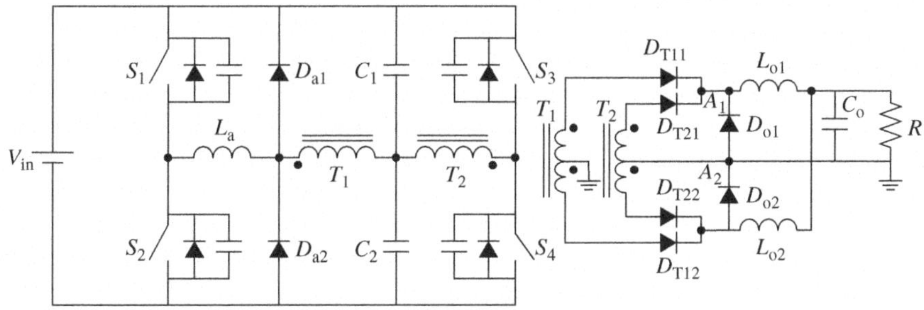

Figure 8.12 ZVS-PWM converter based on dual half-bridge converters.

ringing, like the clamping diodes used in the converter discussed in Section 8.2 of this chapter and shown in Figure 8.3a. The two converters can each be made to operate with a duty-cycle of $D = 0.5$ with a phase shift between the two that varies from 0° to 180°. This PWM method is actually the conventional phase-shift PWM method as this method has the gating signals of the switches in a ZVS-PWM-FB converter leg be $D = 0.5$ and complementary to each other, with voltage regulation made possible by shifting the phase of one leg relative to those of the other leg.

When the phase shift is 0°, the two converters are in phase and operate in parallel and the effective duty-cycle of the secondary is $D = 0.5$, which matches the $D = 0.5$ of each individual half-bridge converter. When the phase shift is 180°, the two half-bridge converters are now out of phase and they superimpose their voltages on the nodes A1 and A2 of the current doubler so that the effective duty-cycle of the secondary is $D = 1$, which makes the output current close to DC. When the phase offset is between 0° and 180°, which makes the effective secondary duty-cycle to be in the range of $0.5 < D < 1$, there will be times when the two half-bridge converters will be working in parallel and times when the secondary voltage of each half-bridge will be added to that of the other. What should be noted is that for any effective duty-cycle in the range of $0.5 \leq D \leq 1$, there is no needless circulating current in any of the two half-bridge converters, as both converters are always working in an energy-transfer mode of operation.

If the converter is operating with a light load and an effective secondary duty-cycle of $D < 0.5$ is desired, then each individual half-bridge is operated with its own duty-cycle of $D < 0.5$ and the phase shift between the two converters is 0°. As explained above, half-bridge converters can be made to operate without circulating current and with fewer switching losses for duty-cycles that are less than $D = 0.5$. As a result, the converter shown in Figure 8.12 can operate without any circulating current throughout the whole effective secondary duty-cycle range $0 \leq D \leq 1$ so that the conduction losses that it would otherwise cause are avoided.

Variations on the dual half-bridge converter concept have been proposed by other researchers. An example of one such variation was proposed in [14]

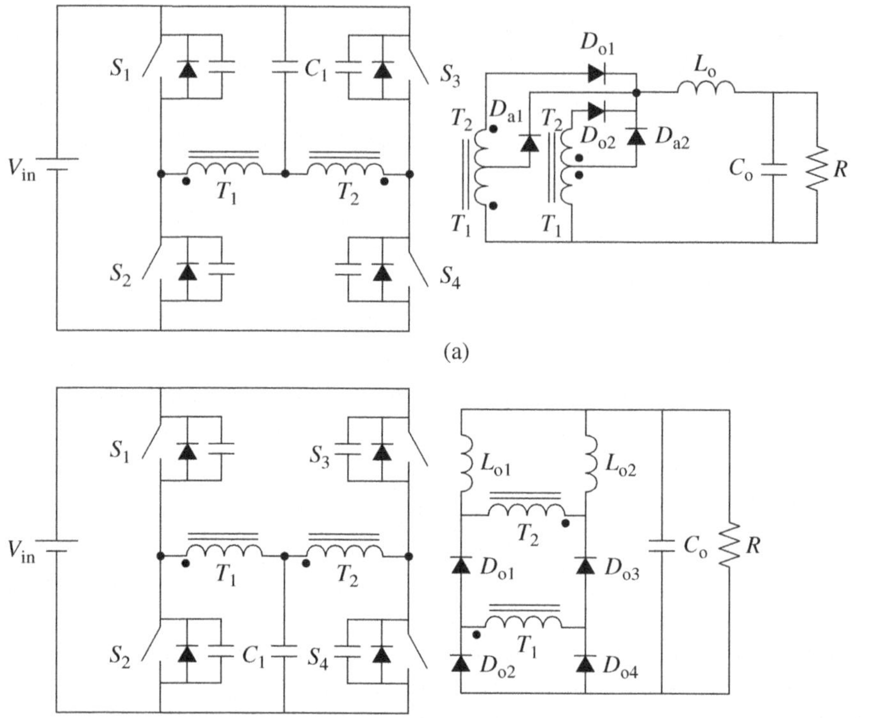

Figure 8.13 ZVS-PWM converters based on dual half-bridge converters: (a) Single secondary inductor implementation. (b) Stacked secondary winding implementation.

and is shown in Figure 8.13a. In this converter, only one capacitor is used to establish a mid-point voltage and the secondary implementation is changed to allow the converter to have an effective secondary duty-cycle of $D < 0.5$ even though its switches operate with $D = 0.5$. The mid-point voltage is always half of the DC bus voltage as each converter switch is always operated with a duty-cycle of $D = 0.5$. Unlike the converter proposed in [13], the duty-cycle of the half-bridge converter is never reduced and the output voltage can be regulated by shifting the phases of the two half-bridges relative to each other. One advantage of operating with the half-bridge always having a duty-cycle of $D = 0.5$ is that there is always voltage at the input of the output LC filter so that the size of the output inductor can be significantly reduced. Another advantage is that the converter switches can be made to work with ZVS, even at lighter loads, as opposed to the loss of ZVS that occurs in the converter shown in Figure 8.12 when its switches are operated with $D < 0.5$. These advantages, however, are achieved at the expense of no-load to very light-load operation. Although the converter can operate with $D = 0.5$ over a greater load range than the converter proposed in [13] due to the way the secondary windings are set up, there is always voltage at the input of the output LC filter that can never be made to go down to zero. As a result, the converter must always operate with some load and its load range is not as wide as that of a conventional ZVS-PWM-FB converter.

The converter shown in Figure 8.13a can be modified to have a current doubler output as proposed in [15] and shown in Figure 8.13b. What this secondary winding arrangement does is that it places the two transformer windings in series with each other so that current can be made to flow through in the same direction relative to the dot notation, which adds voltage, or in opposite directions relative to the dot notation, which subtracts voltage. This extends the range of secondary duty-cycles over which the converter switches can be made to operate with $D = 0.5$ so that $D = 0.5$ operation is possible even under no load conditions.

Although full-bridge converter with dual half-bridges have several advantages over the conventional ZVS-PWM-FB converter, they also have several drawbacks as well. One of these is that their control may not be as simple to implement. For example, the converter proposed in [13] requires a very sophisticated control implementation to make the converter operate with phase-shift control for heavy loads and PWM control for light loads. In some dual half-bridge converters, the converter must be implemented with some sort of control that ensures that the two half-bridges transfer the same amount of power to the output. Another drawback is that a non-standard transformer implementation is required to implement the dual half-bridges. A third drawback is the fact that the secondary implementation is not a simple center-tapped transformer, but is more sophisticated. This results in more secondary output conduction losses as current must flow through more components in its path.

8.8 ZVS-PWM-FB CONVERTERS WITH MODIFIED SECONDARY-SIDE CIRCUITS FOR ZVS OPERATION

Several recent approaches have looked at implementing mechanisms to ensure the ZVS operation of the primary-side switches of ZVS-PWM full-bridge converters at the secondary side of the converter rather than at the primary side. These approaches, however, are limited to applications when the output voltage is high and the output current is relatively low as they involve the placement of additional components that can create conduction losses to the secondary side of the converter. For example, these approaches are not attractive for applications where a 12 V/50 A output is needed, but may be attractive for applications where an isolated DC bus of several hundreds of volts is needed. Within such extreme output requirements, there are no hard-and-fast rules as to whether or not a full-bridge converter with a secondary side ZVS mechanism is suitable for a particular application.

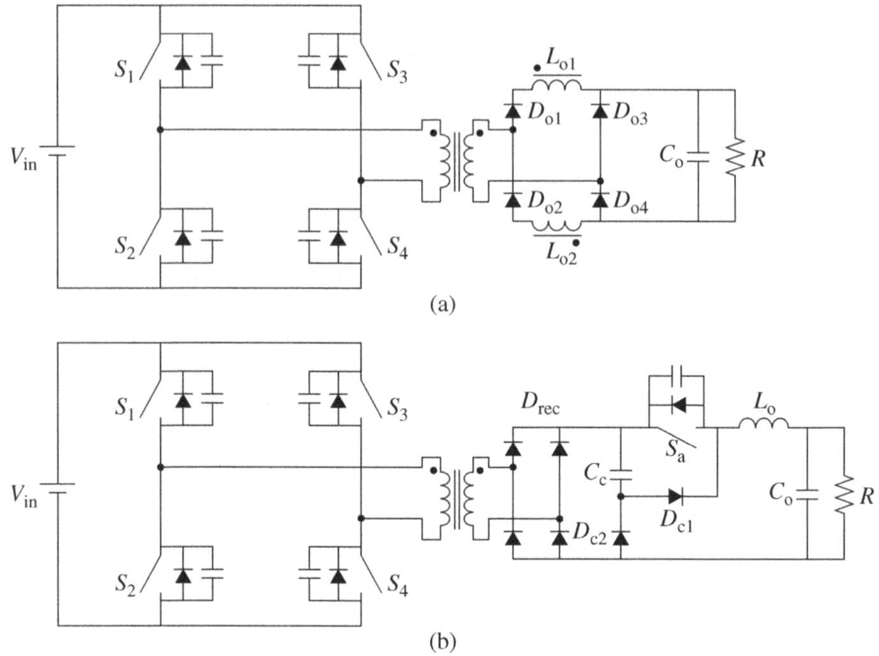

Figure 8.14 ZVS-PWM converters based on dual half-bridge converters with secondary circuit modifications for ZVS operation: (a) Coupled inductor rectifier. (b) Secondary auxiliary switch with passive snubber.

Modifications to the secondary-side can be either passive or active and examples demonstrating both types are presented in this section. An example of a passive secondary-side implementation that helps the primary-side switches operate with ZVS was proposed in [16] and shown in Figure 8.14a. This converter is a full-bridge converter with a coupled inductor secondary rectifier. Current flows through diagonally opposed diodes (D_{o1} and D_{o4} or D_{o2} and D_{o3}) whenever the converter is in an energy-transfer mode and there is voltage impressed across the primary transformer winding, which means that there is voltage across the secondary winding as well. Current flows through diodes D_{o1} and D_{o2} whenever the converter is in a freewheeling mode of operation and there is zero voltage impressed across the primary winding, which means that there is zero voltage across the secondary winding as well.

It is during the transition from an energy-transfer mode to a freewheeling mode of operation that the converter shown in Figure 8.14a differs most from the conventional ZVS-PWM-FB converter. While this transition is happening, there is a negative voltage that is impressed across the transformer secondary winding that extinguishes the current in this winding in a way that is not possible in a conventional ZVS-PWM full-bridge converter. This action also helps to reduce the amount of current flowing in the transformer primary so that ultimately what the coupled inductor implementation does is to reduce the amount of circulating current that flows during a freewheeling mode of operation – current that does nothing but create conduction losses and increase RMS current stresses in the primary-side switches. Reducing this circulating current comes at the cost of reducing the amount of primary-side inductive energy that is available to discharge the lagging legs switches, which are the switches that turn on or off to transition the converter out of a freewheeling mode of operation to an energy-transfer mode. This means that the lagging leg switches cannot be the same devices as the leading leg switches, but must be devices with less output capacitance instead. Such devices, however, can be implemented in the lagging leg as the reduced freewheeling mode circulating current results in there being less current for these switches to turn off. Moreover, the magnetizing inductance current can be made greater, if necessary, to provide additional inductive energy to discharge the lagging leg switch capacitances.

In Section 5.3, which discussed various secondary-side implementations, it was shown how multiple outputs of a multi-output converter can be regulated by some form of post-secondary regulation. One example, which was explained in that section, involved the use of a switch placed in series with a secondary diode to create a buck converter stage that was controlled independently of the primary side. A buck converter is placed in series with each secondary winding and is controlled independently of the other secondary windings to ensure tight regulation of each output. This concept can be extended to ensure the ZVS operation of the primary switches.

Consider the converter proposed in [17] and shown in Figure 8.14b, as an example of an active secondary circuit mechanism that helps ensure the ZVS operation of the primary-side converter switches. This converter is a full-bridge converter with a full-bridge diode rectifier at its secondary and a series switch S_a with a non-dissipative snubber. When the secondary switch is off, no current can flow out of the secondary winding so that energy is trapped in the transformer primary – energy that can be used to discharge switch capacitances. In essence, the ZVS principle of this converter is similar to the ZVS principle of the saturable reactor converter discussed in Section 8.1.2 of this chapter and shown in Figure 8.2a, but a switch is used to block current instead of saturable reactors so that the energy stored in the magnetizing inductance of the transformer can be used to discharge switch capacitances.

The operation of the secondary switch converter shown in Figure 8.14b can be improved if the primary switches are not operated with phase-shift PWM, but are operated with the following sequence instead: S_1 and S_4 on, then S_2 and S_3 on, then S_1 and S_4 on. In other words, only pairs of diagonally opposed switches, each with a duty-cycle of $D = 0.5$, are allowed to be on at any given time so that a square-wave voltage with no zero-voltage states is impressed across the primary of the transformer. This waveform allows the maximum amount of energy to be built up in the magnetizing inductance and thus helps ensure ZVS operation for very wide load ranges. Voltage regulation in this converter is performed by the secondary switch so that this converter is essentially a square-wave full-bridge converter with fixed gating pattern followed by a buck converter. Since the secondary switch interrupts current when it is turned off, a snubber is needed to provide a temporary current path to avoid current being forced through the switch.

8.9 CONCLUSION

A number of variations on the conventional ZVS-PW-FB converter were discussed in this chapter. These variations, for the most part, try to address the following drawbacks of the conventional converter:

- Limited load range for the ZVS operation of the switches caused by the lack of available inductive energy to discharge switch output capacitances at lighter loads.
- Duty-cycle loss caused by the presence of a transformer leakage inductance that must be sufficiently large to contain the energy needed to discharge the output capacitances of the switches before they are turned on.
- Voltage oscillations across the output secondary diodes that are caused by the interaction between their junction capacitance and the transformer leakage inductance.

The use of saturable reactors – small inductors that can be saturated very easily and that are only a factor whenever a switching transition occurs – both in the converter primary and in the secondary, was initially discussed. They are a simple way of increasing the converter's load range for ZVS operation and since they can reduce the need for less leakage inductance in the transformer and can also lead to less duty-cycle loss and less voltage ringing in the converter. They are more effective when used in the secondary as they are stressed less.

Passive auxiliary circuits, which are additional circuits that can be added either in series with the main path of transformer or in parallel, can expand the ZVS load range of a ZVS-PWM-FB converter more than saturable reactors. They can generate additional circulating current in the primary that increases the amount of inductive energy that is available to discharge the output capacitances of the switches. Several passive auxiliary circuits were presented in this chapter, with each having its own strengths and weaknesses. For example, series passive auxiliary circuits with diode clamps are the most effective in suppressing voltage ringing across converter components. Parallel passive auxiliary circuits are smaller in size than series circuits as their components are not in the main current path. Passive auxiliary circuits with transformers offer better performance, but are more sophisticated and expensive. There are a number of trade-offs that need to be considered in any ZVS-PWM-FB converter design and thus there is no one converter that can be said to be the best for all applications.

The same can be said when comparing passive auxiliary circuits with active auxiliary circuits that have at least one active switch. The main drawback with all passive circuits is that they generate a circulating current that flows in the converter throughout the entire switching cycle, instead of just during a switching transition, when it is actually needed to increase the amount of inductive energy available to ensure the ZVS operation of the converter switches. Active auxiliary circuits can be activated just when a switching transition is about to be made in the converter and deactivated shortly afterward so that additional circulating current flows in the converter for only a small fraction of the switching cycle. Since these converters are less dependent on leakage inductance they can have reduced duty-cycle loss and voltage ringing, just like ZVS-PWM FB converters with passive auxiliary circuits. The main drawback with active auxiliary circuits is their expense, as many such circuits require two auxiliary switches. Although it is possible to use only once active auxiliary switch, as was demonstrated in this chapter, this can only be done if conventional phase-shift PWM is not used to control the converter, which is disadvantageous as it does not allow readily available commercial phase-shift PWM controllers to be used and it makes the distribution of conduction losses uneven among the switches. These drawbacks limit the application of ZVS-PWM-FB converters with a single active auxiliary switch.

Dual half-bridge ZVS-PWM-FB converters consider a ZVS-PWM-FB converter to consist of two half-bridge converters, whose switches can be made to operate with a duty-cycle of $D = 0.5$ each. Operating the switches of a half-bridge converter in this manner ensures that the converter is always in some sort of energy-transfer mode so that needless circulating current is avoided. Even if the switches are made to operate with $D < 0.5$, as is the case with some implementations, switching losses can still be reduced as it can be ensured that the converter switches turn-on with just half of the DC bus voltage across them. Dual half-bridge converters are generally the most efficient type of ZVS-PWM-FB converter, but also the most sophisticated. Moreover, some implementations require additional components in the secondary, which results in more conduction losses for low output voltage/high output current applications.

Several, more recent, ZVS-PWM-FB converters have secondary-side auxiliary circuits. These converters can increase the ZVS load range of conventional ZVS-PWM-FB converters, reduce dependence on leakage inductance as a ZVS mechanism, and even reduce freewheeling mode current to some extent. Just like some of the dual half-bridge converters, however, the fact that these converters have additional secondary-side components makes them less attractive in low output voltage/high output current applications and more attractive in high output voltage/low output current applications.

REFERENCES

1. G. Hua, F. C. Lee, and M. M. Jovanovic, "An improved full-bridge zero-voltage-switched PWM converter using a saturable inductor," *IEEE Transactions on Power Electronics*, vol. 8, no. 4, pp. 530–534, Oct. 1993.
2. W. Chen, F. C. Lee, M. M. Jovanovic, and J. A. Sabate, "A comparative study of a class of full bridge zero-voltage-switched PWM converters," *IEEE Applied Power Electronics Conference (APEC)*, 1995, Mar. 5–9, 1995, Dallas, USA, pp. 893–899.
3. R. Redl, N. O. Sokal, and L. Balogh, "A novel soft-switching full-bridge DC/DC converter: analysis, design considerations, and experimental results at 1.5 kW, 100 kHz," *IEEE Transactions on Power Electronics*, vol. 6, no. 3, pp. 408–418, Jul. 1991.
4. G. Moschopoulos, P. D. Ziogas, and G. Joos, "A fixed frequency ZVS high power, PWM SMR converter with zero to rated load variation capability," *IEEE International Communications Energy Conference (INTELEC)*, 1992, Oct. 4–8, 1992, Washington DC, USA, pp. 351–358.
5. P. K. Jain, W. Kang, H. Soin, and Y. Xi "Analysis and design considerations of a load and line independent zero voltage switching full bridge DC/DC converter topology," *IEEE Transactions on Power Electronics*, vol. 17, no. 5, pp. 649–657, Sept. 2002.
6. A. Safaee, P. Jain, and A. Bakhshai, "A ZVS pulsewidth modulation full-bridge converter with a low-RMS-current resonant auxiliary circuit," *IEEE Transactions on Power Electronics*, vol. 31, no. 6, pp. 4031–4047, Jun. 2016.
7. X. Yang, Y. Li, Z. Gao, L. Xi, and J. Wen, "Analysis and design of full-bridge converter with a simple passive auxiliary circuit achieving adaptive peak current for ZVS and low circulating current," *IEEE Journal of Emerging and Selected Topics in Power Electronics*, vol. 9, no. 2, pp. 2051–2065, Apr. 2021.
8. Y. Jang and M. M. Jovanovic, "A new PWM ZVS full-bridge converter," *IEEE Transactions on Power Electronics*, vol. 22, no. 3, pp. 987–994, May 2007.
9. R. Ayyanar and N. Mohan, "Novel soft-switching DC–DC converter with full ZVS-range and reduced filter requirement—Part I: Regulated-output applications," *IEEE Transactions on Power Electronics*, vol. 16, no. 2, pp. 184–192, Mar. 2001.
10. J. G. Cho, J. A. Sabate, and F. C. Lee, "Novel full bridge zero-voltage-transition PWM DC/DC converter for high power applications," *IEEE Applied Power Electronics Conference (APEC)*, 1994, Feb. 13–17, 1994, Orlando, USA, pp. 143–149.
11. G. Moschopoulos and P. Jain, "A PWM full-bridge converter with load independent soft-switching capability," *IEEE Applied Power Electronics Conference (APEC)*, Feb. 6–10, 2000, New Orleans, USA, 2000, pp. 79–85.
12. F. S. Hamdad and A. K. S. Bhat, "A novel pulse width control scheme for fixed-frequency zero-voltage-switching DC-to-DC PWM bridge converter," *IEEE Power Electronics Specialists Conference (PESC)*, 1999, June 27–Jul. 1, Charleston, USA, pp. 262–268.
13. Z. Ye, "Dual half-bridge DC–DC converter with wide-range ZVS and zero circulating current," *IEEE Transactions on Power Electronics*, vol. 28, no. 7, pp. 3276–3286, Jul. 2013.

14. I. Lee and G. Moon, "Analysis and design of phase-shifted dual H-bridge converter with a wide ZVS range and reduced output filter," *IEEE Transactions on Industrial Electronics*, vol. 60, no. 10, pp. 4415–4426, Oct. 2013.
15. L. Zhao, F. Li, Z. Zhuang, Z. Li, and Z. Luo, "A dual half-bridge converter with current doubler rectifier," *IEEE Transactions on Industrial Electronics*, vol. 67, no. 8, pp. 6398–6406, Aug. 2020.
16. J. -K. Han and G. -W. Moon, "High-efficiency phase-shifted full-bridge converter with a new coupled inductor rectifier (CIR)," *IEEE Transactions on Power Electronics*, vol. 34, no. 9, pp. 8468–8479, Sept. 2019.
17. D. D. Tran, H. N. Vu, S. Yu, and W. Choi, "A novel soft-switching full bridge converter with a combination of a secondary switch and a non-dissipative snubber," *IEEE Transactions on Power Electronics*, vol. 33, no. 2, pp. 1440–1452, Feb. 2018.

CHAPTER 9

ZERO-VOLTAGE-ZERO-CURRENT-SWITCHING DC–DC FULL-BRIDGE PWM CONVERTERS

Zero-voltage-zero-current-switching (ZVZCS) DC–DC full-bridge PWM converters are isolated four-switch DC–DC PWM converters that have two converter switches that operate with ZVS and two converter switches that operate with ZCS. The most basic ZVZCS DC–DC converter topologies, such as the ones that are the subject of this chapter, are based on the conventional four-switch PWM DC–DC full-bridge converter structure. ZVZCS-PWM DC–DC converters are popular in higher power applications as they address a key drawback that the conventional DC–DC PWM-FB converter topology has, which is the current that circulates in the converter when it is in a freewheeling mode of operation. Whenever a full-bridge converter is in a freewheeling mode of operation, there is no voltage impressed across the transformer so that current circulates in the converter's primary side and conduction losses are generated without any power transfer taking place; this circulating current increases component current stresses as well. These conduction losses and current stresses become more significant if the converter operates at higher power levels (e.g. >1.5 kW). The main characteristic of isolated ZVZCS-PWM DC–DC full-bridge converters is that they do not have the freewheeling mode circulating current in their primary side that the standard PWM DC–DC full-bridge converter has. This current is extinguished shortly after a freewheeling mode is entered so that conductions losses and current stresses are not increased significantly.

DC–DC Converter Topologies: Basic to Advanced, First Edition. Gerry Moschopoulos.

Published 2024 by John Wiley & Sons, Inc.

ZVZCS-PWM DC–DC full-bridge converters are generally synthesized by making a change of some kind to the standard ZVS-PWM-FB converter, but doing so results in the lagging leg switches losing their ability to operate with ZVS. This is because there is no available current to discharge any switch capacitances before the switches themselves are turned on as it is extinguished. Since these switches are turned on when there is no current in the converter primary and are turned off after current in the primary has been extinguished, the lagging leg switches of a ZVZCS-PWM DC–DC full-bridge converter operate with ZCS. Given that the operation of the leading leg switches is unaffected by extinguishing any freewheeling mode current, these switches still operate with ZVS so that have the converter switches operate with ZVS and half operate with ZCS, thus they are ZVZCS-PWM converters.

In this chapter, several ZVZCS-PWM DC–DC full-bridge topologies are reviewed. Their basic operation is explained as are their strengths and weaknesses. The chapter ends with examples of ZVS-PWM and ZCS-PWM converters that are versions of ZVZCS-PWM converters, but with modifications that make all the converter switches operate with either ZVS or ZCS.

9.1 FUNDAMENTAL ZVZCS-PWM DC–DC FULL-BRIDGE CONVERTER

One of the first ZVZCS-PWM DC–DC full-bridge converter topologies that was proposed in the power electronics literature was the converter that was proposed in [1] shown in Figure 9.1. It will be referred to as the fundamental ZVZCS-PWM-FB converter in this chapter as it is the simplest ZVZCS-PWM-FB converter. The converter is almost identical to the standard full-bridge topology with a four-diode output rectifier except that a

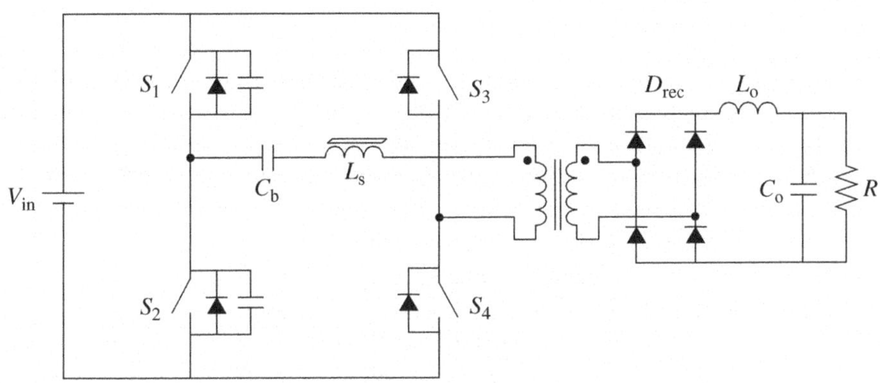

Figure 9.1 Fundamental ZVZCS-PWM DC–DC FB converter.

saturable inductor and a small capacitor, C_b, has been added in series with the transformer primary. The purpose of the capacitor is to act as a voltage source that will extinguish primary-side circulating current when the converter enters a freewheeling mode of operation. In the standard full-bridge converter, primary-side freewheeling mode current cannot be extinguished as there is no counter voltage of any kind that will drive this current down to zero. As a result, no mechanism exists to force the freewheeling current to zero.

The presence of C_b in the ZVZCS-PWM converter shown in Figure 9.1 introduces such a voltage to the S_1–S_3-transformer loop or the S_2–S_4-transformer loop that exists whenever two top switches (S_1 and S_3) or two bottom switches (S_2 and S_4) conduct current. When switches S_1 and S_3 conduct freewheeling current, the polarity of C_b, which is established in the previous mode of operation, is such that it acts against the current. Similarly, when switches S_2 and S_4 conduct freewheeling current and current flows in the opposite direction, the polarity of C_b in the S_2–S_4-transformer loop, which is established in the previous mode of operation, is such that it forces the primary current down to zero. It should be noted that C_b must be implemented with the saturable inductor because the primary current will not remain at zero otherwise, but will try to reverse direction instead. The saturable inductor is not saturated at this instant and provides a sufficiently large inductance to prevent negative current from appearing.

The converter shown in Figure 9.1 has six operating modes within each operating half cycle. The equivalent circuits for these modes of operation are shown in Figure 9.2. The converter is operated with the same standard phase-shift PWM method that is used to operate a ZVS-PWM-FB converter, as was explained in Section 7.2. The switches of each leg turn on and off alternately with nearly 50% duty cycle ratio and the phase shift introduced between the two legs determines the duty cycle of the converter. For the following discussion, it is assumed that all components are ideal that the inductance of the saturable inductor is very large when it is not saturated and zero when it is saturated.

Mode 1 (t_0–t_1): S_1 and S_4 are on and conduct current. The converter is in an energy-transfer mode as a diagonally opposed pair of switches is on and power is being transferred from the input to the output during this mode. Also during this mode, the saturable inductor is saturated and the blocking capacitor voltage increases; this capacitor was charged with the opposite polarity during previous modes.

Mode 2 (t_1–t_2): S_1 is turned off. The primary current discharges the output capacitance of S_2 and charges the output capacitance of S_1. This mode is similar to that of the standard ZVS-PWM-FB converter when a switch has been turned off to take the converter out of an energy-transfer mode.

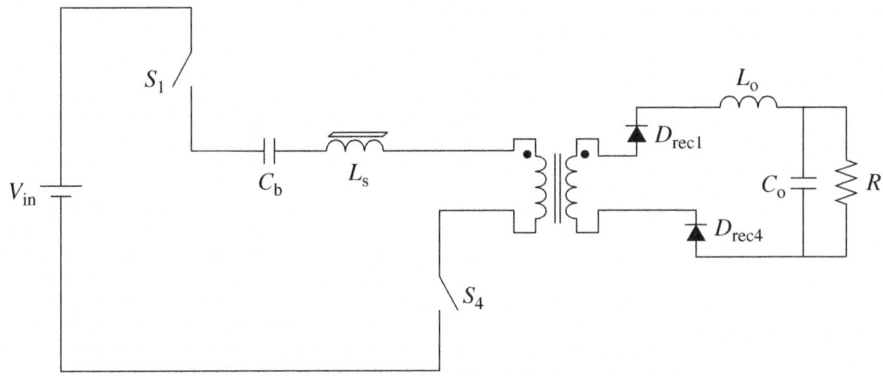

(a)

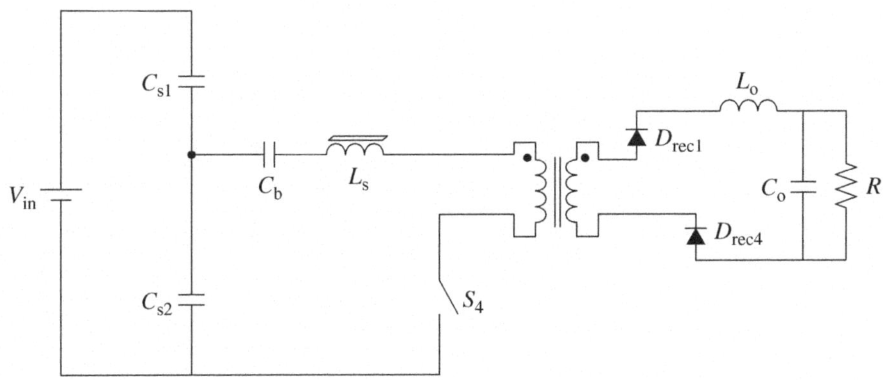

(b)

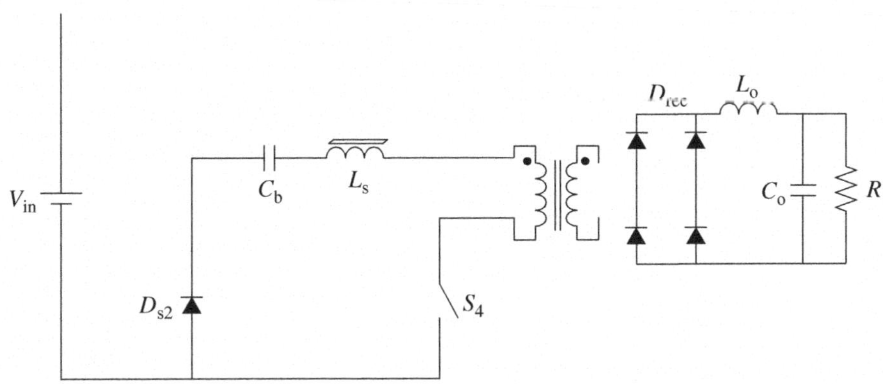

(c)

Figure 9.2 Modes of operation of the fundamental ZVZCS-PWM DC–DC FB converter. (a) Mode 1. (b) Mode 2. (c) Mode 3. (d) Mode 4. (e) Mode 5. (f) Mode 6.

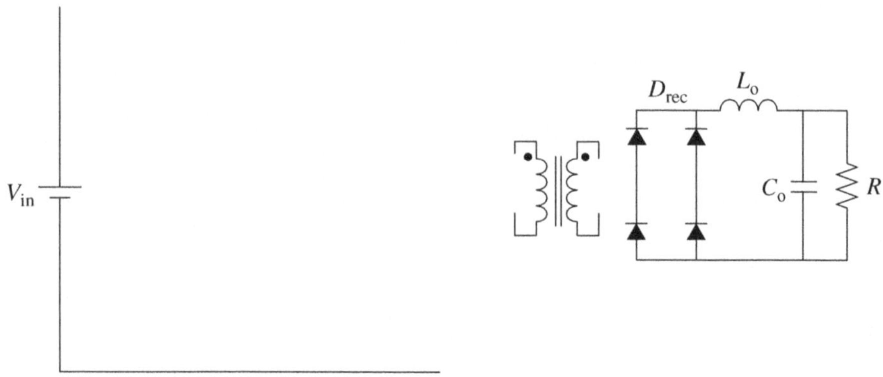

(d)

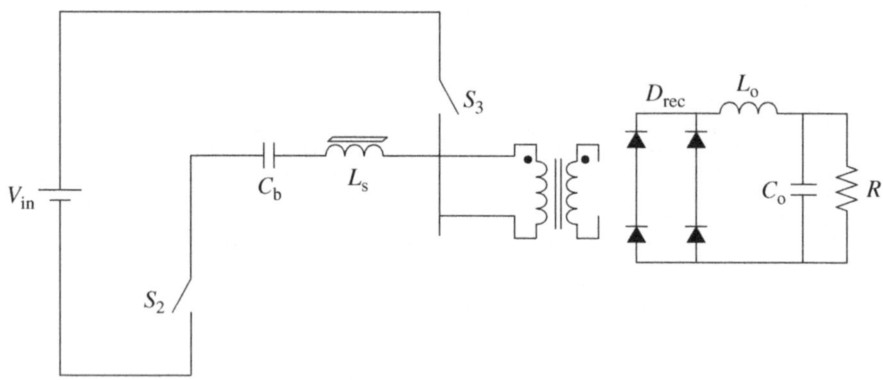

(e)

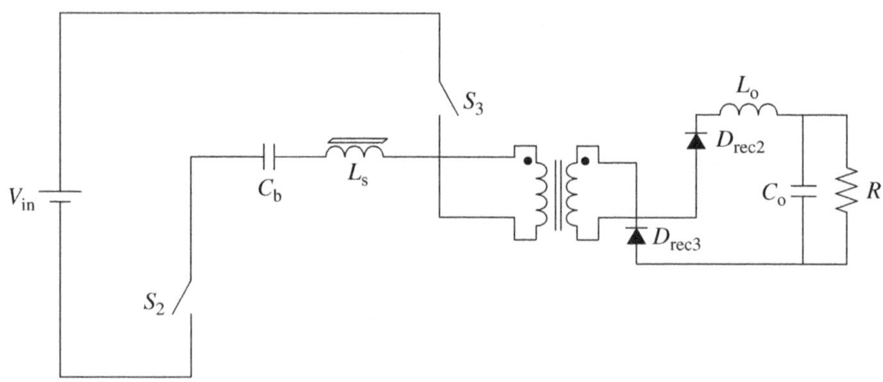

(f)

Figure 9.2 (*Continued*)

Mode 3 (t_2–t_3): This mode begins after the output capacitance of S_2 has been fully discharged and that of S_1 has been fully charged. Current begins to flow through the body diode of S_2. While this is happening, switch S_2 can be turned on with ZVS in the exact same manner as it can be for the standard ZVS-PWM-FB converter. Like a switch in the ZVS-PWM-FB converter after it has been turned on while current is flowing through its body-diode, current may actually flow through the switch in the reverse direction rather than the body-diode if the switching device is a MOSFET with an on-state resistance of $R_{DS,ON}$ that results in a source-drain voltage drop that is lower than that of the body-diode. After the body-diode of S_2 starts conducting, the voltage between the mid-points of the two converter legs, is clamped to zero. The voltage across the blocking capacitor – which is very small compared to the input source voltage, but sufficiently large to act as a constant voltage source during this mode – is applied to the transformer's primary leakage inductance. This causes the primary current to start to decrease during this mode. This also causes the energy stored in the leakage inductance to be transferred to the blocking capacitor; the saturable inductor remains saturated. The difference between the primary current reflected to the secondary and the current in the filter inductor freewheels through the secondary side diodes.

Mode 4 (t_3–t_4): This mode begins when the primary current, which was decreasing during the previous mode, has reached zero. This current will try to reverse direction, but is prevented from doing so by the saturable inductor, which is no longer saturated. All the blocking capacitor voltage is applied to the saturable inductor and this voltage remains constant during this mode. Switch S_4 remains turned on, but no current flows through it.

Mode 5 (t_4–t_5): S_4 is turned off at the start of this mode. The primary side of the converter is open and no current is flowing. Current continues to freewheeling at the secondary side of the converter.

Mode 6 (t_5–t_6): This mode begins with the turning on of switch S_3. This switch turns on with ZCS as current cannot rise abruptly because the saturable inductor is not saturated at this time and there is transformer leakage inductance in the current path. After S_3 is turned on, a voltage that is the sum of the input voltage and the blocking capacitor voltage is applied to the saturable inductor, thus quickly saturating it. The primary current begins to rise at the start of this mode and increases so that it is eventually equal to the reflected output current by the end of the mode. The converter enters Mode 1 at the end of this mode, which is an energy-transfer mode, but with switch S_2 and S_3 on instead of S_1 and S_4. This mode is the final mode of the half-switching cycle.

The ZVZCS-PWM-FB converter shown in Figure 9.1 has the following features:

- *Soft-switching for all the converter switches:* The ZVZCS-PWM-FB converter switches that are turned on to make the converter exit an energy-transfer mode – switches S_1 and S_2 in the above description of the modes of operation, which are considered to be the leading leg switches – operate with ZVS. The leading leg switches can turn on with ZVS during Mode 3 of operation just like the switches in the standard ZVS-PWM-FB converter. Moreover, they are able to turn on even though the load is light because, while the converter is in the process of exiting an energy-transfer mode, the reflected output current is available to help discharge the output capacitances of the switches so that the switches can be turned on when current is flowing their body-diode.

  They can turn off with ZVS if there is sufficient output capacitance (i.e. either the internal output capacitance of the device or a combination of this output capacitance or additional external capacitance) to slow down the rate of rise of voltage across the switch and limit the overlap between the switch voltage and current during this switching transition. Although some turn-off switching loss may exist, it can be considerably less compared what would exist without ZVS.

  The ZVZCS-PWM-FB converter switches that are turned on to make the converter exit a freewheeling mode and enter an energy-transfer mode – switches S_3 and S_4 in the above description of the modes of operation, which are considered to be the lagging leg switches – operate with ZCS. They can be turned on with ZCS as described above for Mode 6, when the converter is in a state where it has no primary current flowing through it and can be turned off with ZCS after primary current has been extinguished as explained above for Mode 5. It should be noted that unlike the lagging leg switches of a conventional ZVS-PWM-FB converter, which typically lose their ZVS turn-on capability when the load is less than 50% full-load, the lagging leg switches of the ZVZCS-PWM-FB converter can operate with ZCS over the whole load range, from no-load to full-load.
- *Reduced conduction losses:* The elimination of the freewheeling mode circulating current in the ZVZCS-PWM-FB converter results in the reduction of conduction losses, especially when the converter is operating at light-load. Under heavy-load conditions, the reduction in conduction losses is not significant because the duration of any freewheeling mode of operation is brief, as the converter is designed to operate with maximum duty-cycle under these conditions. It is when the converter is operating with light loads that the reduction in conduction losses does become significant as the duration of any freewheeling modes is longer.

Given that converters operate mostly with lighter loads for many applications, the reduction in conduction losses is an attractive feature of the ZVZCS-PWM converter.

- *Lack of duty-cycle loss:* The ZVS operation of the ZVS-PWM-FB converter is highly dependent on the transformer primary leakage inductance. This is especially true for the lagging leg switches as it is only the energy from the leakage inductance that is available to discharge switch output capacitances so that converter switches can be turned on with ZVS. As explained in Section 7.3.2, however, there is a trade-off that must be considered in the selection of a leakage inductance value when designing the converter's transformer – increasing the leakage inductance will increase the load range over which the converter can operate with ZVS, but will also result in a reduction in maximum duty-cycle that may force a transformer design with a lower turns ratio. Such a turns ratio will result in more primary current circulating in the converter during energy-transfer modes and thus more conduction losses and more turn-off losses. Moreover, the DC gain of the converter will be very dependent on the load and this makes voltage regulation harder.

  In the ZVZCS converter, issues related to duty-cycle loss can be avoided as the lagging leg switches operate with ZCS and cannot operate with ZVS. Since these switches cannot operate with ZVS, there is no need for leakage inductance energy to discharge their switch capacitances before they are turned on so that this parameter can be minimized. Moreover, as there is no current in the transformer primary during a freewheeling mode, current does not have to change direction, but can immediately begin to flow in the appropriate direction as soon as a pair of diagonally opposed switches are on. With a small, almost negligible leakage inductance and a lack of transformer primary current swing in direction, the ZVZCS-PWM converter has little, if any, duty-cycle loss.
- *If Si-IGBTs are used in the lagging ZCS leg of the converter, then they can be operated with higher switching frequencies:* This is because the ZCS operation of these devices eliminates the current tail, thus shortening the time required to turn them off. This shortening of time results allows the devices to operate with shorter switching cycles thus with higher switching frequencies. For example, a Si-IGBT operating with ZCS can be operated with 100 kHz switching, whereas a Si-IGBT operating without ZCS is limited to about 40 kHz operation. This higher switching frequency operation allows the converter itself to be operated with higher frequency, which, in turn, increases converter power density. This is especially true of silicon devices, but less of a concern for wide bandgap IGBT devices as these devices do not have a current tail.

The ZVZCS-PWM-FB converter has the following drawbacks:

- *Half the converter switches will not operate with the appropriate soft-switching method if all the converter devices are the same:* If all the devices are implemented with MOSFETs, then half the switches will have turn-on switching losses as their output capacitances will not have been discharged beforehand. Moreover, the operation of the converter will be limited in power due to increased conduction losses, given that the drain-source voltage drop of MOSFETs is greater than that of IGBTs for higher power applications. If all the devices are implemented with Si-IGBTs, then half the devices will turn-off with a current tail. The overlap between switch voltage and current can be reduced if the switch output capacitance of the switches in the ZVS leading leg is increased, but this may result in duty-cycle loss that can offset the reduction of duty-cycle loss that would otherwise be caused by a significant leakage inductance as it would take time for switch output capacitances to be charged to the input DC bus level. In addition to duty-cycle loss, turn-on switching losses may be increased as well as more energy will be required to discharge output switch capacitances. ZVS operation will be lost at heavier loads. It should be noted that this is more of an issue when silicon devices are used than when wide bandgap devices are used as switching losses are less significant when wide bandgap devices are used due to their superior switching characteristics. The benefits of reduced conduction losses and RMS current stresses still remain if wide bandgap devices are used.
- *The cost of the devices will increase if different switches are used as cost generally drops significantly with volume:* For example, if a ZVZCS-PWM converter is implemented with all MOSFETs or with all IGBTs, then the cost of having four devices per converter will be cheaper than the cost of having two MOSFETs and two IGBTs per converter. Again, this is less of an issue when wide bandgap devices are used and ZVZCS converters can be implemented with just one type of switching device.
- *The saturable inductor is lossy:* The losses in the saturable reactor can limit the maximum power level of the converter.

A variation to the fundamental ZVZCS-PWM-FB converter that tries to correct the power limiting issue created by the presence of the saturable inductor in the converter is the ZVZCS-PWM-FB converter that was proposed in [2] and shown in Figure 9.3. In this converter, the saturable inductor is replaced by two blocking diodes in series with switches S_3 and S_4. The purpose of these diodes is to block current from flowing through the body-diodes of S_3 and S_4 after the primary current has been extinguished. Since these diodes produce fewer losses than the saturable inductor at higher power levels, the converter can operate with a higher power-limit than before.

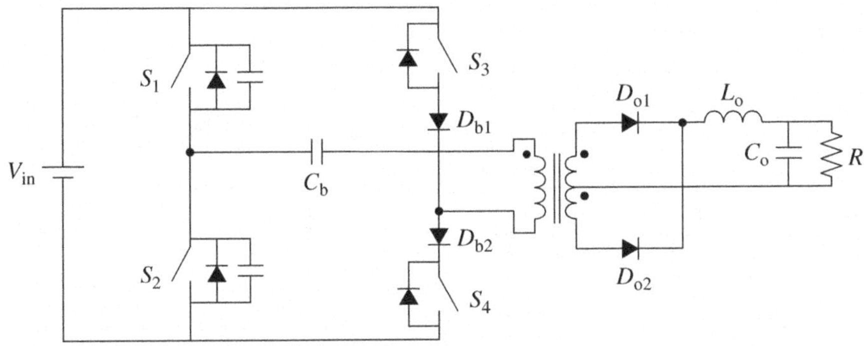

Figure 9.3 ZVZCS-PWM DC–DC FB converter with series lagging leg diodes.

The cost of the converter, however, does increase with the addition of the series diodes, but the use of reverse-blocking IGBTs that have a blocking diode in series with an IGBT in the same device package can partially offset the cost increase.

9.2 ZVZCS-PWM DC–DC FULL-BRIDGE CONVERTERS WITH SECONDARY AUXILIARY CIRCUIT

A more common approach to transforming a conventional ZVS-PWM-FB converter into a ZVZCS-PWM-FB DC–DC converter is to modify the secondary side of the converter rather than the primary. As stated above, the fundamental ZVZCS-PWM-FB converter is limited in power due to losses created by the saturable inductor. Although the modified ZVZCS-PWM-FB converter shown in Figure 9.3 corrects this problem by replacing the saturable inductor with reverse-blocking diodes, these diodes are a source of conduction losses as they are in the main path of current when they are conducting. It would be preferable if ZVZCS can be implemented without the use of any additional circuitry in the primary side of the converter.

A ZVZCS-PWM-FB converter that has a modified secondary instead of a modified primary side is the converter that was proposed in [3] and shown in Figure 9.4a with a four-diode secondary diode bridge. It can be implemented with a center-tapped secondary as shown in Figure 9.4b; for the sake of simplicity, the discussion in this section will focus on the topology in Figure 9.4a. The converter in Figure 9.4a has the same basic topology as the conventional ZVS-PWM-FB converter except for a passive auxiliary circuit that is attached to the output of diode bridge. The auxiliary circuit, which consists of a capacitor C_c and diodes D_{c1} and D_{c2}, is the mechanism by which a counter voltage that can extinguish the primary current when the converter is in a freewheeling mode of operation is produced. The counter voltage is impressed by C_c

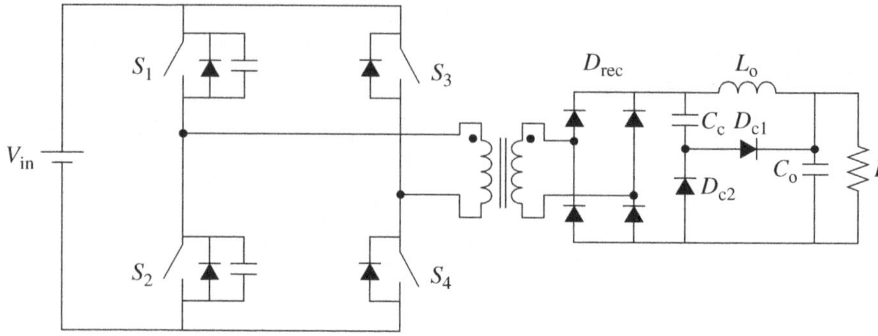

(a)

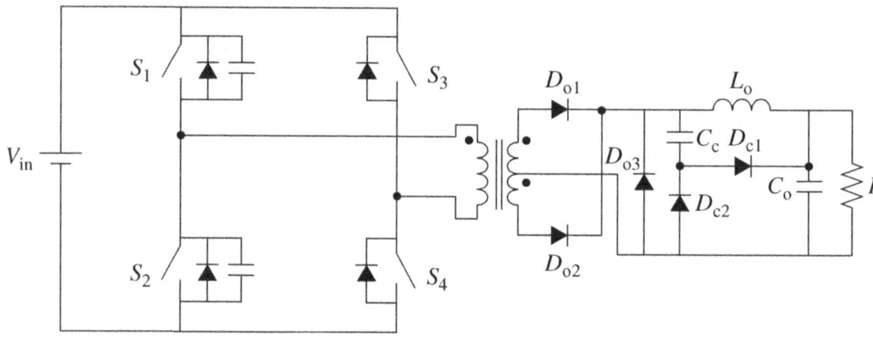

(b)

Figure 9.4 Fundamental ZVZCS-PWM DC–DC FB converter with passive secondary auxiliary circuit. (a) With four-diode secondary diode bridge; (b) with center-tapped transformer secondary.

across the secondary of the transformer then reflected to the primary whenever the converter is in a freewheeling mode of operation. The converter's modes of operation and its operation waveforms for a half switching cycle are shown in Figures 9.5 and 9.6, respectively. The modes of operation are as follows:

Mode 1 (t_0–t_1): Switches S_1 and S_4 are on and are conducting current. As voltage is placed across the transformer primary, the current in the primary rises. In addition to this current, secondary passive circuit capacitor C_c, which was fully discharged during the previous half switching cycle, is charged in a resonant manner during this mode as it interacts with the transformer leakage inductance. Power begins to be transferred from the input to the output during this mode.

Mode 2 (t_1–t_2): This mode begins when secondary passive circuit diode D_{c1} stops conducting current and capacitor C_c reaches its maximum value. With diagonally opposed switches S_1 and S_4 on, the converter is in an

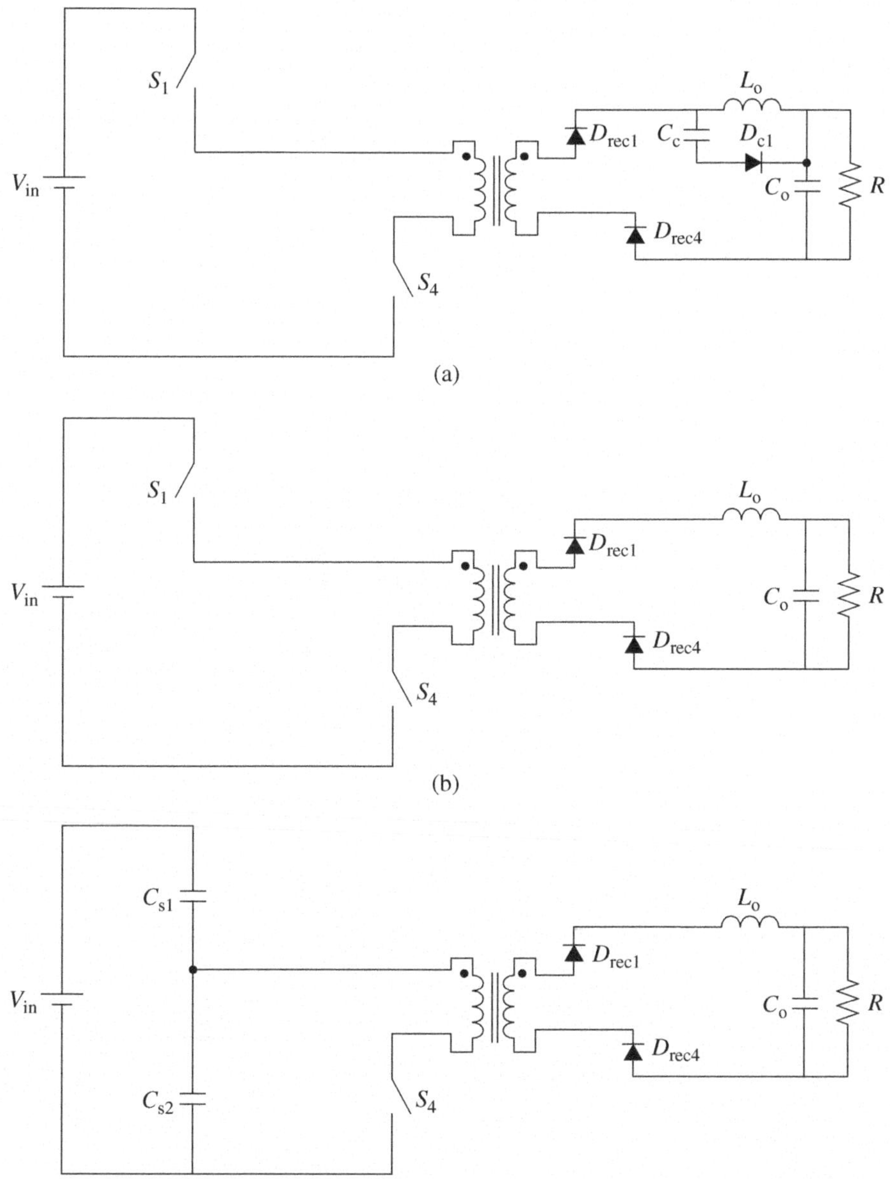

Figure 9.5 Modes of operation of the ZVZCS-PWM DC–DC FB converter with passive secondary auxiliary circuit. (a) Mode 1. (b) Mode 2. (c) Mode 3. (d) Mode 4. (e) Mode 5. (f) Mode 6. (g) Mode 7. (h) Mode 8.

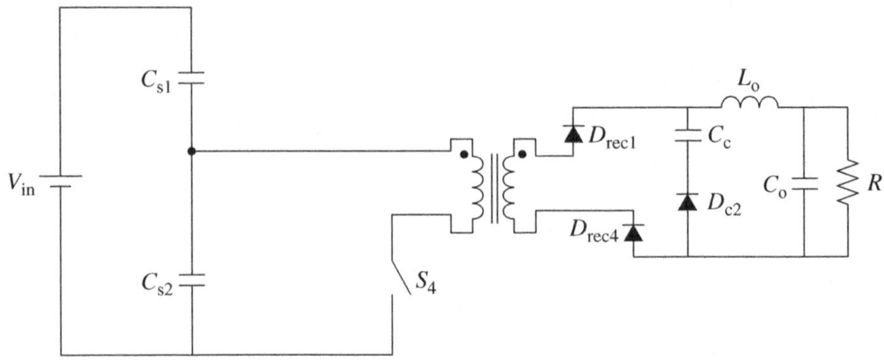

(d)

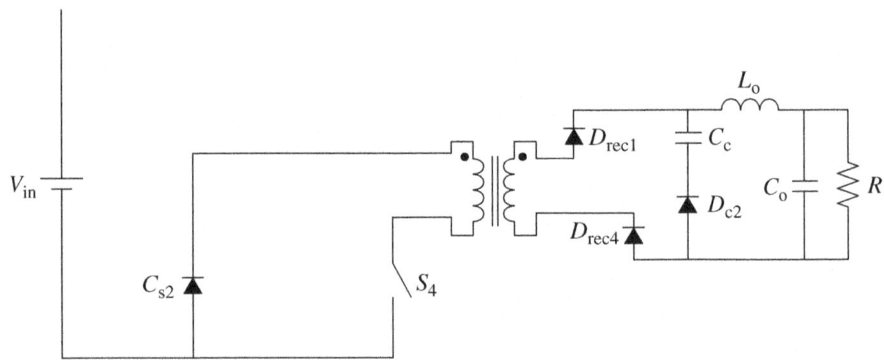

(e)

(f)

Figure 9.5 (*Continued*)

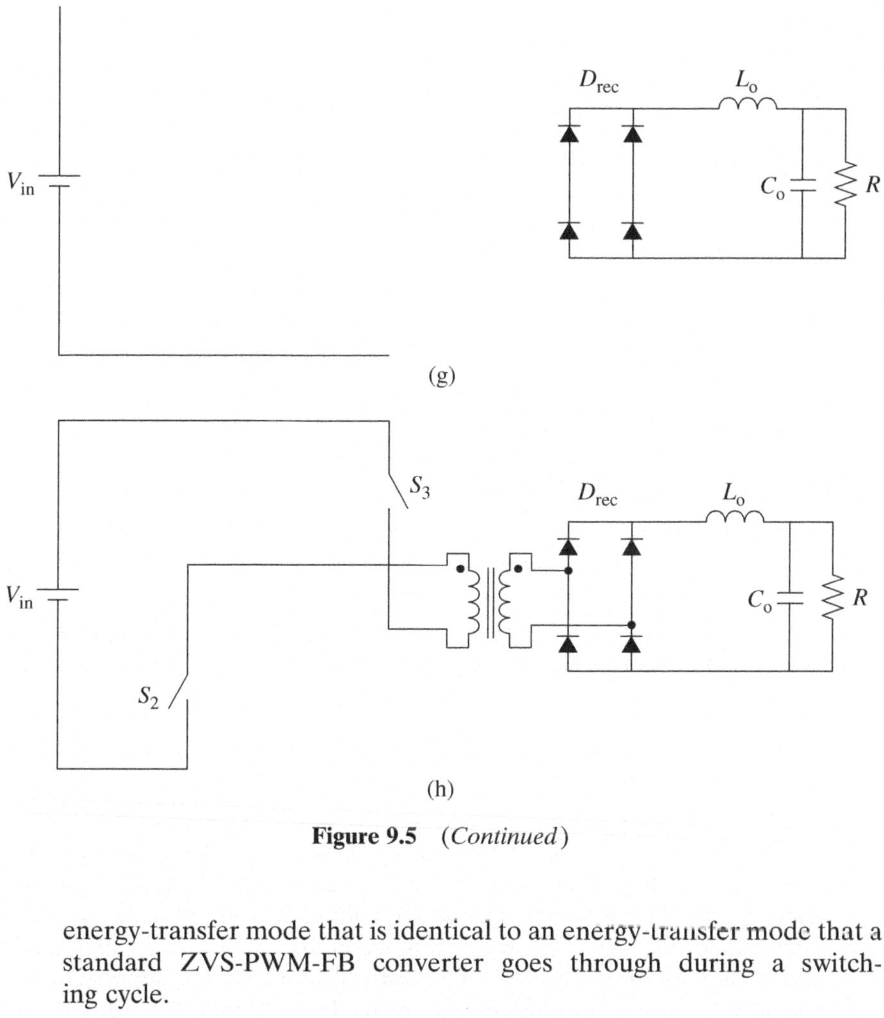

Figure 9.5 (*Continued*)

energy-transfer mode that is identical to an energy-transfer mode that a standard ZVS-PWM-FB converter goes through during a switching cycle.

Mode 3 (t_2–t_3): This mode begins when switch S_1 is turned off. When this happens, the output capacitance of S_1 begins to be charged and that of S_2 begins to be discharged. The primary voltage begins to drop as does the voltage at the output of the secondary diode bridge rectifier. The mode is identical to what a standard ZVS-PWM-FB converter goes through during a switching cycle.

Mode 4 (t_3–t_4): This mode begins when the voltage at the output of the secondary diode bridge begins to drop below the voltage across passive circuit capacitor C_c. This event results in diode D_{c2} starting to conduct as it is no longer reverse-biased. The primary voltage decreases at the same rate as it did during the previous mode, but the voltage across C_c reduces at a slower rate and is applied to the transformer secondary. The difference between the primary voltage and the reflected secondary voltage is

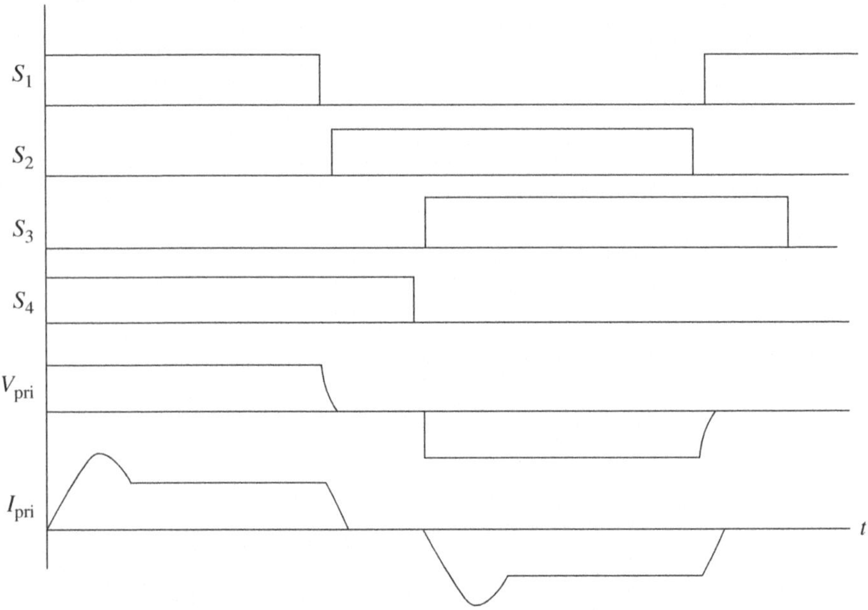

Figure 9.6 Gating signals and primary voltage and current waveforms of the ZVZCS-PWM DC–DC FB converter with passive secondary auxiliary circuit.

applied to the transformer leakage inductance and the primary current start to decrease.

Mode 5 (t_4–t_5): This mode begins when the output capacitance of S_1 has been charged to the input voltage and that of S_2 has been fully discharged. When this happens, current starts to flow through the body-diode of S_3. The reflected secondary voltage is applied to the leakage inductance and the primary current decreases at a faster rate than it did during the previous mode. Capacitor C_c continues to supply current to the load and its voltage continues to drop. It should be noted that switch S_3 can be turned on with ZVS while current is flowing through it.

Mode 6 (t_5–t_6): This mode begins when the primary current has dropped to zero. The only current that flows in the converter during this mode is at the output, which is supplied by capacitor C_c.

Mode 7 (t_6–t_7): The mode begins when capacitor C_c has been completely discharged. When this happens, diodes in the secondary diode bridge that were not conducting current in previous modes start to conduct as the voltage at the output of the bridge is zero. The load current free-wheels through all the diodes in the diode bridge, as shown in Figure 9.5g. Sometime during this mode, S_2 can be turned off with ZCS as there is no current in the device.

Mode 8 (t_7–t_8): This mode begins when switch S_4 is turned on with ZCS. The turn-on is with ZCS because the primary current cannot rise abruptly due to the transformer leakage inductance in the current path. The secondary rectifier output voltage remains zero as the primary current rises and the current-sharing of the diodes in the secondary bridge diodes becomes unequal. This mode ends when two of the diodes in the secondary bridge stop conducting current and the output current goes through two diagonally opposed diodes in the bridge. This concludes the half-switching cycle as the converter enters Mode 1, but with switches S_2 and S_3 conducting current instead of switches S_1 and S_4.

The converter has the same advantageous features that the fundamental and reverse-blocking diode ZVZCS-PWM-FB converters discussed above have soft-switching for all converter switches, reduced conduction losses, lack of duty-cycle loss. Moreover, it has the following advantageous features as well:

- *The auxiliary circuit is active for a small fraction of the switching cycle:* The auxiliary circuit diodes and capacitor conduct current only when capacitor C_c is discharged in Mode 1 and charged in Modes 4 and 5. The duration of this current conduction is significantly smaller than the time that current flows through the saturable inductor in the fundamental ZVZCS-PWM-FB converter or the reverse-blocking diodes of the reverse-blocking diode ZVZCS-PWM-FB converter. This brief duration allows the auxiliary circuit to be implemented with inexpensive, low current-rated components and it does not lead to significant conduction losses that can partially offset the reduction in conduction losses caused by the ZVZCS operation itself.
- *The auxiliary circuit can clamp the ringing caused by the interaction of the junction capacitance of the secondary diodes and the leakage inductance of the transformer*: Unlike the converters previously discussed in Section 9.1, the converter shown in Figure 9.4a can clamp any voltage ringing across the secondary diodes because of the presence of capacitor C_c at the output of the diode bridge. If the voltage across the secondary output diode bridge tries to oscillate, the peak oscillation will be clamped at a voltage level that is the sum of the output voltage and the voltage across C_c. In addition to this voltage clamping, the duration of the ringing can be reduced substantially as energy that would otherwise be transferred back and forth between the transformer's leakage inductance and the junction capacitances of the bridge diode is transferred to the output through C_c.

Two important factors that need to be considered in the design of the ZVZCS-PWM-FB converter shown in Figure 9.4a are the counter voltage setting and the rate that C_c discharges. If the counter voltage is too low, it will take time for the primary current to be extinguished during a freewheeling mode of operation and thus conduction losses and RMS current component stresses will be increased. If C_c is discharged too quickly (i.e. the value of C_c is too low), then the freewheeling mode primary current will not be extinguished and current will continue to flow in the converter primary so that the lagging leg switches will not be able to turn on and off with ZCS. If the counter voltage is too high or C_c is too large, then the operation of the converter will be affected as it will take time for C_c to discharge.

In addition to the passive secondary auxiliary circuit for the ZVZCS-PWM-FB converter shown in Figure 9.4a, a number of other secondary auxiliary circuit have been proposed in the literature, some of which are shown in Figure 9.7. These circuits have different counter voltage settings or different auxiliary capacitor discharge rates than the circuit shown in Figure 9.4a.

The auxiliary circuit shown in Figure 9.7a, which was proposed in [4], is an alternative to the passive circuit shown in Figure 9.4. The main drawback of this circuit, however, is that the peak voltage stress on the secondary diodes may be increased as the two auxiliary circuit capacitors are placed in series, but are charged individually. A more sophisticated version of this circuit that exposes the secondary diodes to lower voltage stress than the circuit in Figure 9.4 or Figure 9.7a was proposed in [5] and is shown in Figure 9.7b, but the reduction of diode voltage stress comes at the cost of more components and more conduction losses, especially if the output current is high and the output voltage is low.

For the ZVZCS-PWM-FB converters shown in Figure 9.4, there is little flexibility in how the level of the counter voltage that is needed to drive the primary current down to zero is set. This level is determined by the transformer turns ratio and by the transformer leakage inductance. The turns ratio, however, is set by DC output/input gain considerations, which limits what the counter voltage level should be, and the leakage inductance should be as small as possible to minimize duty cycle loss, which restricts the range of counter voltage level even more. In order to make the setting of the counter voltage level more flexible, several passive auxiliary circuits use auxiliary transformers or coupled inductor arrangements, such as the ones shown in Figure 9.7c,d. For the auxiliary circuit shown in Figure 9.7c, which was proposed in [6], the counter voltage can be set independently of the main transformer as it can be fine-tuned by the selection of the auxiliary transformer turns ratio, thus providing greater flexibility in the design. For the auxiliary circuit shown in Figure 9.7d, which was proposed in [7], the output filter is implemented as a coupled inductor that can help set the counter voltage and the discharge rate of C_c, based on the reflected output current.

Greater control of the counter voltage and the discharge rate of the auxiliary circuit capacitor can be achieved if the secondary auxiliary circuit is active instead of passive. This is especially needed when the converter is operating with higher loads and thus subject to a wider range of load variations. Examples of ZVZCS-PWM-FB converters with active secondary auxiliary circuits are shown in Figure 9.8. Figure 9.8a shows an active auxiliary circuit that was proposed in [8]. This circuit is essentially an active clamp that allows the

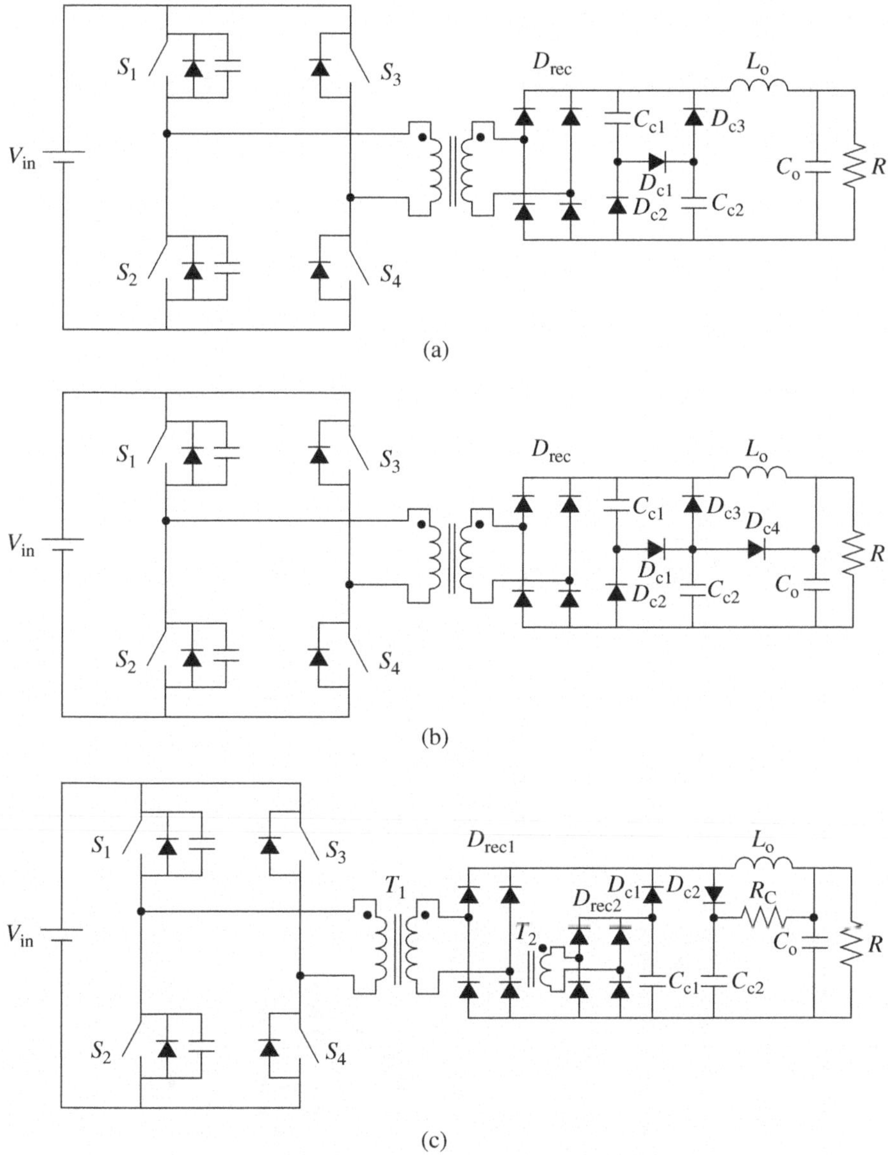

Figure 9.7 Passive auxiliary secondary circuits for ZVZCS-PWM DC–DC FB converters: (a) High voltage clamping circuit. (b) Reduced stress clamping circuit. (c) Auxiliary transformer circuit. (d) Coupled inductor circuit.

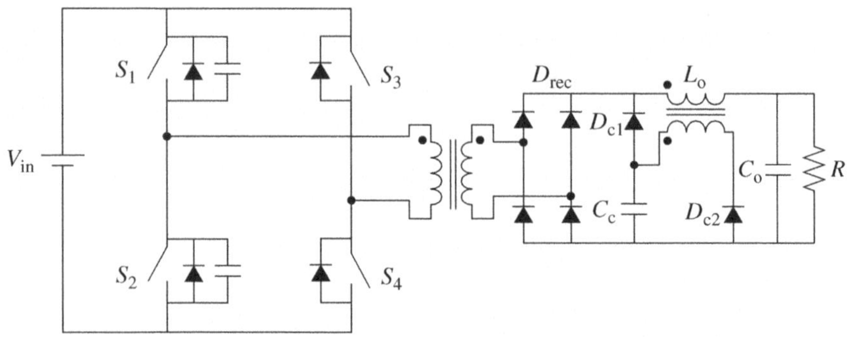

(d)

Figure 9.7 (*Continued*)

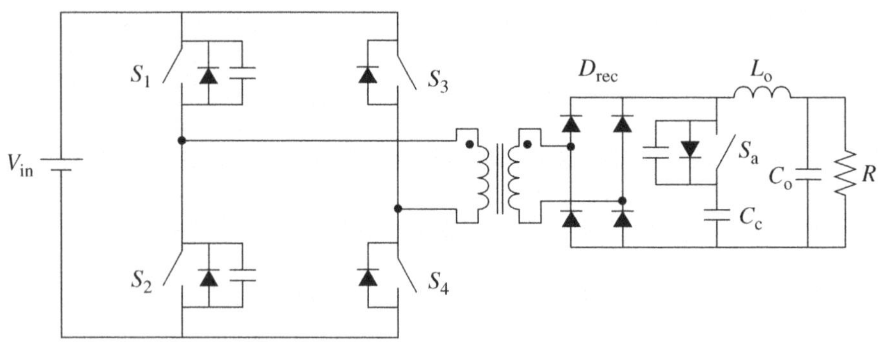

(a)

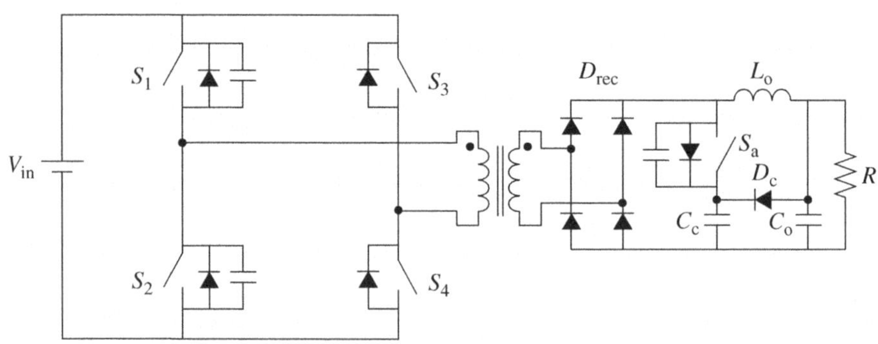

(b)

Figure 9.8 Active auxiliary secondary circuits for ZVZCS-PWM DC–DC FB converters: (a) Simple active clamp circuit. (b) Modified active clamp circuit.

counter voltage and to be controlled by the on-time of the switch – the longer the switch is on, the more the capacitor will discharge and thus the lower the counter voltage will be. Turning off the active clamp switch removes the auxiliary active clamp capacitor from the converter altogether so that the capacitor need not be fully discharged, as is the case for passive auxiliary circuits. In other words, using an active auxiliary circuit allows for the full control of the level of the counter voltage that is needed to extinguish the primary current.

The active auxiliary circuit shown in Figure 9.8b was proposed in [9] and is a modified version of the one shown in Figure 9.8a. The circuit is such that the auxiliary switch is exposed to less peak voltage stress, thus making the overall converter more suitable in applications where the DC output voltage is high. The reduced auxiliary switch voltage stress comes at the price of additional components and the converter becomes less suitable for applications when the output current is high as conduction losses are more of an issue for such applications.

9.3 VARIATIONS OF ZVZCS CONVERTERS FOR FULL ZVS OR FULL ZCS OPERATION

A ZVZCS-PWM-FB converter has half of its switches operating with ZVS and half of its switches operating with ZCS. As stated above, this is a drawback as it requires that two different devices be used in the converter in order to take advantage of the converter's features. Otherwise, half the converter switches are not operating with the soft-switching type that is best suited for them. This is more of an issue if silicon devices are used than if wide bandgap devices are used. It is, however, possible to convert a ZVZCS-PWM-FB converter into a ZVS-PWM or a ZCS-PWM-FB converter by adding some auxiliary circuitry. The main advantage of these converters compared to more traditional ZVS-PWM and ZCS-PWM converters such as the ones discussed in other chapters in this book is that the freewheeling mode circulating current is eliminated, thus resulting in less RMS switch current stress and fewer conduction losses.

9.3.1 ZVS Converters

An example of a ZVZCS-PWM-FB converter that has been modified to be a ZVS-PWM-FB converter is the converter that was proposed in [10] and shown in Figure 9.9. This converter is the same as the ZVZCS-PWM-FB converter shown in Figure 9.4b except that an active auxiliary circuit consisting of switches S_{a1} and S_{a2}, blocking diodes D_{a1} and D_{a2}, inductor L_a, and capacitor C_a is attached to the lagging leg of the converter. The converter operates in the same manner as the ZVZCS-PWM-FB converter in Figure 9.4b, with one notable exception. After the converter has entered a freewheeling mode of

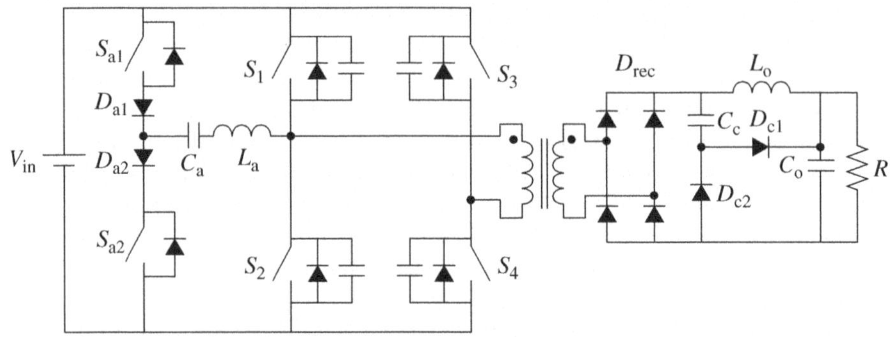

Figure 9.9 ZVS-PWM converter based on ZVZCS-PWM converter.

operation and its primary current has been extinguished by the secondary passive auxiliary circuit, the voltage across each switch of the lagging leg becomes half the input voltage. Just before one of these switches is turned on to force the converter into an energy-transfer mode, the auxiliary switch that is associated with this switch is turned on first to discharge the output capacitance of that switch. For example, if switch S_2 is to be turned on, auxiliary switch S_{a2} is turned on first, thus discharging the output capacitance of S_2 so that its voltage falls to zero; S_2 can then be turned on with ZVS. The current in L_a is eventually extinguished by the voltage across capacitor C_a, which acts as a counter voltage to drive the current to zero.

9.3.2 ZCS Converters

An example of a ZVZCS-PWM converter that has been modified to be a ZCS-PWM converter is the converter that was proposed in [11] and shown in Figure 9.10. In this converter, it is the leading leg of switches that has an additional auxiliary circuit added to it, not the lagging leg switches as these can be turned off with ZCS naturally. The purpose of the auxiliary circuit is to divert current away from a leading leg switch that is on so that it can be turned off with ZCS. Auxiliary switch S_a is turned on whenever S_1 or S_2, is about to be turned off. Regardless of which of the two switches in the leg, S_1 or S_2, is about to be turned off, capacitor C_a has a negative voltage across it from previous modes of operation. When the bottom switch S_2 is about to be turned off, S_a is turned on and current begins to transfer away from S_1 to S_a. S_2 can then be turned off with ZCS after the current transfer has been completed. When the top switch S_1 is about to be turned off, S_a is turned on to initiate a resonant cycle that will ultimately inject current through the body diode of S_1. S_1 can be then turned off with ZCS while current is flowing through its body diode. Given that all four converter switches can be turned off with ZCS, the converter is a ZCS-PWM-FB converter.

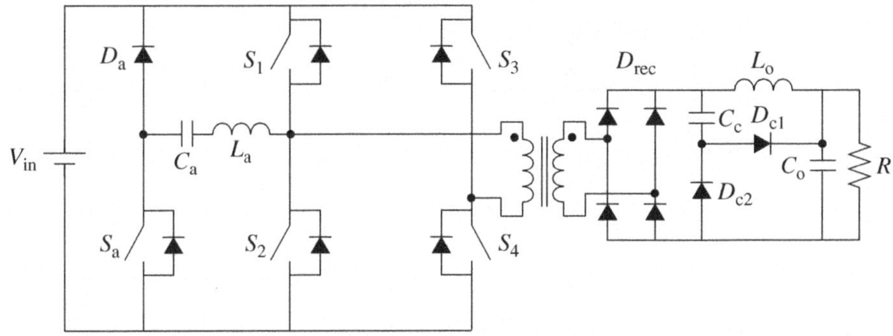

Figure 9.10 ZCS-PWM converter based on ZVZCS-PWM converter.

9.3.3 ZVS-PWM Converters Based on ZVZCS-PWM Converters with Triangular Primary Current Waveform

A different approach to implementing ZVZCS-PWM converters than what has been discussed in this chapter thus far is the full-bridge converter that was proposed as part of a plug-in AC-DC battery charger for electric vehicles in [12] and is shown in Figure 9.11a, along with the gating signals of its switches and the transformer primary current in Figure 9.11b. This converter is essentially a ZVS-PWM full-bridge with passive auxiliary circuits to help the main converter switches operate with ZVS, as was explained in Section 8.3. Capacitor C_b is placed in series with the transformer to ensure that the voltage across the transformer primary is purely AC with no DC voltage component that would saturate the transformer. The output is a battery with an output inductor L_o that acts as a filter.

The most important difference between the converter shown in Figure 9.11a and a standard ZVS-PWM full-bridge converter is the placement of output capacitor C_o, which is placed across the output of the secondary diode bridge, before the output inductor instead of after it. Unlike the passive auxiliary circuit capacitor C_c in the ZVZCS-PWM-FB converter shown in Figure 9.4, this capacitor is never fully discharged, but provides a stiff voltage across the output of the secondary diode bridge. Whenever the converter enters an energy-transfer mode with two diagonally opposed switches on, the lagging leg switch that is turned on does so with ZCS as current in the primary has been extinguished during previous modes of operation. Current in the primary rises and is the result of the difference between the primary voltage and the reflected voltage from the secondary across the leakage inductor of the transformer. Whenever the converter enters a freewheeling mode of operation by the turning on of a leading leg switch, the voltage across

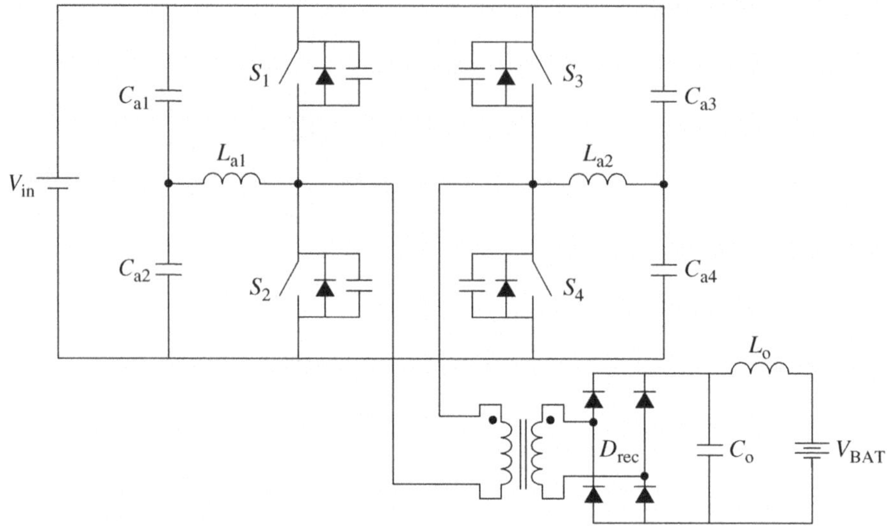

(a)

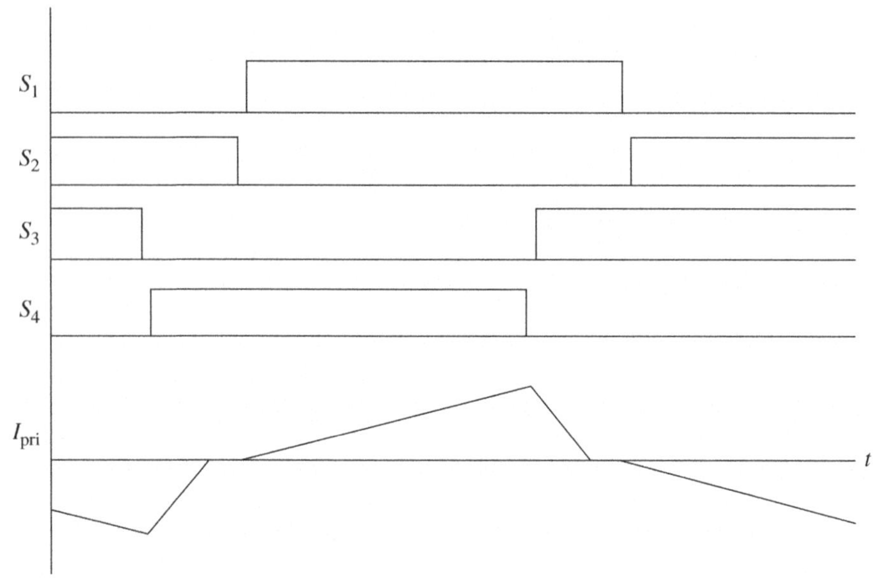

(b)

Figure 9.11 ZVS-PWM converter based on ZVZCS-PWM converters with triangular primary current: (a) Converter topology. (b) Gating signals and transformer primary current waveform.

capacitor C_o is reflected to the primary and acts as a counter voltage to extinguish the primary current. It should be noted that the presence of C_o across the output of the secondary diode bridge forces the primary current to be triangular, as shown in Figure 9.11b. This causes the current to have high triangular peak that increase the peak current stress of the converter switches. If, however, the input voltage is high (e.g. >400 Vdc), then the primary current will be such that the high peak current in the primary becomes less of a concern.

This converter is a ZVS-PWM-FB converter because, like the ZVZCS-PWM-based ZVS-PWM-FB converters discussed in the previous section, auxiliary circuits have been added to the converter. In this converter, passive auxiliary circuits consisting of an inductor and two capacitors have been added to both the leading and the lagging legs of the converter. The addition of the auxiliary circuit to the lagging leg ensures that its switches can be turned on with ZVS. In the case of the leading leg, even though its switches can already operate with ZVS, the auxiliary circuit ensures that it can do so even at very light loads. The converter shown in Figure 9.11a was designed to be used in electric vehicles, which is an especially sensitive application where losing ZVS at no-load can create noise in the control and gate drive circuit due to ringing, which, in turn, may cause shoot-through and the ultimate failure of the power MOSFETs used.

The efficiency of the converter shown in Figure 9.11a can be improved if the amount of current that is generated by the passive auxiliary circuits in the primary is reduced. Although this current is less that the freewheeling current that would be typically generated in a standard ZVS-PWM-FB converter, it does contribute to conduction losses and thus any reduction of this current would be beneficial. A modified version of the converter shown in Figure 9.11a that does just that is the one that was proposed in [13] and shown in Figure 9.12a. The topology is slightly different that the one in Figure 9.11a as the two auxiliary circuits in Figure 9.11a have been combined into one with just one split capacitor arrangement used instead of two and the two inductors combined into a single coupled inductor, with a small inductance added for fine-tuning. The main difference in the two converters, however, is in the way that they are operated. The converter shown in Figure 9.12a was operated with an asymmetrical PWM method (APWM), as shown in Figure 9.12b instead of conventional phase-shift PWM. With APWM, the auxiliary circuit current is no longer fixed, regardless of load, but can be made adaptable to suit the operating conditions of the converter. For example, if the top switches operate with $D = 0.2$, then less current is generated in the auxiliary circuit inductors than if $D = 0.5$. The converter can be designed so that just enough current is available to discharge switch output capacitances when operating under light-load conditions, given that the top two switches are made to be on for a short amount of time instead of $D = 0.5$. Doing so, however, means that the mid-point voltage of the split capacitor arrangement is no longer half the input DC, which means that the capacitors must be rated with different voltages.

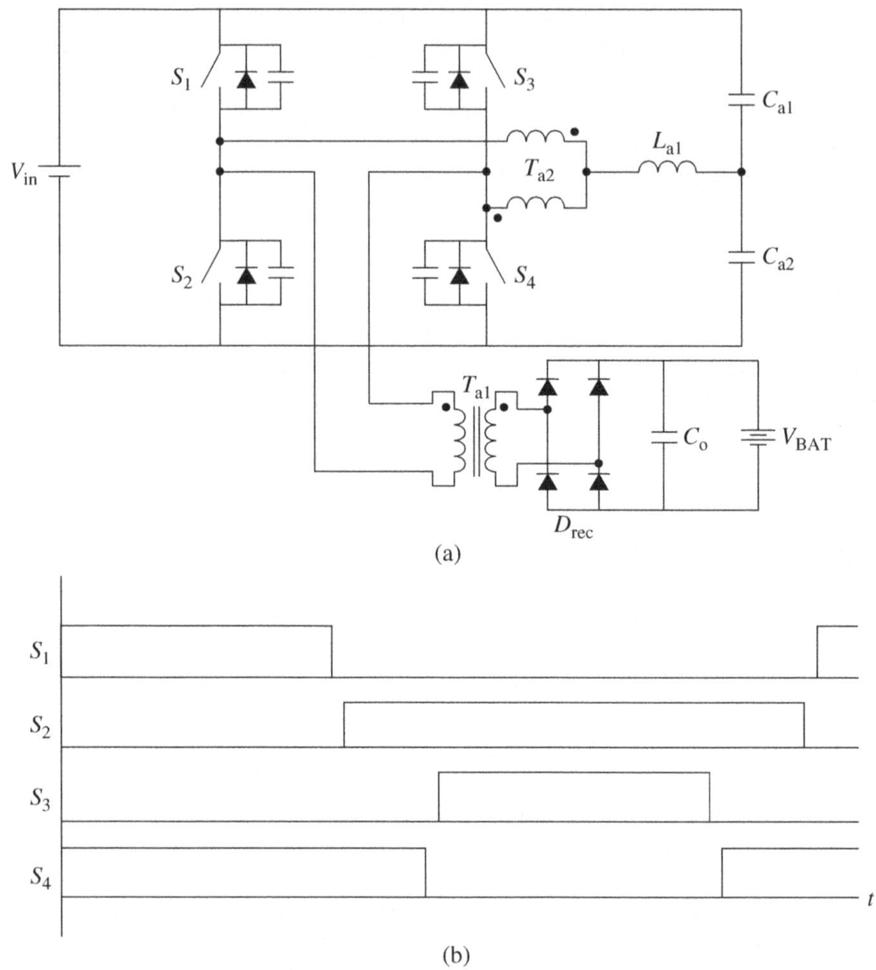

Figure 9.12 Modified ZVS-PWM converter based on ZVZCS-PWM converters with triangular primary current: (a) Converter topology. (b) Converter gating signals based on APWM.

9.4 CONCLUSION

ZVZCS-PWM-FB converters have been proposed as a way to overcome a fundamental drawback of the conventional ZVS-PWM-FB converter topology – the presence of circulating current in the primary side of the converter whenever it is in a freewheeling mode of operation. When the converter is in a freewheeling mode of operation, no power is transferred from the input DC source to the output of the converter and the circulating current just increases conduction losses and RMS current component stresses. ZVZCS-PWM-FB converters can extinguish this current and thus reduce conduction losses; they can reduce turn-off losses and RMS current stresses as well. ZVZCS-PWM-FB converters are especially attractive in higher power converters where these power losses are significant.

The fundamental principle behind ZVZCS-PWM-FB converters is that some sort of counter voltage that can drive freewheeling mode circulating current down to zero is established in the primary side of the converter. One of the first ZVZCS-PWM converters to have been proposed, which was referred to as the fundamental ZVZCS-PWM-FB converter in this chapter, has a capacitor and a saturable reactor in series with its transformer. The voltage across the capacitor is the counter voltage that drives the freewheeling mode circulating current to zero and the saturable inductor keeps current from reversing direction after it has reached zero. If blocking diodes are used in series with the converter's lagging leg switches, then the saturable inductor, which is a source of power losses, can be removed from the converter and the converter can be operated at higher power levels.

The next generation of ZVZCS-PWM converters implemented the counter voltage mechanism at the secondary side rather than the primary side of the converter in the form of a clamp. The advantage of having a voltage clamp at the converter secondary, placed across the output of the secondary diode bridge, is that it can clamp any voltage oscillations caused by the resonant interaction between the leakage inductance of the transformer and the junction capacitance that may exceed the peak voltage stress of the diodes. Passive auxiliary circuit clamps are simple to implement and effective and examples of such clamps were presented in this chapter. The variations in these clamps are related to the setting of the counter voltage level, which impacts the operation of the main converter. Further control of the counter voltage level can be achieved by using active clamps that contain an active switch.

A drawback of ZVZCS-PWM converters, especially when they are implemented with silicon devices, is that half their switches operate with ZVS and half operate with ZCS. This means that these converters must either be implemented with two different sets of switching devices, which increases cost, or, if they are implemented with only MOSFETS or only IGBTs, half the devices do not operate with the optimal type of soft-switching. ZVZCS-PWM-FB converters can be converted to either ZVS-PWM-FB converters or ZCS-PWM-FB converters by adding auxiliary circuits to the converter legs. Several examples of such circuits were presented in this chapter.

REFERENCES

1. J. -G. Cho, J. A. Sabate, G. Hua, and F. C. Lee, "Zero-voltage and zero-current-switching full bridge PWM converter for high-power applications," *IEEE Transactions on Power Electronics*, vol. 11, no. 4, pp. 622–628, Jul. 1996.
2. X. Ruan and Y. Yan, "A novel zero-voltage and zero-current-switching PWM full-bridge converter using two diodes in series with the lagging leg," *IEEE Transactions on Industrial Electronics*, vol. 48, no. 4, pp. 777–785, Aug. 2001.

3. J. -G. Cho, J. -W. Baek, C. -Y. Jeong, and G. -H. Rim, "Novel zero-voltage and zero-current-switching full-bridge PWM converter using a simple auxiliary circuit," *IEEE Transactions on Industry Applications*, vol. 35, no. 1, pp. 15–20, Jan.–Feb. 1999.
4. E. -S. Kim, K. -Y. Joe, M. -H. Kye, Y. -H. Kim, and B. -D. Yoon, "An improved soft-switching PWM FB DC/DC converter for reducing conduction losses," *IEEE Transactions on Power Electronics*, vol. 14, no. 2, pp. 258–264, Mar. 1999.
5. E. -S. Kim and Y. -H. Kim, "A ZVZCS PWM FB DC/DC converter using a modified energy-recovery snubber," *IEEE Transactions on Industrial Electronics*, vol. 49, no. 5, pp. 1120–1127, Oct. 2002.
6. J. -G. Cho, J. -W. Baek, C. -Y. Jeong, D. -W. Yoo, and K. -Y. Joe, "Novel zero-voltage and zero-current-switching full bridge PWM converter using transformer auxiliary winding," *IEEE Transactions on Power Electronics*, vol. 15, no. 2, pp. 250–257, Mar. 2000.
7. H. -S. Choi, J. -W. Kim, and B. H. Cho, "Novel zero-voltage and zero-current-switching (ZVZCS) full-bridge PWM converter using coupled output inductor," *IEEE Transactions on Power Electronics*, vol. 17, no. 5, pp. 641–648, Sept. 2002.
8. J. G. Cho, G. H. Rim, and F. C. Lee, "Zero voltage and zero current switching full bridge PWM converter using secondary active clamp," *IEEE Power Electronics Specialists Conference (PESC)*, 1996, June 23–27, 1996, Baveno, Italy, pp. 657–663.
9. T. Chen and S. Cheng, "A novel zero-voltage zero-current switching full-bridge PWM converter using improved secondary active clamp," *IEEE International Symposium on Industrial Electronics (ISIE)*, 2006, July 9–13, 2006, Montreal, Canada, pp. 1683–1687.
10. D. S. Wijeratne and G. Moschopoulos, "A ZVS-PWM full-bridge converter with reduced conduction losses," *IEEE Transactions on Power Electronics*, vol. 29, no. 7, pp. 3501–3513, Jul. 2014.
11. A. Mousavi and G. Moschopoulos, "A new ZCS-PWM full-bridge DC–DC converter with simple auxiliary circuits," *IEEE Transactions on Power Electronics*, vol. 29, no. 3, pp. 1321–1330, Mar. 2014.
12. M. Pahlevaninezhad, P. Das, J. Drobnik, P. K. Jain and A. Bakhshai, "A novel ZVZCS full-bridge DC/DC converter used for electric vehicles," *IEEE Transactions on Power Electronics*, vol. 27, no. 6, pp. 2752–2769, Jun. 2012.
13. V. R. K. Kanamarlapudi, B. Wang, P. L. So, and Z. Wang, "Analysis, design, and implementation of an APWM ZVZCS full-bridge DC–DC converter for battery charging in electric vehicles," *IEEE Transactions on Power Electronics*, vol. 32, no. 8, pp. 6145–6160, Aug. 2017.

CHAPTER 10

ISOLATED CURRENT-FED DC–DC PWM CONVERTERS

The isolated DC–DC converters that have been covered in previous chapters of this book have been mostly step-down converters that can step down an input DC voltage to a desired DC voltage that is much lower. These isolated converters can step down voltage significantly as their topologies as a combination of a step-down converter and a step-down transformer with more primary windings than secondary windings. Similarly, isolated DC–DC step-up converters are needed in applications where an input DC bus voltage must be significantly stepped up in order to satisfy the demands of particular applications such as

- Renewable energy systems that are fed by low-voltage DC sources such as solar panels and fuel cells. For example, low-voltage solar panels that feed grid-connected solar energy systems typically produce output voltages in the range of about 30–40 V. This voltage is stepped up, then fed to grid-connected inverters that convert the high-level DC bus voltage into an AC voltage that can be fed into a utility grid.
- Energy systems with batteries that have low voltages (12–48 V). For example, in some automotive power systems schemes, the DC voltage produced by a battery needs to be stepped up to much higher DC voltage that is then fed to DC–AC inverters to power auxiliary features.
- Electrostatic chargers, which are used to build up charge on the surface of objects, produce very high voltages (>1 kV). Examples of where

DC–DC Converter Topologies: Basic to Advanced, First Edition. Gerry Moschopoulos.

Published 2024 by John Wiley & Sons, Inc.

electrostatic chargers are used include photocopier and laser printer operation and powder coating – a popular electrostatic process that is widely used to coat metals, such as household appliances, aluminum extrusions, and automobile and bicycle parts.
- Power supplies that must provide a very high voltage to supply medical equipment such as X-ray machines. X-ray machines require very high voltages as part of the X-ray process.

For these applications, a conventional PWM boost DC–DC converter would have to be operated with extremely high duty-cycles that would result in energy transferred from input to output during a small fraction of the switching cycle. Such operation would be very inefficient and would place a tremendous amount of stress on the converter devices. Other possible approaches include the following:

- An isolated converter such as the forward or flyback converter that is typically used as a step-down converter can be designed to step up voltage instead, by making the secondary winding of the converter transformer larger than the primary winding. If the input voltage must be stepped up significantly, operating a converter with a transformer that has a large turns ratio exacerbates the transformer's non-idealities – especially the leakage inductance and the winding capacitance (a winding placed in proximity to another winding results in capacitance and the two windings have charge and are separated by nonconducting material), which can significantly change converter behavior. The transformer's leakage inductance may cause undesirable voltage spikes that can damage converter components, and its winding capacitance may result in current spikes and slow rise times. These nonidealities can lead to greatly increased switching and snubber losses and component voltage stress.
- Two converter stages – a DC–DC boost converter followed by an isolated DC–DC converter – can be used to achieve significant voltage boost. The front-end boost converter can be used to step up the input DC voltage to an intermediate voltage and then the isolated DC–DC converter can be used to step up the intermediate voltage so that the transformer's step-up turns ratio does not have to be very high, thus limiting the above-mentioned issues of leakage inductance and winding capacitance. The drawbacks to this approach are that it requires two DC–DC converters instead of one and it is inefficient as power must be processed by two converters with two sets of losses instead of just one converter.
- A single converter with a topology that is a combination of a step-up converter and an isolated converter can be used. Such a topology combines two voltage step-up mechanisms – that of a voltage step-up converter

and that of a voltage step-up transformer – to achieve high voltage step-up capability. This approach is a popular approach to implementing isolated step-up converters with very high gains as the drawback of the other two approaches are avoided. As the input voltage source is interfaced to the converter with a large inductor that can be considered to be a current source, such a step-up converter is commonly referred to as a current-fed converter in the literature. Isolated current-fed DC–DC converters are the focus of this chapter.

10.1 BASIC CURRENT-FED PUSH–PULL CONVERTER

The most basic isolated current-fed DC–DC converter can be considered to be the current-fed push–pull converter that is shown in Figure 10.1. This converter was first proposed by Clarke in [1], then analyzed in detail by Redl and Sokal in [2, 3]. The converter shown in Figure 10.1 is very similar to the push–pull converter that was presented in Section 4.5.2, but with some fundamental differences:

- A large input inductor is placed between the input DC voltage source and the rest of the converter so that its input matches that of a boost converter.
- There is no output inductor so that its output filter is just a single capacitor, just like a boost converter.
- The input inductor is implemented with a secondary winding that acts as a clamping winding. This winding allows energy stored in the input inductor to be removed whenever both switches are turned off. This mechanism is like the reset mechanism of a flyback converter and allows the converter switches to operate with duty-cycles that are less than 0.5. So that voltage can be stepped down as well as stepped up if desired.

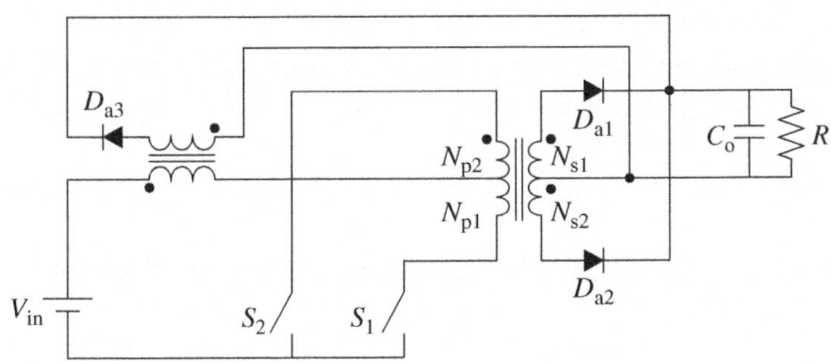

Figure 10.1 Basic current-fed push–pull converter.

Without the secondary winding in the input inductor, there would be no path for inductor current to flow through if both switches are off and energy has been placed in the inductor, which would result in a catastrophic failure of the converter.

- The converter is much less prone to saturation than a conventional push–pull converter as the output capacitor voltage can appear across the transformer at the secondary and be reflected to the primary. There is an inherent reset mechanism to the converter than does not exist in the conventional push–pull converter.

The converter operates as follows: At any given time during a switching-cycle: switch S_1 is on, switch S_2 is on, or both switches are on. Energy is transferred from the primary side of the converter to the secondary side whenever a single switch is on. For example, if S_1 is on, current flows from the source, through the transformer and the switch, then back to the source. While this is happening, secondary diode D_{a1} is conducting so that current can flow to the output. When both S_1 and S_2 are on, current flows through both switches and through both primary transformer windings. Since the polarity of these windings are opposed to each other, the next voltage that is impressed across the transformer's primary is zero, so that no secondary voltage is produced and thus the secondary diode is reverse-biased and does not conduct. When the switches operate with a duty-cycle of $D > 0.5$, the primary section of the converter acts like a boost converter; when $D < 0.5$, the converter acts like a buck converter with input inductor energy transferred directly to the output when both switches are off, as described above.

The main drawbacks of the converter are as follows:

- The converter switches have high peak voltage stress as two times the voltage reflected from the secondary side can appear across a switch if its counterpart switch is turned off.
- Although the gain of the converter can be made high, step-up operation is limited to converter operation with $D > 0.5$. When $D < 0.5$, the converter can actually step-down voltage as energy can be transferred to the output through the secondary diodes.
- A center-tapped transformer must be used and only one half of the transformer primary winding is used at a time. This increases transformer size and cost.

10.2 BASIC TWO-INDUCTOR CURRENT-FED CONVERTER

The two-inductor current-fed converter shown in Figure 10.2, proposed in [4], does not have the drawbacks of the current-fed push–pull converter. The converter consists of two inductors L_1 and L_2, two switches, S_1 and S_2,

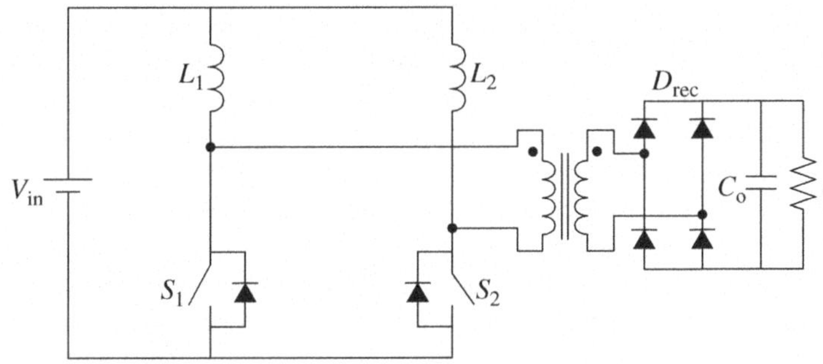

Figure 10.2 Basic two-inductor current-fed converter.

a transformer and a secondary that is the same as the current-fed push–pull converter shown in Figure 10.1. Inductors L_1 and L_2 and switches S_1 and S_2 are arranged in a two-legged H-bridge structure with L_1 in series with S_1, L_2 in series with S_2, and a transformer connected between the midpoints of the two primary-side legs. Like the current-fed push–pull converter, the secondary of the converter can be implemented with a four-diode bridge as shown in Figure 10.2 or with a center-tapped secondary.

The converter operates as follows: The current that flows out of the input DC source is shared equally between the two inductors. If only switch S_1 is closed, then current in L_1 flows through S_1 and current in L_2 flows through the transformer primary and S_1. The current in L_1 rises as the input source voltage is impressed across it while the current in L_2 falls as the transformer primary voltage is greater than the source voltage so that a negative voltage is impressed across L_2. Similarly, if only switch S_2 is closed, then current in L_2 flows through S_2 and current in L_1 flows through the transformer primary, in the opposite direction than when only S_1 is on, and S_2. The current in L_2 rises as the input source voltage is impressed across it while the current in L_1 falls as the transformer primary voltage is greater than the source voltage so that a negative voltage is impressed across L_1. When both switches are on, then the full input voltage is impressed across both L_1 and L_2, the current in both inductors rises, and no voltage is impressed across the transformer primary. It should be noted that the converter's two-inductor structure allows the two inductor currents to be interleaved so that the input current has less ripple than the input current of a current-fed push–pull converter, which has only one input inductor.

The converter is in an energy-transfer mode of operation whenever only one of its two switches is on as current is forced to flow through the transformer primary. The converter is in a short-circuit mode of operation when both converter switches are on and the transformer primary is bypassed altogether. These two modes correspond to the energy-transfer mode of a boost converter when its main boost switch is off and current flows through the boost diode and to the short-circuit/non-energy-transfer mode when the boost switch is on and no current flows through the boost diode and the output. The gating signals of the two converter switches must be such that the converter is never allowed to operate with both switches off. There must always be a path for current to flow in the converter at any instant of the switching, and this can be ensured by having the two gating signals overlap. The amount of voltage step can be increased by increasing the amount of overlap between the two switches.

It should be noted that in Figure 10.2, the converter's secondary is implemented as a four-diode bridge instead of a center-tapped secondary with two output diodes. This implementation is typical when the output voltage is being stepped up to a high voltage level and the output current is low as the peak voltage stress of a diode in the center-tapped configuration is twice the output voltage; this stress takes precedence over conduction losses, which are low. If the converter is used to provide an output voltage that is not as high, then the center-tapped configuration can be used.

The converter shown in Figure 10.2 must be implemented with some sort of snubber that can absorb energy whenever there is a difference between the current flowing in the inductors and the current flowing through the transformer primary. For example, if S_1 is on and it conducts the current flowing through L_1 and the current flowing through L_2 and the transformer primary, then S_2 can be turned on to divert current away from the transformer primary and exit the energy-transfer mode, and S_1 can be turned off sometime afterward. When S_1 is turned off, however, the current in L_1 cannot instantaneously flow through the transformer primary due to the transformer's leakage inductance, which limits the rise in transformer current. There must be some sort of snubber that will allow current to flow through it while current in L_1 transitions to the primary of the transformer. This is true regardless of whether traditional silicon devices or newer wide bandgap devices are used in the converter.

Simple RCD snubbers can be used in the two-switch current-fed converter and, if desired, these snubbers can be supplemented with clamp windings in the two inductors much like the clamp winding of the input inductor of the current-fed push–pull converter. The clamp windings allow energy to be released to the output whenever a switch is off, thus alleviating some of the stress that would otherwise be placed on the RCD snubbers. A popular method of implementing snubbers in the two-switching current-fed converters is to use dual active clamps, as proposed in [5] and shown in Figure 10.3. Although this is a more expensive approach as two additional switches are needed on top of the two main switches, it does have the benefit of allowing the converter to operate with ZVS.

The converter shown in Figure 10.3 operates as follows: If it is assumed that both switches are on, then current in L_1 begins to flow through the body diode of active clamp switch S_{a1} when S_1 is turned off. Sometime later in the switching cycle, before S_1 is to be turned on again, S_{a1} is turned on so that current flowing through the transformer primary becomes the sum of the current

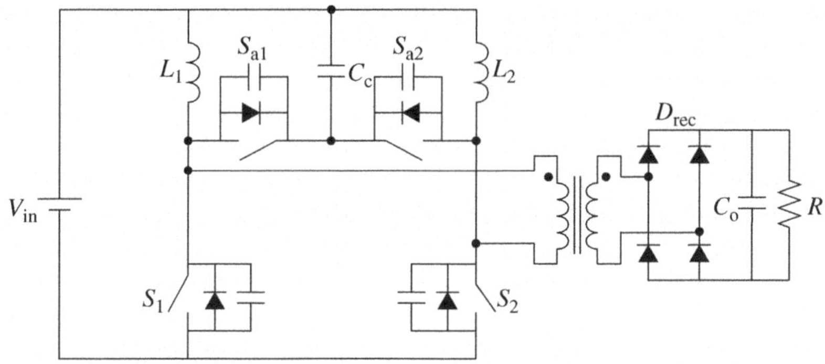

Figure 10.3 Two-inductor current-fed converter with dual active clamps.

through L_1 and the current through S_{a1}. When S_{a1} is turned off, a mismatch between the current in L_1 and the current in the transformer primary current occurs and this mismatch must be accounted for. This is done with the flow of current through the body-diode of S_1 so that the transformer primary current is the sum of that current and the current through L_1. S_1 can be turned on with ZVS while current is flowing through its body-diode. After S_1 is turned on, it eventually conducts current from L_1 and current from L_2 and the transformer primary. S_2 can be turned on with ZVS if the active clamp circuit with S_{a2} is operated in a similar manner.

Compared to the current-fed push–pull converter, the voltage stress of each switch in the two-inductor, two-switch circuit shown in Figure 10.2 is half the voltage stress of the switches of the current-fed push–pull converter. Moreover, the input current can be evenly distributed among the two inductors so that the current ripple in the output capacitor is smaller than that of the current-fed push–pull converter, which has only one input inductor. The main drawback is that additional sensing is needed to ensure this even distribution of current, which increases cost and complexity.

10.3 MODIFIED TWO-INDUCTOR CURRENT-FED CONVERTER WITH AUXILIARY TRANSFORMER

Although the two-inductor current-fed converter is an improvement over the current-fed push–pull converter, it does, however, have a major limitation and that is its inability to regulate the load over a wide range when operated with constant-frequency control. This limitation can be explained as follows: The current in inductor L_1 increases during the entire on-time of switch S_1 and decreases during the entire off-time of switch S_1. Similarly, the current in inductor L_2 increases during the on-time of switch S_2 and decreases during its off-time. As a result, even when the effective converter duty-cycle D, which is defined as the ratio of the overlapping conduction time of the two switches to half of their switching period, is reduced to zero, the inductors continue to store energy since S_1 and S_2 are conducting for half the switching cycle. In other words, current will rise in one of the inductors even though the converter is in an energy-transfer mode. Since at least one of the two switches must be on at any given instant of time, the magnetizing currents of the two inductors cannot be limited, which results in a minimum output power level that the converter must operate with. If the load demands less power than this minimum level, the output voltage increases abnormally because excessive energy has been stored in the inductors. It is possible to extend the load regulation range by increasing the switching frequency to shorten the conduction time of the switches, thus reducing the amount of energy stored in the inductors. The converter in Figure 10.3 can operate with output voltage regulation over a wide load range, but only if variable-frequency control is used. This is disadvantageous in terms of magnetic component design as the converter's magnetics cannot be optimized for a single switching frequency, but must be designed to operate for a wide range of switching frequencies.

The two-inductor current-fed converter can be made to operate with output-voltage regulation from full load to no load over a wide input-voltage range using constant-frequency control if an auxiliary transformer is added to the converter as was proposed in [6] and shown in Figure 10.4. This auxiliary transformer has a unity turns ratio to couple the current paths of the two boost inductors so that both inductors conduct the same current; this allows both inductors to charge and discharge simultaneously. Another way to think about this mechanism is that if just one switch is turned on, the auxiliary winding allows for a counter voltage to appear in series with the inductor that is in the same leg with this switch so that there is a negative voltage across the inductor and its current does not rise. For example, if only S_2 is on, the current

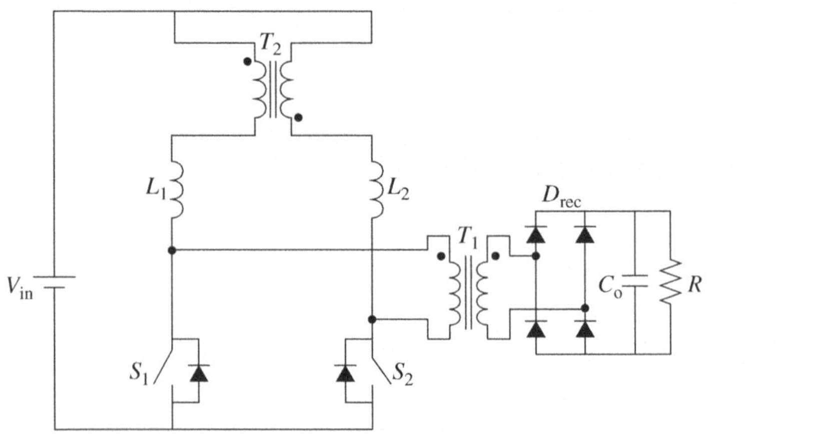

Figure 10.4 Modified two-inductor current-fed converter with auxiliary transformer (snubbers not shown).

in L_2 will not rise as the auxiliary transformer ensures that the voltage across L_2 is negative. It is only when both S_1 and S_2 are on that the current in the inductors can increase.

If the duty-cycle is close to unity, the maximum power is transferred from the input to the output since the maximum energy is stored in the inductors. As the duty-cycle decreases toward zero, less and less energy is stored in both inductors, which enables the output voltage regulation down to very light loads. If the overlapping between two driving signals is small, then the inductor currents become discontinuous. No energy is stored in the inductors when there is no overlapping of conduction times of the two switches. Since the output voltage can be controlled by changing the duty-cycle of the converter, the presence of the auxiliary transformer in the converter allows fixed-frequency control to be used. It should be noted that if the two inductors have the same value and are coupled with a turns ratio of one, then both inductors store and transfer the same amount of energy so that each converter leg processes half the total power. Since the total power is processed in two parallel legs, the conduction loss of the converter is less that for a current-fed push–pull converter, which has just a single power path.

Although the auxiliary transformer helps improve the performance of the two-inductor current-fed DC–DC converter, it also increases cost, size, and weight. It is, however, possible to integrate the main transformer and the auxiliary transformer into one magnetic element as proposed in [7] and shown in Figure 10.5. Given that an explanation of how the transformer core can be designed to accommodate the necessary windings would involve magnetic design concepts that are outside the scope of this book, the reader is referred to [7] for a detailed explanation about the design of the transformer. It should

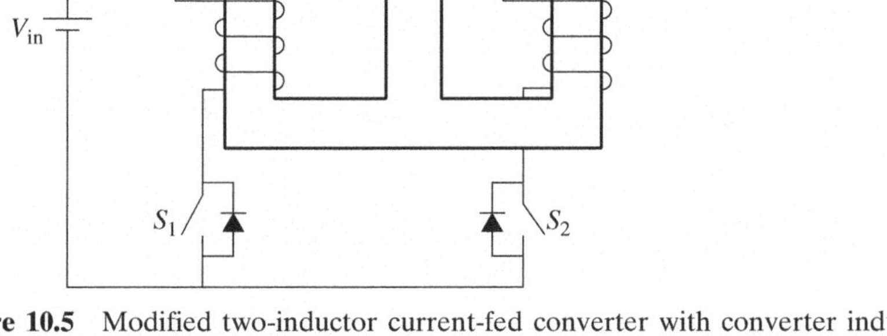

Figure 10.5 Modified two-inductor current-fed converter with converter inductors and auxiliary transformer integrated on one magnetic core (snubbers not shown).

be noted that integrated magnetics are generally difficult for mass production and that expertise in magnetic design is needed to design the transformer for this converter.

10.4 BASIC CURRENT-FED FULL-BRIDGE TOPOLOGY

Four-switch current-fed full-bridge topologies are popular in applications where a converter that can operate with higher power and higher gain than the two-inductor current-fed converter is needed. The basic current-fed full-bridge converter, which is shown in Figure 10.6, operates as follows: During a typical switching cycle, the input current falls and the converter is in an energy transfer mode when only a pair of diagonally opposed switches is on; energy from the primary is transferred to the output through the transformer and the secondary diodes. The input current rises, but no energy is transferred to the output, when all switches are on and the converter is in a boosting mode. The sequence of gating signals that the converter operates within a typical switching cycle is as follows: S_1 and S_4 on, all bridge switches on, then S_2 and S_3 on and again all switches on. In other words, an energy transfer mode when only a pair of diagonally opposed switches is on is always followed by a boosting mode where all the switches are on and no energy is transferred. Operating the converter in this manner mimics the operation of a standard single-switch PWM boost converter where the turning on of the boost switch is analogous to the shorting of the DC bus and the flow of current through the boost diode to the output is analogous to an energy-transfer mode.

It should be noted that the converter can be made to enter a short-circuit mode by having only the two switches of a leg on instead of having all converter switches on. For example, the switching sequence can be made to be S_1 and S_4 on, then S_1 and S_2 on, then S_2 and S_3 on, then S_2 and S_4 on until the start of the next switching cycle when S_1 and S_4 are both on again. Whether the converter is made to operate in short-circuit modes with all the switches on

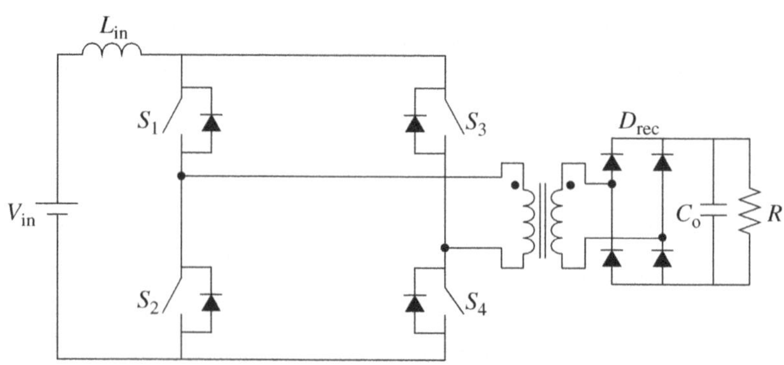

Figure 10.6 Basic current-fed full-bridge converter (snubbers not shown).

or with just the two switches of the same leg is dependent on the operating conditions of the converter and the devices that it is implemented with. There is no clear-cut rule as to which short-circuit mode is advantageous.

In a real-life application, the current-fed converter shown in Figure 10.6 must be modified in some way to account for overvoltage spikes that may occur due to the interaction of the transformer leakage inductance and the output capacitance of the switches. For example, if a transition from a short-circuit mode where all four switches are on to an energy transfer mode where only switch S_1 and S_4 are on is to be made, the switches S_2 and S_3 must be turned off, but doing so would try force current through the transformer so that it could flow through switch S_1 and S_4. Since the transformer has leakage inductance, the transfer of current to the transformer would not be instantaneous so that current would begin to flow through the output capacitances of switch S_2 and S_3. At first, the full input current is split between the output capacitances of these switches; gradually current flowing through the transformer rises and current flowing through the output capacitances of S_2 and S_3 decreases until, eventually, the full input current flows through switches S_1 and S_4, and the transformer.

When a transition from an energy-transfer mode to a short-circuit boosting mode is made, the switches that are turned on to enter the mode may be turned on with ZCS if the leakage inductance of the transformer is sufficiently large. Just as the leakage inductance makes the current transfer from diagonally opposed switches to the transformer gradual when these switches are turned off, so too does the leakage inductance make the current transfer from the transformer to diagonally opposed switches when they are turned on. For example, if the converter is in an energy-transfer mode with switches S_1 and S_4 on and switches S_2 and S_3 are turned on to force the converter into a short-circuit mode, then S_2 and S_3 will be turned on with ZCS as the transformer leakage inductance prevents current from flowing through these switches instantaneously. Such a characteristic can be taken advantage of it ZCS operation of a current-fed full-bridge converter is desired, but the transformer leakage inductance in the transformer should not be made too large, to avoid duty-cycle loss, which would limit the operating range of the converter.

Several actions can be taken in order to reduce the presence of overvoltage spikes across the switches of the current-fed full-bridge converter when they are turned off. These include:

- *Increasing the output capacitances of the switches either by using devices with large output capacitance or by adding additional external capacitance across each converter switch*: Doing so reduces the rate of voltage rise across a switch when it is turned off as more leakage inductance energy (1/2 $L_{lk}I^2$) can be absorbed, which allows any voltage spiking that may occur to be kept to an acceptable level. This approach is only feasible when the primary current is low; otherwise, the amount of output capacitance that would be needed would result in a significant increase in turn-on losses as the energy stored in each output capacitance would be dumped into its associated switch and dissipated when that switch is turned on.

- *Using passive snubbers to provide an alternate path where current can flow through after switches are turned off:* This approach is inexpensive and robust; it is least prone to failure over a wide operating range. Dissipative snubbers are cheap, but lossy; non-dissipative snubbers are more efficient, but more costly and can result in unwanted circulating current flowing in the converter.
- *Using active auxiliary circuit snubbers to divert current away from the main switches before they are turned off or to discharge output capacitances used to limit overvoltage spikes across individual switches before the switches are turned on:* This approach can improve converter efficiency due to ZVS or ZCS, but is the most expensive approach and may not be effective for converters that need to operate over a very wide load range.

The ZVS and ZCS operation of PWM current-fed full-bridge converters are examined in the next sections of this chapter.

10.5 CURRENT-FED DC–DC FULL-BRIDGE CONVERTERS WITH BLOCKING DIODES

Current-fed full-bridge converters can be implemented with blocking diodes placed in series with their switches to prevent their body-diode from conducting or be implemented without these diodes. Converters with series blocking diodes are discussed in this section and converters without blocking diodes will be discussed in the next section. The advantages and disadvantages of placing blocking diodes in series with switches are reviewed at the end of this chapter.

Converters with blocking diodes generally require these diodes because they have some mechanism in their primary that requires that a negative voltage be placed across their switches in order to extinguish the current that flows through them; this enables the switches to turn off softly, with ZCS. Without the series blocking diodes, this negative voltage would generate a significant amount of circulating current in the converter that would flow through the body-diodes of the switches. The purpose of the series blocking diodes is, therefore, to prevent such current from flowing in the converter's primary.

A simple PWM current-fed full-bridge converter that allows its switches to operate with ZCS and turn off softly without overvoltage spikes is the converter shown in Figure 10.7, which was proposed in [8]. This converter is similar to the converter shown in Figure 10.6, but with a few key differences: Capacitor C_r is placed across the transformer, inductor L_r is placed in series with it, and a diode is added in series with each converter switch. Assuming

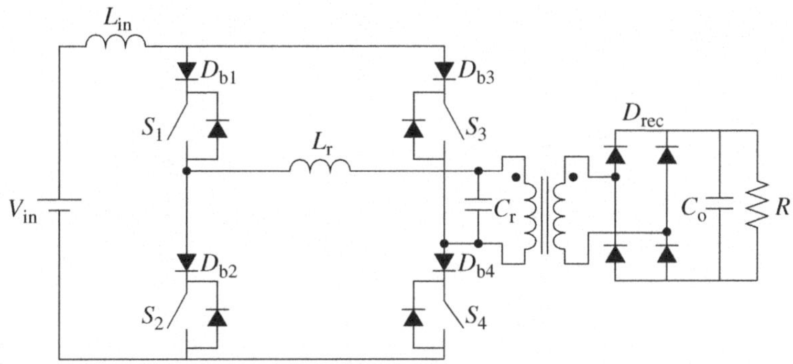

Figure 10.7 Full-bridge converter with blocking diodes and resonant capacitor.

that the converter is in an energy-transfer mode with the input current flowing through switches S_1 and S_4 and capacitor C_r charged so that the terminal connected to S_1 has positive polarity and the terminal connected to S_4 has negative polarity, the converter operates as follows: Switch S_3 is turned on to force the converter into a short-circuit mode. This switch can be turned on with ZCS if L_r is sufficiently large and the transfer of current away from S_1 to S_3 is not instantaneous. S_1 can be turned off with ZCS after the current that was previously flowing through has been fully transferred to S_3. Switch S_2 is turned on when it is time for another energy-transfer mode. Since the terminal of C_r connected to the midpoint of S_3 and S_4 has negative polarity, current begins to be diverted from S_4 to S_2 so that S_4 can be turned off with ZCS after current stops flowing in this switch. S_2 turns on with ZCS as L_r helps slow down the rate of current rise while current is being diverted into this switch. With S_3 and S_4 on, the converter is in an energy-transfer mode and remains so until S_1 is turned on. It should be noted that the polarity of C_r is the opposite of what it was when switches S_1 and S_4 were on and thus the converter is set for the next short-circuit mode/energy-transfer mode change.

Although the PWM current-fed full-bridge converter shown in Figure 10.7 is simple, it has the following drawbacks:

- The resonant capacitor C_r must be large to prevent resonant oscillations.
- The large value of C_r creates duty-cycle loss as it increases the resonant cycle of the interaction with the transformer's leakage inductance. This means that either the operating range of the converter must be limited, or the transformer turns ratio must be results in more primary current and thus more conduction losses.
- The resonant capacitor causes the converter switches to have a peak resonant current that increases the peak current stress of the converter switches.

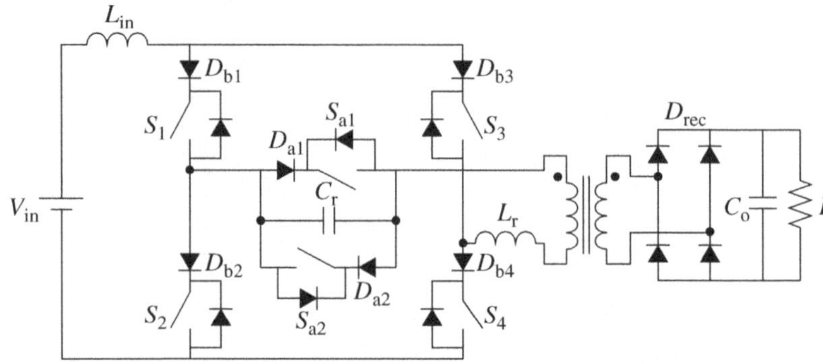

Figure 10.8 Modified full-bridge converter with blocking diodes and active auxiliary circuit.

A variation on this converter was proposed in [9] and is shown in Figure 10.8. For this converter, two switches are placed in parallel to form a bidirectional switch that is placed across the resonant capacitor, C_r. This bidirectional switch is turned on whenever the voltage across C_r is zero, which, in effect, removes the capacitor from the circuit. When it is time to turn off a converter switch, the bidirectional switch is turned off with ZVS and C_r begins to resonate in a way that diverts current away from the switch that is about to be turned off, thus ensuring that it does so with ZCS. Since C_r does not have to be attached to the circuit for all loads at all times, it does not have to be designed to ensure that the converter operates satisfactorily over the full load range. It can instead be designed by considering only what needs to happen during a switching transition so that the converter switches operate with ZCS. The addition of the bidirectional switch allows the converter in Figure 10.8 to avoid the drawbacks that the converter in Figure 10.7 has:

- The resonant capacitor can be considerably smaller in physical size. Any ringing that may occur in the original converter due to the interaction between this capacitor and the primary circuit inductances is avoided as the bidirectional switch removes this capacitor from the converter when it is not used.
- A smaller capacitor means that the duty-cycle loss that the original converter has become negligible as the resonant cycle is reduced. This also results in lower conduction losses as the transformer turns ratio does not have to be lowered to make up for the duty-cycle loss.
- A smaller capacitor also reduces the peak resonant current that the switches are exposed to when a switching transition is made as the resonant impedance of the primary circuit inductances and C_r is increased.

The main drawbacks of the modified converter shown in Figure 10.8 are that two auxiliary switches are needed in the converter and additional sensing is needed to ensure that the auxiliary switches are turned on when the voltage across C_r is zero. These drawbacks increase converter cost, size, and complexity.

10.6 CURRENT-FED DC–DC FULL-BRIDGE CONVERTERS WITHOUT BLOCKING DIODES

The presence of blocking diodes in ZCS-PWM current-fed full-bridge converters such as the ones shown in Figures 10.7 and 10.8 – regardless of whether a blocking diode is placed in series with each switch or the converter switches are implemented with reverse-blocking IGBTs that have the blocking diode in series with the switch inside the device package – is disadvantageous for several reasons: One reason is that the conduction losses of the converter increase as current must flow through more elements at any given time. Another reason is that more devices or more expensive devices must be used, which increases converter cost.

Given that the main reason why the body-diodes of the switches are blocked is to prevent current from flowing through them and that the main reason why current can flow through them is the mechanism for soft-switching operation that is placed in the topology (i.e. capacitor C_r in the converters in Figures 10.7 and 10.8 that is used to divert current away from switches when its voltage is negative), the use of blocking diodes can be avoided if the mechanism for soft-switching is placed "outside" the topology, parallel to it. Converters implemented in such a manner are discussed in this section.

10.6.1 ZVS-PWM Active-Clamp Full-Bridge Converter

A popular current-fed full-bridge converter is the current-fed ZVS-PWM active-clamp full-bridge converter that was proposed in [10, 11] and shown in Figure 10.9. This converter is like a standard PWM current-fed full-bridge

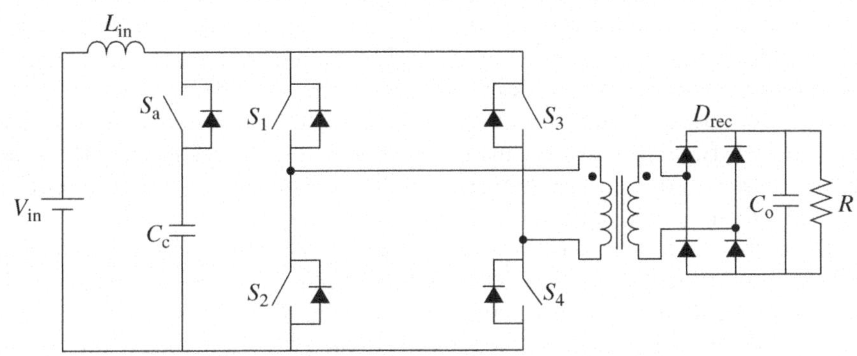

Figure 10.9 ZVS-PWM active-clamp full-bridge converter.

converter but with an active clamp consisting of C_c and S_a attached in parallel to the main full-bridge. The capacitor acts as a clamp that keeps excessive voltage spikes and ringing that can be caused by the interaction of the output capacitances of the bridge switches and the transformer primary leakage inductance when the switches are turned off from appearing across the switches. Energy that would otherwise resonate between the switch output capacitances and the transformer leakage inductance can be transferred to the clamp capacitor and then used to help the main converter switches to turn on with ZVS.

Energy is released from the active clamp capacitor whenever S_a is turned on during an energy-transfer mode of operation when diagonally opposed switches are on. S_a is never on when the converter is in a short-circuit boosting mode when all four switches are on to avoid catastrophic switching device failure. For example, S_a can be turned on soon after S_1 and S_4 have been turned on and the converter has just entered an energy transfer mode. Doing so discharges the energy stored in the clamp capacitor into the leakage inductance of the transformer so that the current in the leakage inductor of the transformer becomes greater than the input current. The active-clamp switch is then turned off just before the other pair of diagonally opposite switches S_2 and S_3 are to be turned on and the converter is about to enter a boosting mode. Once S_a is turned off, the difference between the current in the leakage inductor and the input current discharges the output capacitors of S_2 and S_3 and flows through their body diodes. This reduces the voltage in the active-clamp branch to zero and the corresponding switches (S_2 and S_3) can now be turned on with ZVS. When the current through the transformer leakage inductance drops to zero, the input current starts flowing through the four bridge switches and the converter enters a boosting mode. The input inductor current rises and no energy is transferred from the input of the converter to its output.

The boosting mode ends when the other two diagonally opposite switches, S_1 and S_4, are turned off. After these two switches are turned off, the input current flows through the active-clamp branch through the body diode of S_a and the voltage across the active-clamp capacitor starts rising. The clamp capacitor reduces the rate of voltage rise and limits the magnitude of the spikes across switches S_1 and S_4. S_a is turned on during the time the clamp capacitor is charging up. Once the clamp capacitor voltage exceeds the reflected output voltage at the transformer primary, the current in the transformer primary begins to grow until it becomes greater than the input current and transfers energy to the output. When the current in the leakage inductor exceeds the input current, it changes direction and starts flowing through S_a. Energy continues to be transferred to the output and the input inductor gets discharged. Sometime while the clamp current is flowing through the switch S_a and the current in the leakage inductor is greater than the input current, switch S_a is turned off and it marks the beginning of next half of the switching cycle.

The active-clamp ZVS full-bridge boost converter is an attractive option when isolated voltage step-up power conversion is needed as it is a fixed-frequency ZVS current-fed converter that uses a very simple auxiliary circuit to create ZVS over an extended range of load. Its main drawback is that its main converter switches and its active-clamp switch have a significant amount of conduction losses since current flows either through the active-clamp switch or through its body diode whenever any two diagonally opposite bridge switches are on. This occurs during a significant portion of the switching cycle.

10.6.2 ZCS-PWM Full-Bridge Converter with Parallel Auxiliary Circuit

An alternative approach to the ZVS-PWM active clamp current-fed full-bridge converter is to consider the main circuit (consisting of the S_1–S_4 switches) to be like an equivalent boost converter switch, then place an auxiliary circuit across the bridge. An example of this approach was proposed in [12] and is shown in Figure 10.10. Figure 10.10 shows the auxiliary circuit from the ZCS-PWM boost converter discussed in Section 2.3.2 placed across the main full-bridge. In this converter, the main full-bridge switches operate like those of a standard current-fed PWM full-bridge converter. Energy-transfer modes are followed by short-circuit modes in a manner that corresponds to the turning on and off of a boost converter switch. For example, a typical switching sequence over a full switching cycle can be S_1 and S_4 on, then all switches on, then S_2 and S_3 on, then all switches on; this is Option #1. Another option, Option #2, can be S_1 and S_4 on, then S_1 and S_2 on, then S_2 and S_3 on, then S_2 and S_4 on. The following should be noted:

- For both options, the main full-bridge switches can be turned on with ZCS due to the presence of leakage inductance in the full-bridge transformer. For example, if the converter is operating with the switching

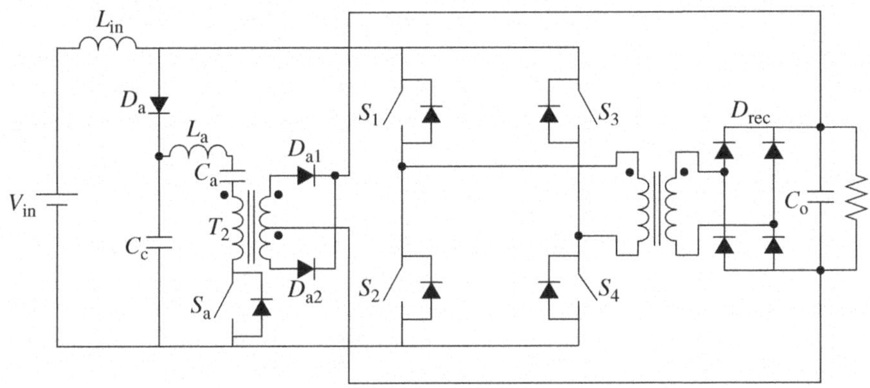

Figure 10.10 ZCS-PWM full-bridge converter with parallel auxiliary circuit.

sequence of Option #1 with S_1 and S_4 on, then S_2 and S_3 can be turned on with ZCS as the leakage inductance delays the transfer of current to switches S_2 and S_3. If this leakage inductance is insufficient, then additional inductance can be added in series with the transformer to slow down the current transfer away from the transformer primary.

- If the converter is operating with the switching sequence of Option #1 and all four switches are on, then current needs to be diverted away from the full-bridge somehow before the appropriate diagonally opposed switches can be turned off with ZCS so that the converter can enter an energy-transfer mode. This can be done by activating the auxiliary circuit.
- Similarly, if the converter is operating with the switching sequence of Option #2 and, for example, the converter is in a short-circuit mode with switches S_1 and S_2 on, then switch S_3 and be turned on with ZCS given that no current will flow through this switch as long as both S_1 and S_2 are on and the converter remains in a short-circuit mode. When it is time for the converter to enter an energy-transfer mode with diagonally opposed switches S_2 and S_3 on, the auxiliary circuit can be activated and used to divert current away from the main full-bridge, which allows S_1 to be turned off with ZCS.

Regardless of which switching sequence option is used in the converter shown in Figure 10.10, activating the auxiliary circuit by turning on switch S_a causes the polarity of the voltage across snubber capacitor C_r to become negative. As a result, current is transferred away from the full-bridge to auxiliary circuit as it tries to flow through a path of least voltage potential. Inductor L_r is placed in series with the full-bridge in order to slow down the rate of current transfer away from the full-bridge switches to allow them to turn off with ZCS – L_r is particularly needed when the converter switches are IGBT devices, which have current tails, as described in Chapters 1 and 3. Moreover, L_r limits the rise in current that may occur when C_r has a negative voltage across it. Without L_r, current in this case can flow through the body-diodes of the full-bridge switching devices and a catastrophic short-circuit would result. Although some current may flow through the body-diodes of the devices, the use of blocking diodes or reverse-blocking IGBTs is unnecessary as negative voltage is present across C_r for a very short time.

It should be noted that, in theory, any auxiliary circuit that can be used in a ZCS-PWM single-switch boost converter can be used in a ZCS-PWM current-fed full-bridge. In reality, however, the auxiliary circuit must have a snubber capacitor/diode combination that can be connected across the bridge like the auxiliary circuit in the converter shown in Figure 10.10 if voltage spikes and ringing across the switches are to be avoided.

10.7 CONCLUSION

Current-fed converters are used in applications where a high output voltage/input voltage gain is required. They are essentially boost-type converters with step-up transformers integrated into their topologies so that there are two mechanisms by which voltage can be stepped up and it is the combination of these two mechanisms that allows these converters to achieve a high voltage gain.

The current-fed push–pull converter was the first isolated current-fed DC–DC converter to be proposed and is the simplest; moreover, it can step down voltage as well. Its clamp, however, limits its operation to low-power applications and its switches have high peak voltage stress. The two-inductor current-fed converter has a higher gain than the current-fed push–pull converter, its switches experience less peak voltage stress, and it can be operated with more power, but its magnetic structure is either bulky or intricate and it is harder to control. It also needs to be implemented with some sort of snubber that can be passive or active. Implementing the converter with dual active clamps is a popular option, especially since the active clamps enable the converter switches to operate with ZVS, but two auxiliary switches are needed, and this increases cost and complexity.

For higher power applications, four-switch current-fed full-bridge-type converters are used. These converters are almost always implemented with some sort of soft-switching, given the power losses that would otherwise accumulate due to the switching losses of four switches; they can be implemented with blocking diodes or without. If they are not implemented with blocking diodes, then some sort of parallel auxiliary circuit is needed to generate the soft-switching mechanism, whether it is ZVS or ZCS. These converters are typically implemented with MOSFETs, thus ZVS is generally the preferred method of soft-switching, except when the input voltage is very low and the input current is very high, as it is in some renewable energy applications. In that case, ZCS would be the preferred method of operation even though the converter may be implemented with MOSFETs.

Whether a current-fed converter for a particular application should be implemented with blocking diodes or without blocking diodes and with a parallel auxiliary circuit will depend on the load and the amount of current that will be flowing in the converter's primary. If the converter is operating mostly with light loads with occasional heavier load operation, then converters with blocking diodes are to be preferred as they will not have any parallel auxiliary circuit losses. If, however, the converter is to operate only with light loads, then a simpler option with fewer switches such as the current-fed push–pull converter is preferable. If the converter is to be operated with heavier loads, then the conduction losses that the blocking diodes would generate would make converter topologies with parallel auxiliary circuits and no blocking diodes preferably. If the converter is to be operated with very heavy loads, then converters with blocking diodes was preferable as parallel auxiliary circuits would generate a considerable amount of circulating current in the converter primary, on top of the primary current that already exists, and this combined current would generate losses that would be on top of the losses generated by the parallel auxiliary circuit. In other words, for any particular application where higher-power current-fed topologies are to be considered, there is generally a range of operating conditions where converters without blocking diodes are preferable, but converters with blocking diodes are to be preferred for operation above and below this range.

REFERENCES

1. P.W. Clarke, "Converter regulation by controlled conduction overlap," U.S. Patent 3,938,024, Feb. 10, 1976, assigned to Bell Telephone Laboratories, Inc., Murray Hill, NJ 07974, U.S.A.
2. R. Redl and N. O. Sokal, "Push-pull current-fed multiple-output [regulated] DC/DC power converter with only one inductor and with 0 to 100% switch duty ratio," in IEEE Power Electronics Specialists Conference (PESC), 1980, pp. 341–345.
3. R. Redl and N. O. Sokal, "Push-pull current-fed multiple-output regulated wide-input-range DC/DC power converter with only one inductor and with 0 to 100% switch duty ratio: operation at duty ratio below 50%," *IEEE Power Electronics Specialists Conference (PESC)*, 1981, June 29–Jul. 3, 1981, Boulder, USA, pp. 204–212.
4. P. J. Wolfs, "A current-sourced DC-DC converter derived via the duality principle from the half-bridge converter," *IEEE Transactions on Industrial Electronics*, vol. 40, no. 1, pp. 139–144, Feb. 1993.
5. S. -K. Han, H. -K. Yoon, G. -W. Moon, M. -J. Youn, Y. -H. Kim, and K. -H. Lee, "A new active clamping zero-voltage switching PWM current-fed half-bridge converter," *IEEE Transactions on Power Electronics*, vol. 20, no. 6, pp. 1271–1279, Nov. 2005.
6. Y. Jang and M. M. Jovanovic, "New two-inductor boost converter with auxiliary transformer," *IEEE Transactions on Power Electronics*, vol. 19, no. 1, pp. 169–175, Jan. 2004.
7. L. Yan and B. Lehman, "An integrated magnetic isolated two-inductor boost converter: analysis, design and experimentation," *IEEE Transactions on Power Electronics*, vol. 20, no. 2, pp. 332–342, Mar. 2005.
8. C. Iannello, S. Luo, and I. Batarseh, "Full bridge ZCS PWM converter for high-voltage high-power applications", in *IEEE Transactions on Aerospace and Electronic Systems*, vol. 38, no. 2, pp. 515–526, Apr. 2002.
9. X. Zhang, H. S. -h. Chung, X. Ruan, and A. Ioinovici, "A ZCS full-bridge converter without voltage overstress on the switches," *IEEE Transactions on Power Electronics*, vol. 25, no. 3, pp. 686–698, Mar. 2010.

10. R. Watson and F. C. Lee, “A soft-switched, full-bridge boost converter employing an active-clamp circuit,” *IEEE Power Electronics Specialists Conference (PESC)*, 1996, June 23–27, 1996, Baveno, Italy, pp. 1948–1954.
11. V. Yakushev, V. Meleshin, S. Fraidlin, “Full-bridge isolated current fed converter with active clamp,” *IEEE Applied Power Electronics Conference and Exposition (APEC)*, 1999, Mar. 14-18, 1999, Dallas, USA, pp. 560–566.
12. A. Mousavi, P. Das, and G. Moschopoulos, “A comparative study of a new ZCS DC–DC full-bridge boost converter with a ZVS active-clamp converter,” in *IEEE Transactions on Power Electronics*, vol. 27, no. 3, pp. 1347–1358, Mar. 2012.

CHAPTER 11

RESONANT CONVERTERS PART I – FUNDAMENTALS

Resonant converters are converters that are implemented with some sort of resonant circuit in their topologies. This circuit consists of capacitors and/or inductors and is used to shape switch voltage or switch current so that switching devices can turn on and off softly. Resonant converters have been a mainstay of modern power electronics since its inception and can be found in many applications, including telecom and server power supplies, electric vehicles, induction heating, lighting, space, and medical power supplies. They have several advantages and disadvantages with respect to PWM converters. Their main advantage is that they offer a simple means for converter switches to operate with soft-switching, as this can be done with just a few passive elements instead of more sophisticated auxiliary circuits, such as the one that are presented in previous chapters of this book. Moreover, when operated with high switching frequencies, resonant converters can absorb parasitic elements into their topologies in a way that is not possible with PWM converters, so that converter size can be further reduced. The main drawback of resonant converters is that they have higher voltage and current stresses than PWM converters.

The popularity of resonant converters has risen and fallen throughout the history of modern power electronics. Resonant converters have become in vogue in recent years due to the development of wide bandgap devices that have better switching characteristics than traditional silicon devices. Such devices allow power converters to operate more efficiently and with

DC–DC Converter Topologies: Basic to Advanced, First Edition. Gerry Moschopoulos.

Published 2024 by John Wiley & Sons, Inc.

considerably higher switching frequency than silicon devices. This allows power electronic designers to use parasitic elements advantageously as part of resonant networks to make switches operate with ZVS or ZCS, mostly with ZVS.

The focus of this chapter is on the basic principles behind resonant power conversion for DC–DC converters. The characteristics of the three fundamental DC–DC resonant converters – series-resonant, parallel-resonant, and series-parallel resonant – are then presented and discussed. Finally, an overview of resonant converters with other types of resonant networks is given, with emphasis on the LLC resonant converter, which is currently among the most popular converters used in industry.

11.1 RESONANT POWER CONVERSION FUNDAMENTALS

Resonant power conversion is based on certain fundamental principles that are related to the impedance of resonant networks. To begin, Figure 11.1 shows a simple series-resonant circuit consisting of a sinusoidal AC voltage source, a capacitor C, an inductor L, and a load resistor R. Each of the passive components has an impedance: the impedance of R is $Z_R = R$, the impedance of L is

$$Z_L = j\omega L \tag{11.1}$$

where

$$\omega = 2\pi f \tag{11.2}$$

where f is the frequency of the AC voltage source, and the impedance of C is

$$Z_C = \frac{1}{j\omega C} \tag{11.3}$$

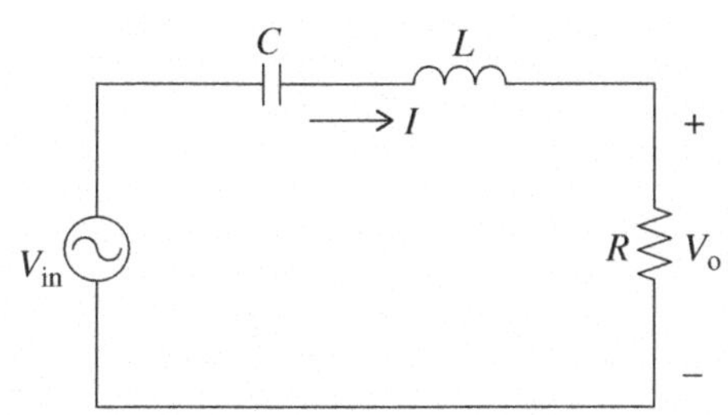

Figure 11.1 Series-resonant circuit with sinusoidal AC source voltage.

The parameter j is an operator that indicates a phase shift of 90° relative to some reference and has a magnitude of $\sqrt{-1}$, according to standard complex number theory. If the numerator and denominator of (11.3) are both multiplied by j, then (11.3) can be expressed as

$$Z_C = \frac{1}{j\omega C} \times \frac{j}{j} = -j\frac{1}{\omega C} \tag{11.4}$$

The overall impedance of the circuit shown in Figure 11.1, Z_T, can be expressed as

$$Z_T = \frac{V_{\text{in}}}{I} = R + j\left(\omega L - \frac{1}{\omega C}\right) \tag{11.5}$$

where V_{in} is the AC input source voltage and I is the current flowing in the circuit. Since the impedances of the inductor and the capacitor are dependent on the frequency of the input AC source, so too is the impedance of the overall circuit and thus Z_T changes as frequency is changed. Given that the impedance of the inductor is subtracted from that of the capacitor, there is a frequency – generally referred to as the resonant frequency in the literature – at which these two impedances will cancel each other out. This resonant frequency, ω_r, can be derived from the fact that $Z_L = Z_C$ at this frequency,

$$\omega_r L = \frac{1}{\omega_r C} \tag{11.6}$$

so that

$$\omega_r = \frac{1}{\sqrt{LC}} \tag{11.7}$$

or

$$f_r = \frac{1}{2\pi\sqrt{LC}} \tag{11.8}$$

The output voltage across the load resistor, V_{o}, can be expressed as

$$V_{\text{o}} = V_{\text{in}} \times \frac{R}{R + j\left(\omega L - \frac{1}{\omega C}\right)} \tag{11.9}$$

and thus the voltage gain – the ratio of output voltage to input voltage can be expressed as

$$\frac{V_o}{V_{in}} = \frac{R}{R + j\left(\omega L - \dfrac{1}{\omega C}\right)} \tag{11.10}$$

Using (11.10), a graph of voltage gain vs frequency ω for the circuit in Figure 11.1 can be plotted as shown in Figure 11.2. It can be seen from this graph that the gain of the circuit varies with the frequency of the AC source and is equal to 1 when the frequency is ω_r, with the gain falling the further away the source voltage frequency is from the resonant frequency. When $\omega = \omega_r$, the impedance of the inductor and the impedance of the capacitor cancel out each other so that (11.10) reduces to 1.

The circuit acts like a pure resistor at the resonant frequency. As the frequency of the input source is reduced, then the impedance of the capacitor increases and that of the inductor decreases so that the impedance of the capacitor is dominant when the AC source voltage frequency is low. Similarly, as the frequency of the input source is increased, then the impedance of the capacitor decreases and that of the inductor increases so that the impedance of the inductor is dominant when the AC source voltage frequency is high. The fact that the circuit becomes more capacitive at low frequencies and more inductive at high frequencies can be used advantageously to generate soft-switching conditions for power circuits.

Consider the circuit shown in Figure 11.3. This circuit is identical to the circuit shown in Figure 11.1 except that the AC sinusoidal source is replaced by

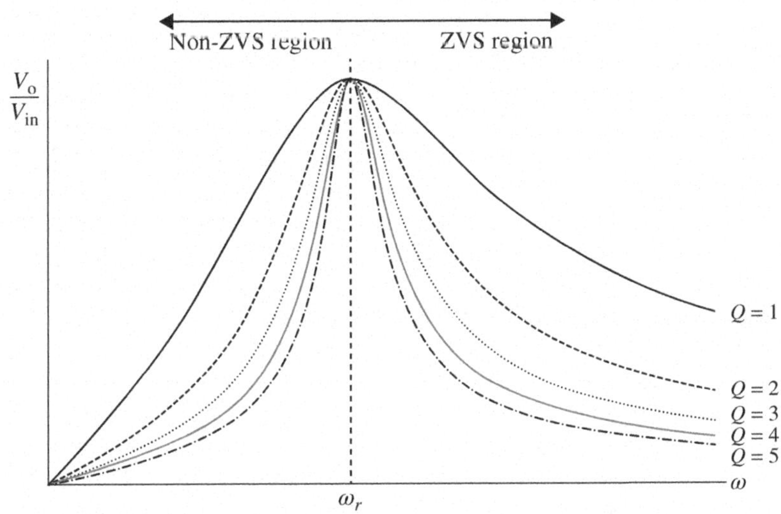

Figure 11.2 Series-resonant circuit gain characteristic.

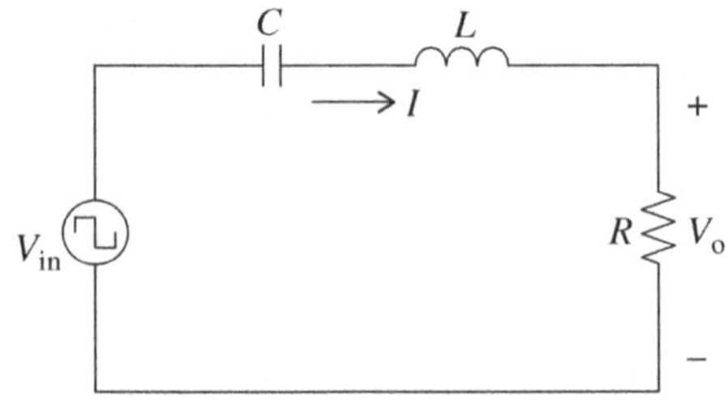

Figure 11.3 Series-resonant circuit with square-wave AC source voltage.

an AC square-wave source. Fourier Series Theory stipulates that any periodic, non-sinusoidal waveform can be expressed as the summation of an average value component and individual sinusoids with frequencies that are multiples of the frequency of the original, periodic, non-sinusoidal waveform. According to Fourier Series Theory, the AC square-wave source can be considered to have a sinusoidal component (harmonic) at its frequency, which is considered to be the fundamental frequency, and sinusoidal components at odd multiples of this frequency. The amplitude of these components or harmonics as a function of the amplitude of the sinusoidal harmonic at the fundamental frequency can be expressed as

$$V_n = \frac{V_1}{n} \qquad n = \text{odd} \tag{11.11}$$

For example, the AC square-wave voltage has a third harmonic V_3 that has an amplitude that is 1/3 that of the fundamental V_1, a fifth harmonic that has an amplitude that 1/5 of V_1, etc. The source voltage does not have any even harmonics, nor does it have an average or DC value.

If the frequency of the input AC square-wave voltage is above the resonant frequency of the circuit and the total impedance of the circuit is such that only the fundamental harmonic of the current flows through it and the other current harmonics have negligible amplitude, then the circuit will have the waveforms shown in Figure 11.4. It can be seen in Figure 11.4 that the current waveform is a sinusoid that lags the square-wave voltage waveform. This is due to the fact that the circuit is more inductive than capacitive when it is operating with a source voltage frequency that is above the resonant frequency. If the source voltage frequency is below the resonant frequency, then the current leads the input voltage. As for the shape of the current waveform, this stems from the fact that the amplitudes of the voltage harmonics of the AC square-wave voltage source diminish as the harmonic frequency increase and the fact that the impedance of the resonant circuit, especially that of the inductor, increases with increasing harmonic frequency. This makes the resulting current harmonics negligible, although some third harmonic may appear in an actual circuit, which creates some distortion in the current.

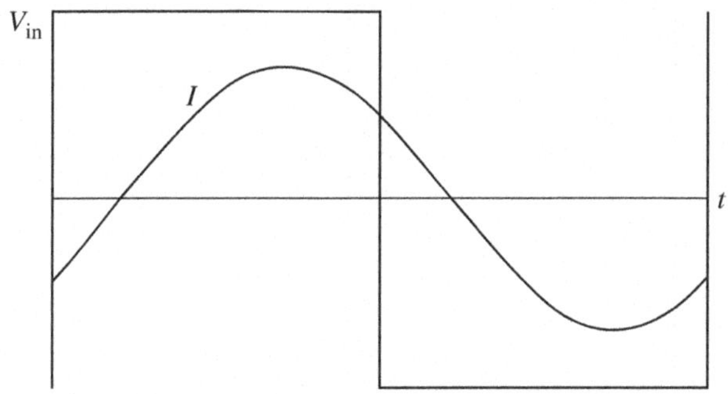

Figure 11.4 Series-resonant circuit waveforms.

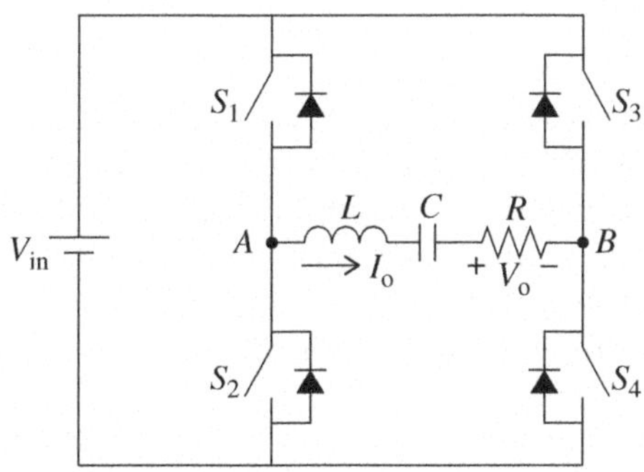

Figure 11.5 Series-resonant full-bridge inverter circuit.

A square-wave voltage such as the one shown in Figure 11.4 can be produced by using a full bridge circuit, as shown in Figure 11.5. In Figure 11.5, four switches are arranged in a full-bridge configuration with each end of the R-L-C circuit connected to the mid-point of a converter leg. A square-wave voltage can be generated across the R-L-C circuit simply by turning diagonally opposed switches in an alternating manner, i.e. turning on S_1 and S_4 and leaving them on for half of the switching cycle, turning off S_1 and S_4, turning on S_2 and S_3 and leaving them on for the other half of the switching cycle, then turning them off. The frequency of the square-wave voltage across the R-L-C circuit can be adjusted by adjusting the switching frequency of the switches.

In order for the circuit shown in Figure 11.5 to work properly, each switch must have an anti-parallel diode connected across it to provide a path for current in the R-L-C circuit to flow through after a pair of switches has been turned off. For example, if switches S_1 and S_4 are on and are then turned off, the current in L must have a path to flow through and this path is provided by diodes D_2 and D_3. While D_2 and D_3 are conducting, switch S_2 and S_3 can be turned on with ZVS. Similarly, when S_2 and S_3 are turned off, current flows

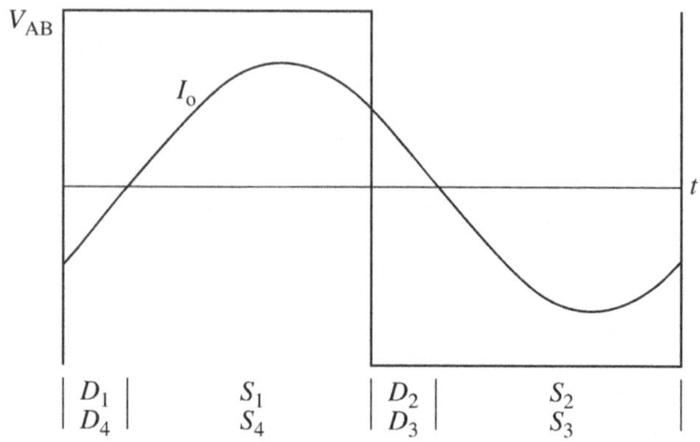

Figure 11.6 Series-resonant full-bridge circuit waveforms with conducting components indicated.

through anti-parallel diodes D_1 and D_4 and thus S_1 and S_4 can be turned on with ZVS as well. Taking the flow of current through the anti-parallel diodes as the start, the sequence of conducting devices is as follows: D_1–D_4, S_1–S_4, D_2–D_3, S_3–S_4, then back to D_1 and D_4 to start a new switching cycle. If the voltage and current waveforms in Figure 11.5 are divided into lengths to time that correspond to this sequence, then the result is the diagram shown in Figure 11.6.

From the diagram shown in Figure 11.6, it can be seen that all the converter switches can operate with ZVS as long as the current is made to lag the voltage. This lag in current, relative to the voltage placed across the R-L-C branch, is necessary for ZVS operation since this is how current is forced to flow into the anti-parallel diode across a switch before it is turned on. The switches cannot operate with ZVS if current is made to lead this voltage. Lagging current is ensured if the switching frequency of the circuit is greater than the resonant frequency of the R-L-C branch. As explained above, when this happen, the R-L-C branch becomes more inductive and current always lags voltage in inductive circuits.

11.2 FUNDAMENTAL RESONANT DC–DC CONVERTERS

The three fundamental resonant DC–DC converter topologies are the series resonant, parallel resonant, and series-parallel-resonant converters. These three fundamental converters are shown in their half-bridge versions in Figure 11.7 and their resonant circuits are shown in Figure 11.8. The following should be noted about the converters in Figure 11.7:

- The series-resonant converter has a capacitive output filter while the parallel resonant and series-parallel-resonant converters have LC

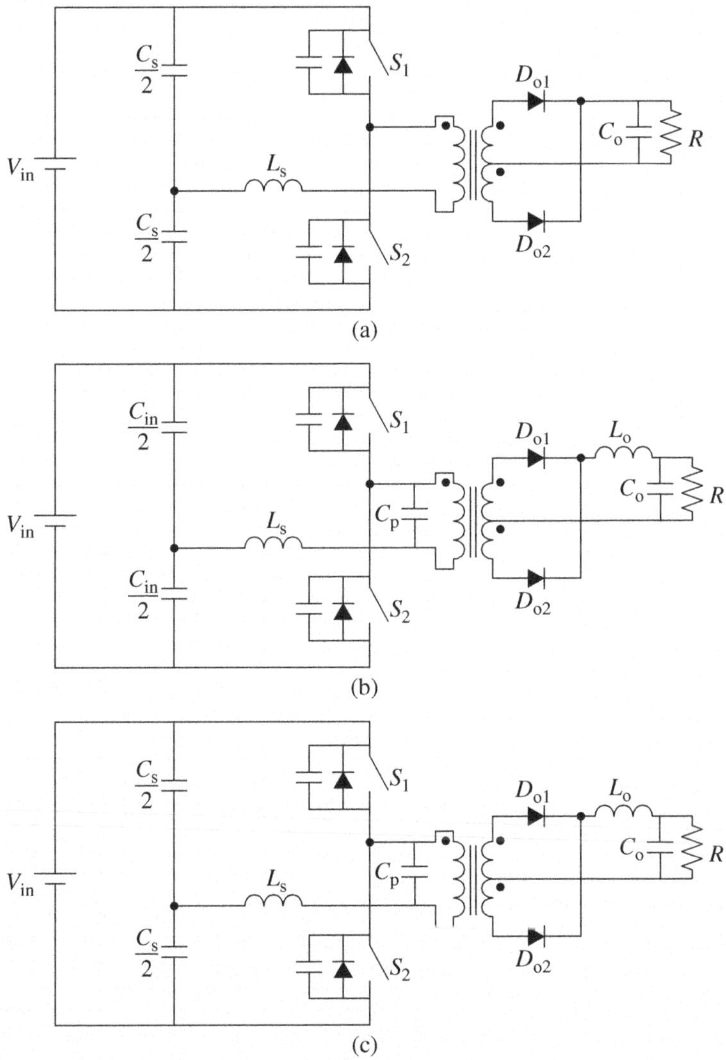

Figure 11.7 Fundamental half-bridge resonant converters: (a) Series-resonant. (b) Parallel-resonant. (c) Series-parallel resonant.

output filters. This is because for the series-resonant converter, there is inductance in the form of L_r that is in series with the transformer so that the secondary must have a capacitor across it. The transformer secondary should not be connected in series with an inductor as there will be a mismatch between the resonant inductor at the primary and this output inductor. Although implementing a series-resonant converter can be

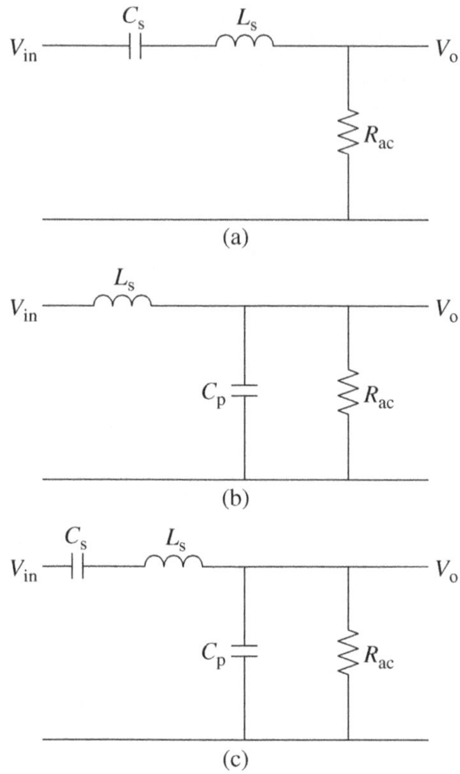

Figure 11.8 Equivalent resonant circuits for the fundamental half-bridge converters: (a) Series-resonant. (b) Parallel-resonant. (c) Series-parallel resonant.

implemented with an LC output filter, as shown in [1], this changes the characteristics of the converter and makes the primary current more like a square-wave, which increases the harmonics that are placed into the transformer, thus requiring that a larger transformer be used. As a result, implementations of series-resonant converters with LC output filters are less common than those with C output filters. In the case of the parallel and series-parallel converters, there is a capacitor across the transformer primary so that there must be an inductor connected to the transformer secondary to avoid a mismatch of voltages.

- For series and series-parallel half-bridge converters, the series capacitor can be embedded in the DC bus capacitors. If the DC bus capacitors are bulk capacitors and a series-resonant capacitor is connected to their midpoint, then the series inductor will see an oscillatory voltage. With this being the case, the series-resonant capacitor can be split into two with one capacitor being the upper DC bus capacitor and the other being the lower DC bus capacitor. The value of each capacitor is half of what the value of the series-resonant capacitor would be if connected to the mid-point of two bulk capacitors. The DC bus capacitors of the parallel-resonant half-bridge converter are bulk capacitors as the resonant inductor in this converter sees a stiff DC voltage. In Figure 11.7, bulk capacitors are indicated as C_{in}, lower value resonant circuit capacitors are indicated as C_s.

11.2.1 Resonant Converter Analysis Using First Harmonic Approximation Method

The basic characteristics of each converter can be determined using an equivalent circuit model that is based on the fundamental harmonic analysis method developed in [2], which is generally referred to as the First Harmonic Approximation Method (FHA) in the literature. Regardless of whether the converter is a full-bridge or a half-bridge, the converter can be essentially reduced to an R-L-C circuit with an input AC sinusoidal source voltage that represents the fundamental harmonic of a square-wave voltage waveform. This fundamental component will be called V_{in} in the discussion that follows as the other harmonics will be neglected. The resistor R in each model represents the load resistor as seen by the primary of the transformer. According to [2], if R_L is a load resistor and $R = R_{ac}$ is the equivalent resistance as seen by the transformer primary, then the relationship between R and R_L for a converter with a capacitive output filter is

$$R = R_{ac} = \frac{8}{\pi^2} R_L \tag{11.12}$$

while that for a converter with an LC output filter is

$$R = R_{ac} = \frac{\pi^2}{8} R_L \tag{11.13}$$

The basic characteristics of each fundamental resonant DC–DC converter can be determined based on the equivalent circuit shown in Figure 11.8. With these characteristics, the basic properties of each converter can be determined, then compared to those of the other converters. As an example, the gain characteristics for a series-resonant converter are derived here.

Taking the gain equation that was derived for the circuit shown in Figure 11.1, Eq. (11.10), and dividing the numerator and denominator by R gives

$$\frac{V_o}{V_{in}} = \frac{1}{1 + j\left(\dfrac{\omega L_s}{R} - \dfrac{1}{\omega C_s R}\right)} \tag{11.14}$$

where $L = L_s$ and $C = C_s$ from Eq. (11.10) and V_{in} is the fundamental component of the square wave of voltage applied to the resonant network by the converter switches, as explained above. Since the input to the resonant circuit is actually a square-wave voltage, the output is also a square-wave voltage. In an actual series-resonant converter, regardless of whether the converter is a full-bridge converter or a half-bridge, the voltage that appears across the transformer secondary is a square-wave. This is because the transformer secondary is connected directly across the output capacitor and the secondary voltage is positive for half the switching cycle when the primary current is positive and negative for the other half cycle. The converter gain in terms of actual converter values is given by the same expression as (11.14), as the factor relating the fundamental components to the square wave amplitudes cancel so that

$$\frac{V_o}{V_{in}} = \frac{1}{1 + j\left(\dfrac{\omega L_s}{R_{ac}} - \dfrac{1}{\omega C_s R_{ac}}\right)} \tag{11.15}$$

Substituting

$$R_{ac} = \frac{8}{\pi^2} R_L \tag{11.16}$$

and defining

$$Q = \frac{\omega_r L_s}{R_L} \tag{11.17}$$

and

$$\omega_r = \frac{1}{\sqrt{L_s C_s}} \tag{11.18}$$

into this equation results in

$$\frac{V_o}{V_{in}} = \frac{1}{1 + j\dfrac{\pi^2}{8} Q\left(\dfrac{\omega}{\omega_r} - \dfrac{\omega_r}{\omega}\right)} \tag{11.19}$$

This equation of voltage gain can be plotted for varying values of Q as shown in Figure 11.9. It should be noted that these curves are based on a first-harmonic approximation and become more accurate as switching

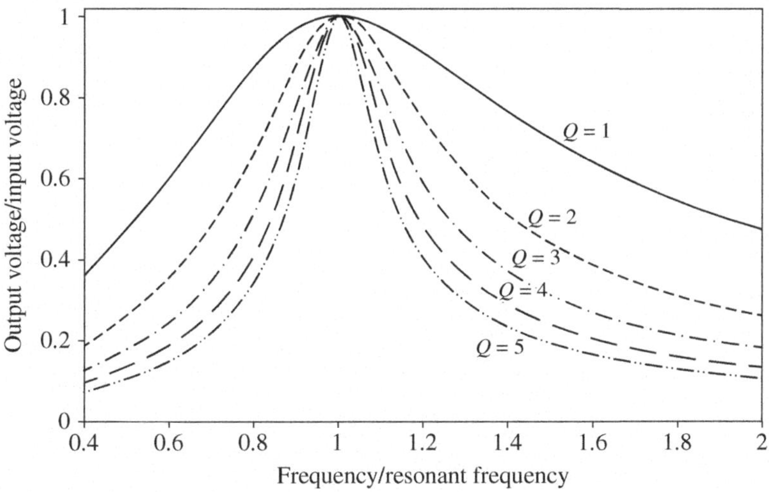

Figure 11.9 Series-resonant converter gain curves based on first harmonic approximation.

frequency is increased and is further away from the resonant frequency, as there is more filtering. They are less accurate at lower switching frequencies that are below the resonant frequency as there is less filtering of current harmonics due to the lower impedance of the resonant circuit at these frequencies.

Voltage gain equations for the parallel-resonant and the series-parallel-resonant converter can be derived in a similar manner. For the parallel-resonant converter, the converter gain can be expressed as

$$\frac{V_o}{V_{in}} = \frac{1}{\frac{\pi^2}{8}\left(1-\left(\frac{\omega}{\omega_r}\right)^2\right)+j\frac{\omega}{\omega_r}\left(\frac{1}{Q}\right)} \tag{11.20}$$

where

$$Q = \frac{R_L}{\omega_r L_s} \tag{11.21}$$

and

$$\omega_r = \frac{1}{\sqrt{L_s C_p}} \tag{11.22}$$

The converter gain equation is plotted for varying values of Q in Figure 11.10.

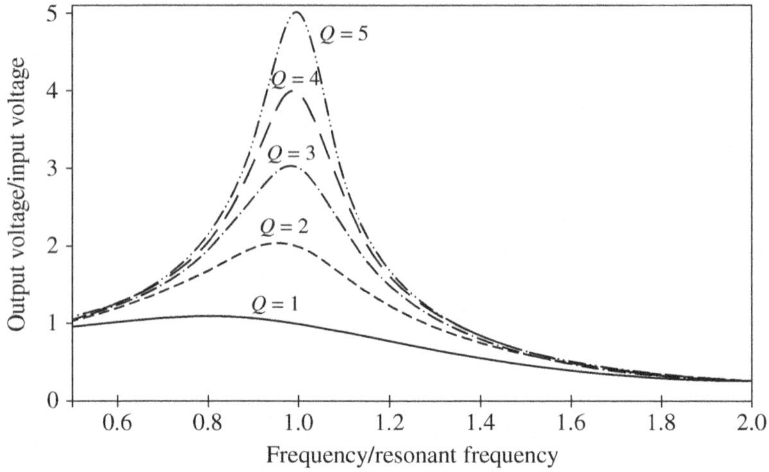

Figure 11.10 Parallel-resonant converter gain curves based on first harmonic approximation.

11.2.2 Series-Resonant Converter vs Parallel-Resonant Converter

Based on the graphs shown in Figures 11.9–11.10 and on their circuit topologies, the following can be stated about the series-resonant and parallel-resonant converters, relative to one another:

- *Voltage step-up/step-down:* The series-resonant converter can be considered to be analogous to a buck-type converter as the maximum output voltage can never exceed the input voltage. The parallel-resonant converter can be considered to be analogous to a boost-type converter as its output voltage can be greater than the input voltage. Although its output voltage can be less than its input voltage, this is only possible when the converter is operated with a switching frequency that is much greater than the resonant frequency, which is not a typical mode of operation.
- *Output filtering:* The output DC filter capacitor of the series-resonant converter must carry high ripple current, about half of the DC output current, which is a major drawback for low output voltage/high current applications. Consequently, the series-resonant converter is generally considered unsuitable for low-output-voltage high-output-current converters, but suitable for high-output-voltage low-output-current applications where output current ripple is less of an issue and filter inductors not needed on the high-voltage side of the converter. Unlike the series-resonant converter, the parallel-resonant converter is suitable for low-output-voltage, high-output-current applications because it does have an output filter inductor that limits the ripple current carried by the output capacitor.

- *Light-load efficiency:* The current in the power devices of the series-resonant converter decreases as the load decreases, so that conduction losses and turn-off losses decrease as well. This is not quite so for the parallel-resonant converter as current flows through the resonant capacitor no matter what is the load so that there will always be a fixed amount of circulating current flowing in the converter that contributes to losses. Moreover, this circulating current increases as the input DC voltage to the converter increases so that the parallel-resonant converter is not well-suited for applications that have a large input voltage or even a large input voltage range.
- *No-load regulation:* The series-resonant converter has the main disadvantage that the output voltage cannot be regulated for the no-load case. As can be seen from the curves of Figure 11.9, the gain curve becomes flatter as Q is decreased (R_L is increased, Q is decreased, and the load is lightened) and is actually a flat horizontal line at no-load, which means that the converter's gain is unaffected by any change in switching frequency. No-load regulation in a series-resonant converter if the converter is occasionally turned on and off in a time-ratio control manner a frequency, but such control is difficult to implement. In contrast to the series-resonant converter, the parallel-resonant and series-parallel resonant converters are able to control the output voltage at no load if operated at a frequency above the resonant frequency. It should be noted that the output voltage of the parallel-resonant converter can rise to very high values at no load (R_L is increased, Q is increased, and the load is lightened) when it is operated right at the resonant frequency, thus operation at this frequency should be avoided.
- *Short-circuit operation:* The parallel-resonant converter is naturally short-circuit proof. If a short is applied directly across the resonant capacitor, the entire square-wave voltage at the input of the resonant network is applied is directly across the resonant inductor and, therefore, current is limited by this impedance. For the series-resonant converter, if the converter is operating near resonance with heavy load and a short-circuit is applied to the converter output, the current will rise to high values as the impedances of the resonant inductor and resonant capacitor cancel each other. To control the output current under such conditions, the frequency of the converter needs to be raised by the converter controller. As it takes a few resonant cycles for the current to rise due to the inductance in the resonant network, the controller will have enough time to take action.
- *Transformer saturation:* The series-resonant converter has a capacitor on its primary side that is connected in series with its transformer. This capacitor acts as a DC blocking capacitor and prevents any volt-second unbalance due to device switching times or forward voltage drops from appearing in the transformer. This characteristic makes series-resonant converter attractive in higher-power applications where a full-bridge

converter is used. The parallel-resonant converter does not have a series DC blocking capacitor connected to its transformer and is thus more susceptible to transformer saturation, unless the converter is a half-bridge converter, where the half-bridge topology naturally prevents transformer saturation from occurring.

11.2.3 Series-Parallel-Resonant Converter

Performing a First Harmonic Approximation analysis similar to that of the series-resonant and parallel-resonant converters, as presented in [2], the converter gain for the series-parallel-resonant converter can be expressed as

$$\frac{V_o}{V_{in}} = \frac{1}{\frac{\pi^2}{8}\left(1 + \frac{C_p}{C_s} - \omega^2 L_s C_p\right) + jQ_s\left(\frac{\omega}{\omega_s} - \frac{\omega_s}{\omega}\right)} \tag{11.23}$$

where

$$Q_s = \frac{\omega_r L_s}{R_L} \tag{11.24}$$

and

$$\omega_s = \frac{1}{\sqrt{L_s C_s}} \tag{11.25}$$

is the resonant frequency based on series capacitor C_s. It can be seen from (11.23) that the gain will depend on the ratio of C_p to C_s, which also determines the series- or parallel-resonant characteristics of the circuit. In Figure 11.11, the gain curves for two cases are plotted. For $C_s = C_p$, the gain is given by

$$\frac{V_o}{V_{in}} = \frac{1}{\frac{\pi^2}{8}\left(\left(2 - \left(\frac{\omega}{\omega_s}\right)^2\right)\right) + jQ_s\left(\frac{\omega}{\omega_s} - \frac{\omega_s}{\omega}\right)} \tag{11.26}$$

This expression is plotted in Figure 11.11a. For $C_s = 2C_p$,

$$\frac{V_o}{V_{in}} = \frac{1}{\frac{\pi^2}{16}\left(\left(3 - \left(\frac{\omega}{\omega_s}\right)^2\right)\right) + jQ_s\left(\frac{\omega}{\omega_s} - \frac{\omega_s}{\omega}\right)} \tag{11.27}$$

and the expression is plotted in Figure 11.11b

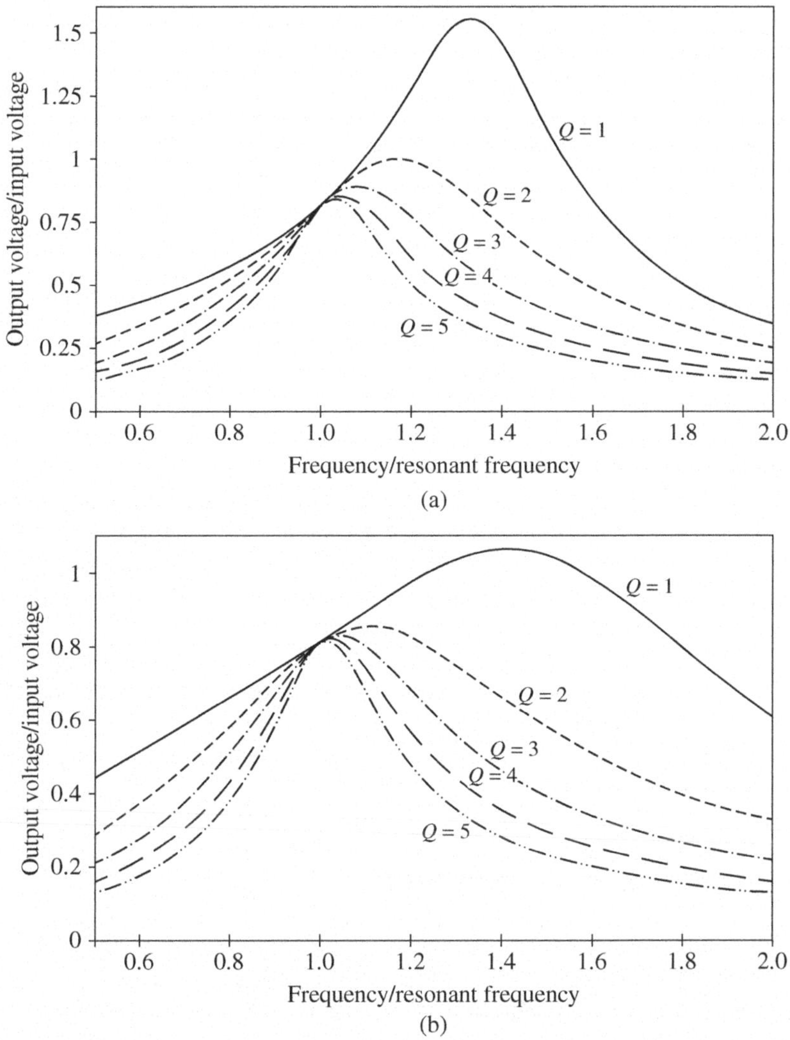

Figure 11.11 Series-parallel converter gain curves based on first harmonic approximation: (a) $C_s = C_p$ (b) $C_s = 2C_p$.

The series-parallel resonant converter is analogous to a buck-boost converter as it can step voltage up or down, as can be seen by its gain characteristics. It tries to take advantage of the best characteristics of the series and the parallel converter while eliminating their weak points. It has the series DC blocking capacitor of a series-resonant converter and the LC output filter of a parallel-resonant converter, but without the drawbacks of lack of no-load regulation for the series-resonant converter and light-load circulating current for the parallel-resonant converter. These two drawbacks can be mitigated with the proper selection of the resonant components, but the converter needs to be operated over a wider range of switching frequencies.

From the characteristic gain curves of Figure 11.11, it can be seen that the converter can operate and regulate at no load provided that the parallel-resonant capacitor C_p, is not too small (if C_p, is zero, then the circuit reverts to the series-resonant converter). It can also be seen that the smaller C_p is, the flatter the gain curves are and the converter resembles a series converter more and more as C_p gets smaller and smaller. For appropriate values of C_p, however, the converter can operate with no load, which removes the main disadvantage of the series-resonant converter.

It is desirable that the series-parallel-resonant converter not have a fixed circulating current flowing through C_p that is independent of load, unlike the parallel-resonant converter. This can be done by decreasing C_p, but doing so makes the gain curves flatter according to Figure 11.11. As C_p gets smaller, the upper frequency needed at light loads increases, which creates more losses. This is the limiting factor in reducing C_p, to reduce circulating current. To have the circulating current decrease with load so that a high part-load efficiency is achieved, it is desirable to select the converter components so that the full-load Q_s is in the neighborhood of 4 or 5. For these values of Q_s, the converter appears essentially as a series-resonant converter and the circulating current decreases as the load decreases. As the load decreases further, however, the converter takes on the characteristics of a parallel-resonant converter, and the circulating current no longer decreases with load, so that that conduction and turn-off switching losses become greater than those of a series-resonant converter.

11.3 LLC RESONANT CONVERTER

At the time of this writing, the most popular resonant converter topology is arguably the LLC resonant converter. Although the converter topology and its properties were established in papers such as [3–5] in the early 1990s, it was not until the 2000s that the topology began to gain in popularity. This converter has some similarities with the fundamental series-parallel-resonant converter, which can also be called an LCC converter, as it has a parallel branch that consists of an inductor (L_p) placed across the primary of the transformer and a series branch that consists of an inductor (L_s) and a capacitor (C_s) placed in series with the inductor/transformer combination. The half-bridge version of this converter, which is the most popular version, is shown in Figure 11.12.

One reason for the popularity of the LLC resonant converter is that it can step voltage up or down like the fundamental series-parallel-resonant converter, but with its third resonant element, the second L, L_p, integrated into the main power transformer. In fact, L_p can be the magnetizing inductance of the main transformer, which reduces converter size and cost. Perhaps the

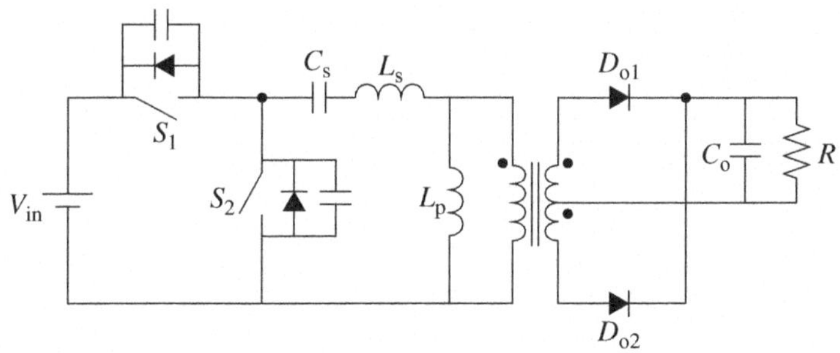

Figure 11.12 LLC resonant converter.

main reason for the converter's popularity, however, is that it can be designed to operate with high input DC voltage in a way that the three fundamental resonant converters cannot. This characteristic makes LLC resonant converter popular as the back-end converter in two-stage AC–DC converters that have a front-end AC–DC converter stage. This converter is typically a boost converter that performs power factor correction (PFC) by shaping the input AC current so that it is sinusoidal and in phase with the input voltage. Since the front-end converter is a boost converter, the output of this converter is high, typically around 400 V in many industrial applications. This high voltage is fed to the input of the DC–DC back-end converter so that a DC–DC converter that can operate with high voltages is needed.

Looking at the gain characteristics of the series-resonant and parallel-resonant converters and considering the fact that these converters can only work with ZVS if their switching frequency is above the resonant frequency, it can be seen the gain can only be reduced if the switching frequency is increased, if ZVS operation is to be maintained. If the load is decreased, then these converters are forced to operate at higher switching frequencies with less current available to discharge switch capacitances. If Si MOSFETS are used, current cannot flow through the body-diodes of switches before they can be turned on with ZVS. The loss of ZVS at high switching frequencies results for a converter with high input voltage results in high switching losses at light loads. This situation may be avoided is the resonant circuit is designed so that there is additional circulating current in the primary, which is especially true for the parallel resonant converter, but this increases conduction losses.

The series-parallel converter is somewhat different from the other two fundamental converters as it actually has two resonant frequencies: a low resonant frequency that is set by the series-resonant tank components L_s and C_s and a high resonant frequency that is set by L_s and the equivalent capacitance of C_s and C_p in series. The graph of characteristic DC gain curves for various loads shown in Figure 11.13 is normalized to the low resonant frequency, which is the series-resonant frequency, and the reflected output voltage. It can be seen that the gain is the same no matter what the value of Q is,

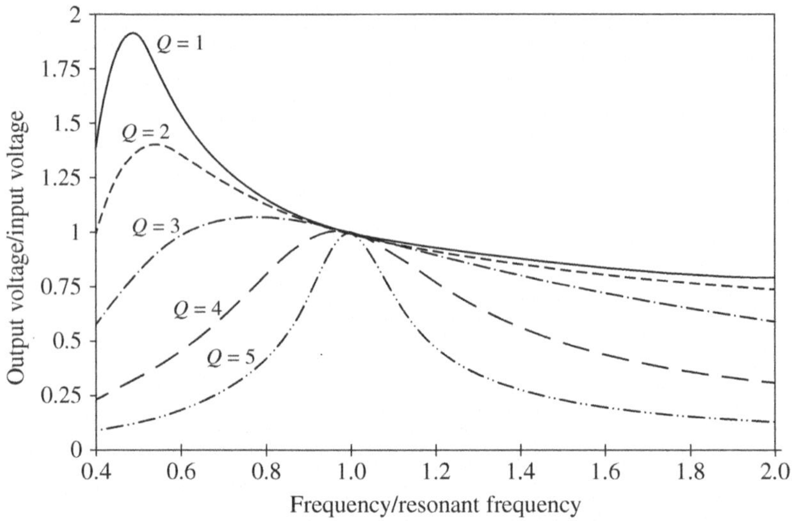

Figure 11.13 LLC-resonant converter gain curves based on first harmonic approximation.

where Q is related to the load as defined by Eq. (11.24). The high resonant frequency is the frequency at which the converter has maximum gain and it changes when Q and the load change. The issue with the series-parallel resonant converter is that operating it around the lower resonant frequency in order to reduce gain means operating it with ZCS so that ZVS operation is lost and the issues that exist for the series-resonant and parallel-resonant converters must be considered as well.

The voltage characteristic of the LLC converter, shown in Figure 11.13, is a "flipped" over version of that of the fundamental series-parallel-resonant converter. There are still two resonant frequencies: a higher frequency, which is the series branch resonant frequency, that is determined by L_s and C_s and a lower frequency, which is the parallel branch frequency, that is determined by the interaction of C_s and the combined inductance of L_p and L_s. In the graph of DC gain curves shown in Figure 11.13, the series-resonant frequency is the same frequency for all the curves and the converter has the same gain for all the curves, whereas the parallel-resonant frequency, which is the frequency at which the converter has the highest gain, is different for each curve. Of the two resonant frequencies, the series-resonant frequency is always the highest. The curves in the graph are normalized with respect to the series-resonant frequency and the output voltage.

The key difference between the LLC converter and the series-parallel resonant converter is that the series-resonant tank component frequency is the higher frequency and the converter can be made to operate with ZVS around this frequency. In other words, the LLC converter can be made to operate with ZVS around both resonant frequencies while the series-parallel converter cannot be made to operate with ZVS around the series-resonant tank component frequency. As a result, the issues that have been described for the three fundamental resonant converters can be avoided as the converter can be operated with ZVS at light loads and with lower switching frequency.

Another way of considering the LLC resonant converter is to examine the impedance of the element that is parallel to the transformer. In the LLC converter, it is an inductor with an impedance that reduces as switching frequency is reduced; in the series-parallel converter, it is a capacitor with an impedance that reduces as switching frequency is increased. Since ZVS operation is related to the amount of current flowing in the converter's primary side, the LLC converter can better maintain ZVS as switching frequency is reduced than can the series-parallel-resonant converter and can actually operate with ZVS even with no load.

The following should be considered when designing a LLC resonant converter:

- The operating frequency must always fall between the series-resonant and parallel-resonant frequencies if the converter is to operate with ZVS.
- With smaller magnetizing inductance, a narrower switching frequency range can be achieved as the gain curves will be sharper, but turn-off switching losses and conduction losses will increase because of high magnetizing current.
- If C_s is too small, then the voltage gain of the converter will be severely limited, thus making the converter unsuitable for particular applications.
- With a larger C_s, the voltage stress on C_s will be smaller, but the short circuit current will be higher and higher switching frequency will be needed to limit the output current.
- The value of C_s should be sufficiently small to get enough voltage gain at heavy load, then L_p should be as large as possible to get the voltage gain within a desired switching frequency range.

Given the popularity of the LLC converter, numerous papers and application notes that demonstrate in detail how the converter can be designed have been published. Examples of such publications such as [4–9] can be found at the end of this chapter in the Reference Section.

11.4 OTHER RESONANT DC–DC CONVERTERS

In addition to the three fundamental resonant DC–DC converters and the LLC resonant converters that have been presented in this chapter, there are many other resonant converters that have been proposed in the literature.

Most of these use three or more resonant elements that are combined to form a resonant network that is placed near the main converter transformer. Each of these resonant converters has its own set of properties and characteristics and thus are well suited or ill suited for particular applications. Given the sheer number of possible resonant converter topologies, it is nearly impossible to discuss each non-fundamental topology in any detail and thus what is presented here is a brief summary:

- In general, the more resonant elements there are in a resonant network, the more likely it is that the converter will require less switching frequency variation under line and load changes [3, 4, 10]. A resonant converter with a fourth-order resonant network generally requires less switching frequency variation under line and load changes than does a converter with a third-order network, which requires less switching frequency variation than does a converter with a second-order network. Although adding more resonant elements may increase converter size and cost, this is may not be the case with some more sophisticated resonant networks that can absorb parasitic components into their network.
- More sophisticated third-order and fourth-order networks must be designed in a way that avoids unwanted multiple resonant interactions. For example in the converter that was proposed in [11] and shown in Figure 11.14, the converter has a series branch that consists of an inductor and capacitor and a parallel branch that consists of a second inductor/capacitor set. If the resonant network is not designed properly, then the current flowing into the network may have sub-oscillations and multiple zero crossings so that ZVS operation is lost. In order to ensure that the converter switches can turn on with ZVS, this current must have a near-sinusoidal shape with not sub-oscillations and just a single zero crossing (from positive to negative) during a switching cycle and

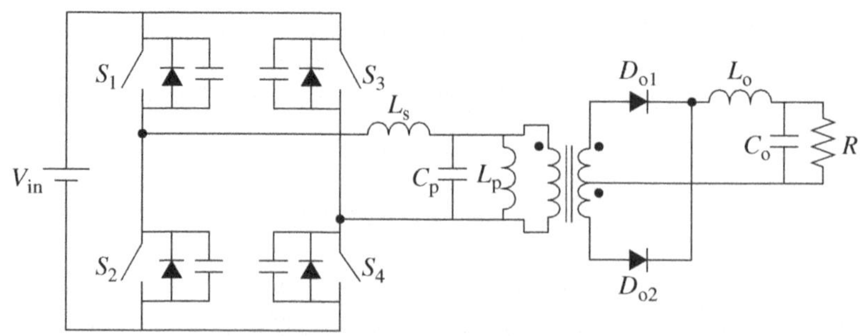

Figure 11.14 Resonant DC–DC converter with three-element resonant circuit.

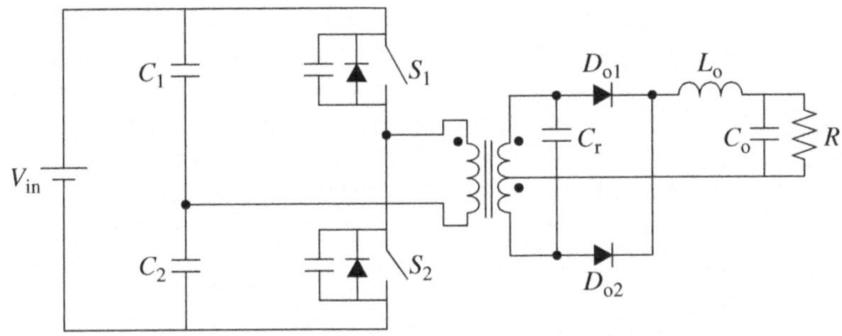

Figure 11.15 Parallel-resonant half-bridge converter with secondary-side resonance.

must lag the square-wave voltage that is placed across the network, as described above.

- Sometimes, it is advantageous to place resonant elements on the secondary side of the transformer in addition to placing some resonant elements on the primary side, as was proposed in [12, 13] and shown in Figure 11.15. Doing so allows for the reduction of the size of certain resonant elements as their effective value is multiplied from the point-of-view of the primary. Putting resonant elements on the secondary side should only be considered for low-output current applications due to the conduction losses that may be caused.
- In topologies where there is a parallel capacitor connected across the transformer, like a parallel-resonant converter, this capacitor can be removed from the transformer primary and placed across a dedicated tertiary winding, as proposed in [14, 15] and shown in Figure 11.16, along with its equivalent resonant circuit. This is done to reduce the size of the capacitor as this tertiary-winding connected capacitor can be smaller, as its impedance is amplified when reflected to the primary. In some applications, the voltage across a parallel connected capacitor

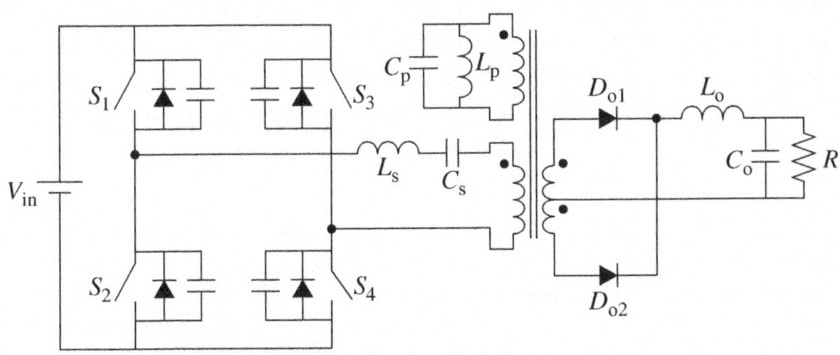

Figure 11.16 Resonant converter with tertiary winding.

can be large, but replacing it with a tertiary winding connected capacitor can reduce this voltage. The use of tertiary windings is limited, however, and should be limited to relative low switching frequencies (<150 kHz) as parasitics components become more prominent when switching frequencies are high. These parasitics become part of the resonant network, making it into a very high-order network (>5 elements) so that multiple zero crossing of the current entering the network is likely to occur. The resonant network may become too sophisticated to be properly used.

- The resonant converters presented in this chapter are ZVS converters as resonant converters are mainly used today as high-frequency, lower power converters where MOSFETs, which work best with ZVS. Resonant converters operated with other devices such as IGBTs, thyristors and thyristor variations are operated with ZCS due to the nature of these devices[1] [16–18]. If ZCS operation if desired, resonant converters can be made to work with ZCS if the current flowing in the resonant network is discontinuous so that it starts at zero at the start of a half-cycle and ends at zero near the end of a half-cycle so that it is at zero at the start of the next half-cycle when current flows in the opposite direction. A discontinuous resonant current waveform ensures that the converter switches can turn on and off with ZCS. Another way of operating a converter with ZCS is to operate it with a switching frequency that is below the resonant frequency. In this case, a negative voltage appears across a switch that helps extinguish the current going through it. If a resonant converter is operated in such a manner where the resonant current leads the voltage across the resonant network, then the switches should have blocking diodes placed in series or be unidirectional IGBTs that have blocking diodes built in the device. Blocking diodes are used to prevent excessive current from flowing through the body-diode of switches when a negative voltage is impressed across them.
- The resonant converters discussed in this chapter are voltage-fed converters as they are far more popular than current-fed converters, such as the ones proposed in [19, 20] and shown in Figure 11.17. Current-fed resonant converters, however, are used in applications where additional voltage boosting is required or in applications where the input current should be continuous with little ripple as in some solar energy system converters. A duality exists between many voltage-fed and current-fed converters;

[1] Thyristors, which are not discussed in this book, are semiconductor devices that are used for high power applications such as for utility-type applications and that can only be turned off is they are starved of current before-hand, which corresponds to ZCS turn-off operation. Variations of the basic thyristor, which are numerous, allow the thyristor to be turned off even when current is flowing through them, but ZCS operation is still preferred. ZCS is preferred for silicon IGBTs due to the current-tail that they have when they are in the process of being turned off.

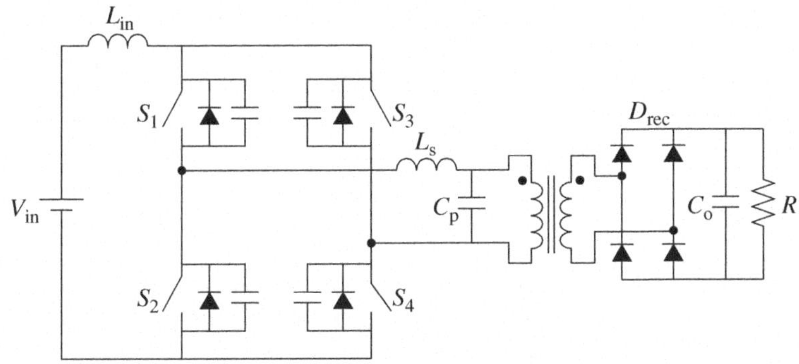

Figure 11.17 Current-fed parallel-resonant converter.

for example, instead of applying a square-wave voltage across a resonant network, a square-wave current can be applied. The underlying principles of current-fed converters are thus very similar to those voltage-fed converters.

11.5 CONCLUSION

Resonant converters use resonant networks consisting of capacitors and inductors to shape voltage to shape switch voltage or switch current so that their switching devices can turn on and off softly. Although the stresses placed on the components of these converters may be higher than those of PWM converters, their ability to operate with soft-switching makes them attractive in many applications, especially in applications where high-frequency power conversion is desired. Resonant converters can absorb parasitics elements that hamper the operation of PWM converters in their resonant networks.

In this chapter, the basic principles behind resonant power conversion for DC–DC converters were explained. The characteristics of three fundamental DC–DC resonant converters – series-resonant, parallel-resonant, and series-parallel resonant – were presented, and it was shown that each converter is analogous to a PWM converter: series-resonant/buck, parallel-resonant/boost, series-parallel-resonant/buck-boost. In addition to the three fundamental converters, there are a vast number of resonant converters with other resonant networks that have been proposed. At the time of this writing, the most popular resonant converter is the LLC resonant converter due to its various properties, which include an ability to step voltage up or down, the integration of one of its resonant tank inductors with the main power transformer, and its ability to be used as the back-end of a two-stage AC–DC converter. These properties were discussed in this chapter and general guidelines for the design of LLC resonant converters were given. The properties and characteristics of other converter were discussed as well. These included converters with more sophisticated resonant networks with more than two resonant elements, converters with tertiary windings, converters with ZCS operation, and current-fed resonant converters.

REFERENCES

1. S. Sooksatra and C. Q. Lee, "Series resonant converter with inductive output filter," *IEEE Industry Applications Society (IAS) Conference*, 1989, Oct. 1–5, 1989, San Diego, USA, pp. 1135–1140.
2. R. L. Steigerwald, "A comparison of half-bridge resonant converter topologies," *IEEE Transactions on Power Electronics*, vol. 3, no. 2, pp. 174–182, Apr. 1988.
3. R. P. Severns, "Topologies for three-element resonant converters," *IEEE Transactions on Power Electronics*, vol. 7, no. 1, pp. 89–98, Jan. 1992.
4. I. Batarseh, "Resonant converter topologies with three and four energy storage elements," *IEEE Transactions on Power Electronics*, vol. 9, no. 1, pp. 64–73, Jan. 1994.
5. A. K. S. Bhat, "Analysis and design of LCL-type series resonant converter," *IEEE Transactions on Industrial Electronics*, vol. 41, no. 1, pp. 118–124, Feb. 1994.
6. R. Lin and C. Lin, "Design criteria for resonant tank of LLC DC-DC resonant converter," *IEEE Industrial Electronics Conference (IECON)*, Nov. 7–10, 2010, Glendale, USA, 2010, pp. 427–432.
7. H. Huang, "Designing an LLC Resonant Half-Bridge Power Converter," *Texas Instruments Power Supply Design Seminar*, SEM1900- Topic 3, TI Literature Number: SLUP263, 2010.
8. A. Lind, "LLC converter design note," *Infineon Technologies Application Note*, AN 2013-03, V1.0, 2013.
9. S. De Simone, "LLC resonant half-bridge converter design guideline," *STMicroelectronics Application Note*, AN2450, 2014.
10. M. T. Outeiro and G. Buja, "Comparison of resonant power converters with two, three, and four energy storage elements," *IEEE Industrial Electronics Conference (IECON)*, 2015, Nov. 9–12, 2015, Yokohama, Japan, pp. 1406–1411.
11. P.K. Jain, H. Soin and M. Cardella, "Constant frequency resonant DC/DC converter topologies with zero switching losses," *IEEE Transactions on Aerospace and Electronics Systems*, vol. 30, no. 2, pp. 534–544, Apr. 1994.
12. K. Liu and F. C. Lee, "Secondary-side resonance for high-frequency power conversion," *IEEE Applied Power Electronics Conference (APEC)*, 1986, Apr. 28–May 1, 1986, New Orleans, USA, pp. 83–89.
13. M. M. Swamy and A. K. S. Bhat, "Analysis and design of a parallel resonant converter with secondary-side resonance," *IEEE International Symposium on Circuits and Systems (ISCAS)*, 1989, May 8–11, 1989, Portland, USA, pp. 1847–1850.
14. A. K. S. Bhat, "Analysis and design of a parallel resonant converter with the resonating capacitor on a tertiary winding," *IEEE Transactions on Industry Applications*, vol. 29, no. 6, pp. 1069–1075, Nov.–Dec. 1993.

15. P. Jain, P. Jain and J. Quaicoe, "Tertiary side resonant DC/DC converter with integrated magnetics," *IEEE Transactions on Magnetics*, vol. 32, no. 5, pp. 5016–5018, Sept. 1996.
16. F. P. Dawson and P. Jain, "A comparison of load commutated inverter systems for induction heating and melting applications," *IEEE Transactions on Power Electronics*, vol. 6, no. 3, pp. 430–441, July 1991.
17. P. Jain, F. P. Dawson and S. B. Dewan, "A near-zero current-switching series resonant inverter using GTOs," *IEEE Transactions on Industrial Electronics*, vol. 39, no. 4, pp. 351–358, Aug. 1992.
18. G. Ivensky, I. Zeltser, A. Kats and S. Ben-Yaakov, "Reducing IGBT losses in ZCS series resonant converters," *IEEE Transactions on Industrial Electronics*, vol. 46, no. 1, pp. 67–74, Feb. 1999.
19. V. Caliskan, J. Gegner and C. Q. Lee, "Current-driven zero-voltage-switched resonant converter with capacitive output filter: analysis and experimental results," *IEEE Applied Power Electronics Conference (APEC)*, 1992, Feb. 23–27, 1992, Boston, USA, pp. 211–218.
20. S. Jalbrzykowski and T. Citko, "Current-fed resonant full-bridge boost DC/AC/DC converter," *IEEE Transactions on Industrial Electronics*, vol. 55, no. 3, pp. 1198–1205, Mar. 2008.

CHAPTER 12

RESONANT CONVERTERS PART II – PWM CONTROLLED, QUASI-RESONANT, AND ULTRAHIGH-FREQUENCY CONVERTERS

This chapter is a continuation of the previous chapter, which introduced fundamental concepts related to resonant power conversion. The focus of this chapter is on the following topics:

- *PWM operation of resonant converters*: This is the operation of resonant converters with fixed switching frequencies as opposed to variable switching frequencies.
- *Quasi-resonant converters*: These are converters that can be synthesized from PWM converters by adding resonant elements so that they are partially PWM and partially resonant converters.
- *Ultrahigh-switching frequency resonant converters*: These are resonant converters that incorporate parasitic elements in their topologies and can be operated with extremely high switching frequencies (>1 MHz).

The basic principles related to each topic are discussed and example converters are presented to demonstrate these principles.

DC–DC Converter Topologies: Basic to Advanced, First Edition. Gerry Moschopoulos.

Published 2024 by John Wiley & Sons, Inc.

12.1 FIXED FREQUENCY RESONANT CONVERTERS

The discussion in the previous chapter assumed that resonant converters are operated with variable switching frequency control. Variable switching frequency control is typically used in two-switch half-bridge resonant topologies (<500 W) as the gating signals of each switch has a fixed duty-cycle of $D = 0.5$ and is complementary to that of the other switch. Operating with $D = 0.5$ ensures that the switches turn on with ZVS. The switches in a half-bridge converter cannot turn on with ZVS if D is <0.5 because the current flowing through the body-diode of a switching device will be extinguished and the output capacitance of the device will be charged to a nonzero voltage by the time the device is turned on. In addition to ZVS, operating with $D = 0.5$ helps to minimize the harmonic content of the voltage that is fed to the output as the symmetrical square-wave voltage that is placed across the converter's resonant network has a better non-fundamental harmonic content than does a voltage with zero-voltage gaps. With the duty-cycle of the half-bridge converter switches fixed at $D = 0.5$, varying the converter's switching frequency and thus varying the impedance that the resonant network presents to the voltage across it is the only way that the voltage can be regulated. Similar arguments can be made for operating full-bridge converters so that their primary voltage has no zero-voltage states and only diagonally opposed switches are on at any given time during the switching cycle.

The main disadvantage with variable switching frequency control is that it becomes less attractive when a converter is forced to operate with heavier loads (>500 W) over a wide load range. In such cases, it becomes difficult to optimize the design of converter magnetics as they must be used for a wide range of switching frequency; this is not an issue for lower power converters with narrower switching frequency variation. From the point-of-view of magnetics, it is easier to design transformers and inductors for one specific frequency, thus the benefits of using fixed-frequency PWM for higher-power full-bridge converters outweigh any benefit that may be achieved by operating a resonant converter with its switches having $D = 0.5$ and with variable switching frequency control. Several methods of fixed-frequency control are presented in this section.

12.1.1 Full-Bridge Resonant Converters Operated with Phase-Shift PWM

The operation of the four-switch ZVS-PWM full-bridge converter was discussed in detail in Chapter 7 of this book. There, it was shown how the converter can be controlled by using the phase-shift PWM technique, which involves establishing complimentary gating signals of $D = 0.5$ for each switch of each of the two legs (minus some dead time to allow ZVS operation to occur) then shifting the gating signals of the switches of one leg relative to those of the other leg. It was shown in papers such as [1, 2] how four-switch full-bridge resonant DC–DC converters can be operated with phase-shift PWM so that the width of the pulses of the square-wave voltage waveform that appears across the resonant network can be varied. The narrower the voltage pulses are, the less fundamental harmonic component there will be, which results in lower output voltage after this fundamental component is stepped down by the transformer, then rectified and filtered.

As stated above, PWM control is advantageous as it allows the design of converter magnetics to be optimized for one switching frequency instead of having to work for a range of switching frequencies; it simplifies the control of the converter as well. In [1], phase-shift PWM was used to operate a series-resonant full-bridge converter, but better performance can be ensured if more sophisticated resonant networks are used, as was proposed in [2] and shown in Figure 12.1. Considering the converter shown in Figure 12.1a, assuming that switches S_1 and S_3 have been turned on in previous modes and that the resonant tank current flows through S_3 and the body-diode of S_1, turning off S_3 results in current charging and discharging the output capacitances of S_3 and S_4, C_{s3} and C_{s4}, respectively. When C_{s4} is discharged, current then flows through the body-diode of S_4, allowing this switch to be

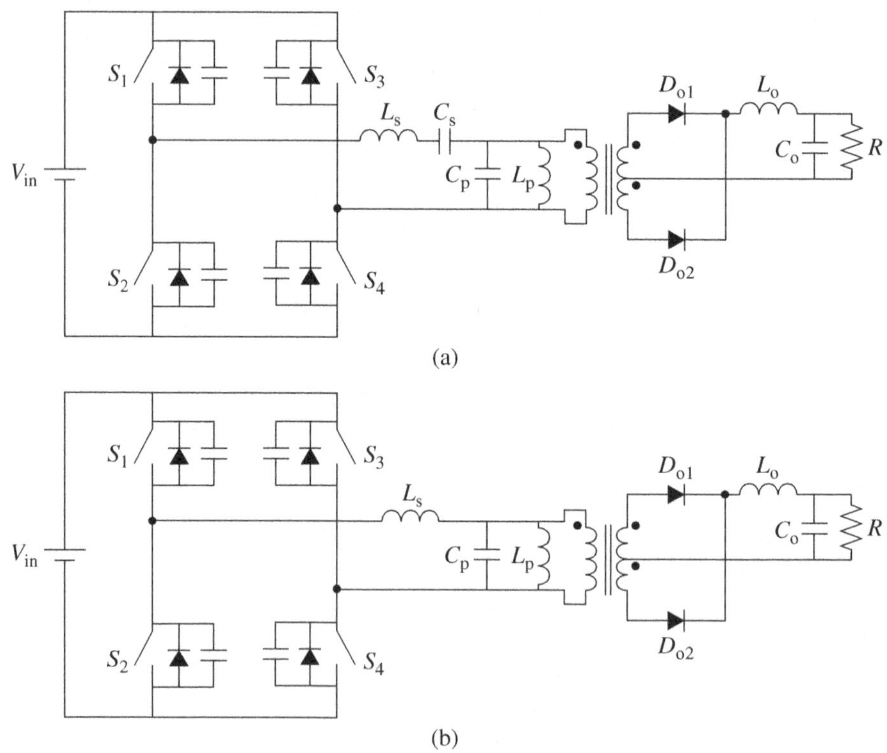

Figure 12.1 Full-bridge resonant converters operated with phase-shift PWM: (a) Series-tuned parallel off-tuned. (b) Parallel-tuned.

turned on with ZVS. With S_1 and S_4 both on, the resonant tank current reverses direction and now flows through both switches. If S_1 is turned off a little later in the cycle, this initiates a series of events that ends with S_2 turning on with ZVS as current flows through its body-diode. With S_2 and S_4 both on, the voltage across the resonant tank drops to zero and current flows through the body-diode of S_2 and switch S_4.

A little later in the switching cycle, S_4 is turned off and current eventually flows through the body-diode of S_3, allowing this switch to be turned on with ZVS. With both S_2 and S_3 on, the polarity of the voltage differs from what it was near the start of the switching cycle with S_1 and S_4 on and current reverses direction. A negative voltage is impressed across the resonant tank until S_2 is turned off and S_1 is turned on with ZVS. The converter then operates as it did at the start of the switching cycle with no voltage across the resonant tank and with current flowing through S_3 and the body-diode of S_1.

The ZVS operation of a resonant converter operating with phase-shift PWM can only be ensured if the resonant tank current lags the voltage across the tank. The more this current lags the voltage across the resonant tank, the wider the input voltage and output load voltage range for ZVS operation will be, but this comes at the cost of greater conduction losses and more current will be circulating in the converter. If the power transferred to the resonant tank can be approximated as being $P = VI\cos\theta$ where V and I are the fundamental rms components of the voltage across the resonant tank and the resonant tank current, respectively, and θ is the phase difference between the two, then increasing θ to make the resonant current more lagging will require that I be increased as well to deliver the same amount of power.

What is needed then is a resonant tank that will allow voltage to be regulated by using a narrow range of phase-shift control so that θ can be made small. For the converter shown in Figure 12.1a, this can be achieved by a proper design of the resonant components in the two branches of the resonant circuit. The components of the series branch should be tuned to the switching frequency so that the branch provides zero impedance at this frequency. The components of the parallel branch should be off-tuned so that this branch provides an effective inductive impedance at the switching frequency. If the resonant circuit is designed in such a manner, then the output voltage becomes less dependent on the load so that the converter can be controlled by a narrow range of phase-shift PWM. It also allows the converter to have a high gain, which results in lower conduction losses.

The series-tuned parallel off-tuned resonant converter shown in Figure 12.1a has one key drawback and that is that the voltage across the series resonant capacitor can become very high when the input voltage is high. As a result, this converter is best suited for low-voltage, high-current applications. To address this drawback for high-voltage, low-current applications, the resonant converter shown in Figure 12.1b was proposed in [2]. The converter is similar to the one shown in Figure 12.1a except that the series resonant capacitor has been removed. The series inductor L_s provides high impedance to the harmonic components that are present in the output voltage of the bridge inverter, limits discharging from the parallel capacitor to the switches of the full bridge inverter during a zero-voltage interval, and assists in the ZVS operation of the converter. The components of the parallel branch provide a stiff, near-sinusoidal voltage across the transformer and confine circulating current within the parallel branch. The components L_p and C_p of the parallel branch are tuned at the switching frequency of the converter. The value of the series inductor L_s is chosen so that the resonant frequency f_{rs} of L_s and parallel capacitor C_p is higher than the resonant frequency of parallel inductor f_{rp} L_p and parallel capacitor C_p. As explained in [2], it is possible for the converter to operate in multiple-zero-crossing mode (MZC mode) or single-zero-crossing mode (SZC mode). In MZC mode of operation, the resonant frequency f_s is such that the series-resonant current has multiple zero crossings in each half cycle, unlike SZC mode where the switching frequency is such that the current has just a single zero-crossing in each half-cycle. Since MZC mode of operation does not always ensure that the converter switches operate with ZVS, unlike SZC mode, the converter should always be operated in SZC mode with a ratio of $1.5 < f_{rs}/f_{rp} < 1.7$, as recommended in [2].

12.1.2 Resonant Converters Operated with Asymmetrical PWM

Resonant converters can be operated with fixed-frequency control by using asymmetrical PWM methods. Unlike variable switching control where a symmetrical square-wave voltage with no zero-voltage states in impressed across a resonant circuit and the output voltage is regulated by varying the switching frequency, or phase-shift PWM where a symmetrical square-wave voltage with zero-voltage states is placed across the resonant circuit and the output voltage is regulated by adjusting the width of the voltage pulse, asymmetrical PWM methods impress an asymmetrical voltage across the resonant circuit. Asymmetrical PWM for two-switch and full-bridge resonant converters is discussed here.

12.1.2.1 Half-Bridge Resonant Converters Operated with Asymmetrical PWM

Asymmetrical PWM can be used to achieve fixed-switching frequency operation for lower power two-switch resonant converters that can be used as point-of-use power supplies. For example, consider the series-resonant two-switch converter that was proposed in [3] and is shown in Figure 12.2. With this converter, fixed-frequency operation is possible if the two switches are operated with different and complementary duty cycles. If switch S_1 is made to operate with a duty cycle of D and S_2 with a duty cycle of $(1 - D)$ so that S_2 is on when S_1 is off and vice versa, a square-wave voltage with levels of V_{in} and 0 will appear across the series resonant network. The resonant capacitor, C_r, blocks the DC component of the square-wave voltage, which is zero when $D = 0.5$, but increases as D is decreased. The current through

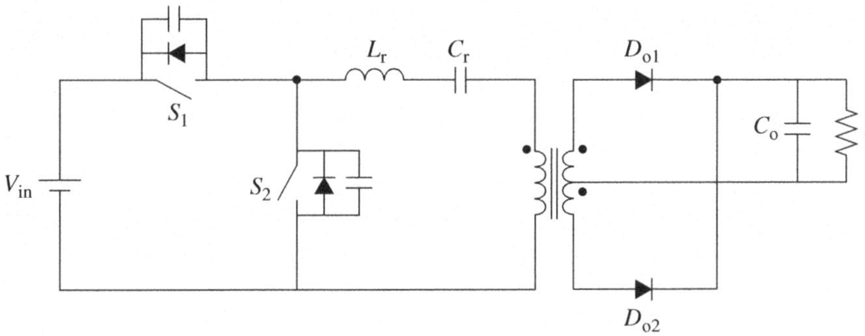

Figure 12.2 APWM series-resonant converter.

the resonant network, which is based on the fundamental AC component of the square-wave voltage, is AC, and this AC current is passed through the transformer, rectified, and then filtered to become the output DC voltage. The maximum output voltage appears at the output when $D = 0.5$. Decreasing D or increasing D past 0.5 decreases the fundamental AC component of the square-wave voltage seen by the resonant network so that the output DC voltage is decreased as well.

The APWM converter shown in Figure 12.2 works as follows: Assuming that switch S_1 is off and current is flowing through S_2, S_2 is turned off at some time $t = t_0$ and current flows through capacitors C_{s1} and C_{s2}, thus discharging and charging these capacitors, respectively. Eventually, voltage across C_{s1} falls to zero, which enables the body-diode of S_1 to conduct. While current is flowing through the body-diode of S_1, S_1 can be turned on with ZVS. Since the voltage across L_r is negative late in this mode, the current through the body-diode of S_1 decreases until it becomes zero. When current starts to flow through S_1, at some time, $t = t_1$, input voltage V_{in} appears across the $L_r - C_r -$ transformer circuit and power is transferred from the input to the output. When S_1 is turned off at some time $t = t_2$, C_{s1} and C_{s2} begin to charge and discharge, respectively, until C_{s2} is fully discharged. Current then flows into the body-diode of S_2, allowing this switch to be turned on with ZVS. Late in this mode, the current in L_r begins to decrease and continues to do so until it becomes zero. When current begins to flow through S_2 at some time $t = t_3$, zero voltage appears across the $L_r - C_r -$ transformer circuit and power is transferred from C_r to the output. This mode continues until S_2 is turned off to initiate the next switching cycle.

The following should be noted about the APWM converter shown in Figure 12.2:

- Both C_{s1} and C_{s2} discharge into the resonant circuit and thus the energy that they store when switches S_1 and S_2 are off, respectively, gets transferred to the load. This allows a larger capacitor value to be used, which helps reduce turn-off losses.

- The converter can be made to operate at no-load if the transformer magnetizing inductance is reduced. Less magnetizing inductance results in more current available to discharge switch capacitors, but also more conduction losses. A trade-off must be considered between ZVS range and conduction losses.
- Operating the converter with an asymmetrical PWM method complicates the implementation of synchronous rectifiers (SRs) at the secondary as the flow of current through the output diodes becomes asymmetrical as well. This is especially true under light-load conditions when the duty-cycle is low and the square-wave voltage that is impressed across the resonant tank circuit is very asymmetrical. The asymmetrical distribution of current in the output rectifiers is seen as a drawback by some who prefer to operate resonant converters with variable switching frequency control, which places symmetrical square-wave voltages across their resonant tanks. This makes the distribution of current in the output rectifying diodes symmetrical as well.
- The input voltage range over which ZVS can be achieved is narrow. Although this range can be made wider by designing the resonant tank so that its voltage gain characteristic is narrow, this will increase the voltage across the resonant capacitor and lead to more losses.

The input voltage range over which ZVS can be achieved can be expanded if an auxiliary circuit consisting of two capacitors, C_1 and C_2, and an inductor L_a are added to the converter as was proposed in [4] and shown in Figure 12.3. The auxiliary circuit allows the switches, especially S_1, to operate with ZVS, regardless of the input voltage and output load. This is because when one switch is on, current is made to flow in L_a, and this current can be used to discharge the output capacitance of the complementary switch in the same leg after the switch is turned off. Since the ZVS operation of the converter is less reliant on the resonant tank, the series inductor can be reduced in size to the

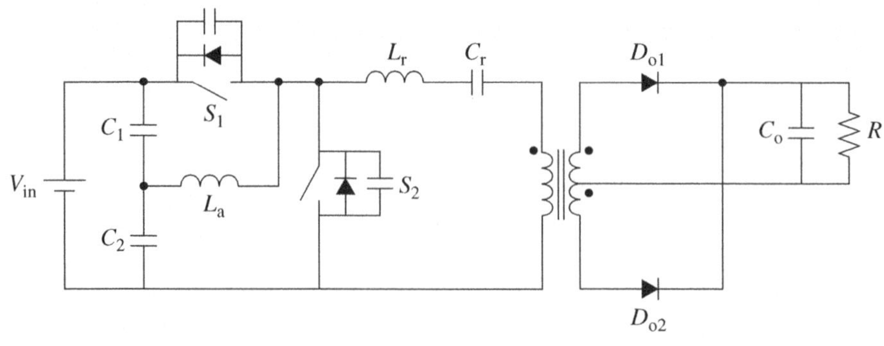

Figure 12.3 APWM series-resonant converter with passive auxiliary circuit for ZVS operating range extension.

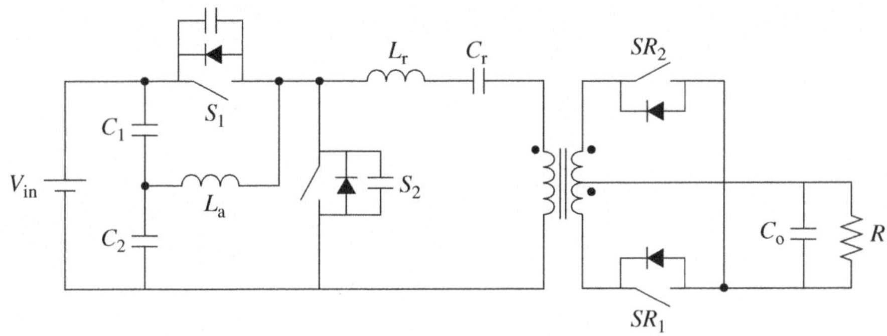

Figure 12.4 APWM converter with passive auxiliary circuit and synchronous rectifiers.

extent that it can be integrated with the transformer as a single magnetic element.

The operation of the APWM converter shown in Figure 12.3 can be further improved so that it can operate with ZVS under even wider input voltage and load conditions if the secondary diodes are replaced with active synchronous rectifier switches, as was proposed in [5] and shown in Figure 12.4. With two sets of switches, a set at the primary side of the converter and a set and the secondary side, two control loops can be used instead of one. One control loop can be implemented for the primary-side switches so that they can be controlled to take care of input voltage variations. A secondary control loop can be used for the secondary side switches so that they can be controlled to take care of output load variations. Details of this overall converter control scheme are presented in [5].

A hybrid-control strategy that can further improve the efficiency of the converter under heavy loads, light loads, and very light loads was proposed in [6]. With this control scheme, the converter works as follows:

- For heavy loads, the output voltage is regulated by the primary-side controller and the secondary-side controller is inactive.
- For light loads, the secondary-side controller is used to deal with load variations and the primary-side controller is used to deal with input voltage variations.
- For very light loads, a pulse-skip primary side-control technique is used, whereby gating signals are fed to the converter during some switching cycles, but not for all switching cycles. If the load is very light, gating signals are only fed to the switches occasionally and not often.

With the modifications that have been discussed thus far, a series-resonant APWM converter can be made to operate with ZVS over a wide line and load range. Light-load ZVS can be achieved using the passive auxiliary circuit

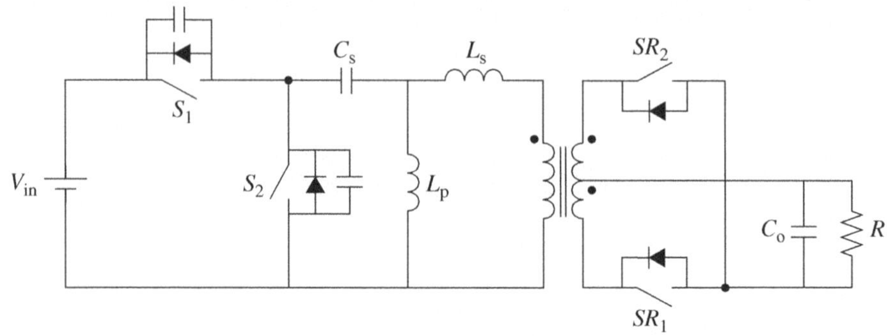

Figure 12.5 CLL resonant converter with synchronous rectifiers.

connected to the primary-side switches, but the current this circuit produces can be relatively excessive for light loads and thus cause an increase in conduction losses. The need for an auxiliary circuit can be avoided and these conduction losses can be reduced if the resonant tank is changed so that it is a CLL circuit, as was proposed in [7] and shown in Figure 12.5.

The CLL APWM converter is like the series-resonant APWM converter, but with an additional inductor L_p connected in parallel, at the node of series capacitor C_s and series inductor L_s. In a series-resonant APWM converter with a passive auxiliary circuit, the current of the series-resonant tank is solely responsible for achieving ZVS. In the CLL APWM converter, the voltage between C_s and L_s varies with load so that when L_p is placed across this voltage, it provides another current path to increase the lag of the switch current, thus increasing the ZVS range of the converter. The addition of L_p provides an additional degree of freedom in the design of the converter so that the efficiency of the converter over a specified range of input voltage can be maximized with proper design.

12.1.2.2 Full-Bridge Resonant Converters Operated with Asymmetrical PWM

Asymmetrical PWM can also be used in full-bridge resonant converters as a fixed-frequency alternative to phase-shift PWM. Considering the full-bridge series-resonant converter shown in Figure 12.6, the converter can be operated with an asymmetrical PWM technique like the one used to operate the APWM converter. With this PWM method, switches S_1 and S_4 are each made to operate with a duty cycle D and are both on simultaneously while switches S_2 and S_3 are made to operate with a duty cycle of 1-D and are both on simultaneously. A third, more sophisticated method is to use so-called asymmetrical clamped-mode control (ACM) as was proposed in [8, 9]. With this method, again considering the series resonant converter shown in Figure 12.6, the switches in one leg are made to operate with a duty cycle of 0.5 and are complementary to each so that both switches in the leg are never

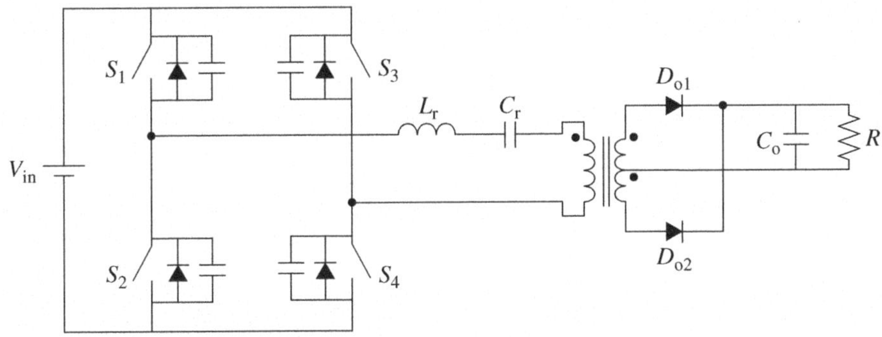

Figure 12.6 Series-resonant full-bridge converter.

on at the same time. The switches in the other leg, however, are made to operate in an asymmetrical manner with one switch operating with a duty cycle of D and its counterpart in the same leg operating with a duty cycle of 1-D. For example, in the series resonant converter shown in Figure 12.6, S_1 can have a duty cycle of D, S_2 can have a duty cycle of 1-D, and S_3 and S_4 can each have a duty cycle of 0.5. The output voltage can be regulated by adjusting D and by phase-shifting the gating signals of one leg related to those of the other so that the PWM method has two degrees of freedom – the amount of phase shift and D – compared to phase-shift PWM and APWM, which have only one. A comparison of these three methods of fixed-frequency PWM was performed in [10, 11] for a full-bridge series resonant converter, and it was determined that using the ACM PWM method results in the best efficiency because it has two degrees of freedom.

It should be noted that more sophisticated methods of controlling resonant converters have been proposed in the literature, such as the ones reviewed in [12]. The literature that is related to the topic of resonant converter control, whether it is variable switching control or PWM control, is vast and deep and well beyond the scope of this book.

12.1.3 Adding Variable Resonant Components

Another approach to operating a converter with fixed switching, but one that maintains symmetrical current distribution and stresses for the switches is to operate the converter with a duty-cycle of $D = 0.5$ regardless of the input voltage and load and change the resonant network itself. One way to do so is to add additional active and passive elements to the converter, as was proposed in [13] and shown in Figure 12.7. The topology shown in Figure 12.7 has a variable capacitor that consists of two switches connected in series with their body-diodes arranged so that the switch combination forms a bidirectional switch. A capacitor is connected

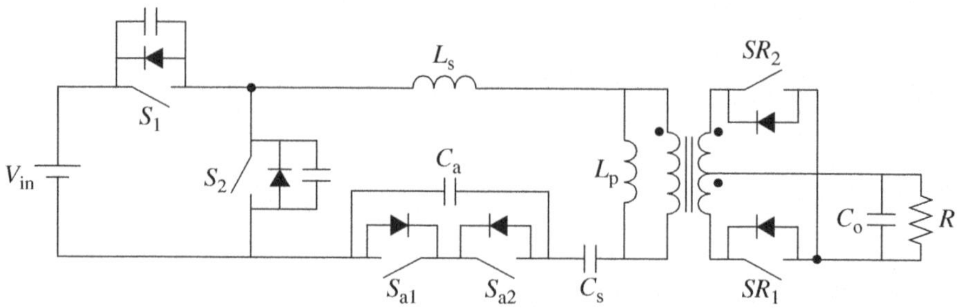

Figure 12.7 LLC-resonant half-bridge converter with variable resonant capacitor.

parallel to this bidirectional switch and the whole switch/capacitor arrangement is connected in series with the resonant capacitor of the LLC resonant network. The general principle of operation is that the longer the switches are on during a switching cycle, the small the effective capacitance of the LLC network should be. Varying the resonant capacitance on-time changes the equivalent impedance of the resonant network, which changes the gain of the converter without changing the switching frequency or the duty-cycle of the switches; the duty-cycle is kept fixed at $D = 0.5$. Each switch is turned on whenever the voltage across C_a is negative relative to its polarity so that current does flow through the switch immediately, but only after it has been discharged. This approach works best with some sort of current sensing that makes the turning on of each switch be relative to the zero-crossing points of the resonant network current.

12.2 QUASI-RESONANT CONVERTERS

Quasi-resonant converters (QRCs) derive their name from the fact that they are a mix of PWM and resonant converters – they are not fully resonant converters, but quasi-resonant. They are partially resonant converters as their topology contains resonant elements that shape switch voltage and current waveforms to allow their switches to turn on with ZVS or with ZCS. They are not fully resonant, however, as they are typically synthesized by adding resonant elements to some standard PWM-controlled topology such as a boost or buck converter and thus have some of the features of the original PWM topology. They have the same advantages and drawbacks of fully resonant converters – quasi-resonant converters can operate with soft-switching with just a few additional resonant components – but they need to be operated with variable switching frequency control and their devices are subjected to high voltage and/or current stresses.

Much of the initial work on QRCs was done by F. C. Lee and his research group during the mid 1980s and a summary of this work during this time was published in [14]. An example of a QRC is the ZVS-QRC buck converter shown in Figure 12.8. This converter has the same topology as a standard PWM buck converter except that resonant elements C_r and L_r have been added to the converter – C_r is added across the switch, in parallel, and L_r is added in series with the switch. If the switching frequency is sufficiently high (>1 MHz), the output capacitance of a MOSFET may be used as C_r and L_r may be the parasitic inductance of wiring or a printed-circuit-board track. Circuit diagrams that show the modes of operation that the converter goes through during a switching cycle are shown in Figure 12.8b–e while the switch voltage and current waveforms are shown in Figure 12.9. It is assumed that the current through output inductor L_o is a pure DC current, I_o, with no ripple.

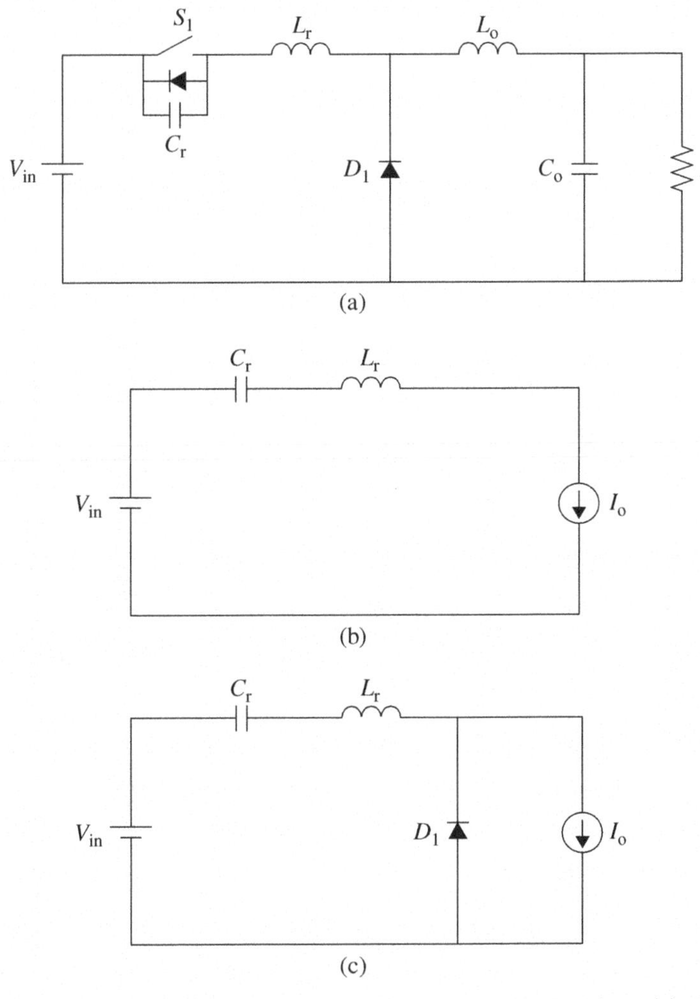

Figure 12.8 ZVS-QRC buck converter and modes of operation. (a) Converter. (b) Mode 1. (c) Mode 2. (d) Mode 3. (e) Mode 4.

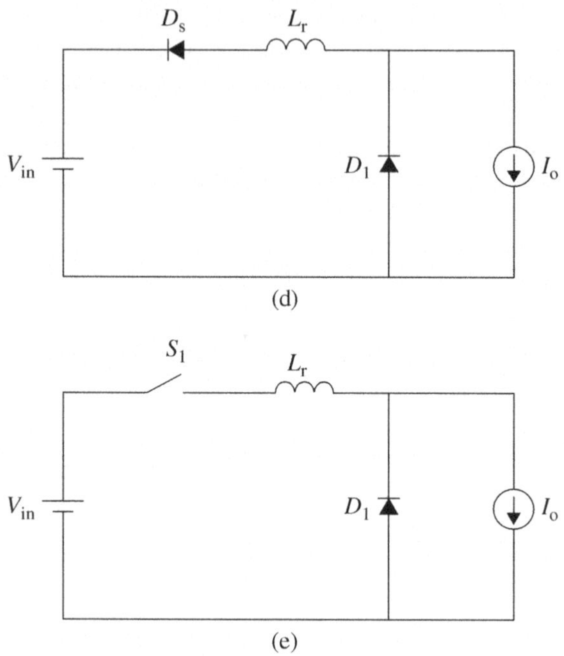

Figure 12.8 (*Continued*)

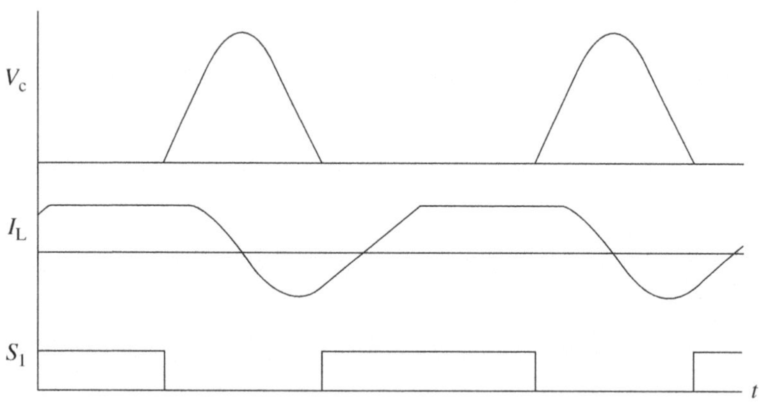

Figure 12.9 ZVS-QRC buck converter waveforms.

The converter's modes of operation for an entire switching cycle are as follows:

Mode 1 (t_0–t_1): At $t = t_0$, the switch is turned off and current flows through capacitor C_r as there is current in L_r. C_r limits the rate of rise of the voltage across the switch so that it can turn off with ZVS.

Mode 2 (t_1–t_2): This mode begins at $t = t_1$ as soon as C_r has been charged to the input voltage V_{in}, which drops the voltage at the cathode of the diode to zero (if the diode is assumed to be ideal with no forward voltage drop) so that the diode becomes forward biased and begins to conduct. The current in the diode is the difference between the current flowing through L_o, I_o and the current flowing through L_r, I_{Lr} with $I_{Lr} < I_o$. During this mode, the current in L_r reaches zero, then reverses direction, and begins to discharge C_r. This mode ends when C_r has been fully discharged and the voltage across it is zero.

Mode 3 (t_2–t_3): This mode begins at $t = t_3$ when current begins to flow through the body-diode of the switch. While this is happening, the current in L_r is flowing toward the source and is decreasing as the voltage across L_r is negative. Moreover, the switch can be turned on with ZVS as the voltage across it is clamped to zero.

Mode 4 (t_3–t_4): This mode begins when the current in L_r reaches zero, then reverses direction, and flows through the switch. It continues to increase until the end of this mode. The converter then begins to operate in Mode 1 again.

The ZVS-QR buck converter shown in Figure 12.8 can be transformed into a ZCS-QR buck converter simply by changing the placement of C_r, as shown in Figure 12.10. Instead of C_r being placed in parallel with the switch, it is placed in parallel with the main buck diode. Circuit diagrams that show the modes of operation that the converter goes through during a switching cycle are shown in Figure 12.10b–e and the switch voltage and current waveforms are shown in Figure 12.11. It is assumed that the current through output inductor L_o is a pure DC current with no ripple, I_o, and that the switch is a unidirectional switch. In the case of a MOSFET, this would mean that the body-diode of the device is blocked with a series diode. The converter's modes of operation for an entire switching cycle are as follows:

Mode 1 (t_0–t_1): Before $t = t_0$, the switch is off and current freewheels through the main buck diode. At $t = t_0$, the switch is turned on and current in L_r begins to rise. Since L_r is in series with the switch, the switch turns on with ZCS.

Mode 2 (t_1–t_2): This mode begins at $t = t_1$ when the current in L_r, I_{Lr} is equal to the current in L_o, I_o. When this happens, the diode stops conducting and since I_{Lr} continues to rise, the difference in current between I_{Lr} and I_o begins to charge C_r. Sometime during this mode, the voltage across C_r, V_{Cr} becomes greater than the input voltage V_{in} so that the current in L_r begins to drop.

Mode 3 (t_2–t_3): This mode begins at $t = t_2$ when the current through the switch and L_r is zero. When this happens, the switch can be turned off with ZCS. C_r began to discharge sometime during Mode 2 and it continues to discharge during this mode.

Mode 4 (t_3–t_4): This mode begins at $t = t_3$ when C_r has been fully discharged and I_o begins to flow through the diode. This mode ends when the switch is turned on to initiate another switching cycle.

It should be noted that regardless of whether ZVS or ZCS is used, the stresses on the converter components can be high compared to those of a PWM converter. For example, the peak voltage stress of a switch in a ZVS-QR buck converter may be as much as 2.5 times that of a switch in PWM buck

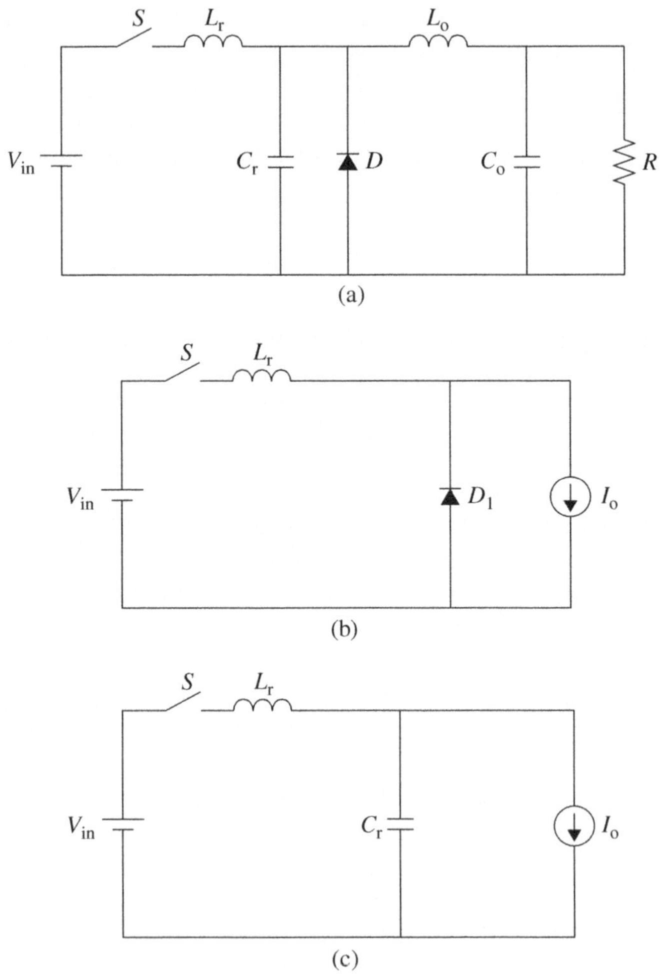

Figure 12.10 ZCS-QRC buck converter and modes of operation. (a) Converter. (b) Mode 1. (c) Mode 2. (d) Mode 3. (e) Mode 4.

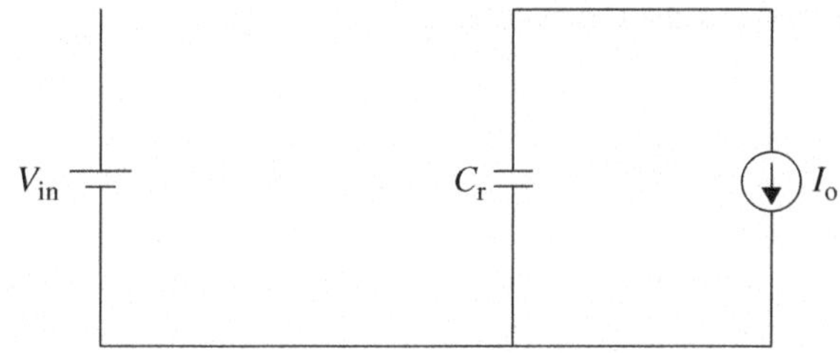

(d)

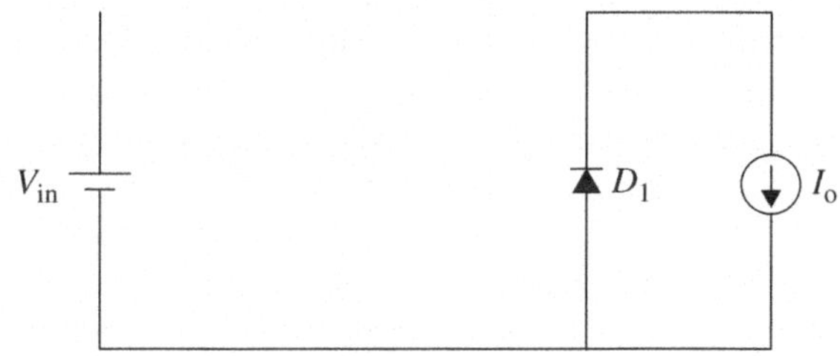

(e)

Figure 12.10 (*Continued*)

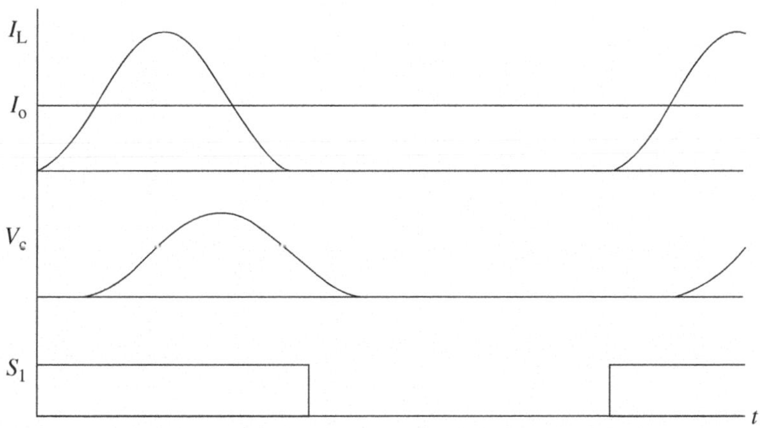

Figure 12.11 ZCS-QRC buck converter waveforms.

converter. Also, it is possible to implement ZCS-QR converters with bidirectional switches, devices with body-diodes. In this case, there will be a small amount of negative current that will flow through the body-diode after Mode 2. The decision as to whether or not to place a blocking diode in series with the device to keep current from flowing through the body-diode depends on the operating conditions of the converter, which dictates the design of the quasi-resonant network and thus the magnitude of this current.

12.2.1 Resonant Pulse Converters

The main principle of quasi-resonant converters – taking PWM converters and adding resonant elements to shape their waveforms – can be extended to full-bridge converters. One such example is the current-fed ZCS resonant boost converter that was proposed in [15] and is shown in Figure 12.12. For this converter, the structure of the resonant circuit matches that of the resonant circuit of the ZCS-QRC converter. Although the converter shown in Figure 12.12 has the same topology as a parallel resonant converter, the design of the resonant circuit components is fundamentally different. In a current-fed parallel-resonant converter, the resonant circuit components are selected in a way that makes the primary current sinusoidal. For the converter shown in Figure 12.12, the resonant circuit components are smaller so that any resonance that occurs is only for a fraction of the switching cycle. This can be seen from the switch voltage and current waveforms that are shown in Figure 12.12b and thus converters such as the one shown in Figure 12.12 are sometimes referred to as resonant pulse converters. Although the waveform is essentially a square-wave with some resonant humping, the converter

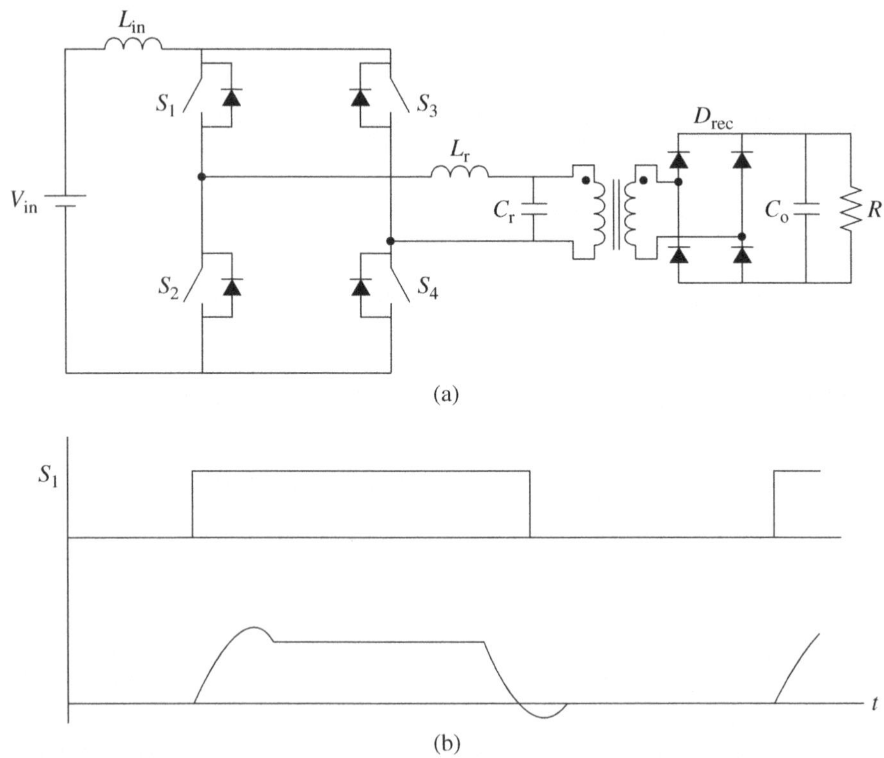

Figure 12.12 Current-fed ZCS resonant-pulse converter: (a) Topology. (b) Switching waveforms.

must be controlled by using a variable switching frequency constant on-time, variable off-time method, much like the QRC-ZVS buck converter shown in Figure 12.10. A ZCS turn-off of a converter switch can only occur when the current in the switch is forced to zero this, the on-time of the switch must be fixed. The frequency is changed by varying the off-time of the converter switches.

12.2.2 Fixed-Frequency Quasi-Resonant Converters

A quasi-resonant converter can be made to operate with fixed switching frequency if a switch is added either in series or in parallel to one of the resonant elements, as was proposed in [16] and shown in Figure 12.13, along with some of its waveforms. In this converter, which is a ZCS-QRC buck converter,

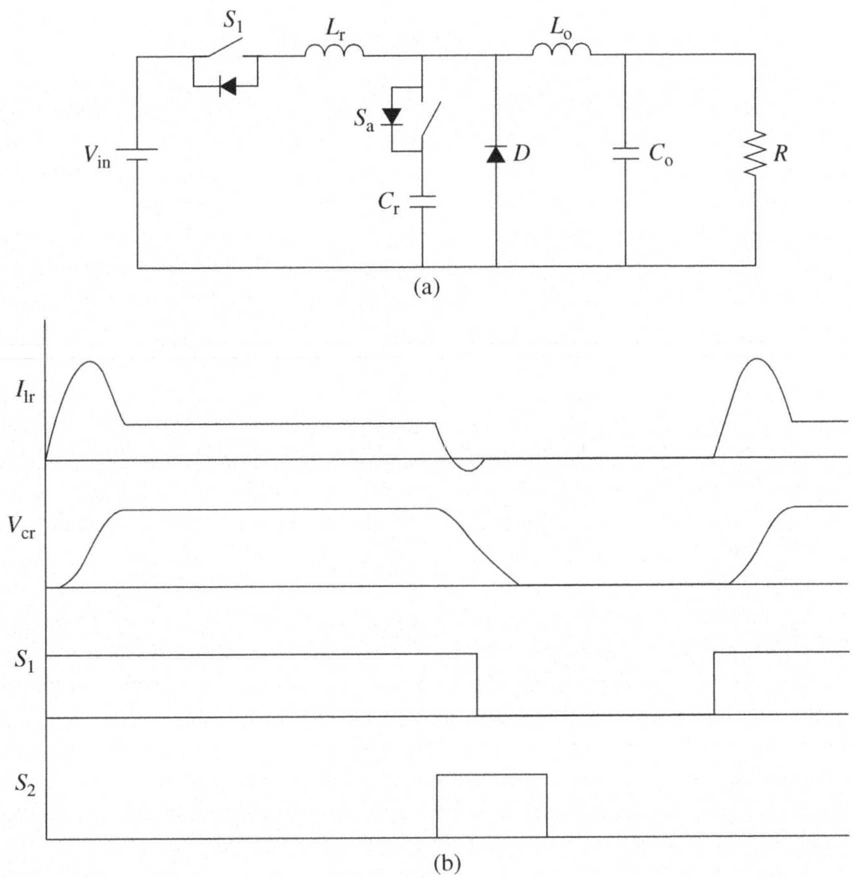

Figure 12.13 Fixed-frequency quasi-resonant converter: (a) Topology (b) Key converter waveforms.

an auxiliary switch S_a has been added in series with resonant capacitor C_r to delay the resonant cycle caused by the interaction of L_r and C_r. This delay occurs during Mode 2 when C_r is about to discharge; with the switch in series, this discharge is postponed until the switch is turned on. It is this action that triggers the resonant mechanism by which the main buck converter switch can be turned on with ZCS, as described above. Since the length of time for the delay is controllable, the converter can thus be made to operate with a fixed switching frequency.

Fixed frequency QRCs such as the one shown in Figure 12.13 are rarely used as they have been mostly superceded by the ZVS-PWM and ZCS-PWM converters that were presented in Chapters 2 and 3, mainly due to their high peak component stresses. Nonetheless, they can be useful occasionally as their topologies are relatively simple compared to many of these converters.

12.3 ULTRAHIGH FREQUENCY CONVERTERS

Since the development of silicon MOSFETs in the late 1970s and early 1980s, there has been interest among power electronic researchers in developing converter topologies that are well suited for ultrahigh switching frequency operation (>1 MHz). This has been driven by the desire to make power converters that are as small and light as possible and the fact that the size of power converters can be reduced by increasing their switching frequency so that the size of capacitors and magnetic elements such as transformers and inductors can be reduced. The main difference between ultrahigh frequency topologies and topologies for lower, more typical switching frequencies is that parasitics such as transformer leakage inductance, transformer winding capacitance, diode junction capacitance, switch output capacitance, and inductance from wiring and printed circuit board (PCB) tracks become much more prominent and cannot be ignored. Ultrahigh frequency converters are resonant converters that can incorporate parasitic elements in their topologies and thus use these elements advantageously.

Interest in ultrahigh frequency converters has risen in recent years due to the development of GaN-type devices that are based on wide bandgap semiconductor technology. GaN devices can be operated at higher switching frequencies than traditional silicon devices as they can be turned on and off much more quickly. This reduces the overlap between voltage and current during a switching transition, which reduces switching losses. Improved semiconductor technology, however, is just one consideration in the operation power converter at ultrahigh switching frequencies; magnetics and EMI are other considerations that need to be taken into account. Nonetheless, the development of devices such as GaN has made the development of ultrahigh switching converters more of a reality than ever before.

12.3.1 Multi-Resonant Converters

Multi-resonant converters, which were first proposed in [17, 18], are more sophisticated versions of quasi-resonant converters as they require more resonant elements, hence their name. An example of a multi-resonant (MR) ZVS buck converter is the converter proposed in [17] and shown in Figure 12.14. The outstanding feature of multi-resonant converters is that they can absorb switch output capacitance and diode junction capacitance at ultra-high frequencies so that their size can be reduced. In the converter shown in Figure 12.14, the converter incorporates the output capacitance of the switch, the junction capacitance of the diode, and the parasitic inductance of the PCB track that connects the switch to the diode so that no additional resonant elements need to be added.

Most multi-resonant converters can be synthesized from PWM converters with the following steps:

(i) One resonant capacitor is placed in parallel with the active switch, regardless of whether it is bidirectional (current can flow through the body-diode of the device) or unidirectional (the body-diode of the device is blocked by a series diode).

(ii) Another resonant capacitor is placed in parallel to the main converter diode.

(iii) An inductor is inserted in the loop containing the switch and the diode, which can also contain voltage sources and filter and blocking capacitors.

The additional resonant elements in multi-resonant converters help reduce switch voltage stress and improved load range significantly, compared to quasi-resonant converters. It should be noted, however, that multi-resonant converters should not be operated at lower switching frequencies as the converter parasitic elements are no longer sufficient to be included in resonant networks and the size of the resonant elements that need to be added makes the converter large.

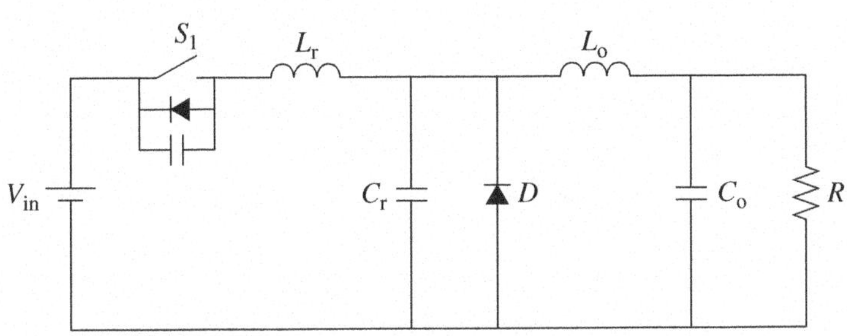

Figure 12.14 Multi-resonant buck converter.

12.3.2 Ultrahigh Frequency Converters Based on Radio-Frequency Amplifier Circuits

Another approach to ultrahigh frequency power conversion stems from radio-frequency (RF) amplifier circuits. Some of the first ultrahigh frequency converters were derived from radio-frequency amplifier theory, where circuits with LC elements are commonly used to produce ultrahigh frequency oscillators. One of the first converters of this type was the Class E amplifier boost converter that was proposed in [19, 20] and shown in Figure 12.15. The core of the topology is switch and the series LC resonant network formed by L_1 and C_1. The elements to the right of this resonant network form a matching network in line with transmission-line electromagnetic theory to try to maximize gain. The matching network can be implemented in any number of ways, as long as there is an output diode in the end to rectify voltage and make it DC. A detailed explanation of electromagnetic theory and transmission lines is beyond the scope of this book and what follows is a general explanation based on circuit principles that have been discussed throughout this book.

The way this converter works is as follows: The various resonant elements interact with each other so that the voltage across the switch can be made to be oscillatory with zero-voltage points. If properly designed, the converter can be made to have a single zero-voltage switch voltage crossing point so that if the switch is turned on at that point, then it can do so with ZVS. The converter can have a variable on-time, but its off-time is almost fixed. The switch voltage rises, then falls in a resonant manner after the switch is turned off and the time that it takes for the voltage to return to zero after this turn-off is almost fixed. This time is not quite fixed because it is dependent on the current in the converter, but the load range is small, thus the variability of time is minimal. In essence, the converter has a variable on-time and fixed off-time and thus requires variable switching frequency control to operate. If the load range is small, however, then the variation in switching frequency is small as well and thus the converter can be made to operate with an almost fixed switching frequency. If PWM control is used, then the switching frequency can be fixed,

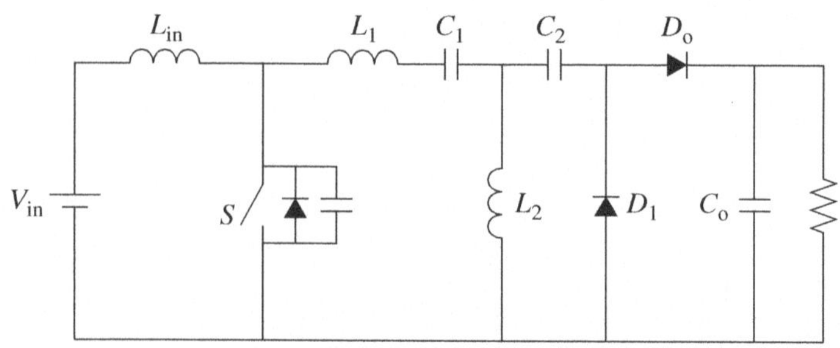

Figure 12.15 Class E converter.

but the switch may not always turn on with ZVS; however, it can be made to turn on with a small amount of voltage across it so that switching losses are minimal.

12.3.3 Ultrahigh Frequency Converters with Air-Core Inductors

Ultra-high frequency converters are limited to low power applications because of issues related to parasitic elements and issues related to magnetics as well. While increasing switching frequency can reduce the size of converter components, this is true only up to a point when it then becomes counter-productive to raise switching frequency further. For example, a 75 W multi-resonant flyback converter that could operate with 22 MHz switching frequency was proposed in [21] in 1988. Although power conversion at 22 MHz was possible, issues related to skin effect in wiring and core utilization in the transformer meant that the desired converter size reduction did not actually occur.

Ultrahigh power conversion above 10 MHz is possible, but only if air-core inductors are used. Air-core inductors are coils of wire with no magnetic core so that they have no core losses, but their use is limited to low power converters (<50 W) because of issues related to flux leakage. To date, converters with ultrahigh switching frequencies of 50 MHz and above that use air-core inductors have been reported in the literature. One such converter was the Class φ converter proposed in [22, 23] and shown in Figure 12.16. This converter operates in a manner that is similar to that of a Class E converter, but its components have lower voltage stress and it requires fewer resonant elements, due to matching network considerations. A 110 MHz, 23 W implementation was reported in [24]. Such an implementation would be impossible with if currently existing magnetic cores are used instead of air-core inductors. It should be noted that converters with such high switching frequencies cannot be controlled by using periodic gating signals as the gating signals would be too narrow for the switch to be turned on and off. Instead, several converters like the one shown in Figure 12.16 are placed in parallel and the output is regulated by what is known as ON–OFF control or burst-mode control. As described in [22],

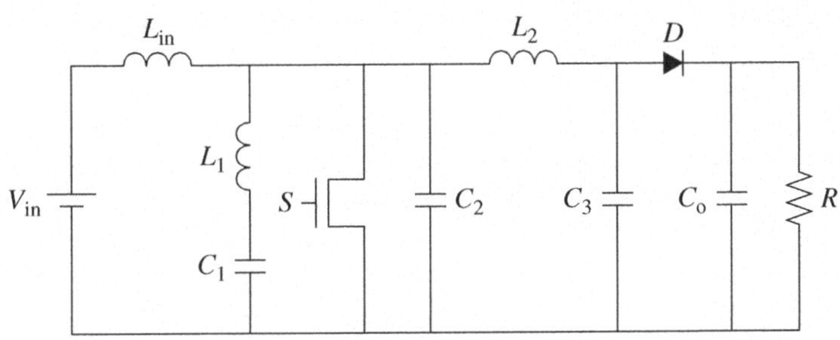

Figure 12.16 Class φ converter.

a base converter is used to regulate the output and the additional paralleled converters are activated or deactivated, depending on the load demand. If the load is heavy, then most of the paralleled converters are operated in burst mode; if the load is light, then only the base converter is operated.

12.4 CONCLUSION

Resonant converters are often operated with variable switching frequency control but can be operated with fixed switching frequency as well. Several PWM methods for the fixed-frequency operation of resonant converters were presented, and it was shown how a variable capacitor can make a half-bridge resonant converter operate with fixed switching frequency. Quasi-resonant converters, converters that can be synthesized from existing PWM converters that are not quite fully resonant converters were then discussed. The chapter concluded with a discussion of ultrahigh frequency resonant converters that can absorb parasitic elements in their topology. Although these converters can operate with very high switching frequencies of at least 1 MHz, the use of air-core inductors can make them operate with ultrahigh frequencies of at least 50 kHz. These converters, however, are limited to low power applications due to various factors related to their magnetics, including core losses and core utilization if magnetic cores are used and flux leakage if air-core inductors are used.

REFERENCES

1. J. A. Sabate and F. C. Lee, "Off-Line application of the fixed-frequency clamped-mode series resonant converter," *IEEE Transactions on Power Electronics*, vol. 6, no. 1, pp. 39–47, Jan. 1991.
2. P. K. Jain, H. Soin and M. Cardella, "Constant frequency resonant DC/DC converter topologies with zero switching losses," *IEEE Transactions on Aerospace and Electronics Systems*, vol. 30, no. 2, pp. 534–544, Apr. 1994.
3. P. K. Jain, A. St-Martin and G. Edwards, "Asymmetrical pulse-width-modulated resonant DC/DC converter topologies," *IEEE Transactions on Power Electronics*, vol. 11, no. 3, pp. 413–422, May 1996.
4. S. Mangat, M. Qiu and P. K. Jain, "A modified asymmetrical pulse-width-modulated resonant DC/DC converter topology," *IEEE Transsctions on Power Electronics*, vol. 19, no. 1, pp. 104–111, Jan. 2004.
5. D. J. Tschirhart and P. K. Jain, "Design procedure for high-frequency operation of the modified series-resonant APWM converter to reduce size and circulating current," *IEEE Transactions on Power Electronics*, vol. 27, no. 10, pp. 4181–4191, Oct. 2012.
6. S. Pan, M. Pahlevaninezhad and P. K. Jain, "Adaptive hybrid primary/secondary-side digital control for series resonant DC-DC converters in 48V VR applications," *IEEE Journal of Emerging and Selected Topics in Power Electronics*, vol. 3, no. 2, pp. 422–429, Jun. 2015.

7. D. J. Tschirhart and P. K. Jain, "A CLL resonant asymmetrical pulse-width-modulated converter with improved efficiency," *IEEE Transactions on Industrial Electronics*, vol. 55, no. 1, pp. 114–122, Jan. 2008.
8. J. M. Burdio, L. A. Barragan, F. Monterde, D. Navarro and J. Acero, "Asymmetrical voltage-cancellation control for full-bridge series resonant inverters," *IEEE Transactions on Power Electronics*, vol. 19, no. 2, pp. 461–469, Mar. 2004.
9. L. A. Barragan, J. M. Burdio, J. I. Artigas, D. Navarro, J. Acero and D. Puyal, "Efficiency optimization in ZVS series resonant inverters with asymmetrical voltage-cancellation control," *IEEE Transactions on Power Electronics*, vol. 20, no. 5, pp. 1036–1044, Sept. 2005.
10. J. M. Burdio, F. Canales, P. M. Barbosa and F. C. Lee, "Comparison study of fixed-frequency control strategies for ZVS DC/DC series resonant converters," *IEEE Power Electronics Specialists Conference*, 2001, June 17–21, 2001, Vancouver, Canada, pp. 427–432.
11. S. Bai, Z. Pantic and S. Lukic, "A comparison study of control strategies for ZVS resonant converters," *IEEE Industrial Electronics Conference (IECON)*, 2010, Nov. 7–10, 2010, Glendale, USA, pp. 256–262.
12. M. Z. Youssef and P. K. Jain, "A review and performance evaluation of control techniques in resonant converters," *IEEE Industrial Electronics Conference (IECON)*, 2004, Nov. 2–6, 2004, Busan, South Korea, pp. 215–221.
13. G. Wen-Jian and K. Harada, "A new method to regulate resonant converters," *IEEE Transactions on Power Electronics*, vol. 3, no. 4, pp. 430–439, Oct. 1988.
14. F. C. Lee, "High-frequency quasi-resonant converter technologies," *Proceedings of the IEEE*, vol. 76, no. 4, pp. 377–390, Apr. 1988.
15. R. Chen, T. Liang, J. Chen, R. Lin and K. Tseng, "Study and implementation of a current-fed full-bridge boost DC–DC converter with zero-current switching for high-voltage applications," *IEEE Transactions on Industry Applications*, vol. 44, no. 4, pp. 1218–1226, Jul.–Aug. 2008.
16. I. Barbi, J. C. O. Bolacell, D. C. Martins and F. B. Libano, "Buck quasi-resonant converter operating at constant frequency: analysis, design, and experimentation," *IEEE Transactions on Power Electronics*, vol. 5, no. 3, pp. 276–283, Jul. 1990.
17. W. A. Tabisz and F. C. Y. Lee, "Zero-voltage-switching multiresonant technique-a novel approach to improve performance of high-frequency quasi-resonant converters," in *IEEE Transactions on Power Electronics*, vol. 4, no. 4, pp. 450–458, Oct. 1989, doi: 10.1109/63.41774.
18. M. M. Jovanovic, W. A. Tabisz and F. C. Y. Lee, "High-frequency off-line power conversion using zero-voltage-switching quasi-resonant and multiresonant techniques," in *IEEE Transactions on Power Electronics*, vol. 4, no. 4, pp. 459–469, Oct. 1989, doi: 10.1109/63.41775.
19. R. J. Gutmann, "Application of RF circuit design principles to distributed power converters," *IEEE Transactions on Industrial Electronics and Control Instrumentation*, vol. IECI-27, no. 3, pp. 156–164, Aug. 1980.
20. R. Redl, B. Molnar and N. O. Sokal, "Class E resonant regulated DC/DC power converters: analysis of operations, and experimental results at 1.5 MHz," *IEEE Transactions on Power Electronics*, vol. 1, no. 2, pp. 111–120, Apr. 1986.
21. W. C. Bowman, F. T. Balicki, F. T. Dickens, R. M. Honeycutt, W. A. Nitz, W. Strauss, W. B. Suiter, and N. G. Ziesse, "A resonant DC-to-DC converter

operating at 22 megahertz," *IEEE Applied Power Electronics Conference (APEC)*, 1988, Feb. 1–5, 1988, New Orleans, USA, pp. 3–11.

22. J. Rivas, J. Shafran, R. Wahby, and D. Perreault, "New architectures for radio-frequency dc/dc power conversion," *IEEE Transactions on Power Electronics*, vol. 21, no. 2, pp. 380–393, Mar. 2006.
23. J. M. Rivas, Y. Han, O. Leitermann, A. D. Sagneri, and D. J. Perreualt, "A high-frequency resonant inverter topology with low-voltage stress," *IEEE Transactions on Power Electronics*, vol. 23, no. 4, pp. 1759–1771, Jul. 2008.
24. R. C. N. Pilawa-Podgurski, A. D. Sagneri, J. M. Rivas, D. I. Anderson and D. J. Perreault, "Very-high-frequency resonant boost converters," *IEEE Transactions on Power Electronics*, vol. 24, no. 6, pp. 1654–1665, Jun. 2009.

CHAPTER 13

THREE-LEVEL DC–DC CONVERTERS

Voltage-fed full-bridge-type converters such as the ones discussed in Chapters 7–9 are typically implemented with switching devices that have peak voltage ratings of 600 V or less. For applications where the input DC bus voltage is higher than 600 V – such as a two-stage AC–DC converter where the front-end AC–DC converter converts AC grid voltage into an 800 V DC bus voltage that is then fed to an isolated DC–DC converter – power electronic designers have two options. The first is to implement full-bridge-type topologies with switching devices that have a peak voltage rating that is higher than 600 V, but these devices are more expensive and less commercially available than lower voltage-rated devices, or they are larger in size and thus take up more space. A more attractive option for such applications, and the one that is more commonly used, is to use some sort of three-level DC–DC converter topology instead of more conventional "two-level" converter topologies such as the ZVS-PWM-FB converter. Three-level converters are converters whose switching devices are subjected to half the input DC bus voltage, which allows devices with peak voltage ratings of 600 V or less to be used in high-voltage applications.

The focus of this chapter is on three-level DC–DC converters. In this chapter, three fundamental three-level DC–DC converters are introduced and their basic operation is explained. Variations on these fundamental topologies that are more efficient are then presented and the extension of the three-level concept to stacked converters is explained. The chapter concludes with a

DC–DC Converter Topologies: Basic to Advanced, First Edition. Gerry Moschopoulos.

Published 2024 by John Wiley & Sons, Inc.

discussion on the use of three-level converters for applications with conventional DC voltage buses as a means of increasing efficiency.

13.1 FUNDAMENTAL THREE-LEVEL DC–DC PWM CONVERTERS

Many three-level DC–DC converters have been proposed by power electronics researchers. Of these converters, the most fundamental are:

- Neutral-point-clamped (NPC) three-level converters
- Flying capacitor three-level converters
- Three-level converters with series blocking capacitor

In this section, the basic operation of these fundamental converters is reviewed and the features of each converter are compared to those of the others.

13.1.1 Neutral-Point-Clamped Three-Level DC–DC Converter

The first three-level converter to have been proposed was the DC–AC inverter proposed in [1] in the early 1980s. This converter is referred to as an NPC three-level converter in the literature because all three-phase converter legs are connected to a common neutral point, which is formed by the mid-point of two DC bus capacitors placed in series. The converter is a three-level converter as it can synthesize three discrete voltage levels at the AC terminal of each phase – - $+\frac{1}{2}V_{in}$, 0 and $-\frac{1}{2}V_{in}$ – instead of just two, 0 and V_{in}. It was not until the early 1990s, however, that the use of three-level converters for DC–DC power conversion was investigated. Most of the initial work on three-level DC–DC converters was done by Ivo Barbi and his research group at the University of Florianopolis in Brazil. At first, three-level DC–AC converter topologies were adapted for use as DC–DC converters, then new three-level DC–DC converters that took advantage of general DC–DC converter properties were proposed. It should be noted that although DC–AC inverters of 5 levels, 7 levels, and even higher number of levels have been proposed and are common, few, if any, multilevel DC–DC converters with more than three levels have been proposed as they are considered to be impractical; therefore, only three-level DC–DC converters are considered in this chapter.

The first three-level DC–DC to have been proposed was the NPC converter proposed in [2] and shown in Figure 13.1a. The converter is a half-bridge-type structure that consists of the following:

- Four switches, S_1–S_4, arranged in a stacked position in a single leg.
- Two capacitors connected in series across the DC bus, C_1 and C_2.

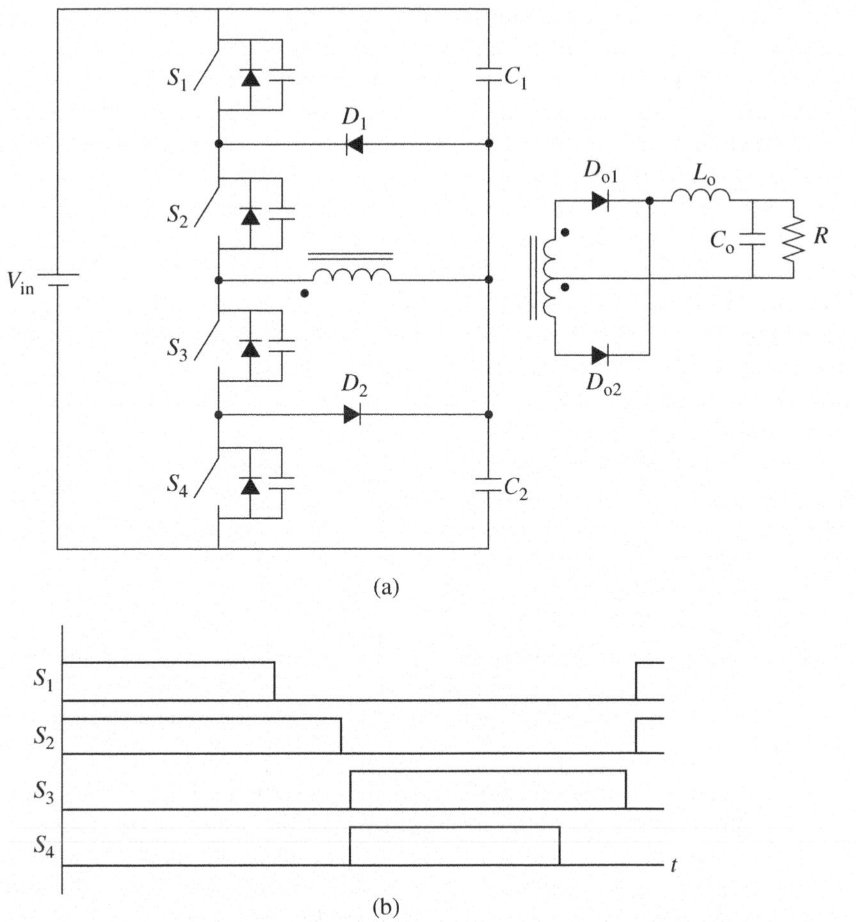

Figure 13.1 Neutral-point-connected three-level DC–DC converter: (a) Topology. (b) Switch gating signals.

- A transformer with one end connected to the mid-point of the capacitors and the other connected to the mid-point of the switch leg.
- Two diodes, D_1 and D_2, with anodes that are connected to the mid-point of the bus capacitors and cathodes that are connected to the mid-point of S_1 and S_2 in the upper part of the converter (D_1) and to the mid-point of S_3 and S_4 in the lower part of the converter (D_2).
- Two diodes, D_{o1} and D_{o2}, that are used to rectify the secondary voltage of the transformer, as in a conventional two-level full-bridge converter.
- An output LC filter.

Typical gating signals for the four converter switches are shown in Figure 13.1b. Unlike a conventional ZVS-PWM-FB converter, the three-level

NPC converter does not operate with the phase-shift PWM method described in Section 7.2. Instead, the gating signals are arranged in the following manner: The inner switches – the switches that do not touch one of the DC bus voltage rails, S_2 and S_3 – operate with a 50% duty cycle and are complementary with each other so that S_3 is off whenever S_2 is on and vice versa. The outer switches, S_1 and S_4, operate with variable duty cycles; it is these switches that control the duty cycle of the converter. S_1 and S_2 are turned on at the same time and the voltage across C_1 is impressed across the primary of the transformer. The converter is in an energy-transfer mode until S_1 is turned off. Current then flows through diode D_1 and S_2 and the converter enters a freewheeling mode of operation with no voltage impressed across the transformer primary, as both its terminals have the same voltage. S_3 and S_4 operate in a similar manner to impress the voltage across C_2 across the transformer primary, with the opposite polarity.

Equivalent circuit diagrams that show the modes of operation for a half-switching cycle for the three-level converter in Figure 13.1a are shown in Figure 13.2. The output filter and the load have been replaced by a current source to simplify the figures. The converter has the following modes of operation for a half-switching cycle:

Mode 1 (t_0–t_1): During this mode, switches S_1 and S_2 are on and half the dc bus voltage, $V_{in}/2$, is impressed across the transformer primary. The converter is in an energy-transfer mode as energy is being transferred to the output.

Mode 2 (t_1–t_2): During this mode, switch S_1 is turned off and the voltage of C_{s1} increases from zero to $V_{in}/2$ while the voltage of $(C_{s3} + C_{s4})$ decreases from V_{in} to $V_{in}/2$. Switches S_3 and S_4 are each exposed to a voltage of $V_{in}/4$. Voltage continues to be impressed across the transformer primary and energy is transferred to the output.

Mode 3 (t_2–t_3): This mode begins after C_{s1} has been fully discharged to $V_{in}/2$. The converter is in a freewheeling mode of operation as current flows through D_1 and S_2 as there is no voltage across the transformer primary, current freewheels in the primary and freewheels in the secondary as well, with the output current ideally split between D_{o1} and D_{o2}.

Mode 4 (t_3–t_4): This mode begins after S_2 is turned off and the converter begins the process of exiting a freewheeling mode of operation. The output capacitor of S_2, C_{s2} begins to be charged while the output capacitors of S_3 and S_4, C_{s3} and C_{s4}, begin to discharge. Current continues to freewheel through output diodes D_{o1} and D_{o2}.

Mode 5 (t_4–t_5): This mode begins as soon as C_{s3} and C_{s4} have been fully discharged and C_{s2} has been charged to half of the input voltage; C_{s1} continues to be charged to half the input voltage. With current flowing through the body diodes of S_3 and S_4, these switches can be turned on with ZVS. Voltage is impressed across the transformer primary and the

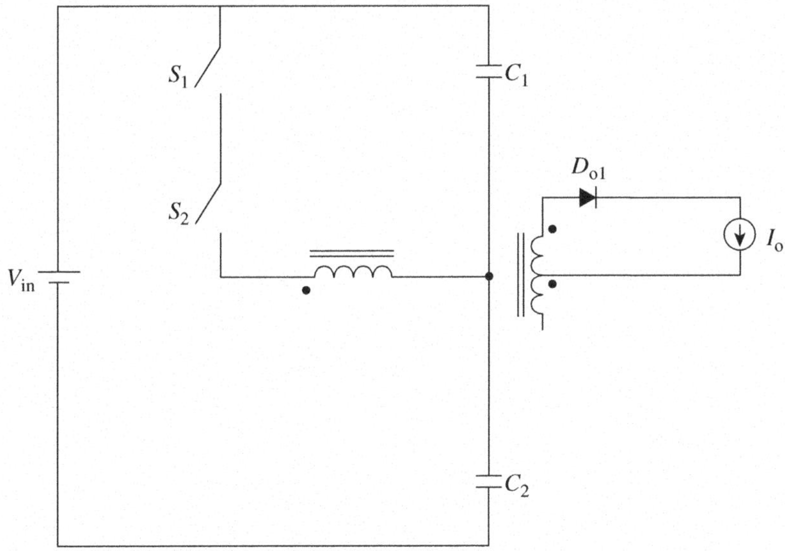

(a)

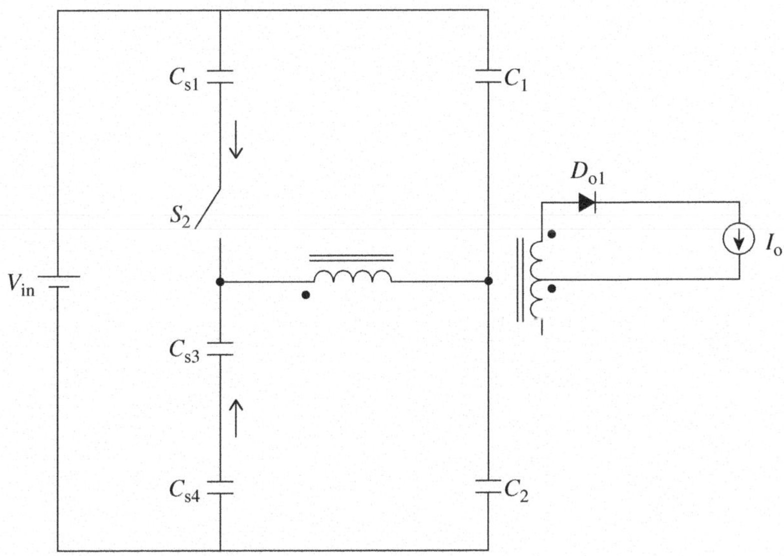

(b)

Figure 13.2 Modes of operation for the neutral-point connected three-level DC–DC converter shown in Figure 13.1. (a) Mode 1. (b) Mode 2. (c) Mode 3. (d) Mode 4. (e) Mode 5. (f) Mode 6. (g) Mode 7.

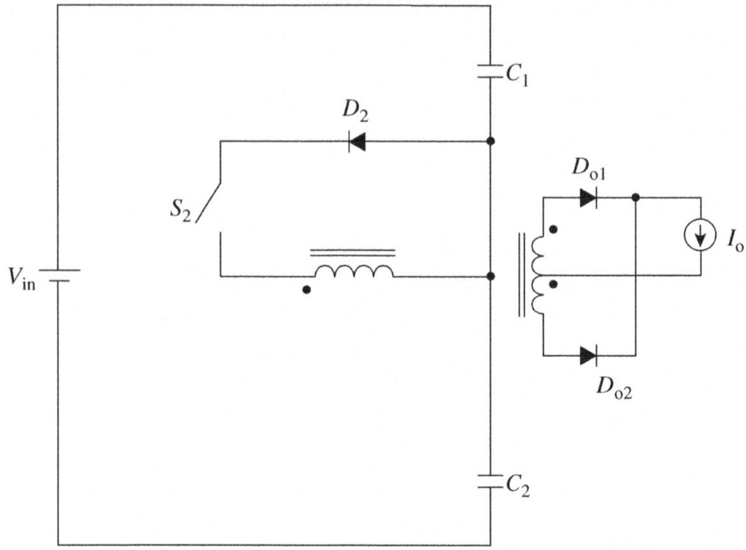

(c)

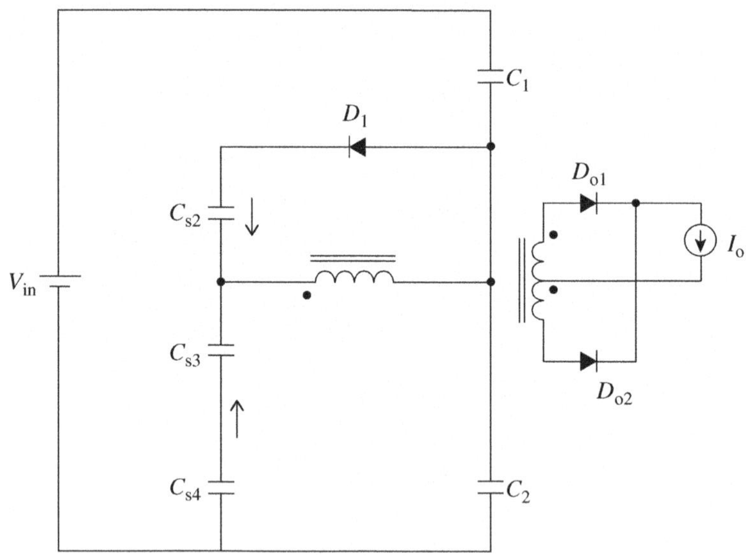

(d)

Figure 13.2 (*Continued*)

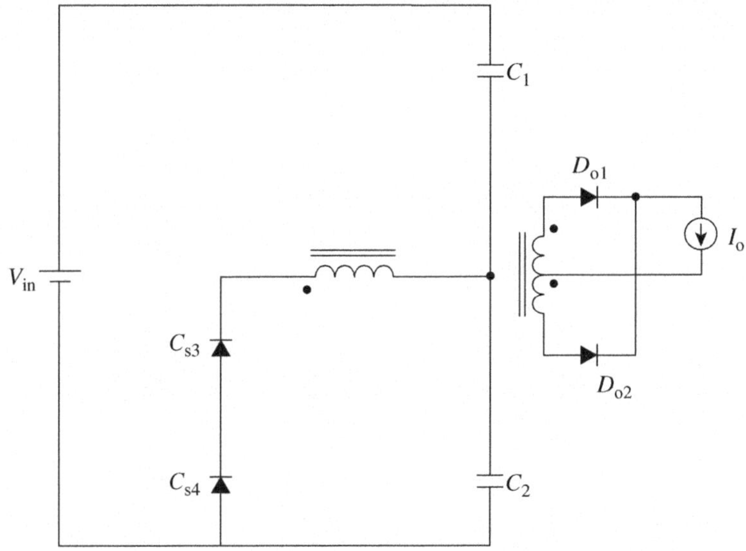

(e)

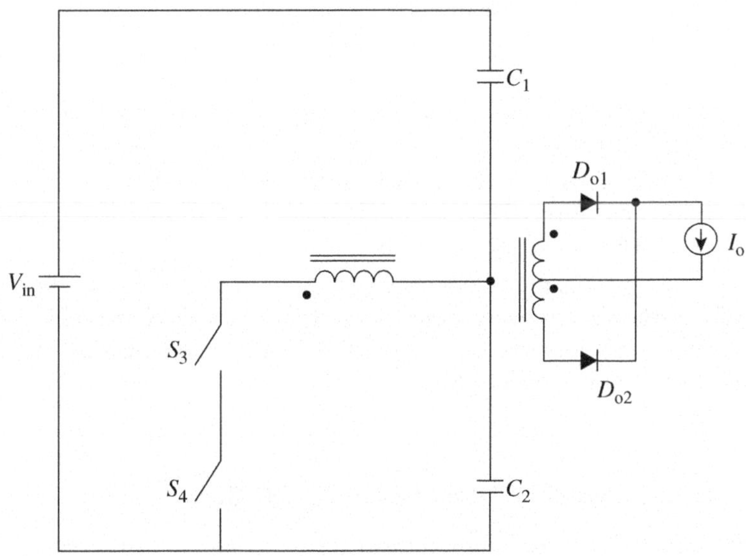

(f)

Figure 13.2 (*Continued*)

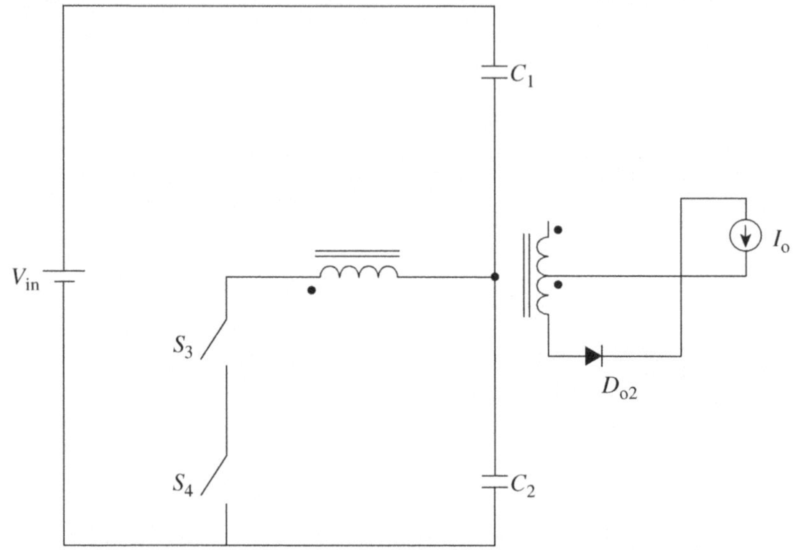

(g)

Figure 13.2 (*Continued*)

process of transferring output current away from D_{o1} so that it all flows through D_{o2} begins.

Mode 6 (t_5–t_6): This mode begins when the current in the primary reverses direction and starts to flow through S_3 and S_4. Output current continues to be transferred away from D_{o1}.

Mode 7 (t_6–t_7): This mode begins when the output current has been completely transferred away from D_{o1} so that it all flows through D_{o2}. The converter is in an energy-transfer mode with current flowing through S_3 and S_4 instead of S_1 and S_2.

13.1.2 Flying Capacitor Three-Level DC–DC Converter

In the mid-1990s, researchers began to investigate an alternative to the NPC three-level inverter, which was the "flying capacitor" three-level converter [3]. This was first adapted for use as a DC–DC converter in [4]; the three-level flying capacitor DC–DC converter is shown in Figure 13.3a. This converter is very similar to the three-level NPC converter, with the most significant difference being that the two primary-side freewheeling diodes of the NPC converter have been replaced by a single capacitor C_f, which gives the converter its name. Typical gating signals are shown in Figure 13.3b. It can be seen the gating signals of S_2 and S_3 are complimentary to each other,

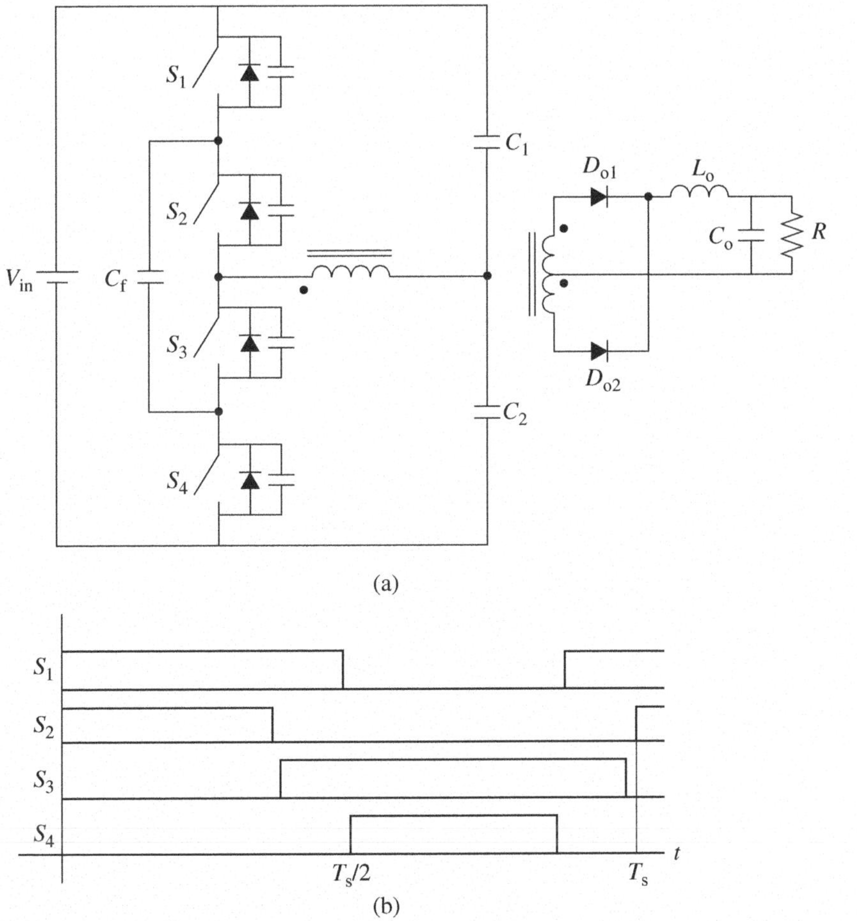

Figure 13.3 Flying capacitor three-level DC–DC converter. (a) Topology. (b) Gating signals.

as are those for S_1 and S_4, and that the S_1/S_4 gating signals are shifted by 180° in relation to those of S_2/S_3. The converter is controlled by changing the duty cycles of S_2 and S_4.

Equivalent circuit diagrams that show the modes of operation for a half-switching cycle for the three-level converter in Figure 13.3a are shown in Figure 13.4. The converter has the following modes of operation for a half-switching cycle:

Mode 1 (t_0–t_1): During this mode, switches S_1 and S_2 are on and the voltage across DC bus capacitor C_1, which is half of the input voltage, is impressed across the transformer primary. Output diode D_{o1} is forward biased and the converter is in an energy-transfer mode of operation.

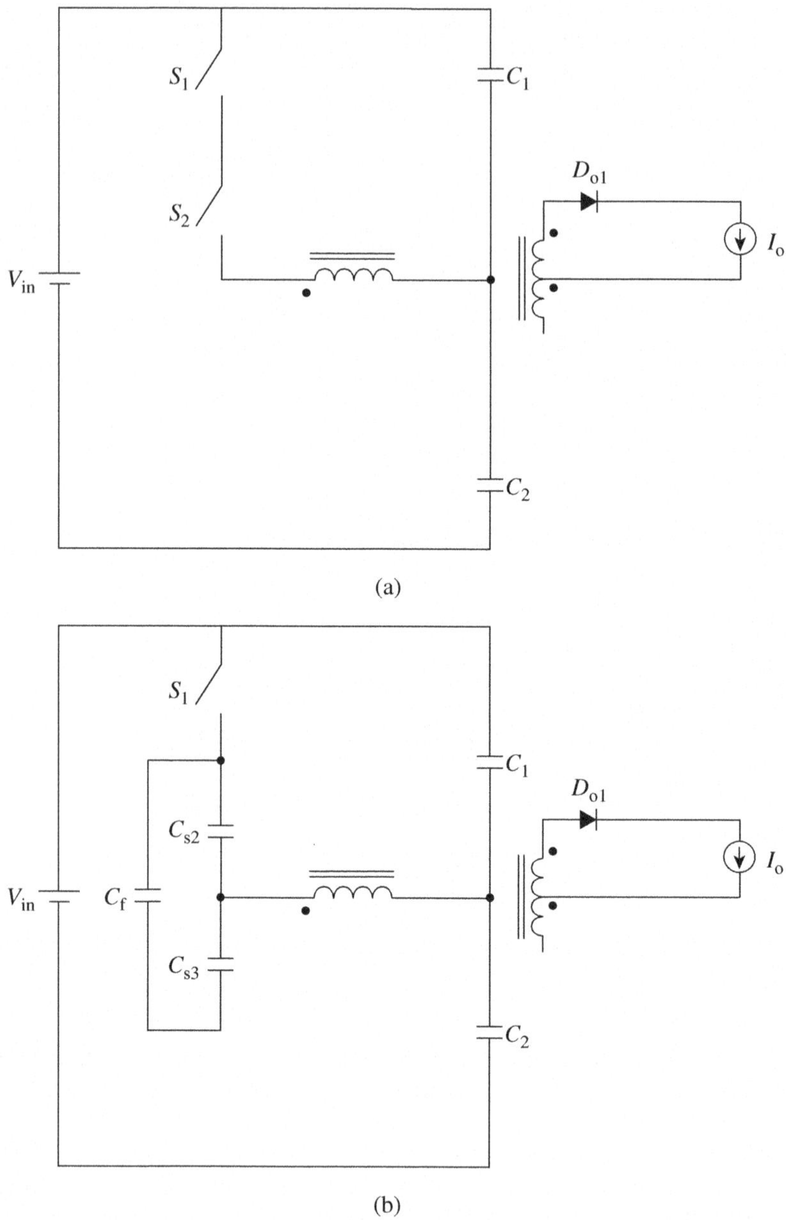

Figure 13.4 Modes of operation for the flying capacitor three-level DC–DC converter shown in Figure 13.3. (a) Mode 1. (b) Mode 2. (c) Mode 3. (d) Mode 4. (e) Mode 5. (f) Mode 6. (g) Mode 7.

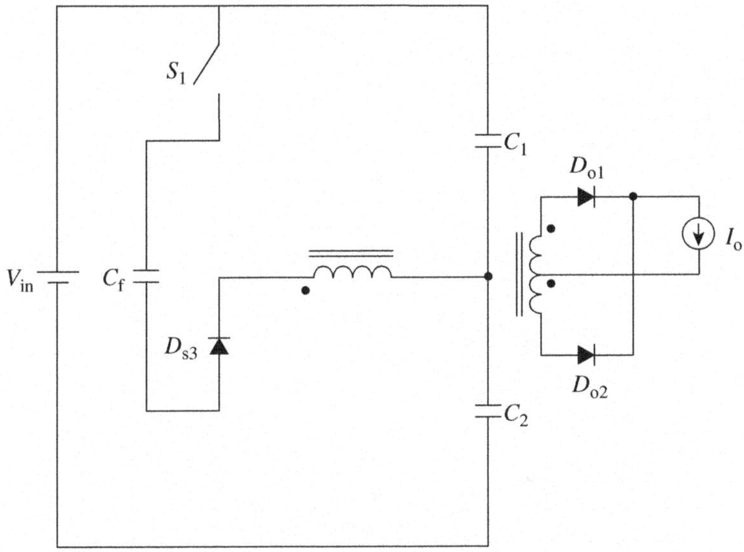

(c)

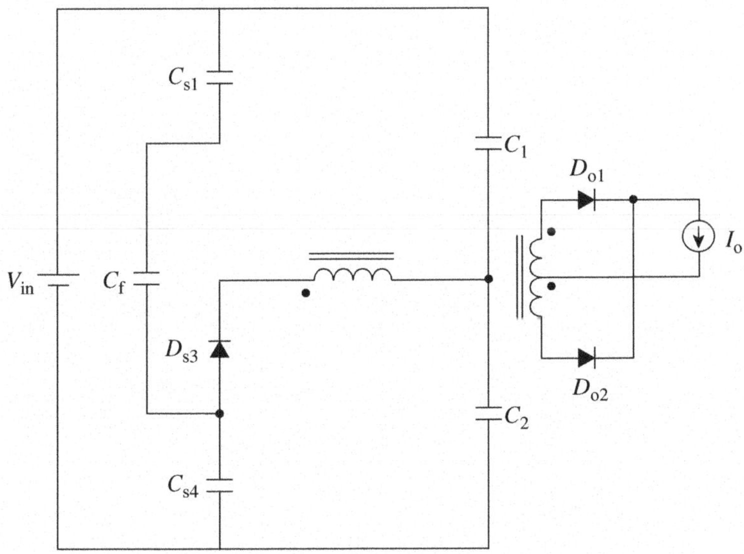

(d)

Figure 13.4 (*Continued*)

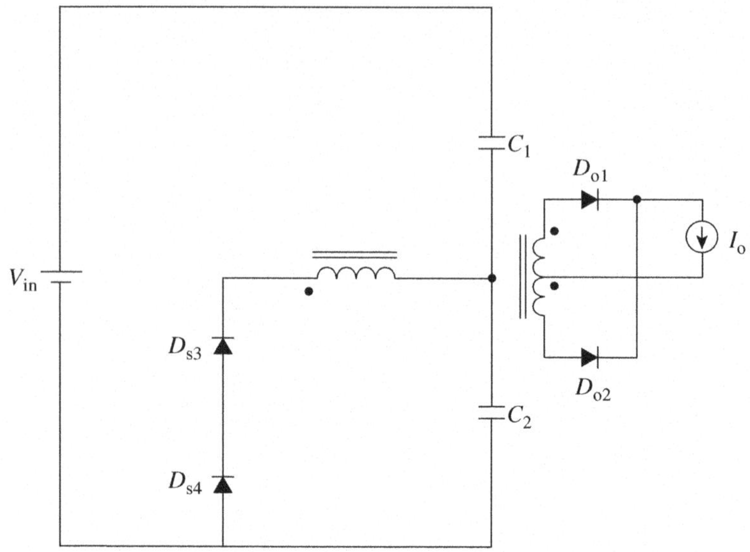

(e)

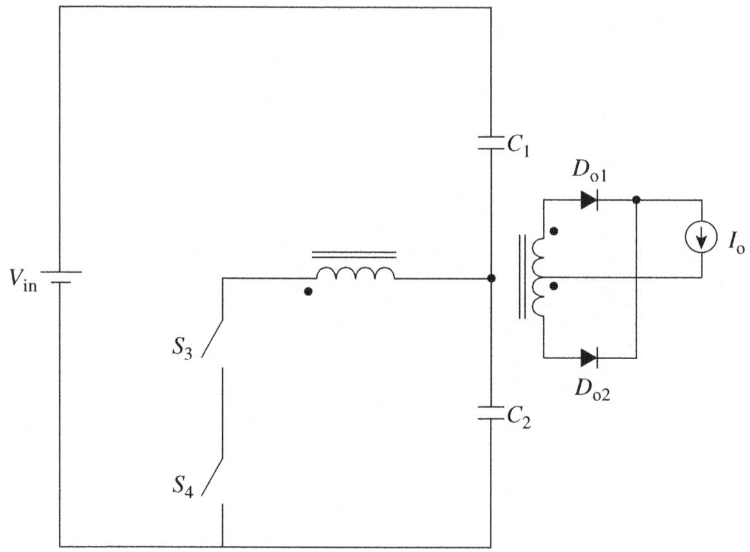

(f)

Figure 13.4 (*Continued*)

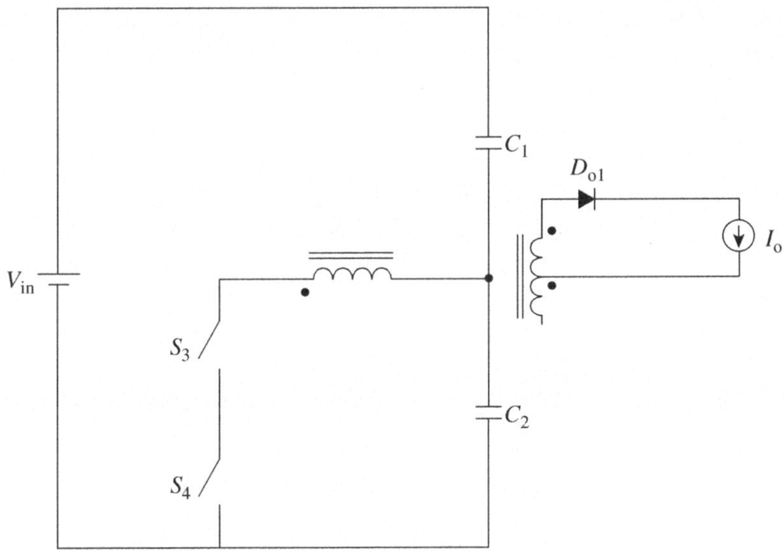

(g)

Figure 13.4 (*Continued*)

Mode 2 (t_1–t_2): This mode begins after S_2 is turned off. When this happens, the output capacitor of S_2, C_{s2} begins to be charged and the output capacitor of S_3, C_{s3} begins to be discharged through the flying capacitor C_f. The converter is in the process of exiting an energy-transfer mode and power continues to be delivered to the output through output diode D_{o1}.

Mode 3 (t_2–t_3): This mode begins after C_{s3} has been fully discharged and C_{s2} has been charged to half the input voltage. The converter is in a freewheeling mode of operation as current flows through S_1, C_f, and the body diode of S_3 and C_1. There is no voltage across the transformer primary as the voltage across C_f, which is $V_{in}/2$, cancels the voltage across C_1, which is $V_{in}/2$ as well. Current freewheels in the primary and freewheels in the secondary as well, with the output current ideally split between D_{o1} and D_{o2}.

Mode 4 (t_3–t_4): This mode begins after S_1 is turned off and the converter begins the process of exiting a freewheeling mode of operation. The output capacitor of S_1, C_{s1} begins to be charged while the output capacitor of S_4, C_{s4} begins to discharge through C_f. Current continues to flow through the body diode of S_3 and freewheel through output diodes D_{o1} and D_{o2} as well.

Mode 5 (t_4–t_5): This mode begins as soon as C_{s4} has been fully discharged and C_{s1} has been charged to half of the input voltage. With current flowing through the body diodes of S_3 and S_4, these switches can be turned on with ZVS; it should be noted that S_3 can be turned on earlier with ZVS during Mode 3 if desired. Voltage is impressed across the transformer primary and the process of transferring output current away from D_{o1} so that it all flows through D_{o2} begins.

Mode 6 (t_5–t_6): This mode begins when the current in the primary reverses direction and starts to flow through S_3 and S_4. Output current continues to be transferred away from D_{o1}.

Mode 7 (t_6–t_7): This mode begins when the output current has been completely transferred away from D_{o1} so that it all flows through D_{o2}. The converter is in an energy-transfer mode with current flowing through S_3 and S_4 instead of S_1 and S_2.

13.1.3 Three-Level DC–DC Converter with Series Blocking Capacitor

The third type of fundamental three-level DC–DC converter is the converter that was proposed in [5] and shown in Figure 13.5a. Like the other

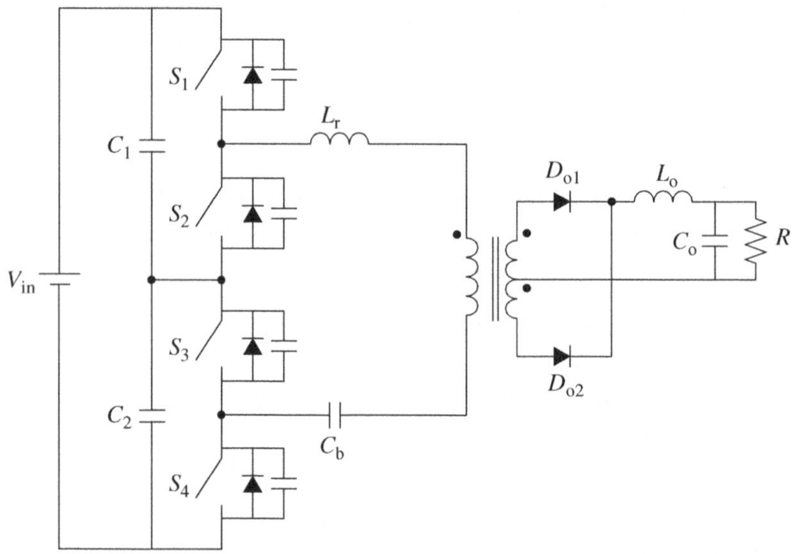

(a)

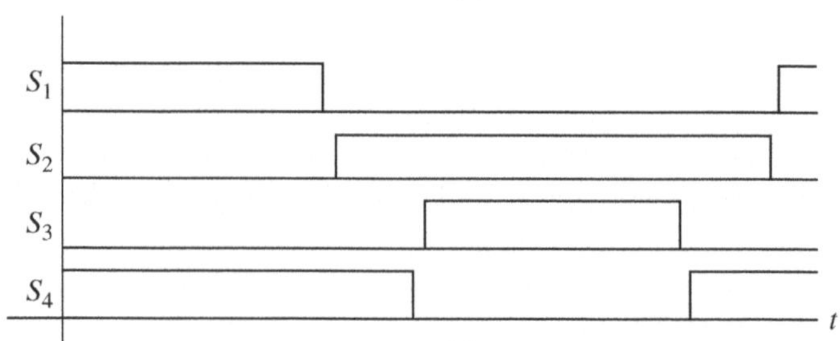

(b)

Figure 13.5 Three-level DC–DC converter with series blocking capacitor. (a) Topology. (b) Switch gating signals.

converters, the converter has four switches aligned in series between the two DC bus voltage rails and two DC bus capacitors with a mid-point connection, but the transformer is connected differently, with one end connected to the connection of S_1 and S_2 and the other end connected to the connection of S_3 and S_4. A small blocking capacitor C_b is placed in series with the transformer to ensure that it does not saturate. V_{Cb} is the voltage that resets the transformer and is ideally $V_{in}/2$. Additional inductance may be added in series to the transformer or built into the transformer in the form of leakage inductance to extend the load range of ZVS; this is shown as L_r in Figure 13.5a. Typical gating signals are shown in Figure 13.5b. It can be seen that the gating signals of S_1 and S_2 are complimentary to each other as are those for S_3 and S_4 and that the gating signals for S_1 and S_3 are the same, but phase-shifted by 180° relative to each other, as are those for S_2 and S_4.

Equivalent circuit diagrams for each mode of operation that the converter goes through during a half-switching cycle are shown in Figure 13.6. The converter has the following modes of operation:

Mode 1 (t_0–t_1): During this mode, switches S_1 and S_4 are on. The voltage that is impressed across the transformer primary is V_{in}–V_{Cb}. Output diode D_{o1} is forward-biased and the converter is in an energy-transfer mode of operation.

Mode 2 (t_1–t_2): This mode begins after S_1 is turned off. When this happens, the output capacitor of S_1, C_{s1} begins to be charged and the output capacitor of S_2, C_{s2}, begins to be discharged.

Mode 3 (t_2–t_3): This mode begins after C_{s2} has been fully discharged and C_{s1} has been charged to half the input voltage. The converter is in a freewheeling mode of operation as current flows through the body diode of S_2, C_b and S_4. There is no voltage across the transformer primary because $V_{in}/2$ voltage is applied to both ends of the transformer. V_{Cb} is $V_{in}/2$ so that Kirchoff's voltage law can be satisfied in the loop where current flows in this mode, assuming that the voltage drop across L_r is negligible. Current freewheels in the primary and freewheels in the secondary as well, with the output current ideally split between D_{o1} and D_{o2}. The transfer of current away from D_{o1} begins earlier in the switching cycle, towards the end of Mode 2.

Mode 4 (t_3–t_4): This mode begins after S_4 is turned off and the converter begins the process of exiting a freewheeling mode of operation. The output capacitor of S_4, C_{s4}, begins to be charged while the output capacitor of S_3, C_{s3}, begins to discharge. Current continues to flow through the body diode of S_2 and freewheel through output diodes D_{o1} and D_{o2} as well.

Mode 5 (t_4–t_5): This mode begins as soon as C_{s3} has been fully discharged and C_{s4} has been charged to half of the input voltage. With current flowing through the body diodes of S_2 and S_3, these switches can be turned on with ZVS; it should be noted that S_2 can be turned on earlier with ZVS during Mode 3 if desired. Voltage is impressed across the transformer

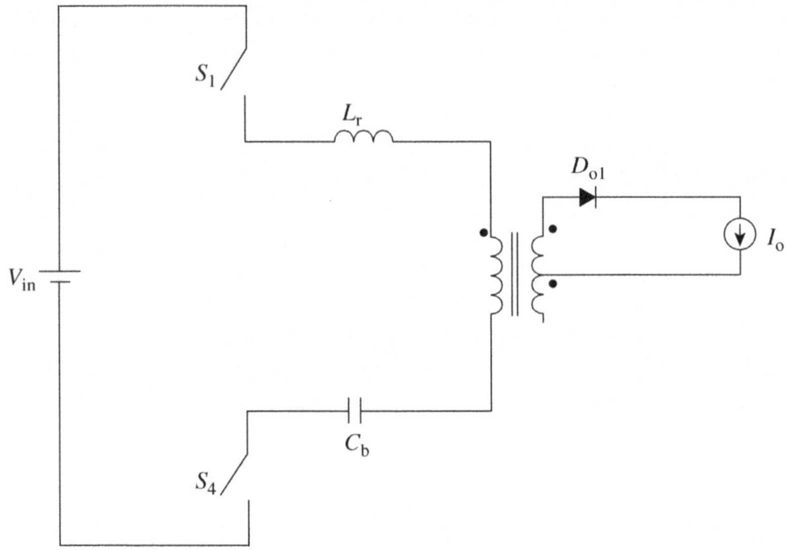

(a)

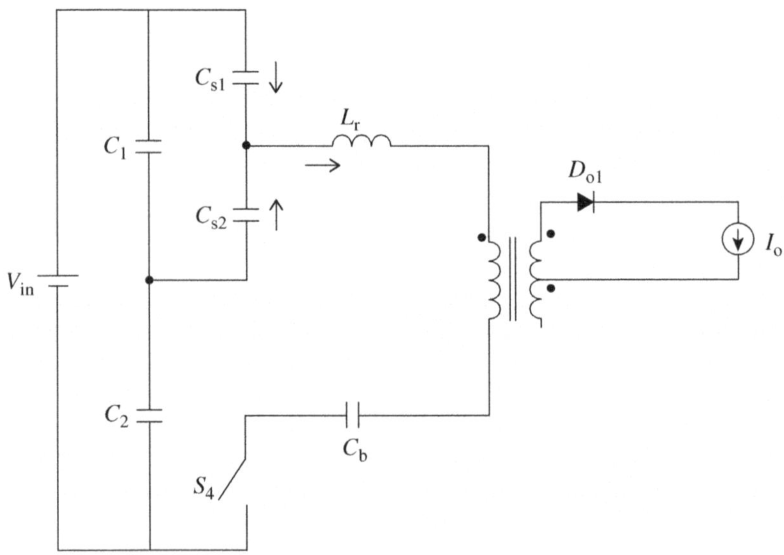

(b)

Figure 13.6 Modes of operation for the three-level DC–DC converter with series blocking capacitor shown in Figure 13.5. (a) Mode 1. (b) Mode 2. (c) Mode 3. (d) Mode 4. (e) Mode 5. (f) Mode 6. (g) Mode 7.

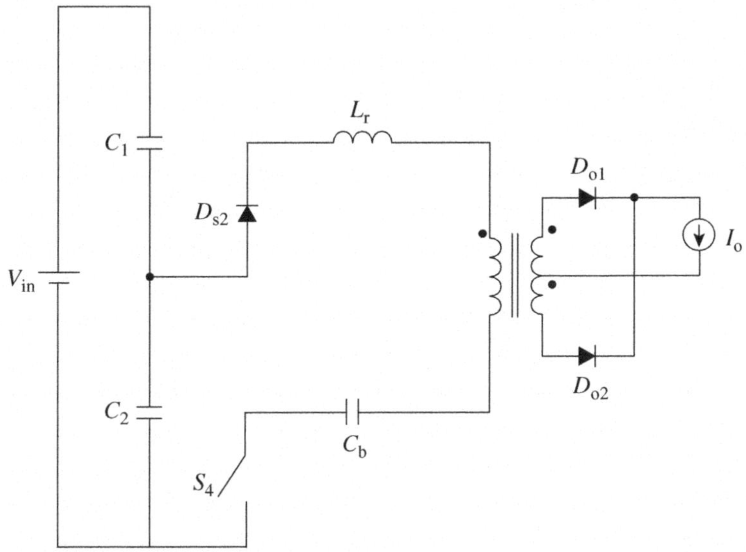

(c)

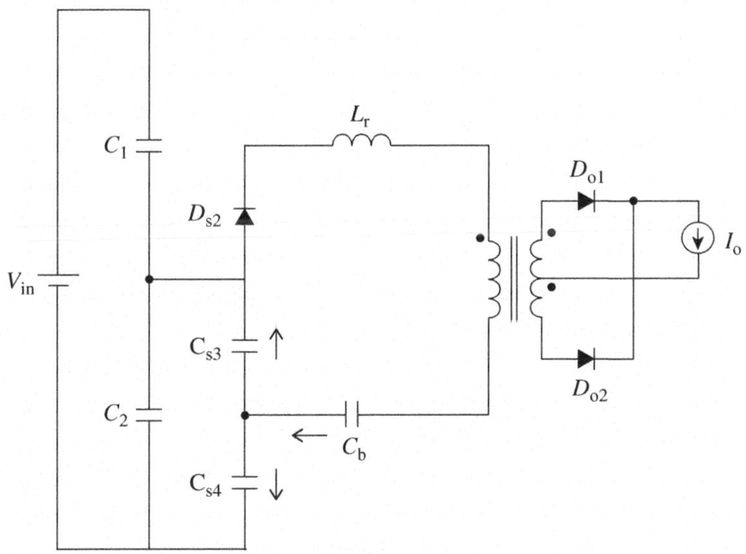

(d)

Figure 13.6 (*Continued*)

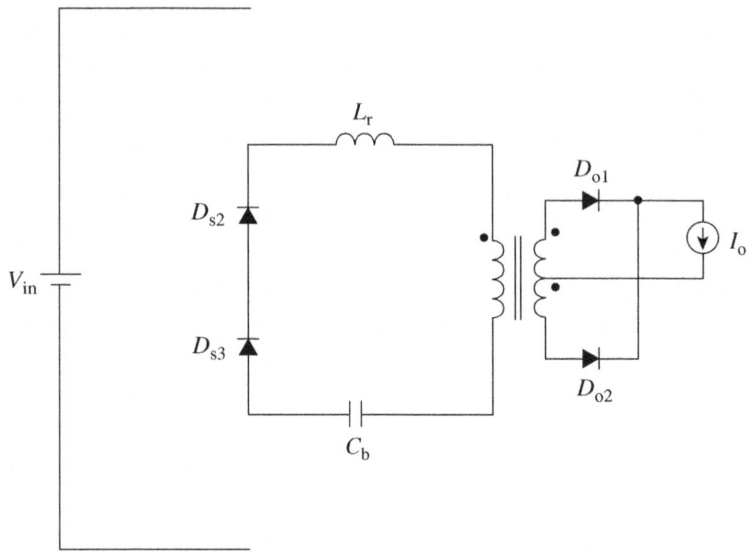

(e)

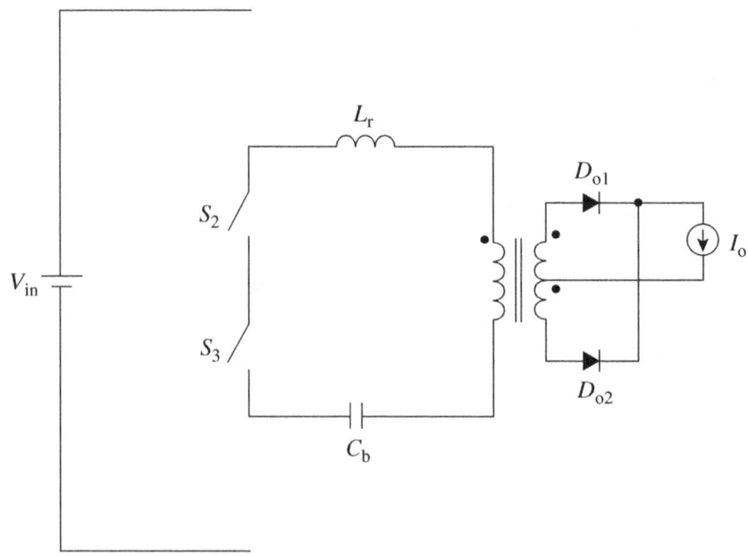

(f)

Figure 13.6 (*Continued*)

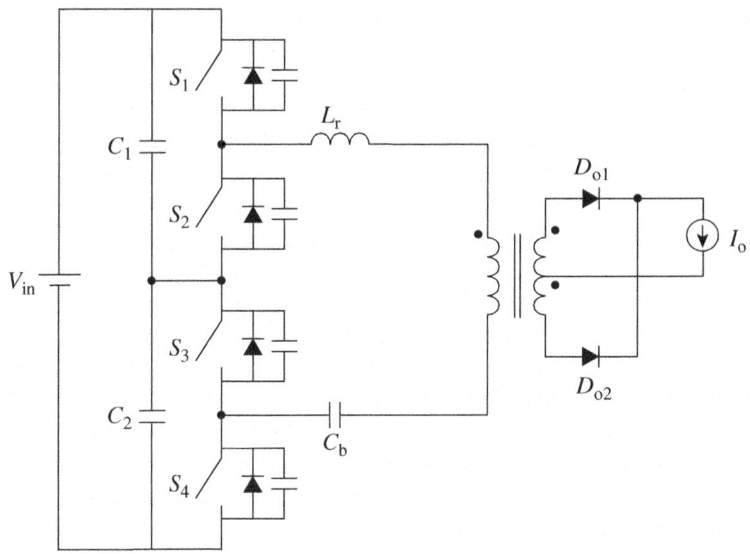

(g)

Figure 13.6 *(Continued)*

primary and the process of transferring output current away from D_{o1} so that it all flows through D_{o2} begins.

Mode 6 (t_5–t_6): This mode begins when current in the primary reverses direction and starts to flow through S_2 and S_3. Output current continues to be transferred away from D_{o1}.

Mode 7 (t_6–t_7): This mode begins when the output current has been completely transferred away from D_{o1} so that it all flows through D_{o2}. The converter is in an energy-transfer mode with current flowing through S_2 and S_3 instead of S_1 and S_4, as $V_{Cb} = V_{in}/2$ is impressed across the transformer primary.

13.1.4 Comparison of Fundamental Three-Level DC–DC Converter Topologies

A comparison of the efficiency of all three fundamental three-level converters was performed in [6]. It was determined that all three converters have approximately the same efficiency, with the NPC converter being the least efficient and the converter with the series blocking capacitor being the most efficient. The converter with the series blocking capacitor is limited to three-levels whereas as the NPC and flying capacitor converters can be implemented with five and seven levels for very high DC bus voltages, if desired.

Of the three converters, the NPC converter is the most expensive because it has two more diodes in its topology than the other converters, but it is also the easiest converter to maintain voltage balancing between the two DC bus capacitors. Voltage balancing, which is the ability to balance the voltages across the

DC bus capacitors so that they are both the same, helps reduce the amount of ripple at the output and ensure that each switch is exposed to half the input voltage at most.

When three-level converters are used in low power applications, the primary current is sufficiently small that any difference in the current that is used to charge and discharge the two bus capacitors is small as well and thus any voltage difference between the two capacitors is also small. When three-level converters are used in higher power applications, then the charging/discharging current of the capacitors becomes more significant and so too does the voltage difference among them unless something is done. It is easier to perform voltage balancing for the NPC converter than the other converters. The issue of voltage balancing is of great concern for high-power DC–AC inverters and there is a vast literature on this topic, one that is beyond the scope of this book. Comparing NPC inverters with flying capacitor inverters (there are no series blocking capacitor inverters), it has been shown that it is easier to perform voltage balancing on NPC inverters than on flying capacitor inverters.

13.2 MODIFIED THREE-LEVEL DC–DC CONVERTERS

Various modifications to the three fundamental three-level DC–DC converters have been proposed by researchers to correct their deficiencies and improve their efficiency. This section covers ZVS and ZVZCS converters that have been modified from the three fundamental three-level DC–DC converters.

13.2.1 ZVS Three-Level Converters

One drawback that all three fundamental three-level converters have is that they must be operated with some sort of asymmetrical PWM scheme and cannot be operated with the phase-shift PWM method described in Section 7.2. This is a drawback because they cannot be implemented with dedicated and readily available integrated-circuit (IC) phase-shift PWM controllers that are popular due to the widespread use of ZVS-PWM-FB converters. There are few, if any, such controllers for the asymmetrical PWM methods required by the three fundamental three-level converters and thus these PWM methods must be implemented with more sophisticated controller hardware, which increases converter size and cost.

A three-level converter that can be operated with conventional phase-shift PWM was proposed in [7] and shown in Figure 13.7 along with its gating signals. It can be seen that the converter is essentially an NPC converter with a flying capacitor added to it. Unlike the fundamental flying capacitor converter, the flying capacitor in this converter is smaller and mainly serves to provide a path for current to flow when need to satisfy the current path requirements forced by using phase-shift PWM. It should be noted that like

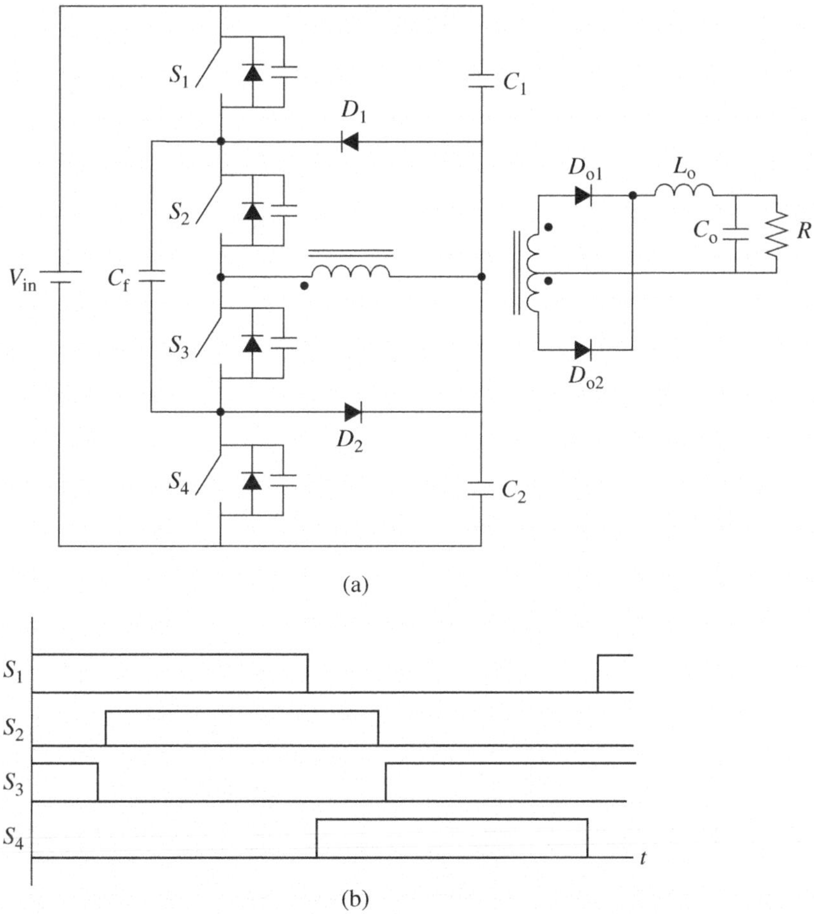

Figure 13.7 Modified three-level DC–DC converter for phase-shift PWM operation. (a) Topology. (b) Switch gating signals.

conventional phase-shift PWM, all the switches have gating signals that are approximately 50% of the switching period in width, minus some small dead-time to prevent any shoot-through or short-circuiting. The gating signals of switches S_1 and S_4 are complementary to each other as are those for switches S_2 and S_3 and the gating signals for the S_1/S_4 switches are phase-shifted by 180° relative to those of the S_2/S_3 switches.

Equivalent circuit diagrams for each mode of operation that the converter goes through during a half-switching cycle are shown in Figure 13.8. The converter has the following modes of operation:

Mode 1 (t_0–t_1): During this mode, switches S_1 and S_2 are on and the voltage across DC bus capacitor C_1, which is half of the input voltage, is impressed across the transformer primary. Output diode D_{o1} is forward-biased and the converter is in an energy-transfer mode of operation.

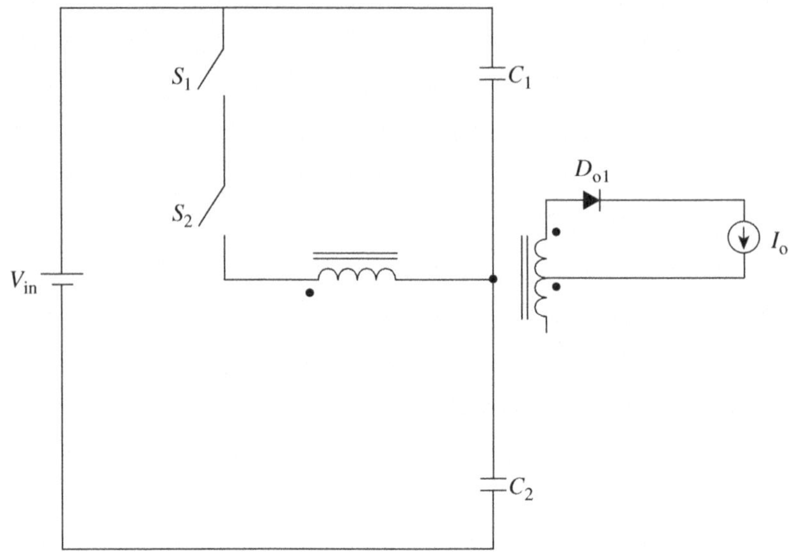

(a)

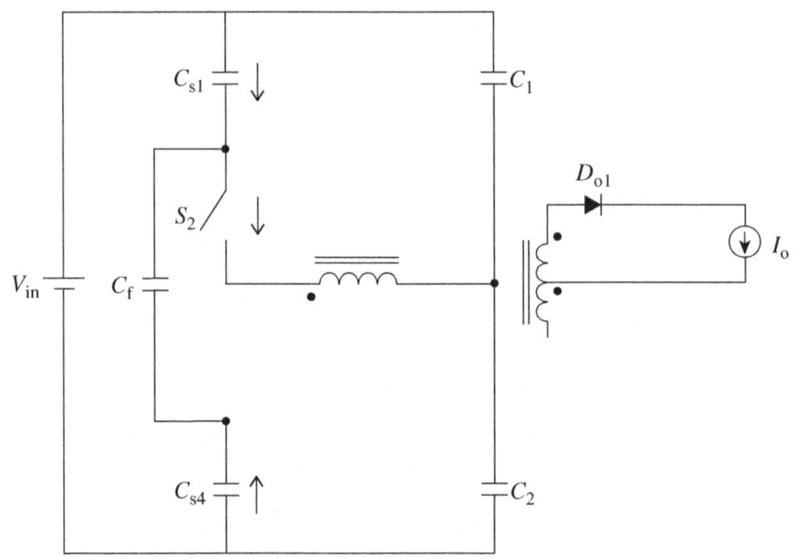

(b)

Figure 13.8 Modes of operation of the modified three-level DC–DC converter shown in Figure 13.7. (a) Mode 1. (b) Mode 2. (c) Mode 3. (d) Mode 4. (e) Mode 5. (f) Mode 6. (g) Mode 7.

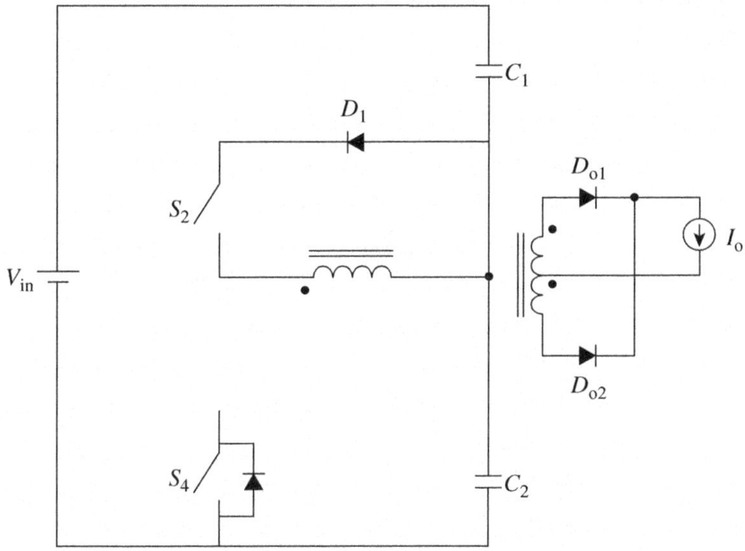

(c)

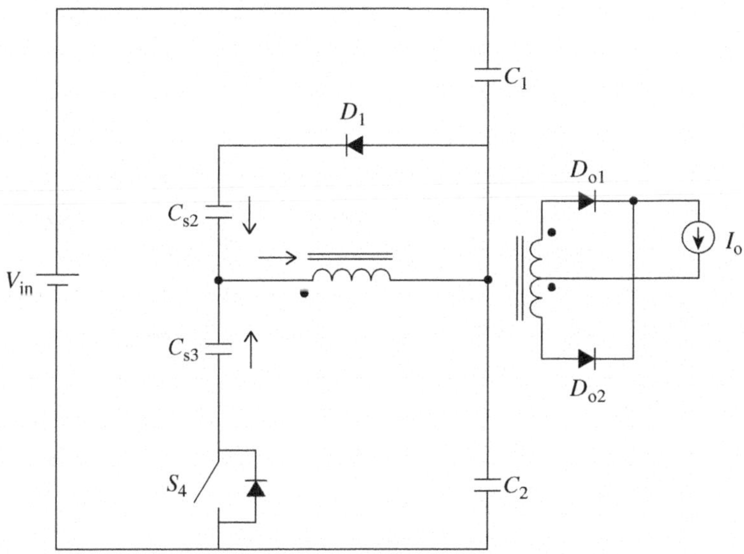

(d)

Figure 13.8 (*Continued*)

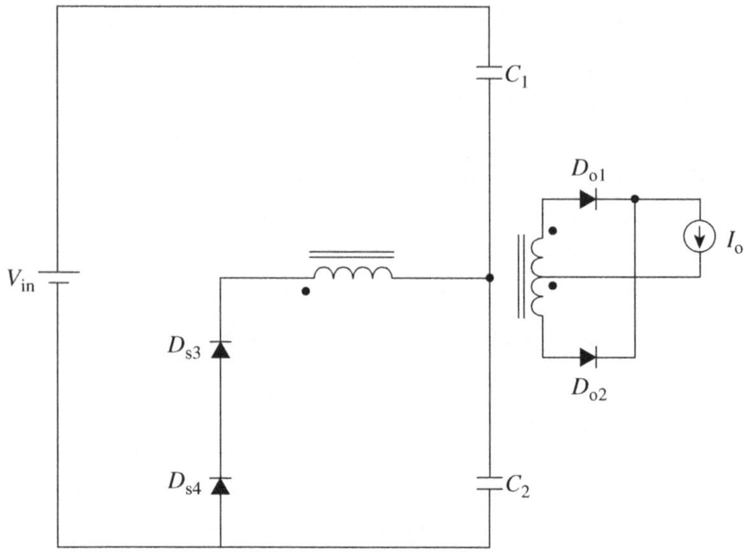

(e)

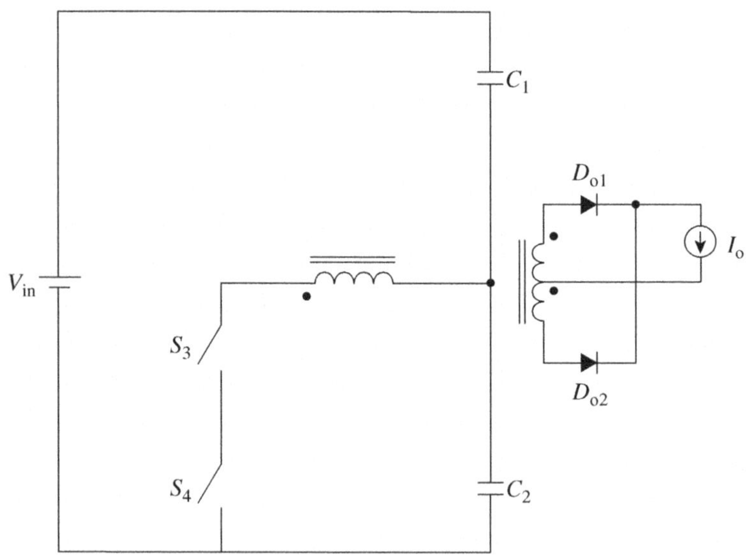

(f)

Figure 13.8 (*Continued*)

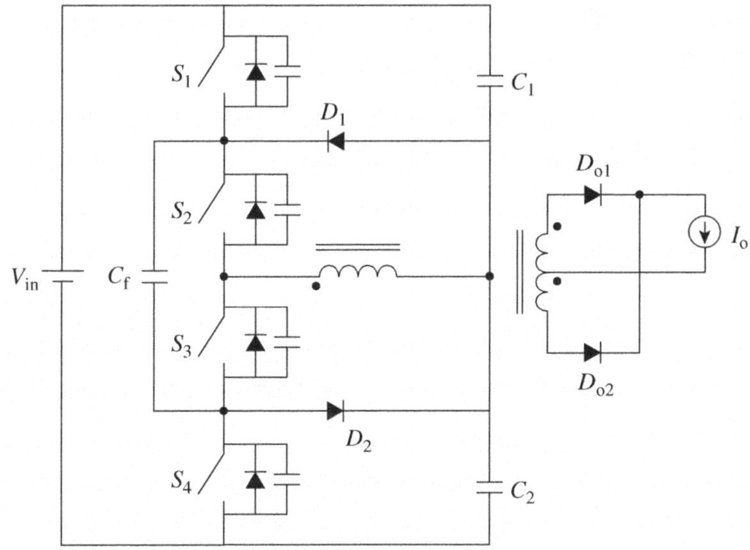

(g)

Figure 13.8 (*Continued*)

Mode 2 (t_1–t_2): This mode begins after S_1 is turned off. When this happens, the output capacitor of S_1, C_{s1} begins to be charged and the output capacitor of S_4, C_{s4}, begins to be discharged through the flying capacitor, C_f. The converter is in the process of exiting an energy-transfer mode and energy continues to be transferred to the output through output diode D_{o1}.

Mode 3 (t_2–t_3): This mode begins after C_{s4} has been fully discharged and C_{s1} has been charged to half the input voltage. Current flows through the body diode of S_4, through flying capacitor C_f, through S_2 and bus capacitor C_2. Capacitor C_f is being discharged during this mode. Also during this mode, S_4 can be turned on with ZVS. At the secondary side, both secondary diodes conduct current with the transformer secondary winding voltage being zero and load current ideally shared equally among the two diodes. The converter is in a freewheeling mode of operation.

Mode 4 (t_3–t_4): This mode begins with the turning off of S_2. Capacitor C_{s2} begins to be charged while the output capacitance of S_3, C_{s3} begins to be discharged. Current continues to freewheel in the output through diodes D_{o1} and D_{o2}.

Mode 5 (t_4–t_5): This mode begins when current begins to flow through the body diode of S_3. S_3 can be turned on with ZVS during this more. The flow of current through the output remains unchanged.

Mode 6 (t_5–t_6): This mode begins when current in the primary reverses direction and starts to flow through S_3 and S_4. Output current continues to be transferred away from D_{o1}.

Mode 7 (t_6–t_7): This mode begins when the output current has been completely transferred away from D_{o1} so that it all flows through D_{o2}. The converter is in an energy-transfer mode with current flowing through S_2 and S_3 instead of S_1 and S_4, as $V_{in}/2$ voltage is impressed across the transformer primary.

A number of modified three-level converters with passive auxiliary circuits that are used to extend the load range over which the converter can operate with ZVS have been proposed. Examples of such converters are shown in Figure 13.9. The converters shown in Figure 13.9a, b, which were proposed in [8, 9], respectively, are implemented with a passive auxiliary circuit placed between one end of the main primary transformer winding and the mid-point formed by two DC bus capacitors in series. The auxiliary circuit helps generate additional current and additional inductive energy that makes it easier to discharge the output capacitances of switches that are about to be turned on. Figure 13.9b has clamping diodes to suppress voltage ringing similar to the converter that was presented in Section 8.2. The three-level ZVS-PWM converter that was proposed in [10] and shown in Figure 13.9c has a mechanism that generates additional inductive energy with the coupled inductor formed by L_{a1} and L_{a2}. The mid-point between C_{a1} and C_{a2} has a voltage that is equal to $V_{in}/4$ and with the mid-point of S_2 and S_3 being $V_{in}/2$ or 0, a voltage of $V_{in}/4$ is generated across L_{a1} and L_{a2}, which can be used to generate more primary current and thus more inductive energy to discharge the output capacitances of the switches. This mechanism is adaptive as more current is generated when the converter is operating under light-load conditions, when more current is needed, than under heavy-load conditions, when the regular primary current is sufficient.

13.2.2 ZVZCS Three-Level Converters

It was demonstrated in Chapter 9 how a conventional full-bridge converter could be converted into a ZVZCS-PWM converter by developing a counter-voltage source that can extinguish the current that circulates in the converter when it is operating in a freewheeling mode; similar techniques can be implemented in three-level converters. For example, the use of a capacitor placed in series with the primary transformer to convert a ZVS-PWM-FB converter into a ZVZCS-PWM converter was discussed in Section 9.1. A similar approach can be used to do the same for the modified three-level converter shown in Figure 13.10 and this is what was proposed in [11, 12]; this ZVZCS-PWM converter is shown in Figure 13.10a. The freewheeling current in this converter is extinguished in the same manner as the two-level ZVZCS-PWM converter presented in Section 9.1 – current circulates in the primary and the counter-voltage due to C_b forces the current down to zero. In this converter, however, instead of using a saturable inductor, blocking diodes in series with the inner switches, S_2 and S_3, are used to ensure that current does not reverse direction and begin to flow through these switches.

Figure 13.10b shows a three-level ZVZCS-PWM converter that was proposed in [13]. This converter uses the same active switch technique to

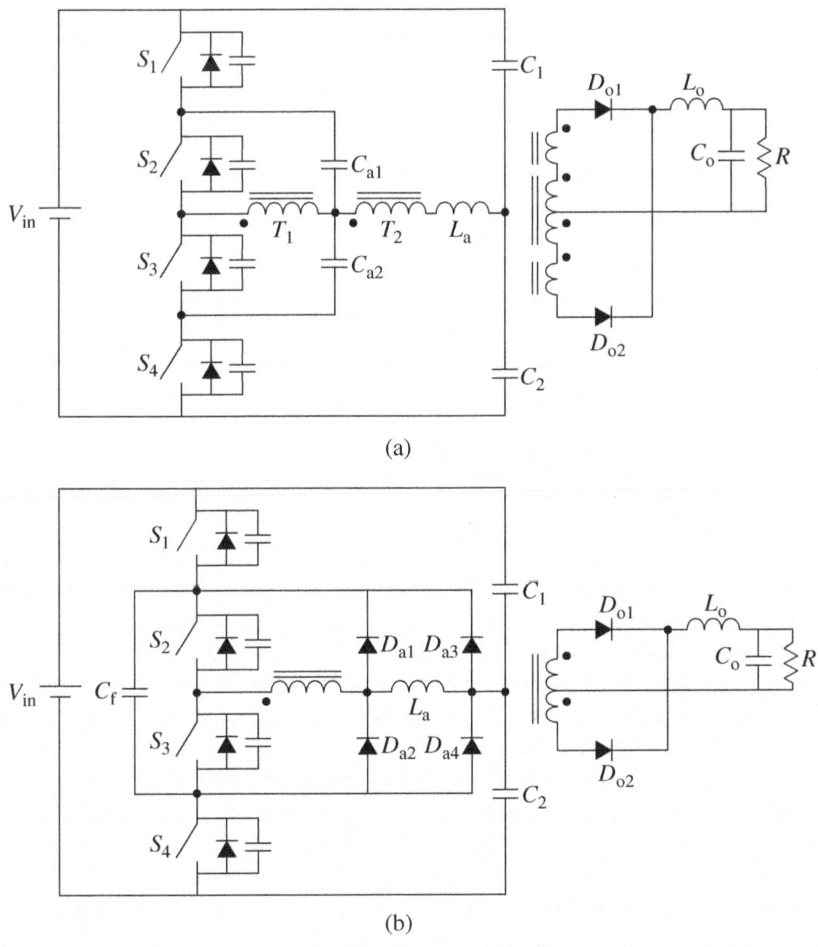

Figure 13.9 ZVS three-level DC–DC converters. (a) Converter with simple passive auxiliary circuit. (b) Converter with clamping diode auxiliary circuit. (c) Converter with adaptive ZVS auxiliary circuit.

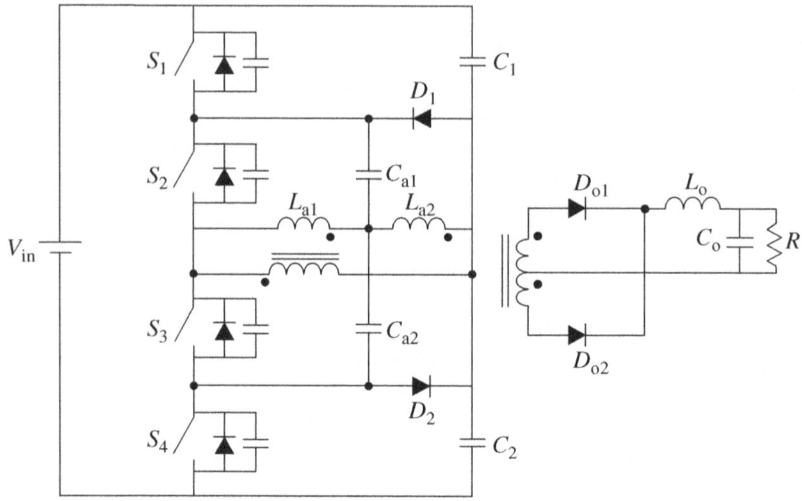

(c)

Figure 13.9 (*Continued*)

extinguish freewheeling current that was described in Section 9.2. It is an example of how ZVZCS techniques for conventional two-level full-bridge converters can be extended to three-level DC–DC converters.

A comparison between the fundamental NPC converter shown in Figure 13.1, the modified NPC converter with the flying capacitor shown in Figure 13.7, the modified converter with extended ZVS range shown in Figure 13.9b, and the ZVZCS-PWM converter shown in Figure 13.10 was made in [14]. The converter specifications that were used for this comparison were: input voltage $V_{in} = 400$ V, output voltage $V_o = 48$ V, maximum output power $P_{o,max} = 700$ W, and switching frequency $f_{sw} = 100$ kHz. A lower input voltage was used to increase the primary current to investigate the effectiveness of using ZVZCS-PWM. It was shown that the modified ZVS converter shown in Figure 13.9b was the most efficient because it can be operated with ZVS at lighter loads than the other converters and that the fundamental NPC converter was the least efficient. Whether the ZVZCS-PWM converter is more efficient than the modified ZVS converter shown in Figure 13.9b depends, however, on the amount of current flowing in the primary. Depending on the specifications, it may be that ZVZCS-PWM is preferable if the converter is operating with a considerable freewheeling current that needs to be extinguished, which can be the case for high power converters operating with heavy loads and low to medium duty-cycles.

Three-level DC–DC PWM converters can be modified to become resonant converters simply by placing resonant components in series or in parallel to the transformer, just as for two-level full-bridge converters. Typically, this

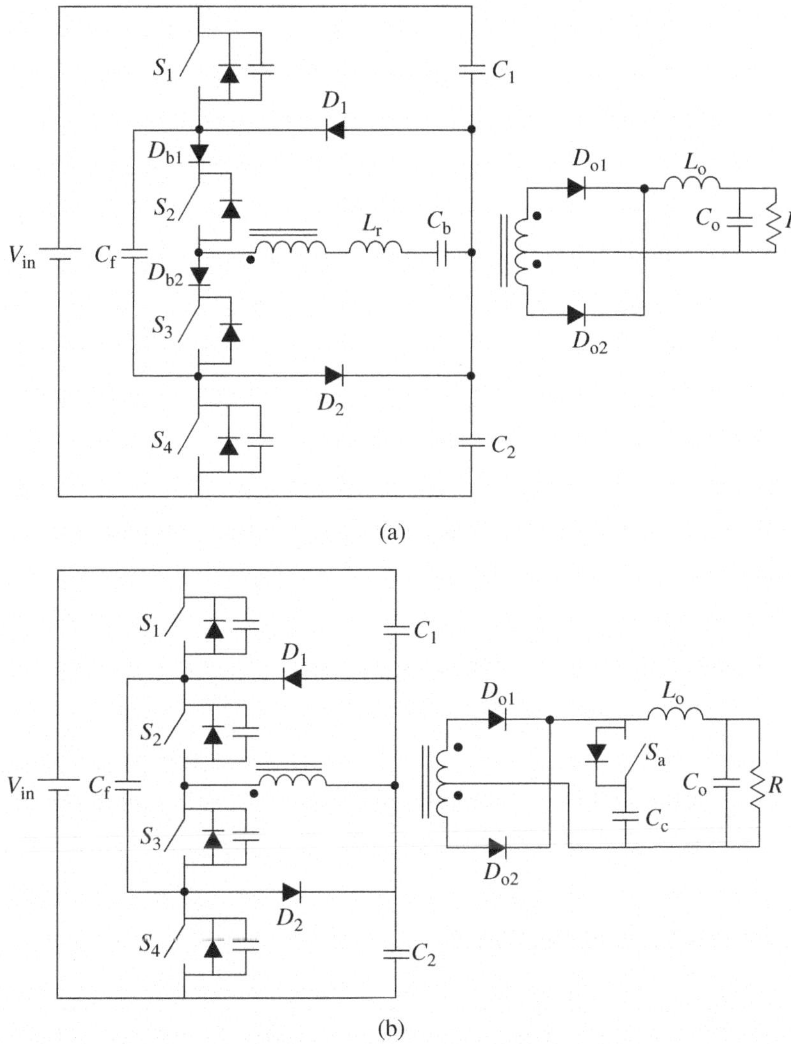

Figure 13.10 ZVZCS three-level DC–DC converters. (a) Converter with primary counter-voltage capacitor and blocking diodes. (b) Converter with secondary-side active auxiliary circuit.

is done by modifying one of the three fundamental converters discussed in Section 13.1 or the phase-shift PWM converter discussed in Section 13.2.1. An example of a three-level series resonant converter that was proposed in [15] is shown in Figure 13.11. This converter has been synthesized by adding the series resonant tank that was described in Section 11.2 to the NPC

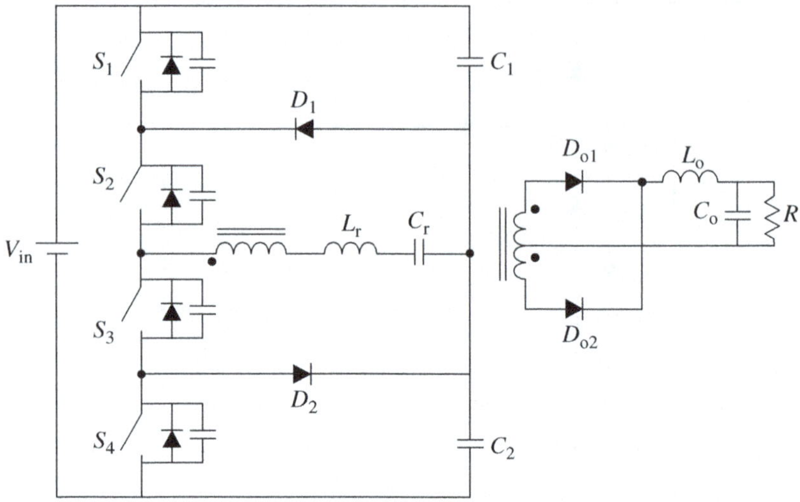

Figure 13.11 Three-level neutral-point connected series resonant DC–DC converter.

fundamental converter. The specific modes of operation are very similar to those of the base fundamental converter so that in the case of the converter shown in Figure 13.11, the modes of operation of this converter are very similar to those of the NPC converter discussed in Section 13.1.1. In general, any resonant tank current that can be used in a two-level converter can be used in a three-level converter as a similar square voltage waveform can be impressed across the tank in both converters.

13.3 STACKED CONVERTERS

The main benefit of three-level DC–DC converters over two-level converters is that it allows switches with lower peak voltage ratings to be used. A simplified variation on the three-level concept is to form DC–DC converters by simply stacking DC–DC converter modules on top of each other; the overall converter is referred to as a "stacked converter" or as a "series-input" converter (since the modules are connected in series between the positive and negative or ground DC bus voltage rail). The main advantage of a stacked converter approach over the three-level approaches that have been discussed in this chapter is modularity; the main drawback is that they require more components.

An example of such a stacked converter is the stacked DC–DC flyback converter shown in Figure 13.12a, which has been created by stacking two flyback converter modules on top of each other. By arranging the two flyback converter modules in this manner, the voltage stress of each switch is half of what

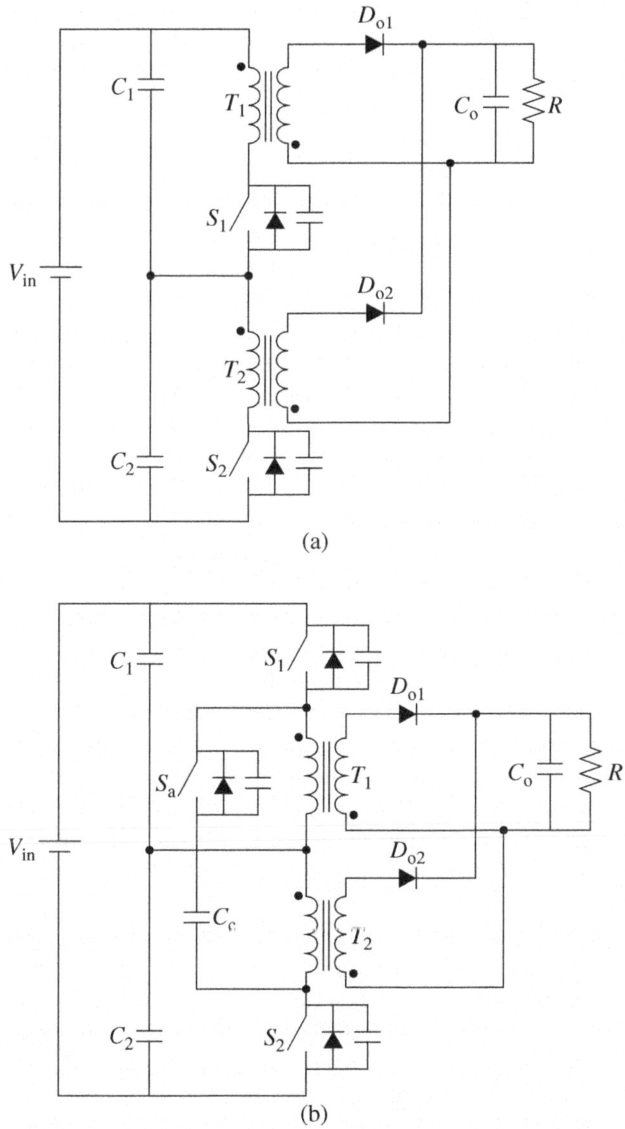

Figure 13.12 Stacked ZVS flyback converter. (a) Basic converter. (b) Converter with common active clamp circuit for ZVS.

it would be if the two modules were connected in parallel in an interleaved manner. The following should be noted:

- It is very simple to stack more than two converter modules to further reduce peak voltage switch stresses.

- The outputs of the two converter modules in the converter shown in Figure 13.12a are interleaved, which doubles the output ripple frequency and thus allows for a reduction in the size of the filter capacitor.

A second benefit to stacking converter modules is that it makes it possible for the modules to share components or work with integrated passive components to reduce size and cost. For example, by modifying the top flyback module in the converter shown in Figure 13.12a so that the drain of S_1 is connected to the top DC input rail, the converter can be implemented with just one magnetic core instead of two separate cores as was proposed in [16]. It should be noted that if only a single magnetic core is used for both modules, then the duty cycle D for each module must be limited to $D < 0.5$ to allow time for the core to demagnetize. Another example of how stacking converter modules can result in the elimination of components is the stacked ZVS active clamp flyback converter that was proposed by [17] and shown in Figure 13.12b. In this converter, stacking allows the active clamp circuit formed by S_a and C_c to be shared by S_1 and S_2 so that the use of two independent active clamps, one for each switch, can be avoided, thus reducing cost and size.

The concept of stacking converters can be applied to more sophisticated converters than flyback or forward converters. One example is the ZVS PWM converter that was proposed by [18] and shown in Figure 13.13a. This converter consists of two half-bridge modules that have been stacked on top of each other. Each half-bridge has been implemented with a simple diode clamping circuit to eliminate any voltage ringing that may be caused by the interaction of transformer leakage inductance and the junction capacitance of output secondary diodes. This clamping circuit is the same as the one discussed in Section 8.2. The output section is a current doubler, as described in Section 5.2. The secondary windings of each half-bridge module are connected in parallel.

The basic principle behind the converter is that a half-bridge converter with its two switches operating with a 50% duty cycle does not have any freewheeling mode of operation – voltage is either impressed across the transformer primary with one polarity when one of the switches is on or impressed with the opposite polarity when the other switch is on. If the duty cycle of the switches is reduced so that it is less than 50%, however, then the converter switches will not operate with ZVS as the current in the primary will have been extinguished by the time a converter switch is about to be turned on so that its output capacitor cannot be discharged beforehand. The duty cycle of the converter switches must be 50% in order for them to operate with ZVS, but fixing the duty cycle means that the converter cannot regulate its output voltage.

If two half-bridge modules are used, then each half-bridge converter module can be operated with a 50% duty cycle and can be phase-shifted from 0° to 180° with respect to the other. This phase-shifting affects the voltage that is delivered to the current doubler filter input nodes V_{Do1}, at the cathode of

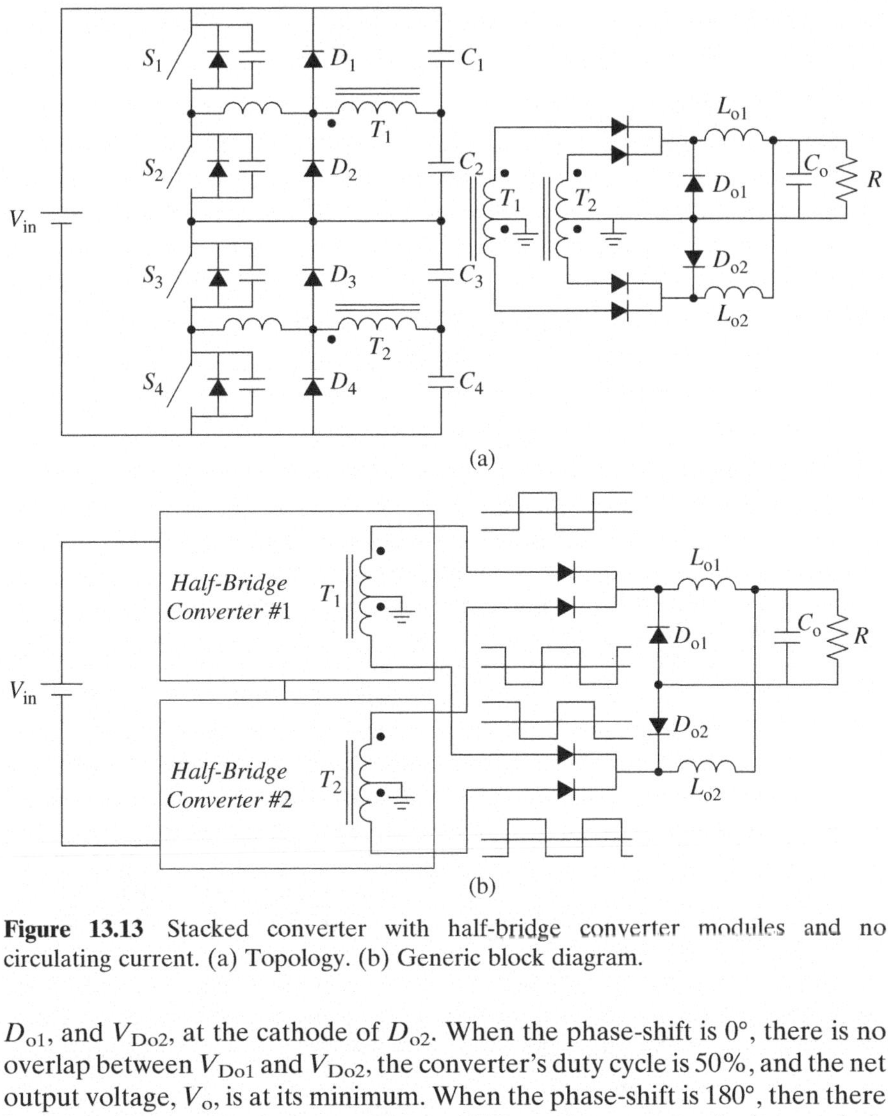

Figure 13.13 Stacked converter with half-bridge converter modules and no circulating current. (a) Topology. (b) Generic block diagram.

D_{o1}, and V_{Do2}, at the cathode of D_{o2}. When the phase-shift is 0°, there is no overlap between V_{Do1} and V_{Do2}, the converter's duty cycle is 50%, and the net output voltage, V_o, is at its minimum. When the phase-shift is 180°, then there is a complete overlap between V_{Do1} and V_{Do2}, the converter's duty cycle is 100%, and V_o is at its maximum.

Figure 13.13b shows a simplified block diagram that illustrates how phase-shifting works. If the converter is operated with a 180° phase-shift, then all secondary winding diodes that are connected to either the cathode of D_{o1} or the cathode of D_{o2} will conduct for half of the switching cycle and there are no gaps in V_{Do1} and V_{D02}. If the phase-shift between the two converters is reduced, then V_{Do1} and V_{Do2} will have gaps and the output voltage will be reduced. The output voltage can be regulated by phase-shifting the two half-bridges from full load to very light load conditions.

13.4 THREE-LEVEL DC–DC CONVERTERS IN APPLICATIONS WITH LOW AND CONVENTIONAL DC BUS VOLTAGE

Although the primary use of three-level DC–DC converter topologies is for applications where the input DC voltage is high (>600 V), these topologies can be used advantageously in lower power applications (<1 kW) where the input voltage is at a more conventional level. Since the switches of a three-level DC–DC converter turn on with half the voltage that the switches of a conventional ZVS-PWM-FB converter turn on with, then the ½ CV^2 turn-on switching losses of a three-level converter with Si MOSFETs will be just 25% of those of a ZVS-PWM-FB converter, thus improving converter efficiency. This property can be used in the following applications:

- *Light-load efficiency improvement:* Power converters are typically designed to operate with maximum efficiency under heavy-load conditions, with the peak efficiency being about 70–90% of full load. Due to the massive proliferation of power converters in society at large and the fact that many converters only operate with heavy loads occasionally, light-load efficiency has become a growing concern among power electronics experts and a number of ways have been proposed to improve it. One such method was investigated in [19] where the efficiency of a three-level NPC was compared to that of a conventional ZVS-PWM-FB converter. It was determined in [19] that the three-level NPC converter does have better light-load efficiency but worse heavy-load efficiency than the ZVS-PWM-FB converter. Since the three-level NPC converter is based on a half-bridge structure, the current flowing in its primary will be twice that of the current flowing in the primary of a ZVS-PWM-FB converter; thus, the three-level converter will have more conduction losses. Although some of these conduction losses can be somewhat offset by the use of lower voltage-rated devices with lower on-state resistance values for the converter switches, nevertheless, there will reach a point in load where the efficiency of a three-level converter will be less than that of a ZVS-PWM-FB converter; this point is dependent on the application and the specifications. As a result, the main conclusion in [19] was that the three-level NPC converter is the better converter for applications where the converter is expected to operate mainly under light-load conditions with occasional heavy-load operation, while the ZVS-PWM-FB converter is the better converter for applications where the converter is expected to operate mainly under heavy-load conditions.
- *Low power/high switching frequency applications:* In applications where a power converter of <100 W operating with a high switching frequency (>500 kHz) is to be used, such as to power small-scale electronics on a motherboard of some type, the use of three-level converter topologies becomes attractive given their improved light-load efficiency. This is

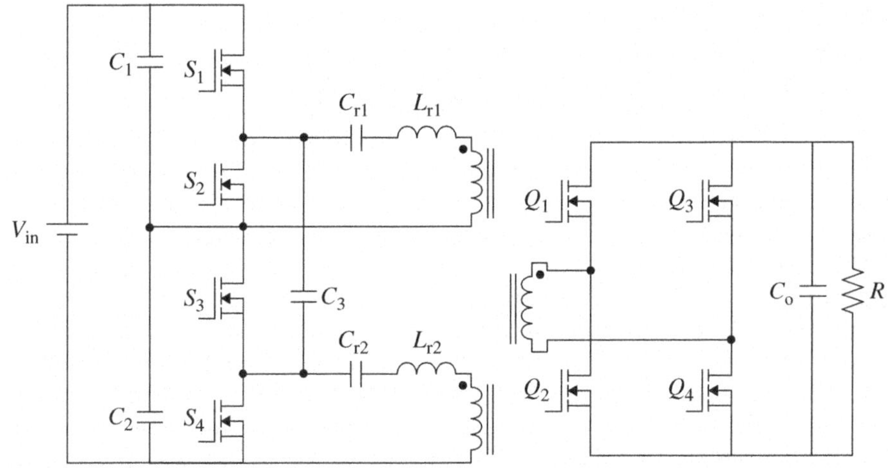

Figure 13.14 Low power three-level resonant converter for high switching frequency applications.

true whether Si or wide bandgap devices are used. An example of such a converter is the 18–80 V input, 5 V output, 75 W, 800 kHz multi-stage converter proposed in [20] and shown in Figure 13.14; only the isolation stage is shown in the figure. The converter is essentially a flying capacitor three-level converter like the one shown in Figure 13.3 with two series resonant tanks – one for S_1 and S_2, the other for S_3 and S_4 – with a synchronous rectifier output that can also be used for voltage regulation if desired. This dual resonant structure allows the load range for ZVS to be extended if Si devices are used as the current that would otherwise be available to discharge the switch output capacitances would be too low. Given that switching losses are related to ½ CV^2f, the fact that a three-level topology has a quarter of the turn-on losses of a two-level topology and that primary current is very low, makes the choice of a three-level topology advantageous. In addition to ZVS operation, the multi-level structure allows the converter to have other advantages such as reduced device ratings and reduced magnetic size and parasitic transformer capacitances; it also allows for devices with better switching characteristics and fewer parasitics to be used. All these advantages are due to the lower voltage that the converter components are exposed to.

13.5 CONCLUSION

Three-level DC–DC converters are DC–DC converters with switches that are exposed to half the voltage that two-level converters like the conventional ZVS-PWM-FB converter are exposed to. As a result, these converters are commonly used in applications where the DC input voltage is high, to avoid the use of more expensive, less available, and larger devices. Initially, they were based on topologies that were developed for DC–AC inverters but have gradually evolved for dedicated DC–DC power conversion. Many three-level topologies are derived from three fundamental three-level DC–DC converter topologies that were presented in this chapter, with each topology having its own advantages and disadvantages relative to the others. These topologies can be modified so that they can operate with an extended load range for ZVS, with ZVZCS instead of ZVS, or as resonant converters. Examples of each of these implementations were been given in this chapter.

The three-level converter concept can be expanded to allow the use of conventional converter modules simply by stacking two modules, one on top of the other; more modules can be stacked if desired. When two modules are used, not only are the voltage ratings of the converter components reduced, but the size and the cost of the overall converter can be reduced as well. Examples of how the stacked converter concept can allow for the reduction of the number of converter magnetics and soft-switching auxiliary circuits were presented in this chapter.

Three-level converters can also be used in applications where the DC bus voltage is not high, but less than 600 V. In these applications, the emphasis is on reducing converter switching losses, given that the converter switches are exposed to half the voltage that the switches in two-level converters are exposed to so that ½ CV^2 related turn-on losses are naturally reduced by 75%. Three-level converters can be used to improve light-load efficiency, which has become of increasing concern in recent years, and to improve the efficiency of low voltage/low power converters operating with very high switching frequencies (>500 kHz). The use of three-level converters to improve efficiency in this manner, however, is restricted to light-load applications as many of these converters are based on half-bridge structures with a primary current that is double that found in two-level converters. Such current results in increased conduction and turn-off losses that will offset any savings in turn-on losses that may occur by using three-level converters.

REFERENCES

1. A. Nabae, I. Takahashi, and H. Akagi, "A new neutral-point-clamped PWM inverter," *IEEE Transactions on Industry Applications*, vol. 17, no. 5, pp. 518–523, Sept. 1981.
2. J. R. Pinheiro and I. Barbi, "The three-level ZVS-PWM DC-to-DC converter," *IEEE Transactions on Power Electronics*, vol. 8, no. 4, pp. 486–492, Oct. 1993.
3. J. S. Lai and F. Z. Peng, "Multilevel converters – A new breed of power converters," *IEEE Transactions on Industry Applications*, vol. 32, no. 3, pp. 509–517, May–Jun. 1996.

4. E. Deschamps and I. Barbi, "A three-level ZVS PWM DC-to-DC converter using the versatile multilevel commutation cell," *IEEE Brazilian Power Electronics Conference (COBEP)*, 1997, pp. 85–90.
5. I. Barbi, R. Gules, R. Redl, and N. O. Sokal, "DC-DC converter: four switches V/sub pk/=V/sub in//2, capacitive turn-off snubbing, ZV turn-on," *IEEE Transactions on Power Electronics*, vol. 19, no. 4, pp. 918–927, Jul. 2004.
6. E. Deschamps and I. Barbi, "A comparison among three-level ZVS-PWM isolated DC-to-DC converters," *IEEE Industrial Electronics Society (IECON)*, 1998, Aug. 31–Sept. 4, 1998, Aachen, Germany, pp. 1024–1029.
7. F. Canales, P. M. Barbosa, J. M. Burdio, and F. C. Lee, "A zero voltage switching three-level DC/DC converter," *IEEE International Telecommunications Energy Conference (INTELEC)*, 2000, Sept. 10–14, 2000, Phoenix, USA, pp. 512–517.
8. E. Deschamps and I. Barbi, "A flying-capacitor ZVS PWM 1.5 kW DC-to-DC converter with half of the input voltage across the switches," *IEEE Transactions on Power Electronics*, vol. 15, no. 5, pp. 855–860, Sept. 2000.
9. X. Ruan, D. Xu, L. Zhou, B. Li, and Q. Chen, "Zero-voltage-switching PWM three-level converter with two clamping diodes," *IEEE Transactions on Industrial Electronics*, vol. 49, no. 4, pp. 790–799, Aug. 2002.
10. P. Das, M. Pahlevaninezhad, and A. K. Singh, "A novel load adaptive ZVS auxiliary circuit for PWM three-level DC–DC converters," *IEEE Transactions on Power Electronics*, vol. 30, no. 4, pp. 2108–2126, Apr. 2015.
11. X. Ruan, L. Zhou, and Y. Yan, "A novel zero-voltage and zero-current-switching PWM three-level converter," *IEEE Power Electronics Specialists Conference (PESC)*, 2001, June 17–22, 2001, Vancouver, Canada, pp. 1075–1079.
12. X. Ruan, L. Zhou, and Y. Yan, "Soft-switching PWM three-level converters," *IEEE Transactions on Power Electronics*, vol. 16, no. 5, pp. 612–622, Sept. 2001.
13. F. Canales, P. Barbosa, and F. C. Lee, "A zero-voltage and zero-current switching three-level DC/DC converter," *IEEE Transactions on Power Electronics*, vol. 17, no. 6, pp. 898–904, Nov. 2002.
14. M. Narimani, G. Moschopoulos, and D. Wijeratne, "A comparative study of three-level DC–DC converters," *IEEE Energy Conversion Congress and Exposition (ECCE)*, 2013, Sept. 15–19, 2013, Denver, USA, pp. 3971–3976.
15. Y. Gu, Z. Lu, L. Hang, Z. Qian, and G. Huang, "Three-level LLC series resonant DC/DC converter," *IEEE Transactions on Power Electronics*, vol. 20, no. 4, pp. 781–789, Jul. 2005.
16. T. Qian and B. Lehman, "Coupled input-series and output-parallel dual interleaved flyback converter for high input voltage application," *IEEE Transactions on Power Electronics*, vol. 23, no. 1, pp. 88–95, Jan. 2008.
17. N. Golbon, F. Ghodousipour, and G. Moschopoulos, "A DC–DC converter with stacked flyback converters," *IEEE Energy Conversion Congress and Exposition (ECCE)*, 2013, pp. 4894–4899.
18. M. Narimani and G. Moschopoulos, "A new DC/DC converter with wide-range ZVS and reduced circulating current," *IEEE Transactions on Power Electronics*, vol. 28, no. 3, pp. 1265–1273, Mar. 2013.
19. M. Narimani and G. Moschopoulos, "An investigation on the novel use of high-power three-level converter topologies to improve light-load efficiency in low

power DC/DC full-bridge converters," *IEEE Transactions on Industrial Electronics*, vol. 61, no. 10, pp. 5690–5692, Oct. 2014.

20. M. Chen, K. K. Afridi, S. Chakraborty, and D. J. Perreault, "Multitrack power conversion architecture," *IEEE Transactions on Power Electronics*, vol. 32, no. 1, pp. 325–340, Jan. 2017.

CHAPTER 14

HIGH GAIN CONVERTERS

High gain DC–DC converters are DC converters that have a high output voltage to input voltage ratio. The popularity of these converters has exploded in recent years due to their use in applications such as renewable solar energy systems, electric vehicles, DC microgrids, fuel cell energy systems, and medical equipment. The common requirement for these applications is power converters with very high voltage gains. This can be for stepping up low input voltages such as those from solar cells or batteries and stepping them up to voltages that can be fed to downstream converters, or for taking conventional DC voltages (200–400 V) and stepping them up to an ultra-high voltage that can be several thousand voltages, as is the case for some medical and electrostatic applications.

Although conventional isolated topologies such as the flyback, forward, half-bridge, and full-bridge converters can be used to generate high voltage gains simply by implementing them with transformers that have high step-up turns ratios, such transformers have considerable parasitic inductance and capacitance that can affect their design and operation (e.g. more snubbing to suppress possible voltage spikes, more challenging ZVS design). Current-fed topologies such as the ones presented in Chapter 10 help alleviate this issue, but to a limited extent, if very high voltage gains are required. Isolated resonant converters such as the parallel-resonant converter are also an option, but high-gain requirements complicate their design and operation (e.g. the impact of parasitics on the resonant network and frequency range).

DC–DC Converter Topologies: Basic to Advanced, First Edition. Gerry Moschopoulos.

Published 2024 by John Wiley & Sons, Inc.

As a result, unless an application requires transformer isolation, non-isolated high-gain converters with a voltage-boosting mechanism of some type or with a combination of various voltage-boosting mechanisms are commonly used in applications where a high-gain converter is needed. Various voltage-boosting mechanisms are presented in this chapter, along with an explanation for each mechanism discussed, in example converters[1]. The chapter concludes by summarizing the voltage-boosting mechanisms presented in this chapter and comparing their various properties.

14.1 VOLTAGE MULTIPLIER CIRCUITS

Voltage multiplier circuits are simple, inexpensive circuits that typically consist of diodes and capacitors. Some researchers include circuits with inductors in this category, while others do not. The outstanding features of these circuits are:

- They are passive circuits that can be added to existing converter topologies or can be used to replace components in existing topologies.
- The main principle used to achieve high voltage gain is voltage multiplication – multiplying the voltage that is placed across capacitors.

These circuits can generally be divided into two categories: voltage multiplier rectifiers that are used at the output stage of converters to boost voltage while simultaneously rectifying AC or pulsating DC waveforms and internal voltage multiplier circuits that are embedded within a converter's topology. Each type is reviewed in this section.

14.1.1 Output Voltage Multiplier Circuits

Figure 14.1a shows a circuit that is referred to as a Greinacher voltage doubler in the literature after its creator H. Greinacher, who proposed it in [2]. This circuit is connected to an AC sinusoidal or square-wave input, typically across the secondary winding of a transformer and works as follows: Capacitor C_1 is charged to the input voltage V_{in} when the input voltage is negative. When the input voltage is positive, the output capacitor C_2 is charged to the input voltage V_{in} plus the voltage across C_1, which is also V_{in}, so that the output voltage becomes $2V_{in}$. The main drawback of this circuit is that the diodes are

[1] At the time of this writing, the most complete summary of high-gain converters could be found in [1], a paper that was co-authored by two former Editors-in-Chief of the IEEE Transactions on Power Electronics and that is now considered to be a classic paper in the field. The author of this book would like to acknowledge this paper in the preparation of this chapter.

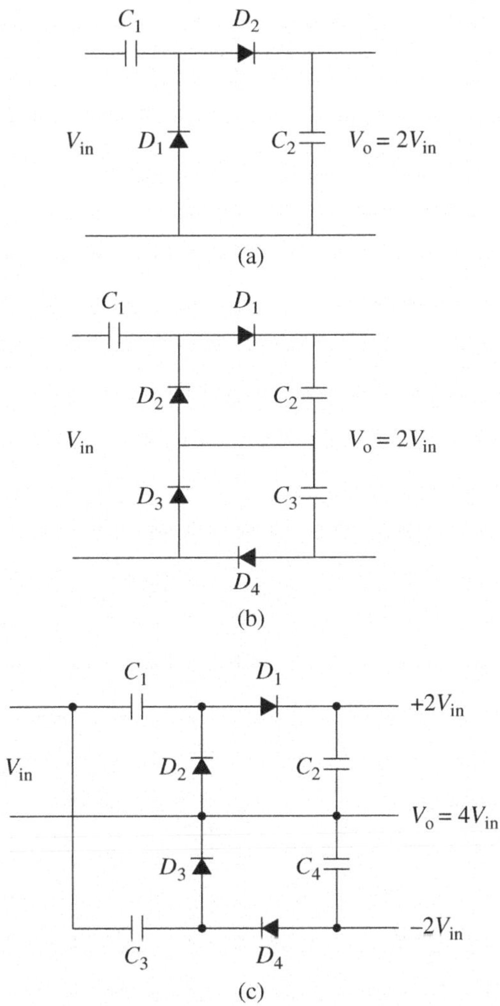

Figure 14.1 (a) Greinacher voltage multiplier circuits: (a) Voltage doubler. (b) Improved voltage doubler. (c) Voltage quadrupler.

exposed to $2V_{in}$ voltage stress so that the circuit must be implemented with more expensive diodes with higher peak voltage ratings. A variation on the classic Greinacher voltage doubler that corrects this drawback was proposed in [3] and is shown in Figure 14.1b. The arrangement of D_2 and D_3 in series ensures that each of these diodes has a peak voltage stress that is half the output voltage, or just V_{in}, while the arrangement of capacitors C_2 and C_3 in series ensures that diodes D_1 and D_4 have a peak voltage stress of just V_{in} as well. Additional voltage gain can be achieved if two Greinacher voltage doublers are stacked on top of each other, as was proposed in [4] and shown in Figure 14.1c. Doing so makes the output voltage to be $4V_{in}$ instead of $2V_{in}$ if the voltage is taken across both capacitors, or $2V_{in}$, if it is taken across each output capacitor, relative to a neutral point formed by the mid-point of C_2 and C_4 in series. This circuit is useful where a center-tapped transformer is used, or in a half-bridge-based transformerless inverter, commonly found in PV systems, where a neutral point terminal is often required.

Another commonly used output voltage multiplier circuit is the Cockcroft–Walton voltage multiplier circuit that was proposed by Cockcroft and Walton in [5] and shown in Figure 14.2a. As with the Greinacher voltage multiplier circuit shown in Figure 14.1a, this circuit is typically connected across the secondary winding of a transformer and works as follows: When the input voltage reaches its negative peak $-V_{pk}$, current flows through diode D_1 to charge capacitor C_1 to a voltage of V_{pk}. When the input voltage reverses polarity and reaches its positive peak $+V_{pk}$, it adds to the voltage across C_1 when current flows from C_1 through diode D_2, thus charging capacitor C_2 to a voltage of $2V_{pk}$. D_1 is reverse-biased as C_2 is being charged. This is an example of a half-wave rectifier as C_2 is charged only during half of the input voltage cycle. The peak voltage stress of each diode is V_{in}. If this rectifier is duplicated and

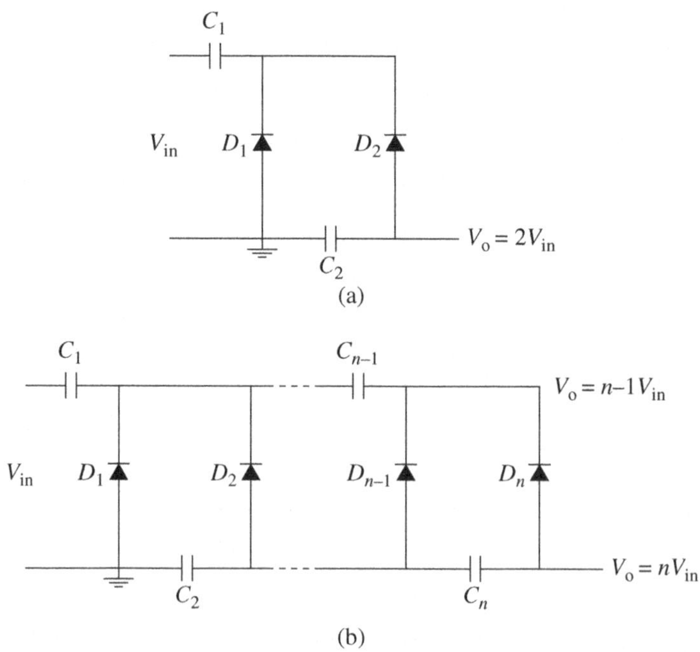

Figure 14.2 Cockcroft–Walton voltage multiplier circuits: (a) Voltage doubler. (b) Generic voltage multiplier.

placed in series as shown in Figure 14.2b, then more voltage boosting can be achieved. The main advantage of Cockcroft–Walton voltage multiplier circuits over Greinacher voltage multiplier circuits is that it is very easy to achieve further voltage multiplication simply by adding voltage multiplier circuits in series. As a result, much higher voltage gains can be achieved with Cockcroft–Walton-based voltage multiplier circuits rather than with Greinacher ones. The main drawback to Cockcroft–Walton voltage multiplier circuits is that they cannot be used with center-tapped transformers or in applications where a neutral point terminal is needed.

Figure 14.3a shows an example of a commonly used voltage-doubler circuit. This circuit operates as follows: When the input voltage reaches its

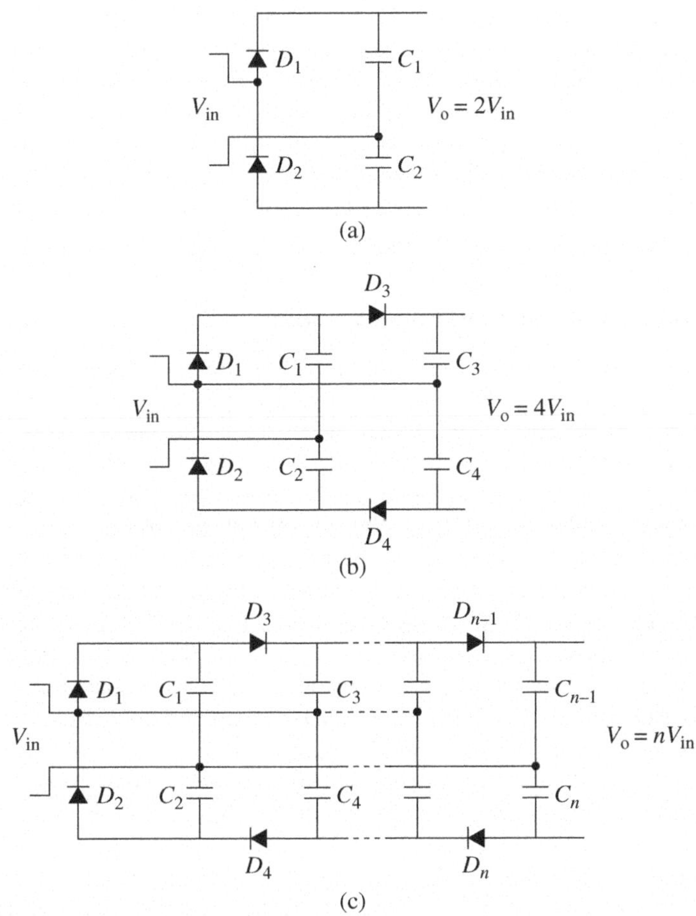

Figure 14.3 Output voltage rectifier circuits: (a) Voltage doubler rectifier. (b) Voltage quadrupler rectifier. (c) Voltage multiplier rectifier.

positive peak $+V_{pk}$, C_1 is charged to a voltage of $+V_{pk}$ through diode D_1 and D_2 is reverse-biased. When the input voltage reaches its negative peak $-V_{pk}$, C_2 is charged to a voltage of $-V_{pk}$ through diode D_2 and D_1 is reverse-biased. The output voltage is thus the sum of the voltages across the two capacitors, $+2V_{pk}$. The voltage doubler shown in Figure 14.3a actually consists of two half-wave rectifiers that are stacked on top of each other and are an example of a full-wave rectifier as either C_1 or C_2 is charged, regardless of the polarity of the input voltage. Similar to the Cockcroft–Walton rectifier, additional stages can be added in series to the voltage doubler shown in Figure 14.3a to further multiply voltage and thus ensure even more voltage boosting, as shown in Figure 14.3b,c. The circuit cannot be used with a center-tapped transformer but does have a neutral point terminal available if needed.

It should be noted that the use of output voltage multiplier circuits is limited to applications where the output current is low. This is due to the fact that current is forced to flow through numerous elements before it can reach the output of the converter; thus, the conduction losses of these circuits become very high unless the current through them is restricted. Nonetheless, as long as the output current is low, output voltage multiplier circuits are a simple and excellent way to boost voltage, whether they are attached to the output section of a conventional converter or to the output of a converter with some other voltage-boosting mechanism.

14.1.2 Internal Voltage Multiplier Circuits

An example of a converter with an internal voltage multiplier circuit is the modified Zeta converter that was proposed in [6]. The conventional Zeta DC–DC converter topology, which was presented in Section 1.3.3, is shown in Figure 14.4a, and the modified high-gain topology is shown in Figure 14.4b. The converter is a Zeta converter, but with its middle section replaced with an internal voltage multiplier circuit that consists of two diodes, D_1 and D_2, and two capacitors, C_1 and C_2. It has two main modes of operation: When the switch is on, as shown in Figure 14.4c, energy is placed in the inductor L_1 as the input voltage source is placed directly across it to form a current loop. A second current loop consisting of the input source, the switch, C_1, C_2, the output LC filter and the load is also formed while the switch is on and energy is transferred to the output during this time. When the switch is off, as shown in Figure 14.4d, the energy that was placed in the boost inductor in the previous mode is transferred to capacitors C_1 and C_2 through diodes D_1 and D_2. Current in the output inductor flows through the same elements, but the direction of this current is opposite that of the current through L_1. In order for the capacitors to be charged, the current in L_1 must be greater than that of L_o, as it is the difference between the current in these two inductors that charges the capacitors. What should be noted is how the voltage at the left-hand side of L_o when the switch is on is the sum of the input voltage and the voltage across C_1 and C_2. It is this fact that ensures that

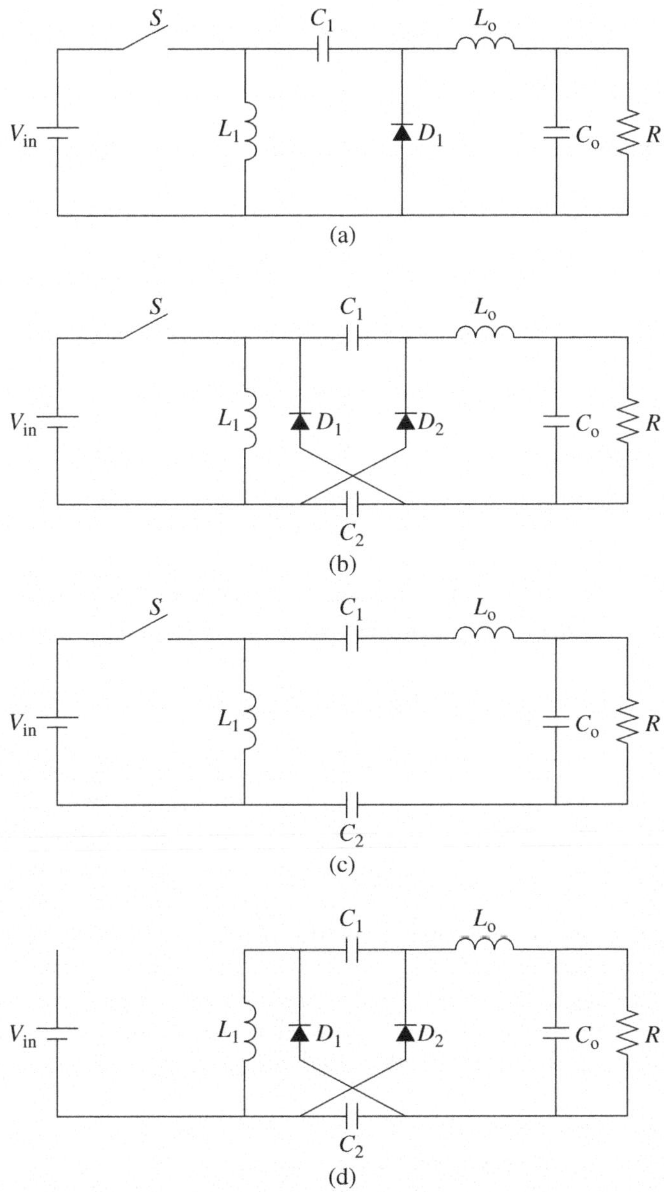

Figure 14.4 (a) Conventional Zeta DC–DC converter. (b) Modified zeta-based high-gain converter with internal voltage multiplier circuit. (c) Equivalent circuit when the high-gain converter switch is on. (d) Equivalent circuit when the high-gain converter switch is off.

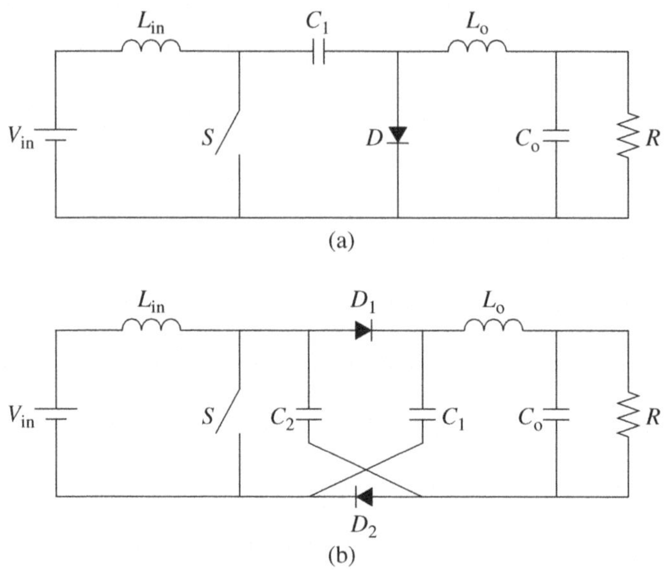

Figure 14.5 (a) Conventional Cuk converter. (b) Modified Cuk-based high-gain converter with an internal voltage multiplier.

greater voltage boosting occurs than what would be possible without the internal voltage multiplier circuit.

Figure 14.5 shows an internal voltage multiplier circuit that was proposed for a Cuk converter in [7]. The conventional Cuk converter topology, which was presented in Section 1.3.1, is shown in Figure 14.5a and the modified high-gain circuit is shown in Figure 14.5b. As with the high-gain Zeta converter shown in Figure 14.4, the high-gain converter is synthesized by replacing the middle components of the circuit with a different voltage multiplier circuit. The general principle behind any of these circuits is that any capacitor in the circuit is charged and then when it is time to transfer energy to the output, this capacitor voltage is added to any voltage that would normally be applied to the output to ensure an increased voltage gain.

14.2 SWITCHED CAPACITOR CONVERTERS

Switched capacitor converters [8–15] are converters that can perform voltage boosting without the use of any magnetic components. They are mostly popular in very low-power conversion applications and can be implemented in integrated circuits. Although integrated circuits are outside the scope of this book, converters that use switched capacitors have been proposed for converters operating with higher power levels and, thus, some of the more basic switched capacitor converters are reviewed here.

A number of switched capacitor converters of varying complexity are shown in Figure 14.6. The fundamental switched capacitor converter is the simple charge pump circuit shown in Figure 14.6a. In this circuit, the two switches are turned on and off in a complimentary manner so that switch S_1 is on whenever switch S_2 is off and vice versa. When S_1 is turned on, energy is pumped into capacitor C_1 and it charges to the input voltage level. When S_2 is turned on, the energy stored in C_1 is transferred to C_2.

The simple charge pump circuit shown in Figure 14.6a cannot perform voltage boosting by itself and must be modified in some way before it can do so. Figure 14.6b shows a basic switched capacitor circuit that uses two sets of switching positions to double input voltage, designated as 1 and 2 in the figure. When the two switches are in position 1, capacitor C_1 is charged to the input voltage. The switches are then switched to position 2 and C_1 is placed in series with the input source so that the two voltage charge C_2 and make it have a voltage that is twice the input voltage. Multiple switched capacitor circuits can be connected in series to achieve higher voltage gains as shown in Figure 14.6c. In this converter, two of the basic switched capacitor circuits of Figure 14.6b are connected in series so that the voltage across C_2 is double that of C_1 and the voltage across C_o is double that of C_2, thus making the output voltage across C_o four times that of C_1 and four times that of the input voltage.

Switched capacitor circuits can be implemented in a series-parallel manner, as shown in Figure 14.6d. The main advantage of this type of circuit is that the capacitors in the circuit all have the same voltage and thus their size can be limited. This concept can be extended to build up the voltage by placing a number of these circuits as shown in Figure 14.6e, which is typically referred to as a ladder-switched capacitor as a ladder of capacitors is charged and the output voltage is the sum of the voltages across the capacitor "steps."

Figure 14.6f,g shows two popular types of switched capacitor circuits. Figure 14.6f shows the Dickson switched capacitor, which achieves voltage boosting by using mostly diodes instead of active switches. Active switches are only used to generate two sets of pulses with an appropriate phase-shift with the charging and discharging of the capacitors synchronized to the pulses. The voltage gain of a Dickson switched capacitor circuit is equal to N times the input voltage, where N is the number of stages.

The Makowski switched capacitor circuit, shown in Figure 14.6g, can provide high-voltage boosting with low-device requirements. It is also known as a Fibonacci switched capacitor circuit because its voltage gain characteristic increases according to the Fibonacci number sequence $\{1, 1, 2, 3, 5, 8, 13, \ldots\}$. Although this circuit can achieve high voltage gain, it is difficult to implement as it has a wide range of different voltages with high capacitor voltage stresses and most of its switches are not ground referenced. The lack of ground-reference switches is an issue with other switched capacitor implementations as well.

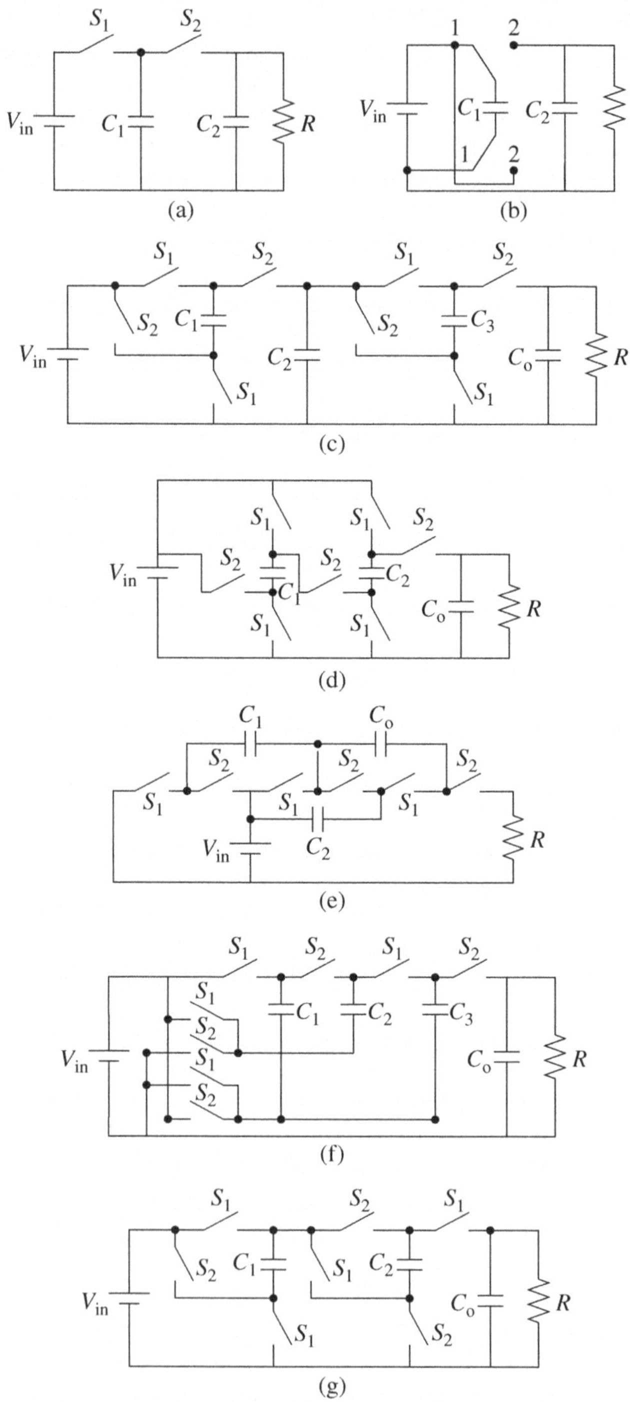

Figure 14.6 Basic switched capacitor circuits. (a) Basic charge pump. (b) Basic switched capacitor. (c) Doubler. (d) Series-parallel. (e) Ladder. (f) Dickson. (g) Markowski or Fibonacci.

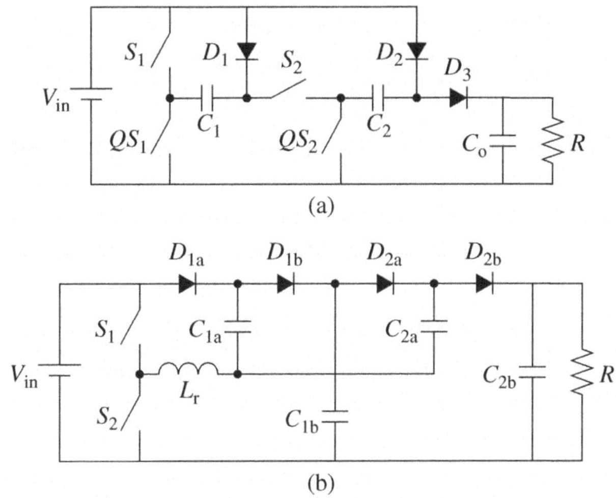

Figure 14.7 Example switched capacitor converters. (a) Converter proposed in [16]. (b) Converter proposed in [17].

Switched capacitor technology has been adapted for higher power applications. An example of such a converter is the one that was proposed in [16] and shown in Figure 14.7a. This topology consists of several individual cells that have been strung together and the basic principle behind the converter is that the capacitor in each cell is charged and the voltage across each capacitor is then added to that of the other capacitors so that a high voltage gain can be achieved. For example, for the first cell consisting of S_1, QS_1, C_1, and D_1, C_1 is charged through D_1 and QS_1 when S_1 and S_2 of the next cell are off. When S_1 and S_2 are turned on later in the switching cycle, the voltage across capacitors C_1 and C_2 is added to V_{in} so that the output voltage becomes a multiple of V_{in}.

The power that can be processed by switched capacitor circuits is limited by their high-current transients as they do not have any inductors in their topology. Current transients are caused by the exposure of capacitors with differing voltage levels to each other. As a result, switched capacitor circuits are modified by adding inductors to their topology to provide impedance against current transients, as was done for the switched capacitor circuit proposed in [17] and shown in Figure 14.7b. Adding inductors to a switched capacitor circuit eliminates current transients and increases efficiency since it eliminates the need to implement switched capacitor circuits with small, lossy, transient-limiting resistances, as is done in many cases.

14.3 VOLTAGE-LIFT AND SWITCHED INDUCTOR CONVERTERS

The voltage multiplier and switched capacitor circuits that have been presented thus far in this chapter do not use inductors in their topologies and achieve their high gain by storing energy in capacitors to charge them, then placing these capacitors in series to boost voltage. Another approach to boosting voltage is to take standard converter topologies with inductors and replace their inductors with inductor structures that allow more inductive energy to be stored when a switch is on so that a greater amount of energy can then be transferred to the output. These inductor structures can either be passive, in which case a converter with such a structure is considered to be a voltage-lift converter in the literature, or active, in which case the converter is considered to be a switched inductor converter. The main drawback to this approach, however, is that the size and weight of the converter are increased due to the use of magnetic elements.

An example voltage-lift converter that was proposed in [7] is shown in Figure 14.8a. This converter is a boost converter with a passive element

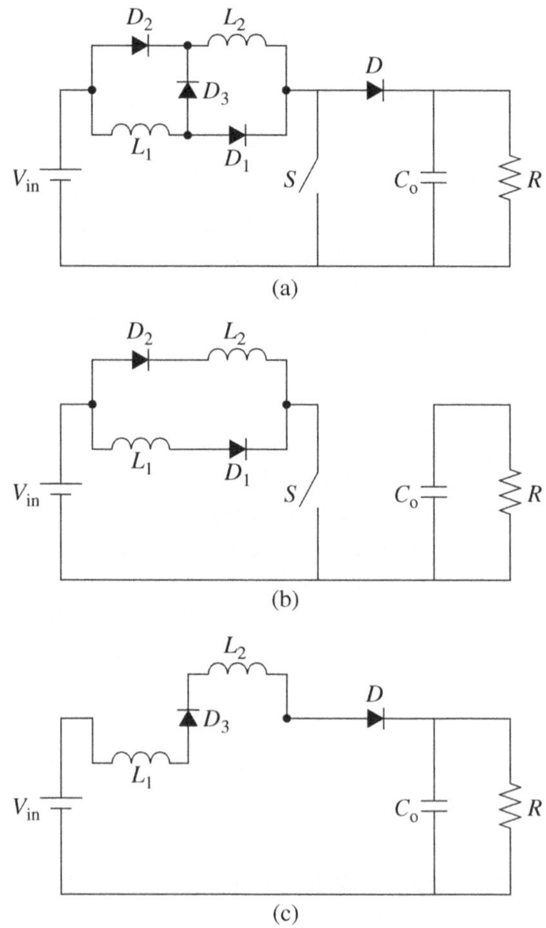

Figure 14.8 Boost-based voltage-lift converter: (a) Topology. (b) Equivalent circuit when the switch is on. (c) Equivalent when the switch is off.

network replacing the input inductor of a standard boost converter. The network can be implemented with multiple windings on a single physical core as the current through each inductor is identical to that of the other. The converter works as follows: When the switch is turned on, as shown in Figure 14.8b, current flows through L_1 and D_1 and D_2 and L_2, but not through diode D_3. When the switch is turned off, as shown in Figure 14.8c, current flows through L_1, D_3, and L_2, but not through D_1 or D_2. Assuming that $L_1 = L_2$, more energy can be transferred to the output than if a boost inductor of $2L_1$ is used in a conventional boost structure. This is because the amount of energy stored in an inductor is related to the square of the current flowing through it, according to $\frac{1}{2}LI^2$, so that if the energy stored in a conventional, single boost inductor is $\frac{1}{2}LI^2$, then the combined energy stored in the inductors of the switched inductor circuit is

$$E = \frac{1}{2}\left(\frac{L}{2} \times (2I)^2\right) + \frac{1}{2}\left(\frac{L}{2} \times (2I)^2\right) = 2LI^2 \tag{14.1}$$

if it is assumed that the current in each inductor of the switched inductor converter is double that of what it would be if a single inductor with twice the value of $L_1 = L_2$ is used. With more energy being transferred from the input to the output, the output voltage is therefore higher than what it would be if a conventional structure is used.

Another approach to voltage lifting is the modified boost converter proposed in [18] and shown in Figure 14.9. The converter works as follows: When the switch is turned on, the voltage source is placed across L_1 and C_1. When the switch is turned off, current flows through the input source, L_1, and C_1 so that the voltage across C_1 is added to the input source to produce a higher output voltage than what would be possible in a standard boost converter. The network formed by L_1 and C_1 can be considered to be a circuit cell and additional voltage boosting can be achieved if such cells are strung together as proposed in [18] and shown in Figure 14.10a. Energy is placed in each inductor of each cell when the switch is turned on, as shown in Figure 14.10b. When the switch is turned off, as shown in Figure 14.10c, a number of elements are placed in series so that voltage is boosted

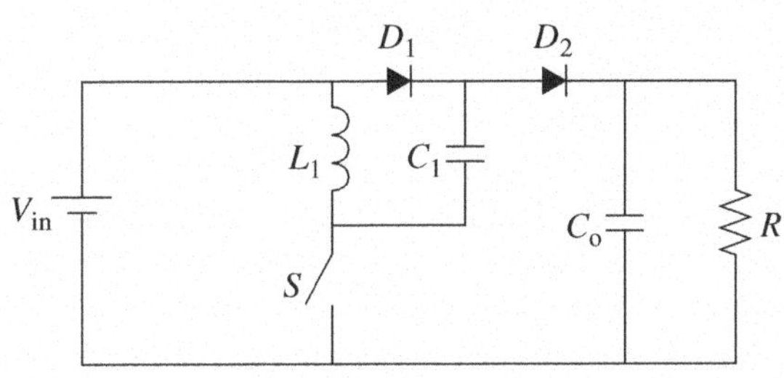

Figure 14.9 Boost-based voltage-lift with voltage-lifting network cell.

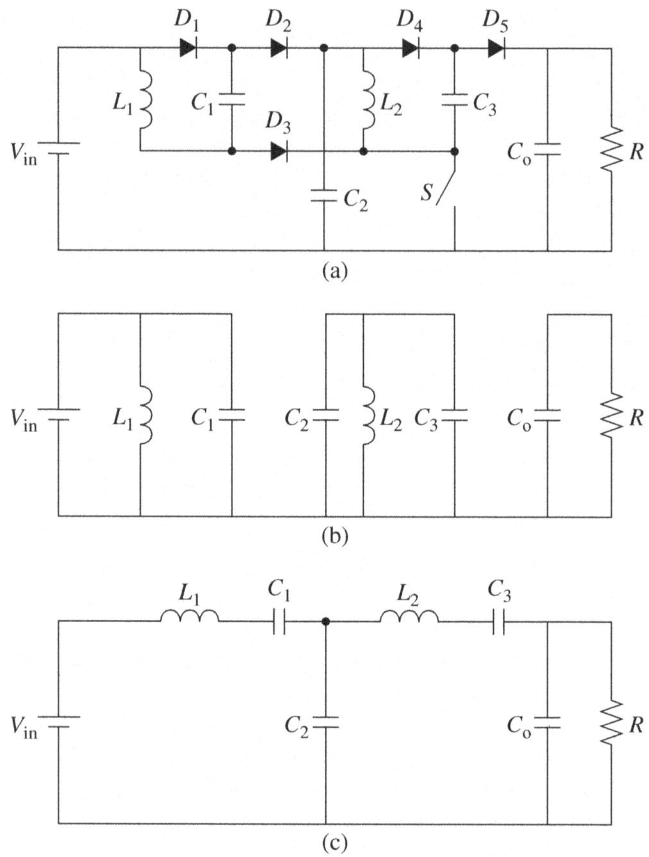

Figure 14.10 Voltage-lift converter with multiple voltage-lifting network cells: (a) Topology. (b) Equivalent circuit when the switch is on. (c) Equivalent circuit when the switch is off.

considerably. The voltage across C_2 in the diagram, which is the result of the voltage-lifting cell that consists of L_1 and C_1, is higher than the input voltage, and the voltage across C_o, which is the result of the voltage-lifting cell of L_2 and C_3, is even higher. If more such capacitors are placed inside the converter and their voltage is added, then a very high converter gain can be achieved.

The very high voltage gain that can be achieved by placing voltage-lifting cells in series is the main advantage that a cellular approach such as what is shown in Figures 14.9 and 14.10 has over a voltage-lifting approach such as the one shown in Figure 14.8. The main drawback is the current transients that can appear as a result of placing capacitors with various voltage levels in parallel. In order to increase the gain of converters that use networks such as the one shown in Figure 14.8, active networks such as the ones that were proposed in [19, 20] and shown in Figure 14.11 are used. These networks have active switches embedded in them and are considered to be switched inductors. They

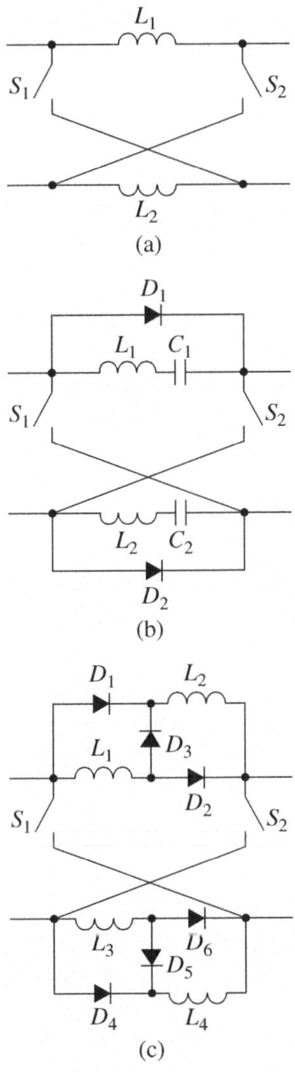

Figure 14.11 Example switched inductor networks: (a) Basic network. (b) Modified network. (c) Modified network with the voltage-lifting circuit of Figure 14.8.

are designed to replace the single boost inductor and single boost switch of a conventional single-switch PWM converter like a boost converter (the left-side terminals are connected across the source, and the right-side terminals are connected to the anode of the boost diode and to ground, respectively). The network shown in Figure 14.11a can be considered to be the basic network. The networks shown in Figure 14.11b,c can be considered to be modified version of the basic network. Converters with such networks can operate with higher conversion ratios and less component stress than the simple switched inductor converter shown in Figure 14.8, but are more sophisticated.

For example, the circuit network shown in Figure 14.11c, which was proposed in [20], combines the active network that was proposed in [19] and shown in Figure 14.11a with the passive network shown in Figure 14.8 to achieve greater voltage gain, but avoid the issue of potentially high current transients that exists in the converters shown in Figures 14.9 and 14.10. Very high voltage gains can be achieved with switched inductor circuits, but the higher the gain, the more sophisticated and expensive the circuit network becomes.

14.4 CASCADED AND QUADRATIC CONVERTERS

A simple way of achieving high voltage gains in converters is to place full converters in series so that the output of one converter is fed to the input of a second converter. This can be done with boost converters, as shown in Figure 14.12a. Assuming that each boost converter stage operates with continuous inductor current so that it has a voltage gain of

$$\frac{V_{\text{ok}}}{V_{\text{sk}}} = \frac{1}{1-D} \tag{14.2}$$

where V_{ox} and V_{sx} are the output and input voltages of a stage k and D is the duty cycle of both stages, then the voltage gain of the overall converter is

$$\frac{V_{\text{o}}}{V_{\text{in}}} = \frac{1}{(1-D)^2} \tag{14.3}$$

Although this is a simple approach, it does require the use of multiple converters, which makes it expensive. So-called quadratic converters can be synthesized by cascading converters in series, then eliminating redundant switches and controllers [21, 22]. They can have the output-to-input voltage conversion ratio of cascaded converters, but at a lower cost, as they have fewer components. They can operate with wider ranges of voltage gain than those of conventional single-switch converters so that they can avoid problems due to extreme current stresses, poor component utilization, and very low efficiency.

An example of a DC–DC quadratic boost converter is shown in Figure 14.12b. The quadratic boost converter is synthesized by cascading two DC–DC boost converters, eliminating the first boost switch, and then connecting the input inductor to the remaining switch. A diode is placed in series with the inductor for this connection to block any circulating current. The converter has been simplified as an active switch and its gate drive circuitry is replaced with a passive diode. The converter works as follows: When the switch is turned on, a positive voltage is impressed across both inductors

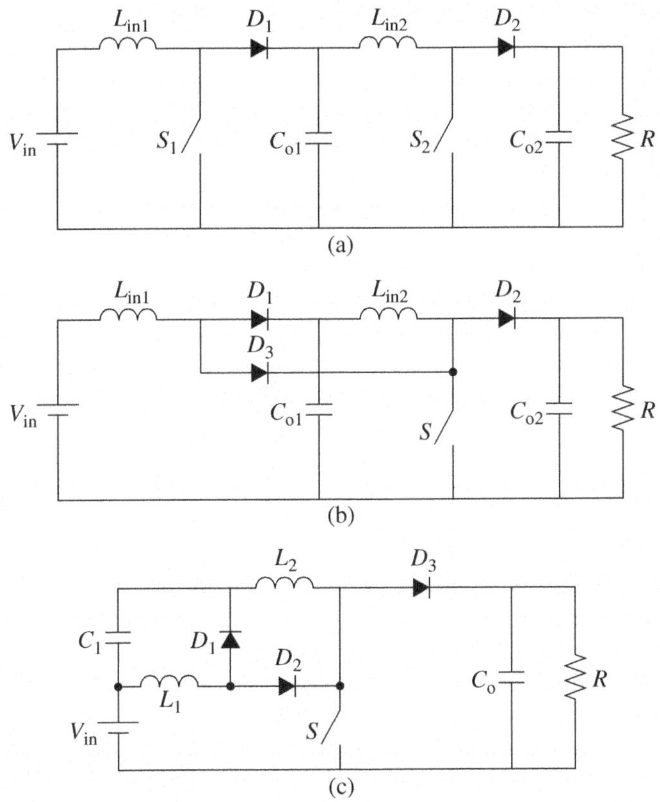

Figure 14.12 (a) Cascaded boost converter. (b) Quadratic boost converter. (c) Modified quadratic converter.

and their currents rise as they flow through the switch. When the switch is turned off, the currents in the two inductors fall as they flow through boost diodes D_1 and D_3, respectively.

There are many variations on the quadratic converter concept, but they are generally of two types. The first type is to cascade two boost converters as shown in Figure 14.12b but with different arrangements of elements to reduce component stress. An example of such a converter is the converter that was proposed in [23] and shown in Figure 14.12c. This converter has been synthesized from two boost converter stages like the quadratic boost converter shown in Figure 14.12b, but the location of the output boost capacitor C_1 of the first stage has been changed so that the input to the second stage is the sum of the voltage C_1 and the input voltage. This allows C_1 to have less voltage across it, thus less stress, which allows a smaller capacitor to be used. The second variation on the quadratic converter concept is to synthesize quadratic converters by using other high-gain mechanisms as part of hybrid converters. The topic of hybrid high-gain converters is discussed later in this chapter.

14.5 CONVERTERS WITH MAGNETIC COUPLING

A popular way to increase the gain of DC–DC converters is to use some type of magnetic coupling, such as tapped inductors, coupled inductors, or transformers. This approach is widely used in applications where no transformer isolation is required between the input and output voltage. The general principle behind these converters is to use a magnetic core with a primary and a secondary winding to step up voltage like a step-up transformer, but to implement the secondary winding in the converter in a way that its voltage is piggybacked on top of some other voltage. What this does is that it reduces the turns ratio of the magnetic core so that the parasitic effects that would occur if a regular transformer is used in an isolated converter that were described at the start of this chapter are reduced. Some examples of high-gain converters with magnetic coupling are presented in this section.

14.5.1 Tapped Inductor Converters

Consider the converter shown in Figure 14.13a. It is a conventional boost converter, but with a tapped inductor instead of a regular inductor. The converter works as follows: When the switch is on and voltage is impressed across the L_1 winding and energy is stored in the magnetic core. When the switch is turned off, the boost diode becomes forward biased and conducts, but given the polarity of the windings of the tapped inductor, the voltage across L_2 is a stepped-up version of the voltage across L_1 so that additional voltage is generated, which ultimately increases voltage gain. The converter shown in Figure 14.13a must be implemented with some sort of snubber for the switch to account for the difference in current between the current flowing in L_1 and the current flowing in winding L_2 after the switch is turned off.

An example of a tapped-inductor boost converter with such a snubber is the converter that was proposed in [24] and shown in Figure 14.13b. In this converter, inductive energy is transferred into capacitor C_c after the switch is turned off. C_c prevents large overvoltage spikes from appearing across the switch. When the switch is turned on again, the energy stored in the capacitor is transferred to C_r through diode D_2 and winding L_2 as there is a current path from C_c to the switch. When the switch is turned off again, the energy is C_r is transferred to the output as this capacitor is discharged first, before current flows through the boost diode D_4. To summarize, energy is placed in C_c when the switch is turned off, transferred to C_r when the switch is turned on again, and then transferred from C_r to the output when the switch is turned off again.

The basic principles of tapped inductor converters can be extended to multiple stages as was done in the converter that was proposed in [25] and shown in Figure 14.13c. This converter uses two-tapped inductors to generate even more voltage than what is possible for a single-tapped inductor converter.

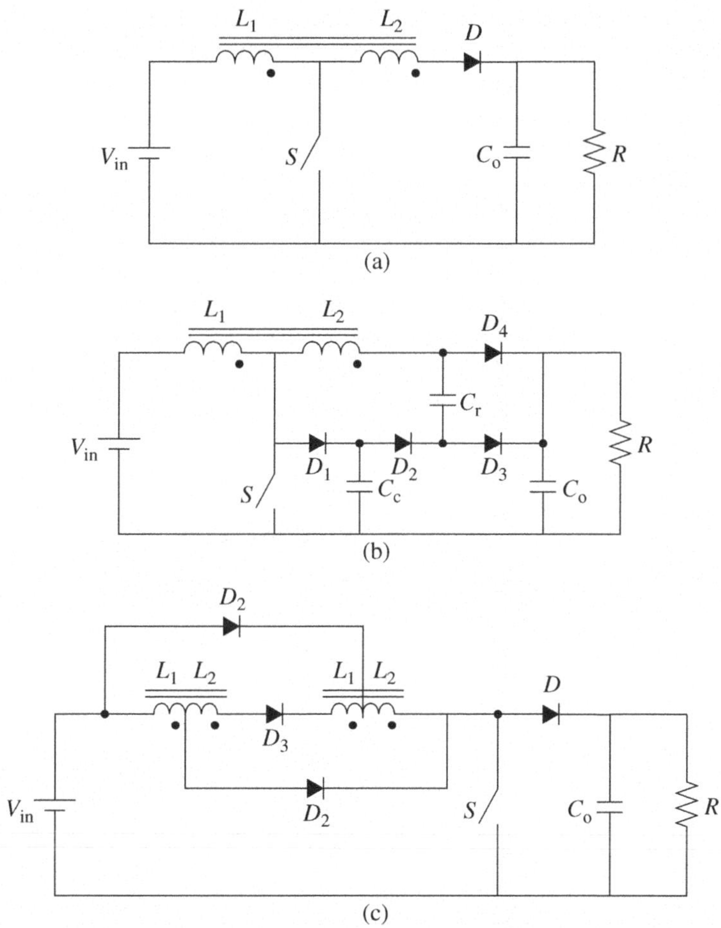

Figure 14.13 (a) Basic tapped inductor boost converter. (b) Tapped inductor boost converter with a snubber. (c) Double-tapped inductor converter.

Even higher voltage gains can be achieved by stringing more tapped inductors in series.

14.5.2 Coupled Inductor Converters

More flexibility in the synthesis of high-gain converters with magnetic coupling, greater gains, and higher converter efficiency can be achieved if inductors are implemented with separate and distinct primary windings instead of tapped windings. Converters that are synthesized by using such an approach are referred to as coupled inductor converters in the literature. An example of a coupled inductor converter is the converter that was proposed in [26] and shown in Figure 14.14a. This converter is essentially a boost converter with a high-gain mechanism that has been implemented with a coupled inductor with separate windings instead of a tapped inductor. A snubber capacitor C_c has been added to suppress any voltage spikes that may occur when the switch is turned off. Just like the tapped inductor, a high voltage gain is achieved by having the primary and secondary windings in series to boost the voltage that is seen by the output.

Additional examples of coupled inductor converter cells are shown in Figure 14.14b and c. Figure 14.14b shows an active clamp-coupled inductor converter that was proposed in [27]. The active clamp capacitor suppresses any voltage spikes that may occur after switch S_1 is turned off. Soon after this turn-off, switch S_2 is turned on to transfer energy into inductor L_r, through the loop with C_c and L_1. Just before S_1 is to be turned on, S_2 is turned off and

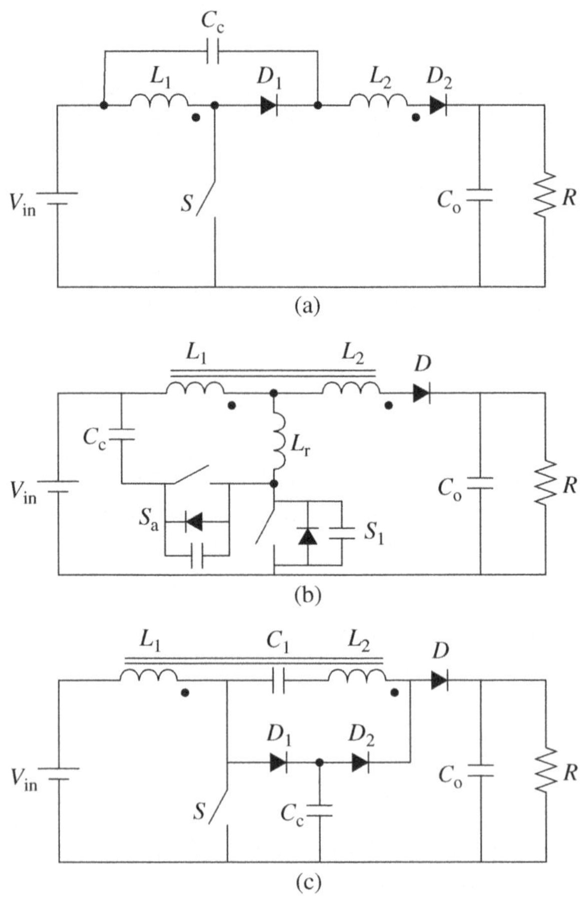

Figure 14.14 Examples of coupled inductor converters: (a) Converter with simple clamp snubber. (b) Converter with an active clamp. (c) Converter with a passive snubber.

current flowing through L_r discharges the output capacitance of S_1, then flows through its body diode, thus allowing this switch to be turned on with ZVS. Figure 14.14c shows a coupled inductor converter that was proposed in [28]. The converter has a capacitor placed between the two coupled inductor windings that are charged with a voltage that is then used to add onto the voltage that is passed along to the output. It has a clamping mechanism that can be used to suppress voltage spikes that would otherwise appear after the switch is turned off.

14.5.3 Transformer-Coupled Converters

In the literature, the term "coupled inductor converter" is generally used to refer to a converter with a topology that has a coupled inductor with two windings that are placed in series with each to build up voltage. There may be elements between the two windings, but the two windings are somehow in series. The term "transformer-coupled converter" is used to refer to a converter with at least one magnetic coupled element where the primary and secondary windings are completely separated and are not serially connected at any time during the converter's operation. Even greater flexibility in the synthesis of high-gain converter topologies can be achieved if these two windings are not serially connected as the secondary winding can be integrated with other voltage-boosting mechanisms to further increase gain.

An example of a transformer-coupled high-gain converter is the one that was proposed in [29] and shown in Figure 14.15a. In this converter, the input inductor of a boost converter is replaced with a flyback transformer and the output of this flyback converter section is stacked on top of the boost converter output. The turns ratio of the transformer is less than what it would be for a transformer in a purely isolated conventional DC–DC flyback or forward converter due to the boost converter mechanism, which reduces the parasitics of the transformer and mitigates any issues that are associated with these parasitics. Additional voltage gain can be achieved if the secondary of the transformer is attached to the voltage doubler circuit shown in Figure 14.3a so that the voltage produced by the transformer is doubled. An example of such an approach was proposed in [30] and is shown in Figure 14.15b. This converter has a boost converter front-end, but a transformer with a series blocking capacitor has been added. The secondary winding of the transformer is attached to a voltage doubler circuit whose output is then stacked on top of the output of a boost converter.

14.6 MULTI-LEVEL AND INTERLEAVED CONVERTERS

The components of many high-gain converters can be exposed to high voltages. This is especially true of a number of multiple-stage cascaded converters where the voltage of each stage has been stepped up from that of previous

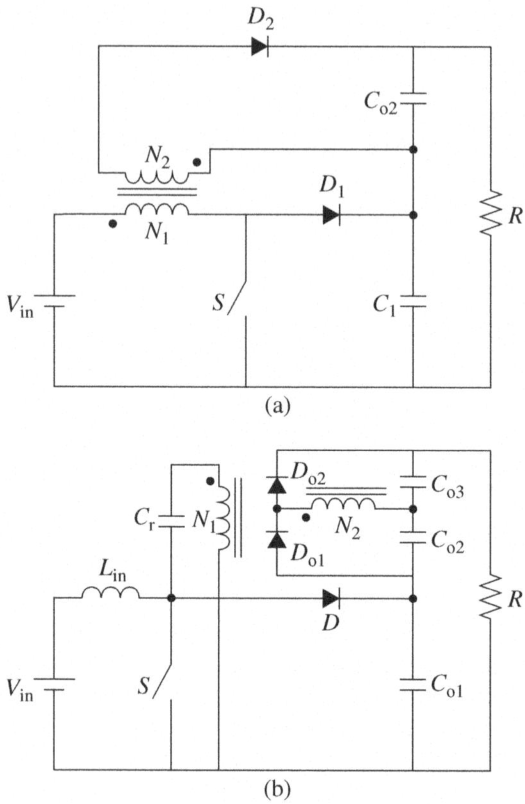

Figure 14.15 Transformer-coupled converters: (a) Converter formed by combining a boost and a flyback converter. (b) Converter with a voltage doubler circuit at the transformer secondary.

stages so that the final output voltage is very high, which exposes the components of the final stages to very high voltages. Similarly, in converters that are supplied with very low input voltages, the input current is very high and this places considerable current stress on the input section components. In this section, approaches for reducing voltage stress and for reducing current stress in the components of high-gain converters are presented.

14.6.1 Multi-Level Converters

A number of researchers have proposed high-gain converters that use multi-level converter concepts to prevent converter components from being exposed to high voltage. Most of these concepts are related to stacking sub-converters or small circuit cells on top of each other so that the voltage across each of these is added to form a very high output voltage, but with the converter components exposed to just the voltage of each sub-converter or cell instead of the full output voltage. Although these converters do have some similarities to stacked multi-level converters such as the ones that were discussed in Section 13.3, there is one key fundamental difference and that is the voltage build-up from sub-converter to sub-converter or from circuit cell to circuit cell that occurs in high-gain multi-level converters, but not in regular multi-level converters.

Examples of high-gain multi-level converter topologies that have been synthesized by stacking sub-converters or circuit cells are the ones shown in Figure 14.16. The converter shown in Figure 14.16a, which was proposed in [31], is a high-gain boost converter with additional stages consisting of capacitors and diodes. The capacitors are intertwined with the diodes so that each capacitor is charged with a fraction of the output voltage, which exposes the diodes to only a fraction of the output voltage as well. Capacitor C_1 is charged before the start of a switching cycle. When the switch is turned on, energy is placed in the input inductor as it is in a standard boost converter. When the switch is turned off, this energy is placed in C_1, again, as it does in a standard

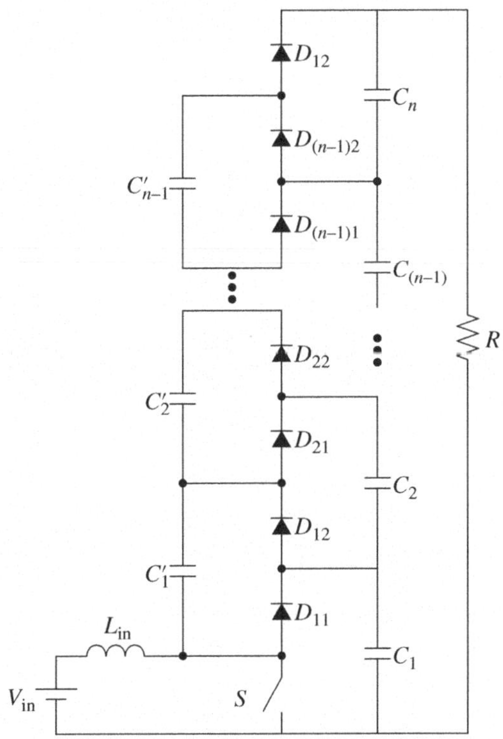

(a)

Figure 14.16 Multi-level high-gain converters: (a) Converter with stacked capacitors. (b) Modular multi-level converter.

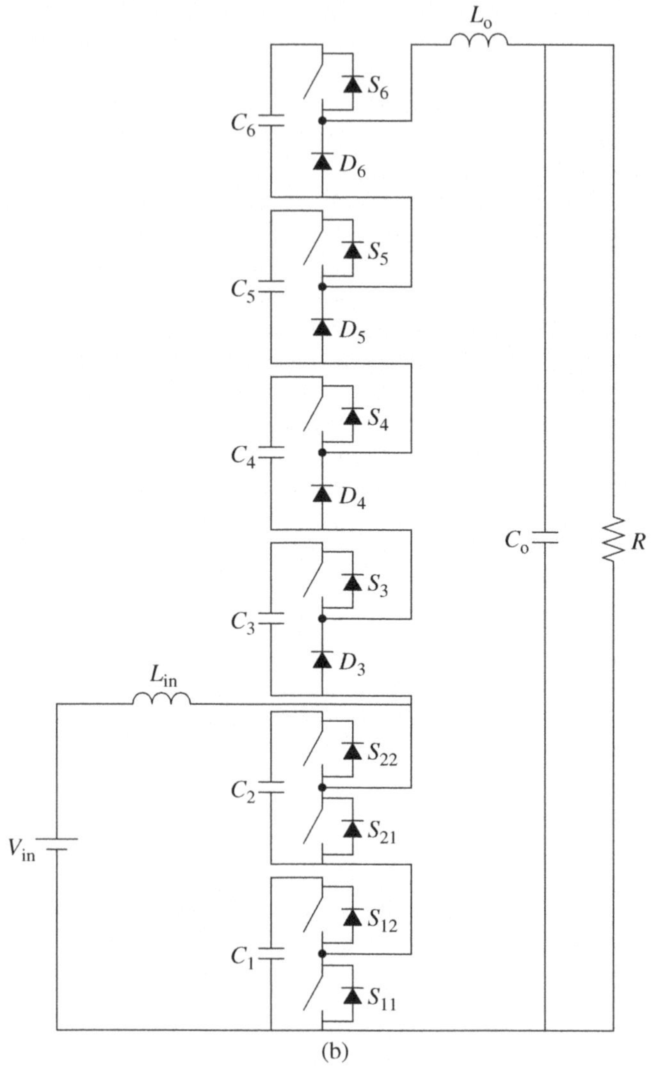

(b)

Figure 14.16 (*Continued*)

boost converter. When the switch is turned on again, then C_1' can be charged by C_1 through diode D_{12}, while energy is placed in the input inductor. When the switch is turned off, C_2 can be charged by C_1' through D_{21}. Eventually, all the capacitors in the converter are charged. In essence, the converter is like a switched capacitor converter but with the placement of the capacitors and diodes being such that they are not exposed to high voltage.

Although the components of this converter are not exposed to the full output voltage, the distribution of voltage among the capacitors is uneven. This is due to differences in the passive components, such as the forward voltage drop of the diodes and the values of the capacitors and their equivalent series resistance, which are not identical for all diodes and capacitors. These differences may be small, according to the tolerances of these components, but these small differences may be magnified so that a converter designer is forced to select components based on the absolute worst possible voltage stress, which makes the converter more expensive.

One way to address the issue of uneven component stresses is to use active switches to balance the capacitor voltage of each stage so that the components of each circuit cell or sub-converter have the same voltage stress. An example of such a converter was proposed in [32] and shown in Figure 14.16b. This converter is based on modular multi-level converters that are commonly used in high-voltage DC–AC inverters and consists of individual capacitor/switch/diode modules that are stacked up on top of each other. Current from the input can either flow through the capacitor and switch of a stage when the switch is on or through the diode, thus bypassing the capacitor whenever the converter is on in a boost mode where the whole input voltage is placed across the voltage. The length of time that the switches in the stage are on helps determine the gain. There may be slight differences between the on-times of the converter switches as adjustments for voltage balancing are made, but the voltage across the capacitor in each active circuit cell can be made to be the same as that of other circuit cells.

14.6.2 Interleaved Converters

Just as there are issues involving the peak voltage stress of components in high-gain converters, there are issues involving current stress as well. In many applications that require a high-gain converter, the input current level is much higher than the output current level, which results in high current stress and high current ripple. To address this issue, interleaving is often used in high-gain converters. Interleaving involves implementing multiple converter modules in parallel to relieve current stress on individual converter components and operating these modules out-of-phase relative to each other to reduce current ripple, as explained in detail in Section 1.4. An example of a high-gain interleaved converter is shown in Figure 14.17. This converter, which was

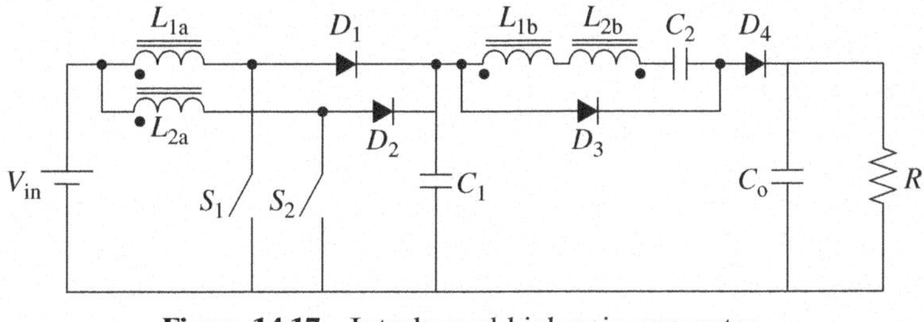

Figure 14.17 Interleaved high-gain converter.

proposed in [33], has two coupled inductor boost converter modules. The secondary windings of each coupled inductor are embedded in a voltage multiplier circuit arrangement along with diode D_3 and capacitor C_2. The voltage multiplier circuit becomes a transformer-coupled voltage multiplier circuit, which is explained in more detail in the next section. Voltage boosting is achieved by having the voltage multiplier circuit placed in series with an interleaved front-end boost converter. Current in the interleaved front end of the converter is split among the two interleaved modules so that the components of each module are not subjected to the same current stress as the components in a conventional single-module boost converter would be. Capacitor C_1, which is the output of the interleaved boost converter, has a voltage that is considerably higher than the input voltage so that the voltage multiplier circuit does not have to be interleaved as the current stress of its components is much less. Additional stages, however, can also be interleaved if desired.

14.7 HYBRID CONVERTERS AND CONVERTER SELECTION

It is very often the case that various high-gain converter mechanisms are combined to synthesize other high-gain converters. These hybrid converters offer better performance than converters of a single type but at greater expense and size as they have more components. Although hybrid converters can be synthesized in any number of ways, the most common way of doing so is either by adding voltage multiplier circuits of some sort or by adding some sort of magnetic coupling to a high-gain DC–DC converter. Several examples of hybrid converters are presented in this section.

The first example of a hybrid converter is the one that was proposed in [34] and shown in Figure 14.18a. It is a boost converter that combines a voltage multiplier circuit that has no magnetic elements with a transformer. The voltage multiplier circuit is seamlessly integrated into a boost converter. The primary winding of the transformer is placed across switch S_1 along with a DC blocking capacitor, while the secondary winding is embedded in the voltage multiplier circuit in place of the inductor in the original voltage multiplier circuit. With the additional voltage boosting mechanism of the transformer, the hybrid converter shown in Figure 14.18a can achieve a higher voltage gain than it can without it.

Figure 14.18b shows another hybrid converter that was proposed in [35]. This converter has been synthesized by combining the quadratic boost converter shown in Figure 14.12b with the voltage multiplier circuit shown in Figure 14.5b. In this converter, the input of the voltage multiplier circuit can be seamlessly matched with the output of the quadratic converter, but an output inductor needs to be added to the output of the voltage multiplier circuit as it cannot be matched to the output filter capacitor. A more sophisticated converter can be synthesized by replacing the simple boost converter front-end of the converter shown in Figure 14.18a with a quadratic boost

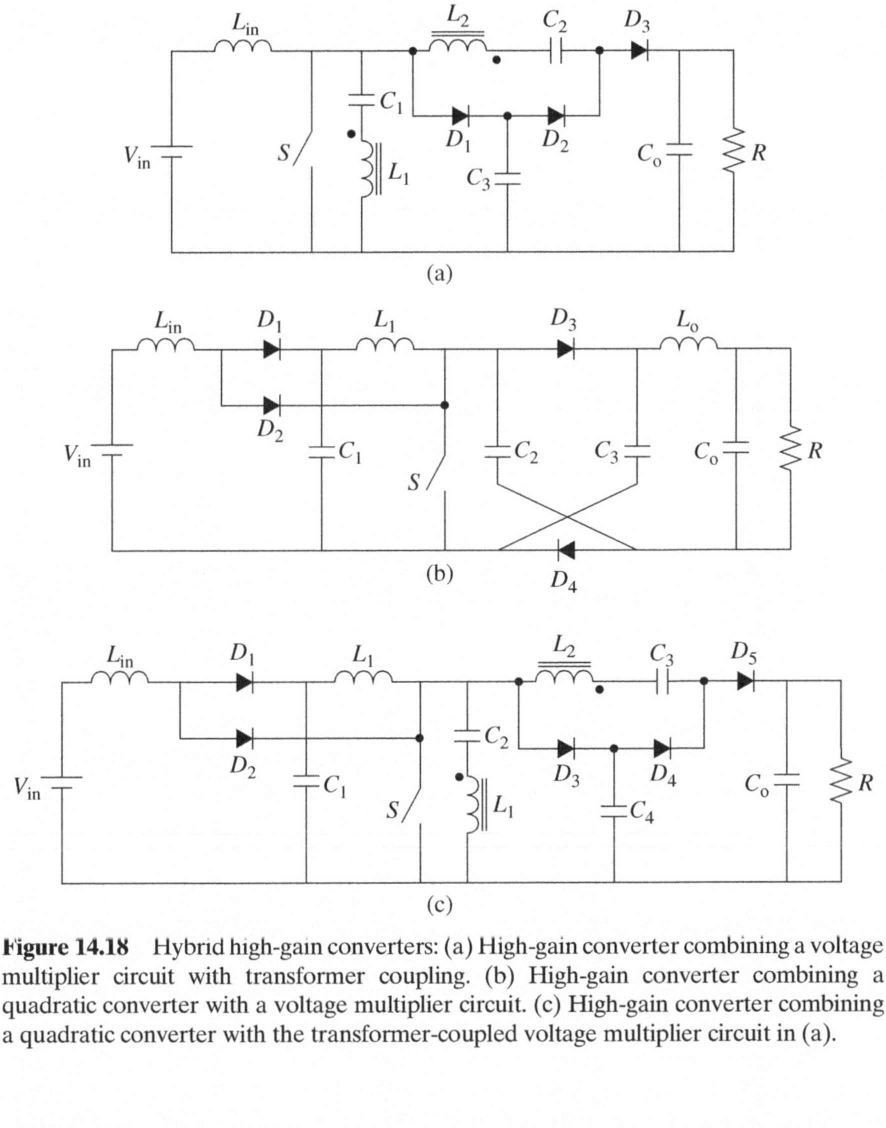

Figure 14.18 Hybrid high-gain converters: (a) High-gain converter combining a voltage multiplier circuit with transformer coupling. (b) High-gain converter combining a quadratic converter with a voltage multiplier circuit. (c) High-gain converter combining a quadratic converter with the transformer-coupled voltage multiplier circuit in (a).

converter, as was done in [36]; the resulting converter is shown in Figure 14.18c. This converter seamlessly integrates three voltage-boosting mechanisms – quadratic boost, voltage multiplier circuit, and transformer – to produce a converter with even greater voltage gain.

In essence, the various high-gain mechanisms that have been presented in this chapter can be considered to be like building blocks that can be combined in a seemingly infinite number of ways to generate new high-gain converters. As a result, the operation of more recently proposed high-gain converters, which are mostly hybrid, can be determined by identifying the various building blocks that they contain and then determining the contribution of each building block to voltage boosting. There is no single high-gain converter that can be said to be clearly superior to all other high-gain converters as there are many factors to consider, such as cost, size, weight, efficiency, component stress, and gain. The selection of a high-gain converter is very much dependent on the particular application that it is to be used for.

The following is a summary of the various high-gain converter types that have been presented in this chapter.

- *Voltage multiplier circuits:* They are simple passive circuits consisting mostly of diodes and capacitors (some have inductors) that can be embedded inside the topology or used as the final voltage-boosting stage at the end. They have a cell-based structure that allows them to be integrated into various structures or strung together in series and can produce a significant amount of voltage-boosting with a simple circuit structure. Their drawbacks include high peak voltage stress of components, especially though of the final stage if several voltage multiplier circuits are placed in series, and conduction losses, as current must flow through a fair number of components. They tend to be used as the final output stage in the secondary of ultra-high-voltage transformer-isolated converters such as those needed for electrostatic and medical applications (e.g. X-ray machine, laser), where the output voltage is very high, in the thousands of volts, and thus output current is very low. They can also be found in numerous low-power hybrid converters where conduction losses can be kept low.
- *Switched capacitor converters:* They are converters that can perform voltage boosting without the use of any magnetic components, just switches and capacitors. Due to the fact that they do not have any magnetic components, they can only be used in very low-power applications because of voltage transients. Although some inductance may be added to limit these transients, their use is generally restricted to implementations in integrated circuits.
- *Switched inductor and voltage-lift converters:* These are converters that perform voltage-boosting with switched inductors instead of switched capacitors. They can be implemented with higher power levels than switched capacitor converters as they do not have issues with voltage transients but are larger in size because they have magnetic elements in their topologies. They are suitable for low-power and medium-power applications but not high-power applications.
- *Quadratic converters and cascaded converters:* These are converters formed by cascading active converter stages. Quadratic converters can be derived from cascaded converters but have redundant elements eliminated. They can achieve high voltage gain without being dependent on coupled inductors or transformers but suffer from conduction losses as

they are derived from topologies with multiple converters in series. These converters can be operated at power levels that are higher than switched inductor and voltage-lift converters, but their power is generally restricted to levels that are lower than other options, such as converters with magnetic coupling.

- *Converters with magnetic coupling:* These converters use some sort of magnetic coupling – a core with a tapped winding (tapped inductor), a core with two separate windings that are placed in series with each other (coupled inductor), a core with two separate windings that are not in series with each other (transformer) – to step up the voltage. The general principle behind these converters is to use a magnetic core with a primary and a secondary winding to step up voltage like a step-up transformer but to implement the secondary winding in the converter in a way that its voltage is piggybacked on top of some other voltage. What this does is that it reduces the turns ratio of the magnetic core so that the parasitic effects that would occur if a regular transformer is used in an isolated converter that were described at the start of this chapter are reduced. These converters can be used in higher power applications and can be made to have good efficiency, especially transformer-coupled converters, as some of the power transferred from input to output can be made to bypass current paths as the secondary winding can be placed at later stages. The voltage stress of the components can be reduced as well in more sophisticated converters. The main drawbacks are the size and weight of the converter magnetics and the negative effects of core leakage inductance, such as voltage spikes, that need to be considered. They can be used in higher voltage applications and as power supplies in higher power applications.
- *Multi-level and interleaved converters:* The components of many high-gain converters can be exposed to high voltages, given the high voltage gain of these converters. This is especially true of a number of multiple-stage cascaded converters where the voltage of each stage has been stepped up from that of previous stages so that the components of the final stages are exposed to very high voltages. They can also be exposed to high currents as the input voltage in many applications where high gain converters are used is low. To reduce the component stresses of these converters, multi-level converters have been developed to reduce peak voltage stress and interleaved converters have been developed to reduce current stresses. Multi-level and interleaved converters are generally modular in structure and can operate with high power, high voltage, and high current. Given the number of components that these converters have, however, they can be bulky, heavy, and costly. Multi-level converters can be used in applications such as in DC microgrids and high-voltage power supplies, while interleaved converters can be used in renewable energy systems and electric and hybrid-electric vehicles.

14.8 CONCLUSION

High-gain DC–DC converters have become very popular in recent years due to the need to step up voltage in many industrial applications, especially for renewable energy systems. Although conventional isolated topologies such as the flyback, forward, half-bridge, and full-bridge converters can be used to generate high voltage gains simply by implementing them with transformers that have high step-up turns ratios, such transformers have considerable parasitic inductance and capacitance that is disadvantageous. As a result, unless an application requires transformer isolation, non-isolated high-gain converters with a voltage-boosting mechanism of some type or with a combination of various voltage-boosting mechanisms are commonly used in applications where a high-gain converter is needed.

Various voltage-boosting mechanisms have been presented in this chapter. For each mechanism, the basic principle behind it was explained and demonstrated with example converters containing the mechanism. A summary of each of these mechanisms and guidelines for their use are given at the end of this chapter.

REFERENCES

1. M. Forouzesh, Y. P. Siwakoti, S. A. Gorji, F. Blaabjerg, and B. Lehman, "Step-up DC–DC converters: A comprehensive review of voltage-boosting techniques, topologies, and applications," *IEEE Transactions on Power Electronics*, vol. 32, no. 12, pp. 9143–9178, Dec. 2017.
2. H. Greinacher, "Greinacher. Erzeugung einer Gleichspannung vom vielfachen Betrag einer Wechselspannung ohne Transformator," *Bulletin des Schweizer Elektrotechnischer Vereins*, vol. 11, pp. 59–66, 1920.
3. W. Li, D. Xu, B. Wu, Y. Zhao, H. Yang, and X. He, "Zero-voltage switching dual-boost converter with multi-functional inductors and improved symmetrical rectifier for distributed generation systems," *IET Power Electronics*, vol. 5, no. 7, pp. 969–977, Jul. 2012.
4. Y. Zhao, X. Xiang, C. Li, Y. Gu, W. Li, and X. He, "Single-phase high step-up converter with improved multiplier cell suitable for half-bridge based PV inverter system," *IEEE Transactions on Power Electronics*, vol. 29, no. 6, pp. 2807–2816, Jun. 2014.
5. J. D. Cockcroft and E. T. S. Walton, "Experiments with high velocity ions," *Proceedings Royal Society London*, vol. 136, pp. 619–630, 1932.
6. B. Axelrod, Y. Berkovich, and A. Ioinovici, "Hybrid switched-capacitor-Cuk/Zeta/Sepic converters in step-up mode," *IEEE International Symposium on Circuits and Systems*, 2005, pp. 1310–1313.
7. B. Axelrod, Y. Berkovich, and A. Ioinovici, "Switched-capacitor/switched-inductor structures for getting transformerless hybrid DC–DC PWM converters," *IEEE Transactions on Circuits and Systems I: Regular Papers*, vol. 55, no. 2, pp. 687–696, Mar. 2008.

8. G. Palumbo and D. Pappalardo, "Charge pump circuits: An overview on design strategies and topologies," *IEEE Circuits and Systems Magazine*, vol. 10, no. 1, pp. 31–45, First Quarter 2010.
9. M. D. Seeman and S. R. Sanders, "Analysis and optimization of switched capacitor DC–DC converters," *IEEE Transactions on Power Electronics*, vol. 23, no. 2, pp. 841–851, Mar. 2008.
10. M. D. Seeman, "A design methodology for switched-capacitor DC–DC converters," *Univ. California, Berkeley, Berkeley, CA, USA, Technical Report* UCB/EECS-2009-78, May 2009.
11. M. S. Makowski and D. Maksimovic, "Performance limits of switched capacitor DC–DC converters," *IEEE Annual Power Electronics Conference (APEC)*, 1995, Mar. 5–9, 1995, Dallas, USA, pp. 1215–1221.
12. J. F. Dickson, "On-chip high-voltage generation in MNOS integrated circuits using an improved voltage multiplier technique," *IEEE Journal of Solid-State Circuits*, vol. 11, no. 3, pp. 374–378, Jun. 1976.
13. M. S. Makowski, "Realizability conditions and bounds on synthesis of switched-capacitor DC–DC voltage multiplier circuits," *IEEE Transactions on Circuits and Systems I: Fundamental Theory and Applications*, vol. 44, no. 8, pp. 684–691, Aug. 1997.
14. J. A. Starzyk, Y.-W. Jan, and F. Qiu, "A DC–DC charge pump design based on voltage doublers," *IEEE Transactions on Circuits and Systems I: Fundamental Theory and Applications*, vol. 48, no. 3, pp. 350–359, Mar. 2001.
15. F. L. Luo, "Switched-capacitorized DC/DC converters," *IEEE Conference on Industrial Electronics and Applications*, 2009, May 25–27, 2009, Xi'an, China, pp. 1074–1079.
16. H. S.-H. Chung, "Design and analysis of a switched-capacitor-based step-up DC/DC converter with continuous input current," *IEEE Transactions on Circuits and Systems I: Fundamental Theory and Applications*, vol. 46, no. 6, pp. 722–730, Jun. 1999.
17. K. K. Law, K. W. E. Cheng, and Y. P. B. Yeung, "Design and analysis of switched-capacitor-based step-up resonant converters," *IEEE Transactions on Circuits and Systems I: Regular Papers*, vol. 52, no. 5, pp. 943–948, May 2005.
18. F. L. Luo and H. Ye, "Positive output super-lift converters," *IEEE Transactions on Power Electronics*, vol. 18, no. 1, pp. 105–113, Jan. 2003.
19. L. Yang, T. Liang, and J. Chen, "Transformerless DC–DC Converters with high step-up voltage gain," *IEEE Transactions on Industrial Electronics*, vol. 56, no. 8, pp. 3144–3152, Aug. 2009.
20. Y. Tang, D. Fu, T. Wang, and Z. Xu, "Hybrid switched-inductor converters for high step-up conversion," *IEEE Transactions on Industrial Electronics*, vol. 62, no. 3, pp. 1480–1490, Mar. 2015.
21. D. Maksimovic and S. Cuk, "Switching converters with wide DC conversion range," *IEEE Transactions on Power Electronics*, vol. 6, no. 1, pp. 151–157, Jan. 1991.
22. D. S. Wijeratne and G. Moschopoulos, "Quadratic power conversion for power electronics: Principles and circuits," *IEEE Transactions on Circuits and Systems I: Regular Papers*, vol. 59, no. 2, pp. 426–438, Feb. 2012.

23. Y.-M. Ye and K. W. E. Cheng, "Quadratic boost converter with low buffer capacitor stress," *IET Power Electronics*, vol. 7, no. 5, pp. 1162–1170, May 2014.
24. J. Kang, J. Han, and S. Han, "Lossless snubber for tapped-inductor boost converter for high step-up application," *IEEE International Conference on Industrial Technology (ICIT)*, 2014, Feb. 26–Mar. 1, 2014, Busan, South Korea, pp. 253–260.
25. N. Vazquez, F. Medina, C. Hernandez, J Arau, and E. Vazquez, "Double tapped-inductor boost converter," *IET Power Electronics*, vol. 8, no. 5, pp. 831–840, 2015.
26. Q. Zhao, F. Tao, Y. Hu, and F. C. Lee, "Active-clamp DC/DC converters using magnetic switches," *IEEE Applied Power Electronics Conference (APEC)*, 2001, Mar. 4–8, 2001, Anaheim, USA, pp. 946–952.
27. T. S. Wu, Y. S. Lai, J. C. Hung, and Y. M. Chen, "Boost converter with coupled inductors and buck-boost type of active clamp," *IEEE Transactions on Industrial Electronics*, vol. 55, no. 1, pp. 154–162, Jan. 2008.
28. R. J. Wai and R. Y. Duan, "High step-up converter with coupled-inductor," *IEEE Transactions on Power Electronics*, vol. 20, no. 5, pp. 1025–1035, Sep. 2005.
29. K. C. Tseng and T. J. Liang, "Novel high efficiency step-up converter," *IEE Proceedings on Electrical Power Applications*, vol. 151, no. 2, pp. 182–190, Mar. 2004.
30. K. B. Park, G. W. Moon, and M. J. Youn, "High step-up boost converter integrated with a transformer-assisted auxiliary circuit employing quasi-resonant operation," *IEEE Transaction on Power Electronics*, vol. 27, no. 4, pp. 1974–1984, Apr. 2012.
31. J. C. Rosas-Caro, J. M. Ramirez, F. Z. Peng, and A. Valderrabano, "A DC–DC multilevel boost converter," *IET Power Electronics*, vol. 3, no. 1, pp. 129–137, Jan. 2010.
32. X. Zhang and T. C. Green, "The modular multilevel converter for high step-up ratio DC–DC conversion," *IEEE Transactions on Industrial Electronics*, vol. 62, no. 8, pp. 4925–4936, Aug. 2015.
33. W. Li, Y. Zhao, Y. Deng, and X. He, "Interleaved converter with voltage multiplier cell for high step-up and high-efficiency conversion," *IEEE Transactions on Power Electronics*, vol. 25, no. 9, pp. 2397–2408, Sept. 2010.
34. Y. Deng, Q. Rang, W. Li, Y. Zhao, J. Shi, and X. He, "Single-switch high step-up converters with built-in transformer voltage multiplier cell," *IEEE Transactions on Power Electronics*, vol. 27, no. 8, pp. 3557–3567, Aug. 2012.
35. J. Leyva-Ramos, R. Mota-Varona, M. G. Ortiz-Lopez, L. H. Diaz-Saldierna, and D. Langarica-Cordoba, "Control strategy of a quadratic boost converter with voltage multiplier cell for high-voltage gain," *IEEE Journal of Emerging and Selected Topics in Power Electronics*, vol. 5, no. 4, pp. 1761–1770, Dec. 2017.
36. N. Zhang, D. Sutanto, K. M. Muttaqi, B. Zhang, and D. Qiu, "High-voltage-gain quadratic boost converter with voltage multiplier," *IET Power Electronics*, vol. 8, no. 12, pp. 2511–2519, Dec. 2015.

CHAPTER 15

THREE-PHASE DC–DC CONVERTERS

Three-phase DC–DC converters are converters that have a three-phase transformer in their topology. A three-phase AC voltage that can be stepped up or stepped down is generated at its primary windings and the resulting three-phase secondary voltage is rectified and then filtered to produce an output DC voltage. They offer several advantages over more conventional topologies that have a standard single-phase transformer, such as reduced input and output filtering due to the effective tripling of the switching frequency, reduced switching stresses for individual devices, and reduced cost compared to three-module interleaved converters that are implemented by combining multiple converters of the same type. Although they were first proposed in the early 1990s, they have not achieved popularity due to limitations in switching device technology and magnetics, but this state of affairs is gradually changing. New wide bandgap devices that can be operated at higher switching frequencies, new magnetic design techniques and core structures, and a greater emphasis on applications such as renewable energy and automotive applications that require higher-power DC–DC converters have ensured that more research and development of three-phase converters is being done by power electronics specialists in recent years.

In this chapter, various three-phase topologies are introduced, their operation is briefly discussed, and their strengths and weaknesses are reviewed. Also discussed in this chapter are various ways in which the three-phase transformer can be implemented, such as using three separate cores, a single standard core, or a core with a special geometry.

DC–DC Converter Topologies: Basic to Advanced, First Edition. Gerry Moschopoulos.

Published 2024 by John Wiley & Sons, Inc.

15.1 FUNDAMENTAL VOLTAGE-FED THREE-PHASE DC–DC PWM CONVERTER

One of the first three-phase DC–DC converters to have been proposed in the literature is the voltage-fed converter that was proposed in [1] and shown in Figure 15.1. This converter, which henceforth will be referred to as the fundamental three-phase DC–DC pulse-width modulation (PWM) converter, has six converter switches at the primary, a three-phase transformer, a three-phase diode bridge rectifier at the secondary, and an output LC filter. The converter can be operated with the symmetrical and asymmetrical PWM methods shown in Figure 15.2, as explained below.

15.1.1 Basic Operating Principles with Symmetrical PWM

The fundamental converter must be operated with a PWM method that ensures that the gating signals of the six converter switches produce a voltage at each of the three transformer primary windings that is the same as the voltage in the other windings but shifted in phase by 120°. This is to ensure that the same amount of power is processed through each winding and to minimize the harmonic content of the output waveform. The gating signals of the fundamental converter with symmetrical PWM, as shown in Figure 15.2a, are relative to the signal for switch S_1, which is referenced to 0°. Following the sequence of gating signals shown in the figure, it can be seen that each signal is shifted by 60° relative to the previous signal in the sequence so that the gating signal for S_2 is 60° behind the signal for S_1, the signal for S_3 is 60° behind the signal for S_2, etc. If these gating signals are examined in conjunction with the converter circuit diagram shown in Figure 15.1, it can be seen that the gating signal for each switch is 180° out of phase with that of the other switch in the same leg so that there is a 180° phase difference between the gating signals of S_1 and S_4, S_2 and S_3, and S_3 and S_6.

If duty cycle D is defined as the ratio of the on-time of a switch over the length of the switching cycle, then the minimum duty cycle that a switch can be operated with is $D = 0.16$, which represents one-sixth of a switching

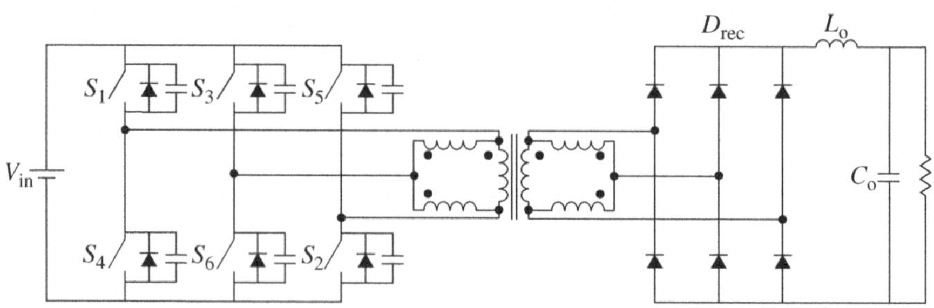

Figure 15.1 Fundamental three-phase DC–DC converter.

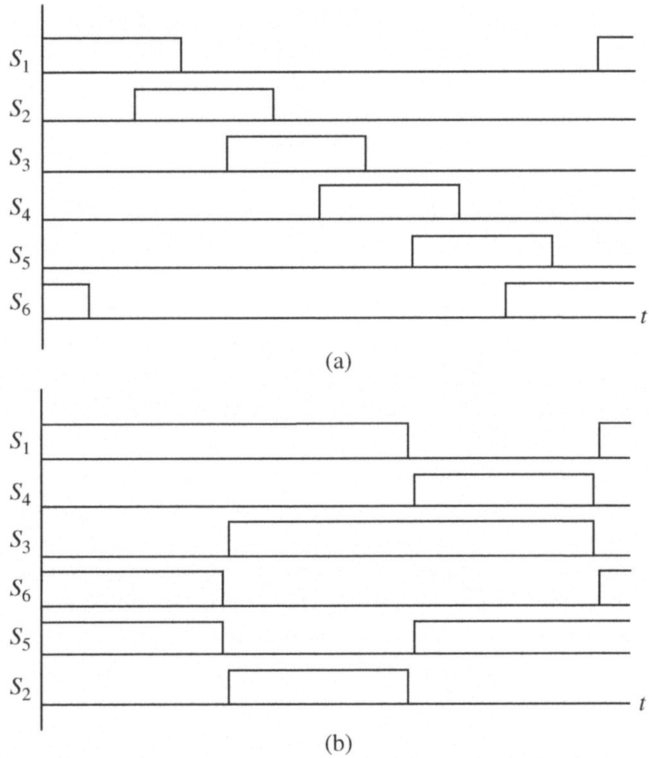

Figure 15.2 Three-phase DC–DC converter PWM patterns: (a) Symmetrical PWM. (b) Asymmetrical PWM.

cycle and the maximum duty cycle that a switch can be operated with is $D = 0.33$, which represents one-third of a switching cycle. Voltage can only be actively impressed across transformer primary windings whenever one top switch (S_1, S_3, or S_5) and one bottom switch (S_2, S_4, or S_6) are on, which allows current to flow from the positive terminal of the voltage source to its negative terminal. If $D < 0.16$, then the converter will operate with only one switch in the converter or none at all. In both cases, no current will flow in the converter so that the output voltage will be zero. The converter switches cannot be operated with $D > 0.33$ as this would result in multiple top switches or multiple bottom switches being on at the same time, which will actually increase the duration of zero-voltage states across the transformer windings, thus reducing the output voltage. The maximum output voltage can be obtained with a duty cycle of $D = 0.33$.

The following should be noted about the operation of the converter:

- The converter can operate with states when $D > 0.16$, but there is no switch on in a particular converter leg. Ideally, this will result in zero-voltage states if the transformer has no leakage inductance. In reality, however, there is leakage inductance in the transformer and the leakage inductance current can flow through the body diodes of the switching devices – from the negative terminal of the voltage source to its positive terminal – so that voltage can be impressed across the transformer windings until this current dies down.

- Operation in the $0.16 < D < 0.33$ range with one top switch and one bottom switch means that just two of the three primary transformer windings can be connected to a fixed voltage rail – either positive or negative – whenever switches are on. This means that the terminal of a winding is floating whenever the other two windings are connected to the voltage rails. This voltage is dependent on the parameters of the converter (e.g. transformer leakage inductance, switch device characteristics, etc.) but never exceeds the voltage of the rails.
- These two facts make the voltage across each transformer winding irregular, but this is not of concern as all that matters is that all three windings are impressed with an identical AC voltage with balanced volt-seconds that is 120° shifted in phase relative to the voltage across the other windings. This voltage is passed through the secondary and filtered so that the actual shape of the primary voltages does not matter, but the irregular nature of the voltage waveform can result in increased output ripple.

15.1.2 Operation with Asymmetrical PWM

A drawback to the fundamental three-phase PWM DC–DC converter is that the switches do not turn on and off softly but do so with switching losses. One way to correct this drawback is to modify the gating signals so that they are asymmetrical, as was proposed in [2] and shown in Figure 15.2b. There are two fundamental differences between the asymmetrical and the symmetrical gating signals shown in Figure 15.2: the gating signals of the switches do not have the same width for all the switches and the gating signals for each switch are complementary to the switch in the same leg so that whenever one of the switches in a leg is on, the other switch in the same leg is off, with the exception of small gaps for dead-time. For example, if switch S_1 is on for about 70% of a switching cycle, then its complementary switch S_4 should be on for about 30% of the cycle. In the asymmetrical PWM scheme shown in Figure 15.2b, the bottom switch in each leg has a duty cycle of D, while the top switch has a duty cycle of $(1 - D)$. As with the symmetrical gating signals shown in Figure 15.2a, the gating signals of the switches in one leg are shifted by 120° in phase relative to those of the switches in the other two legs.

The use of asymmetrical PWM in the fundamental three-phase DC–DC converter results in two fundamental differences in the operation of the converter. The first is that each transformer winding terminal is always connected to either the positive voltage rail or to the ground rail at any instant in time. This results in a better-defined voltage being applied to the transformer as there are no floating voltage states. The other difference is that the switches can now operate with zero-voltage-switching (ZVS), which can be explained as follows: Each of the three transformer terminals is connected to the midpoint of one of the three converter legs. Current will flow into a transformer terminal whenever the top switch of the leg it is connected to is on and out of the same transformer terminal whenever the bottom switch is off. The flow of the current of a transformer terminal is dictated by the voltage that appears across the three windings. If the top switch of a leg is on, there will be one other terminal that will be connected to the ground rail so that current will flow from the source, through the top switch, through the transformer, through a bottom switch and return to the source. Similarly, if the bottom switch of a leg is on, then at least one top switch is on so that current will flow through the bottom switch.

With the direction of current set in this manner, inherently by the implementation of the asymmetrical PWM method, if the top switch of a leg is turned off while it is conducting current, then the current will discharge the output capacitance of the bottom switch of the same leg and the switch can be turned on with ZVS. Similarly, if the bottom switch of a leg is turned off while it is conducting current, then the current will discharge the output capacitance of the top switch of the same leg and the switch can be turned on with ZVS.

ZVS operation can be achieved for a wide load range – from 30% of full load to 100%, according to [2] – but ZVS operation is lost when the converter is made to operate with lighter loads. Of the two switches in a converter leg, it is the top switch with $(1 - D)$ duty cycle that is more likely to lose ZVS. As the load is decreased and D becomes smaller, there is less opportunity to build up energy in the leakage inductance that will allow the output capacitance of the upper switch to be discharged. It should be noted that although asymmetrical PWM may reduce switching losses, it makes the distribution of conduction losses in the converter uneven so that the devices with higher current ratings must be used compared to when the converter is operated with the symmetrical PWM method. Asymmetrical PWM makes the overall conduction losses of the converter greater as well if MOSFETs are used due to $I^2R_{DS,ON}$ on-state resistance losses.

15.1.3 Modified Output Section with Three Output Diodes

The conduction losses of the fundamental three-phase DC–DC converter can be reduced if the three-phase full-bridge diode rectifier that is attached to the secondary of the three-phase transformer is replaced with a diode implementation that has fewer components. With the full-bridge rectifier, the full output current must flow through two diodes in series, which creates conduction losses, especially if the output current is higher and the cost is increased as

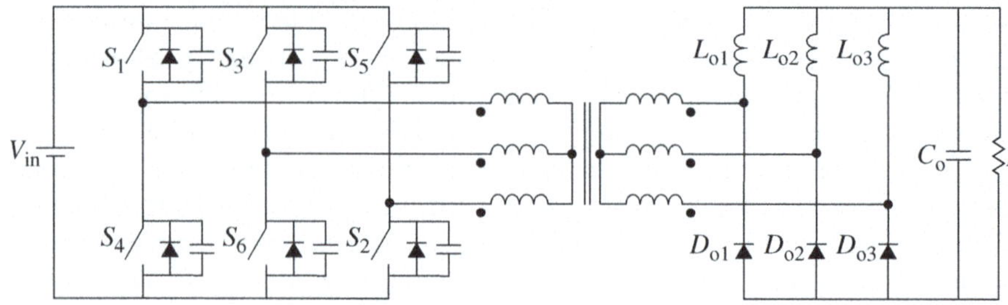

Figure 15.3 Three-phase DC–DC converter with three-diode hybridge secondary.

there are six output diodes at the secondary. If the converter is operated with the asymmetrical PWM method shown in Figure 15.2b, then it becomes possible to modify the converter and reduce the number of output diodes as was proposed in [3] and shown in Figure 15.3. In this implementation, the output section consists of three output diodes and three output inductors.

As was explained in the previous section, when the converter is made to operate with the asymmetrical PWM shown in Figure 15.2b, each of the three terminals of the transformer is connected to one of the two DC voltage rails at any given instant of time so that there is fixed voltage impressed across each of the transformer primary windings. Relative to the neutral point of the wye connection, this voltage may be positive or negative so that the voltage of each of the secondary windings is either positive or negative. In terms of current flow at the secondary, this translates into a situation when at any given time, one phase current may be flowing out of the transformer through one of the three output inductors to the load, then be split into two and flow through two output diodes and into the transformer, or into a situation when two phase currents flow out of the transformer, through two of the output inductors, to the load, then flow through one of the output diodes and back into the transformer. Since the current flows entirely only through one diode or is spilt and flows into two, there is never a situation where the full output current flows through two series diodes; thus conduction losses can be reduced by a full diode drop.

The converter shown in Figure 15.3 has several disadvantages, including the following:

- Three output inductors are needed instead of just one, which increases the size of the converter.
- The converter diodes are subjected to at least 50% more voltage stress than the diodes in the fundamental converter so that more expensive diodes need to be used. This partially offsets any cost savings resulting from the use of fewer diodes unless the output voltage is low when voltage stress is less of an issue.

In the end, the converter shown in Figure 15.3 is best suited for low-output voltage/high-output current applications.

15.2 RESONANT CONVERTERS

A number of researchers have proposed modifying the fundamental three-phase DC–DC PWM converter by adding resonant components to the three-phase transformer so that it becomes a resonant converter that can be operated with soft-switching. Two such converters are discussed in this section.

15.2.1 Parallel Resonant Converter Based on the Fundamental Converter

Prasad, Ziogas, and Manias were the first to propose a three-phase DC–DC resonant converter, a parallel-resonant converter, in [4], as an extension to their work on the fundamental DC–DC converter in [1]. This converter, which is shown in Figure 15.4, is a fixed frequency converter that was operated with the symmetrical PWM pattern shown in Figure 15.2a. The converters' three-phase resonant tank is designed so that the combined tank and load are seen by the three-phase, six-switch bridge to have a lagging power factor so that the converter switches can operate with ZVS. This is ensured by operating the converter with a switching frequency that is greater than the resonant tank frequency, similar to conventional single-phase DC–DC resonant converters such as the ones that were discussed in Chapter 11. As with the converters in Chapter 11, the basic principle is to apply a square-wave-type waveform that contains a fundamental AC harmonic component and harmonics that are multiples of the fundamental across the resonant elements so that the harmonics are filtered out by the resonant tank and the current is a quasi-sinusoidal waveform that lags the voltage.

Under light-load conditions, there may be states when none of the switches in a converter leg is on. The converter is thus more likely to fall into discontinuous modes of operation where the current does not flow in a phase for some time. This creates two issues: The first is that there is no current to

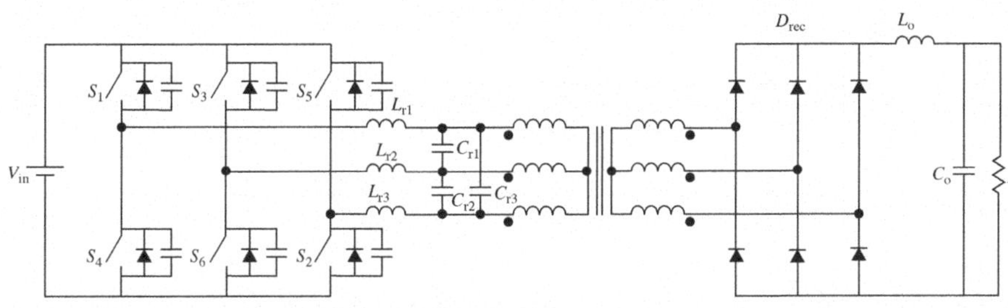

Figure 15.4 Three-phase parallel resonant converter.

discharge the output capacitances of switches before they are turned on. The second is that the various resonant elements can interact with the converter parasitics and cause voltage ringing across the DC bus. A dissipative resistor-capacitor-diode (RCD) snubber was added to the converter in [4] to clamp any such voltage ringing.

15.2.2 Three-Phase Series-Parallel Resonant Converters with Variable and Fixed Switching Frequency Operation

As with conventional single-phase DC–DC resonant converter topologies, three-phase converters can be implemented with various resonant tank arrangements. An example is the three-phase series-parallel resonant converter that was proposed in [5], which is shown in Figure 15.5a along with the gating signals of its switches. The converter topology is the same as that of the fundamental converter, except that resonant elements have been added to the transformer – to its primary and secondary – to make the converter a series-parallel converter. The gating signals for all six switches are 180° wide and the signals of the two switches in each leg are complementary to each other and phase-shifted by 120° relative to the gating signals of the switches in the other converter legs, as shown in Figure 15.5b. The gating signals are

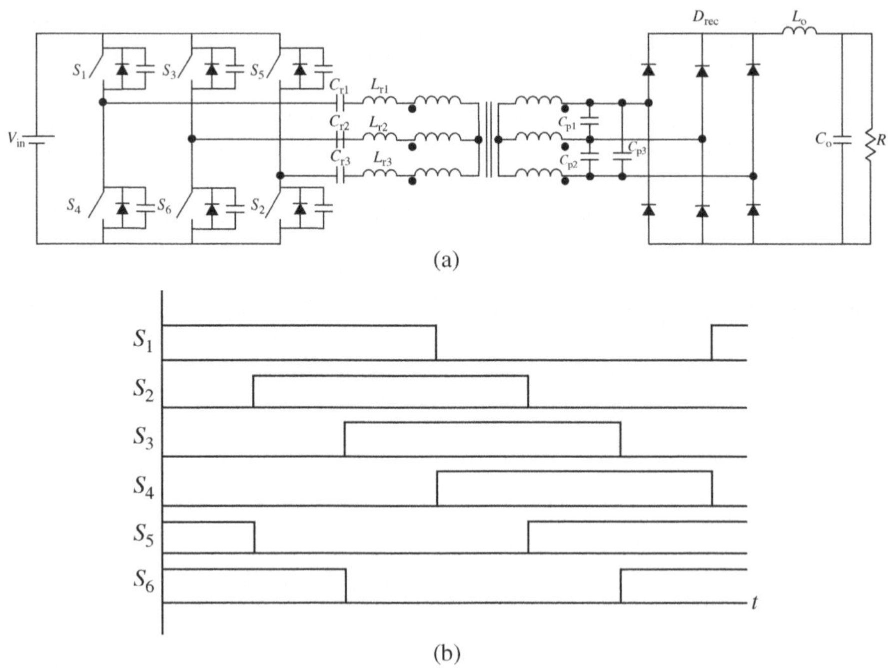

Figure 15.5 Three-phase series-parallel resonant converter: (a) Topology. (b) Gating signals.

fixed and do not change with load; instead, the converter is operated with variable switching frequency control, and thus has all the drawbacks of such operation, such as magnetics that need to be designed to operate with various switching frequencies instead of just one and the need for more sophisticated control.

The converter can be made to operate with fixed switching frequency control if it is operated with the asymmetrical PWM pattern shown in Figure 15.2. If this is done, then the converter behaves as if it were synthesized from three single-phase asymmetrical pulse width modulation (APWM) converters, such as the ones discussed in Section 12.1.2, connected in parallel. The gating signal of one switch of one of these converters operates with a duty cycle of D, the other operates with a duty cycle of $(1 - D)$, and the signals of each converter are phase-shifted by 120° relative to the switches in the other converters. As this converter behaves like three APWM converters, it, therefore, has the same strengths (ZVS with fixed switching frequency control) and weaknesses (asymmetrical current stresses and increased output current ripple at lighter loads) as the APWM converter.

15.3 THREE-PHASE CURRENT-FED DC–DC PWM CONVERTERS

The three-phase DC–DC converter topologies that have been presented thus far in this chapter can be considered to be voltage-fed converters, as there is a direct connection between a voltage source and the converter. These converters can be considered to be analogous to buck-type converters that are used to step down voltage. Although the converters can be made to step up voltage simply by implementing them with transformers that have more secondary windings than primary windings, current-fed boost-type converters are preferable as they require lower turns ratio, thus making the three-phase transformerless susceptible to parasitic effects. A ZVS three-phase current-fed converter and a ZCS converter are presented in this section.

15.3.1 Three-Phase ZVS Active Clamp Converter

One of the first three-phase current-fed DC–DC converters was the converter proposed in [6] and shown in Figure 15.6 along with its gating signals. The converter consists of an input inductor, six switches, a three-phase transformer, a three-phase diode bridge rectifier connected to the secondary of the transformer, an output capacitor filter, and an active clamp consisting of auxiliary switch S_a and clamp capacitor C_c. This converter is essentially the three-phase version of the single-phase current-fed active clamp full-bridge converter that was discussed in Section 10.6.1 and operated in a similar manner. Like that converter, the three-phase converter shown in Figure 15.6a uses the active clamp to clamp any overvoltage spiking that may occur as a result when switches are turned off and a path must be provided for energy trapped in

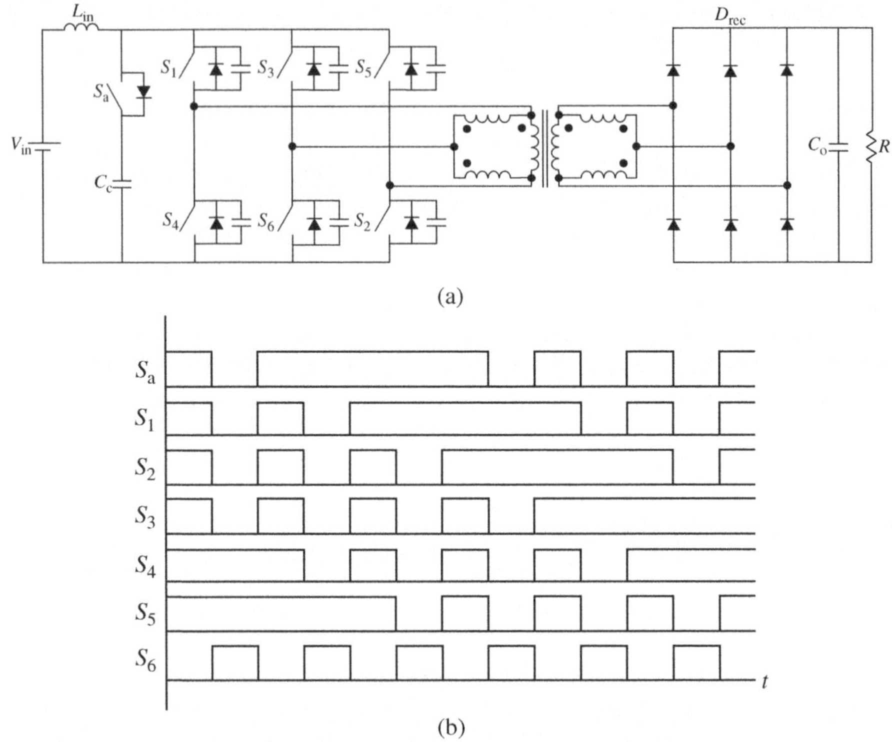

Figure 15.6 Three-phase ZVS current-fed DC–DC converter with active clamp: (a) Topology. (b) Gating signals.

parasitic inductances to be transferred to, instead of being forced upon the turned-off switches. Moreover, again like the single-phase full-bridge converter discussed in Section 10.6.1, energy in the active clamp is used to assist the switching devices in turning on with ZVS by transferring energy to the transformer that can then be used to help discharge the output capacitances of switches before they are turned off. The main difference between the two converters is with the PWM pattern that is used to operate them.

The PWM method used to operate the converter is shown in Figure 15.6b. The gating signals for all six switches are derived in the following manner: Since there are three top switches and three bottom switches, the synthesis of a symmetrical PWM pattern is constrained to the fact that (i) each switch must be on for at least 120° of the switching cycles and (ii) short circuiting or shoot-through boost modes must be intertwined with nonboost modes. As the converter is essentially a boost converter, there must be shoot-through modes of operation if any voltage boosting is to take place.

The synthesis of the PWM pattern starts off with the fact that there must always be at least one top switch and one bottom switch on at the same time so that there is always a path for current to flow through the six-switch bridge at all times. In order to make the pattern symmetrical, each top switch is on for 120° of the switching cycle; as there are three top switches, 3 × 120° covers the full 360° switching cycle. Similarly, each bottom switch is on for 120° while the other bottom switches are off and this ensures that there is a path for current to flow into a bottom switch and back to the source. The sequencing of the gating signals follows the symmetrical PWM sequence shown in Figure 15.2b with each switch being turned on 60° after the preceding switch in the sequence has been turned on. For example, S_2 is turned on 60° after S_1 has been turned on, S_3 is turned on 60° after S_2 has been turned on, etc. This sequence is necessary to ensure that symmetrical three-phase AC waveforms appear across the transformer primary that can then be stepped up, rectified and filtered by the output.

If this switching sequence is maintained, then the only voltage boosting that will occur will be due to the transformer. If additional boosting is to occur, then the six-switch bridge must operate with shoot-through modes. These shoot-through modes can be implemented either by having a top switch and a bottom switch of the same converter leg on at the same time or by turning on all the converter switches, as is shown in Figure 15.6b. The latter strategy is preferred as it distributes the inductor current among all the switches, thus reducing stresses. Moreover, if MOSFETs are used, the net $R_{DS,on}$ of having three converter legs conduct current is a third of what it is if current is forced to flow through only one leg in a shoot-through mode.

The general sequence in the PWM pattern shown in Figure 15.6b is that shoot-through mode and nonshoot-through mode alternate so that one is always followed by the other. Given that the active clamp switch S_a must be off whenever the converter is in a shoot-through and always on otherwise in order to maximize the opportunity for the clamp capacitor to discharge into the converter to prevent the clamp voltage from being excessive, the gating signals for S_a is a train of square-wave pulses, as shown in Figure 15.6b.

15.3.2 Three-Phase ZCS Converter

The active clamp converter shown in Figure 15.6a operates with ZVS. The active clamp can be removed from the converter and the converter switches can be made to operate with ZCS if its topology and PWM method are modified, as was proposed in [7] and shown in Figure 15.7a. The key modification, after the removal of the active clamp, is the addition of three capacitors at the secondary of the converter. These capacitors are resonant capacitors that are used to ensure soft-switching transitions. The gating signals shown in Figure 15.7b are synthesized in a different manner than those shown in Figure 15.6b. The synthesis of the pattern starts off with 120° like those for the active clamp converter, but the pulse width is kept fixed. The converter is controlled by phase-shifting the gating signals of the top switches relative to those of the bottom switches. The amount of voltage boosting that can be done by the converter is determined by the amount of overlap between

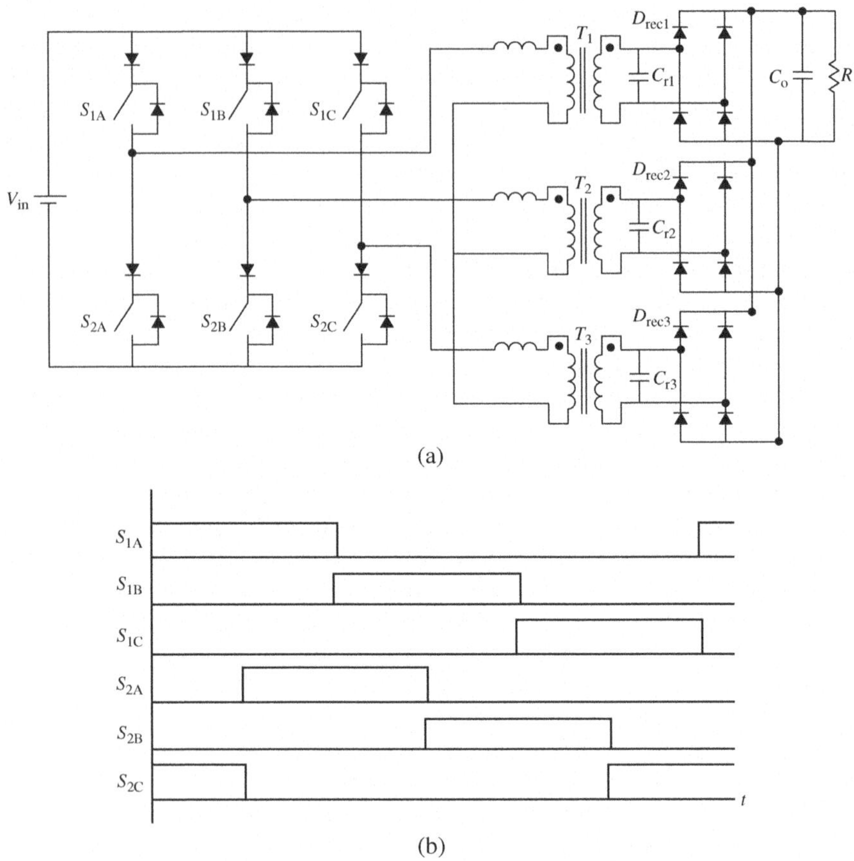

Figure 15.7 Three-phase ZCS current-fed DC–DC converter: (a) Topology. (b) Gating signals.

the top switch of each converter leg and its bottom switch counterpart – more overlap results in more voltage boosting. Since a top switch is always on and a bottom switch is always on, regardless of the amount of phase-shifting or overlap in the pattern, the current will always have a path to flow through in the switch bridge.

Unlike the active clamp converter, which can have shoot-through states when all the converter switches are on, for the ZCS-PWM converter, there is always just one top switch and always just one bottom switch on at all times, except for brief time durations when the current is about to be transferred from a top switch to another top switch, or from a bottom switch to another bottom switch. Whenever a top switch and a bottom switch are on, the resonant capacitors attached to the secondary of the transformer are charged with various polarities. When it comes time to introduce a shoot-through state, say between S_{1A} and S_{2A}, by turning off S_{2C} and turning on S_{2A}, the turning on of S_{2A}

diverts current away from S_{2C}, as it sees lower voltage potential so that S_{2C} can be turned off with ZCS. S_{2A} can be turned on with ZCS as the transformer leakage inductance makes the transition of current into the switch gradual.

The converter can exit the shoot-through mode by turning on S_{1B}. Due to the presence of the resonant capacitors at the secondary of the transformer, the current will see negative voltage if it starts to flow through S_{1B} instead of zero voltage if it continues to flow through S_{1A}. Consequently, S_{1B} can be turned on with ZCS as the transition of current away from S_{1A} is gradual and S_{1A} can be turned off with ZCS after all the current that was flowing through it has been diverted. It should be noted that S_{1A} must be turned off within a certain time, or else the voltage polarities of the resonant capacitances with change and current will begin to flow back through S_{1A}. This time is dependent on the resonant cycle set by the interaction of the resonant capacitances and the transformer leakage inductance.

The ZCS-PWM converter shown in Figure 15.7 is well-suited for higher-power applications where insulated gate bipolar transistors (IGBTs) are used, given that ZCS is the preferred method of soft-switching for these devices. The main drawbacks of the converter are that it requires that reverse-blocking IGBTs are used (IGBTs with a blocking diode in series with an IGBT that has an anti-parallel diode so that current never flows through this diode), that the switches of only one converter leg can be on during shoot-through modes, and that it does not have a snubber at the primary side, which decreases reliability as the converter can fail if there is a sudden misfiring of the converter switches.

15.4 HIGHER-POWER THREE-PHASE DC–DC CONVERTERS

Three-phase DC–DC converters that have been synthesized from three single-phase DC–DC full-bridge converters can be used for higher-power applications. These converters are different from DC–DC converters that are implemented with three paralleled and separate single-phase DC–DC full-bridge converters as they are implemented using three-phase transformers with a three-phase secondary. Although these converters can be expensive and large in size because they contain 3 × 4 bridge switches = 12 switches, they are attractive for use in high-power applications where the switching devices of six-switch three-phase converters would experience significant stress.

15.4.1 High-Power Converter with Three Single-Phase PWM Full-Bridges

An example of such a converter is the converter that was proposed in [8] and is shown in Figure 15.8. The converter works as follows: Each of the four-switch full-bridge converters is operated with standard phase-shift PWM, as explained in Chapter 7. The gating signals of all the switches in the converter

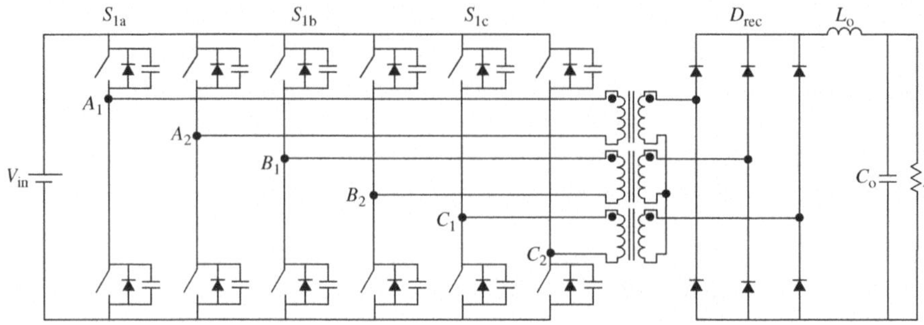

Figure 15.8 High-power three-phase DC–DC converter based on three single-phase PWM full-bridge converters.

have the same pulse width and the signal for an individual switch is shifted in phase by 120° relative to the signals for the other two switches in the other two full-bridge circuits. For example, the gating signal for switch S_{1a} in Figure 15.8 is shifted by 120° in phase relative to the gating signal of S_{1b} and shifted by 120° in phase relative to the gating signal of S_{1c}. The advantages of this converter include the soft-switching operation of all the converter switches, reduced stress for each switch, compared to single-phase isolated DC–DC converters and the fundamental three-phase DC–DC converter, and wide duty cycle range and symmetrical operation that help reduce conduction losses.

15.4.2 High-Power Converter with Three Single-Phase Resonant Full-Bridges

Full-bridge resonant converters can be used as the building blocks for higher-power DC–DC three-phase converters instead of PWM full-bridge converters as was proposed in [9] and shown in Figure 15.9. This converter consists of three single-phase resonant full-bridge converters, each with gating signals that have a duty cycle of D = 50% and shifted 120° in phase relative to the switches in the same position in the other two converters. The main advantage of this arrangement is that it allows PWM control to be performed by using phase-shift PWM in each of the converters so that the gating signals in each converter can be symmetrical, thus avoiding the drawbacks of asymmetrical PWM operation. This is a key advantage for higher-power applications where the additional switch current stresses than some switches may have and the increased conduction losses created by asymmetrical PWM need to be avoided.

15.5 THREE-SWITCH THREE-PHASE DC–DC PWM CONVERTERS

The three-phase DC–DC converters that have been discussed in this chapter have, thus far, contained at least six active switching devices in their topologies. Researchers have proposed converters that have fewer active devices,

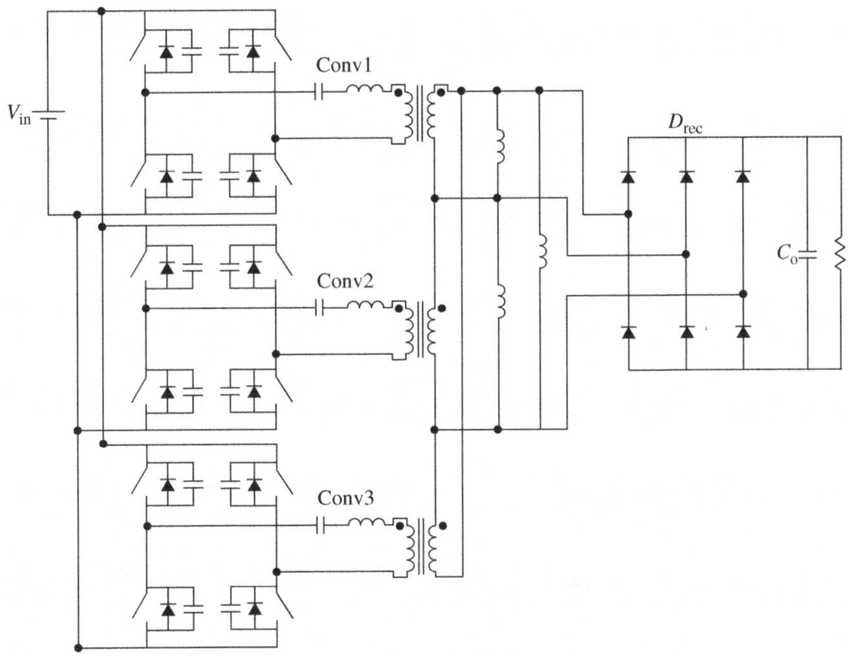

Figure 15.9 High-power three-phase DC–DC converter based on three single-phase resonant full-bridge converters.

as few as three in the main current path, to reduce converter size and cost. The basic principle behind these converters is that it is possible to set the voltage at each of the primary terminals of the three-phase transformer depending on the on state of just one switch so that the number of switches in the overall converter can be reduced. Several of these converters are discussed in this section.

15.5.1 Three-Phase Push-Pull Converter

The first three-switch, three-phase PWM converter was proposed in [10] and shown in Figure 15.10a, along with its gating signals. This converter consists of a three-phase transformer that has just three switches attached to its primary side, just three diodes attached to its secondary side and an output LC filter. Although there are several ways in which the three-phase transformer can be implemented, as will be discussed towards the end of this chapter, the transformer must be implemented with a single E-type core with three legs. The converter works are follows: The switching cycle is divided into three 120° segments and only one of the three switches can be on at any given time. The gating signals of the three switches, shown in Figure 15.10b, are identical, but each signal is shifted 120° relative to that of the other two switches. The

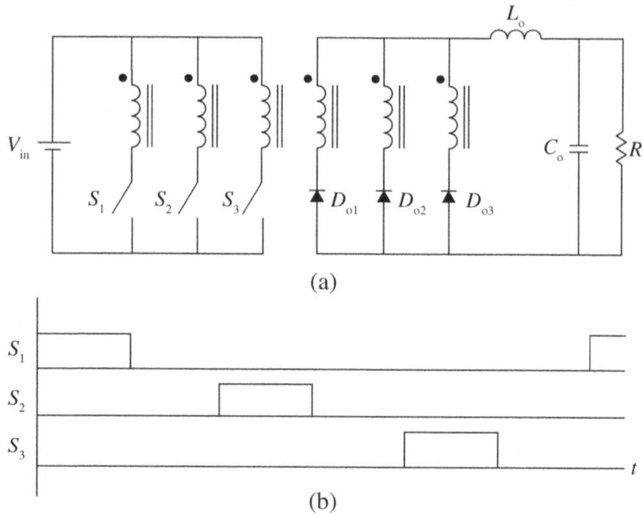

Figure 15.10 Three-phase three-switch DC–DC converter based on push-pull converter: (a) Topology. (b) Gating signals.

duty cycle of the converter is controlled by changing the pulse width of the gating signals, and thus, there are states when none of the three switches are on.

When a switch is on, it generates flux in one of the core legs that splits and flows through the other two core legs in the other direction. If the flux, φ, that is generated when a switch is on, flows through the core leg to which its winding is attached in a positive direction, then this flux is evenly divided among the other two core legs and flows in a negative direction so that a flux of $-\varphi/2$ flows through these two legs. If the other switches are turned on and the same amount of volt-seconds are applied to the transformer, then a core leg will have $+\varphi$ flux flowing through it for 1/3 of the switching cycle and $-\varphi/2$ flux flowing through it for the other 2/3 of the cycle, which makes the net flux zero after a switching cycle. This is similar to the operation of a conventional push-pull converter, where the core is reset by the turning on of switches later in the switching cycle. At the secondary, two of the diodes become forward-biased (the two diodes that are attached to the secondary windings of the phases that do not have a switch that is on) and energy is transferred to the output.

Although the converter shown in Figure 15.10 has a very simple topology, its main drawback is that it is based on the conventional push-pull converter and thus has the main drawback of the push-pull converter. It is easy to saturate because the core reset is dependent on the three gating signals being exactly the same. If there is any difference in the pulse width of any of the gating signals, then flux will accumulate in the core unless some adjustment is made. In a follow-up paper by the authors of [10], the current-fed push-pull

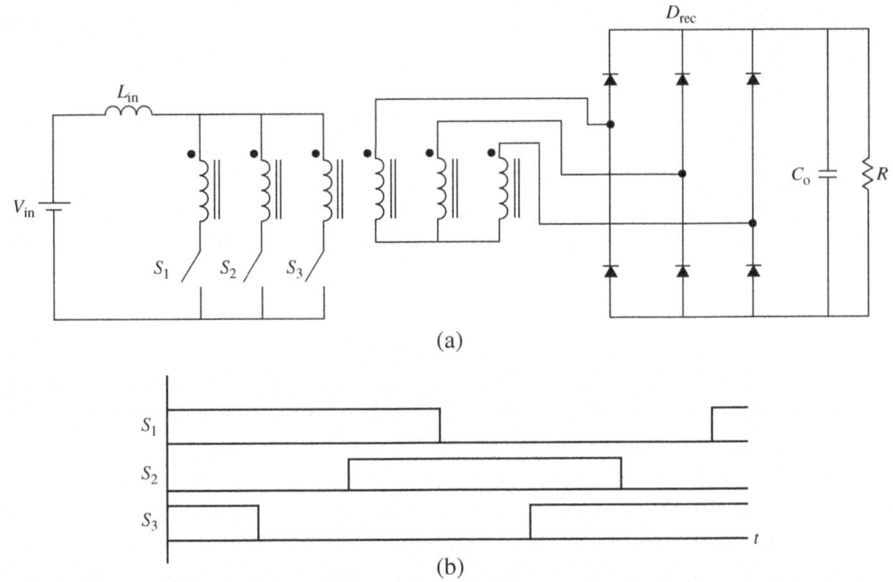

Figure 15.11 Three-phase current-fed push-pull converter: (a) Topology. (b) Gating signals.

converter shown in Figure 15.11, along with its gating signals, was proposed in [11] to correct this drawback. This converter is less prone to saturation because the output capacitor can be placed directly across the secondary transformer windings, much like the output capacitor can be placed across the secondary winding of a flyback converter. With a fixed voltage available to reset the transformer, saturation in the converter shown in Figure 15.11 becomes less likely.

The gating signals shown in Figure 15.11b differ from those shown in Figure 15.10b. As the converter is a current-fed type converter, at least one of the main switches must be on at all times to provide a path for the current in the input inductors to flow through. Failure to provide a path for current to flow in the converter will result in a catastrophic failure. This means that the minimum duty cycle of the main switch relative to the full switching cycle must be $D = 0.33$ as $D < 0.33$ will result in intervals during the switching cycle when no switch is on. The gating signals of a switch are identical to those of the other switches but are shifted by 120° relative to the others.

15.5.2 ZVS Active Clamp Converter

Although the converter shown in Figure 15.11 is less prone to saturation compared to the converter shown in Figure 15.10, the converter switches must be implemented with passive, dissipative snubbers (not shown in Figure 15.11)

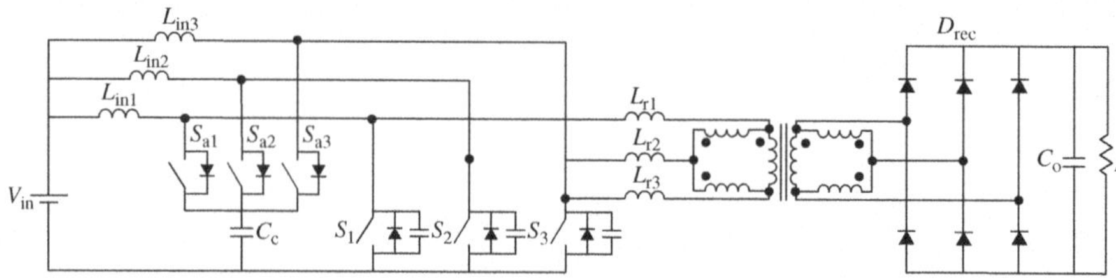

Figure 15.12 Three-phase three-switch ZVS DC–DC converter with three-switch active clamp.

in order to prevent overvoltage spiking from occurring whenever they are turned off. The issue of overvoltage spiking prevention was addressed in the three-switch, three-phase active-clamp DC–DC converter that was proposed in [12] and shown in Figure 15.12. The converter consists of three main power switches (S_1, S_2, S_3), three active clamp switches (S_{a1}, S_{a2}, S_{a3}) placed in series with an active clamp capacitor, a three-phase transformer, and three-phase diode bridge rectifier, and an output capacitor filter.

The PWM pattern of the converter is the same as that shown in Figure 15.11b. Although the PWM pattern is the same, the way the converter is operated is different when $D < 0.33$ and when $D > 0.66$. When $D < 0.33$, a current path through the active clamp capacitor does exist so that the possibility of a catastrophic failure when all three switches are off is avoided. Operation with $D < 0.33$ should generally be avoided, however, as it charges the active clamp capacitor to higher voltage levels, which increases the peak voltage stresses of the converter switches. Operation with $D < 0.33$ should only be considered when the converter is operating under light-load conditions when there is less current available to charge the active clamp capacitor.

Operating the converter shown in Figure 15.12 with $D > 0.66$ actually decreases the gain for the converter, unlike the converter shown in Figure 15.11a. For example, if $D = 0.9$ for each switch, this means that each primary terminal of the three-phase transformer will have 0 V applied to it for 90% of the switching cycle. As a result, all three terminals of the transformer primary will have the same voltage for most of the switching cycle so that the line-to-line voltage that is applied to each primary winding will be zero for most of the switching cycle as well, thus limiting the gain of the converter. The duty cycle of the main converter switches should be limited to $0.33 < D < 0.66$ as much as possible. The limitation of gain is the main drawback that the converter has relative to the converter shown in Figure 15.11a, but it does have several advantages, such as low duty cycle operation for light-loads and interleaving due to the presence of three paralleled input inductors instead of just one. Interleaving is advantageous as it results in input current ripple reduction.

Another benefit is that the converter switches can operate with ZVS due to the three-switch active clamp. The gating signals of each of the three active clamp switches are complementary to those of the main switch that is attached to the same phase. For example, if switch S_1 has a duty cycle of $D = 0.6$, then the duty cycle of active clamp switch S_{a1} will be $(1 - D) = 0.4$. ZVS for each of the main switches is initiated whenever its active clamp switch is turned off. This action cuts off the current path out of the active clamp capacitor and forces current to flow through the body diode of the device that is about to be turned on, thus ensuring ZVS operation. It should be noted that when an active clamp switch is on, it impresses the clamp capacitor voltage on the terminal of the transformer that it is attached to, which ensures that the voltage at this terminal is stiff and not floating. For example, if switches S_1, S_2, and S_3 are on and the clamp capacitor voltage is V_{cl}, then $V_a = 0$, $V_b = V_{cl}$, and $V_c = 0$ so that $V_{ab} = -V_{cl}$, $V_{bc} = V_{cl}$, and $V_{ca} = 0$.

15.5.3 ZCS Converter with Secondary-Side Resonance

A way of allowing the converter to operate with soft-switching without the use of a three-switch active clamp circuit was proposed in [13] and is shown in Figure 15.13. For this converter, there is no auxiliary circuit like an active clamp to help the main converter switches operate with soft-switching as the soft-switching mechanism is similar to the three-phase ZCS current-fed converter that was discussed in Section 15.3.2 and shown in Figure 15.7. Instead of an active clamp circuit, soft-switching is performed.

Although the converter is simple, it has several drawbacks when compared to the active-clamp converter shown in Figure 15.12, including the following:

- Its switches operate with ZCS instead of ZVS. Given that MOSFETs are the device of choice in low input voltage DC–DC converters and that ZVS operation is preferred over ZCS operation due to the need to discharge their output capacitance, ZCS is generally disadvantageous, but it is better than having no soft switching at all. Is may be advantageous if the input voltage is low and the primary current is high.
- The switch currents have a resonant hump that is superimposed on them by the ZCS resonant mechanism. Unlike the ZCS current-fed converter

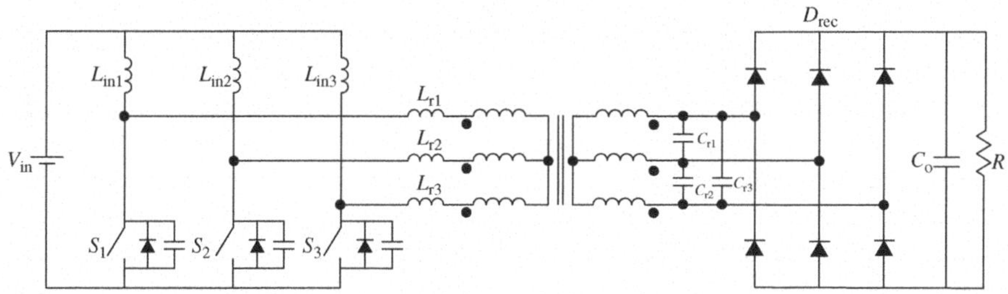

Figure 15.13 Three-phase three-switch ZCS DC–DC converter with secondary resonance.

shown in Figure 15.7, the converter does not have a single stiff current source and individual resonant capacitor clamps at the secondary so that primary inductances and secondary resonant capacitances are essentially floating when it comes to current and voltage, respectively. The resonant current hump that appears in the switch currents just as they are turned on increases the peak current stress of the switches and results in increased conduction losses under light-load conditions as well.

- The converter must be operated with variable switching frequency control to ensure that the switches are turned off with ZCS. Due to the resonant interaction of the secondary resonant capacitors and the inductances in the converter, a switch can only be turned off with ZCS at certain key instants when its current has dropped to zero. Failure to turn a switch off when the ZCS resonant mechanism has forced its current down to zero means that the switch current will soon begin to rise so that when the switch is turned off, the turn-off will not be soft. Variable switching frequency operation complicates the design of the converter magnetics and control as they cannot be done for just one fixed switching frequency.

Nonetheless, in spite of these drawbacks, the converter shown in Figure 15.13 is the simplest and least expensive three-phase DC–DC converter implementation.

15.5.4 Converter with Mini-Flyback Snubber

A modification to the three-switch, three-phase active clamp converter was proposed in [14] and is shown in Figure 15.14. This converter is similar to the active clamp converter and the three main converter switches are operated in the same way as the three main switches in the active clamp converter. The main difference between the two converters is that the three-switch active clamp is replaced by a flyback snubber. The snubber capacitor clamps any overvoltage spiking that may occur when switches are turned off; energy stored in the capacitor is discharged through the flyback transformer. The main advantage of the converter shown in Figure 15.14 is that it requires fewer switching devices and gate drives in total and has no floating gate drives. Moreover, the operation of the flyback snubber is independent of the operation of the main converter switches and, thus, time is not required in the main PWM switching pattern for the auxiliary switch to operate, unlike the three auxiliary active clamp switches in the active clamp converter. The main drawback is that converter switches do not operate with ZVS, but the converter can be used in many renewable energy applications where a low DC voltage from a set of PV panels or fuel cells needs to be stepped up. Operating with low voltages helps reduce switching losses when silicon devices are used. ZVS is less of a concern if wide bandgap devices are used, as these have fewer

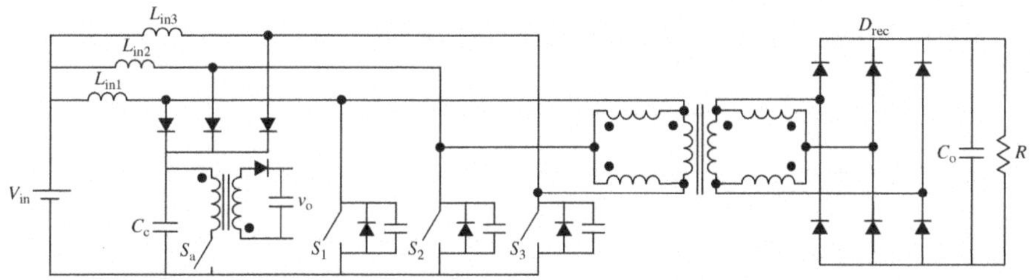

Figure 15.14 Three-phase three-switch DC–DC converter with mini-flyback converter snubber.

switching losses. For applications where wide bandgap devices are required, the converter becomes especially attractive as only four switches are used.

15.6 MISCELLANEOUS THREE-PHASE CONVERTER EXAMPLES

In general, most single-phase four-switch DC–DC converters can be implemented as three-phase six-switch DC–DC converters simply by adding another converter leg so that there are three legs instead of two. In this section, a couple of three-phase converters that do not fit this categorization are presented as examples of how certain three-phase converter properties can be used advantageously.

15.6.1 Three-Phase DC–DC Multi-Level Converter

The first example is the multi-level converter that was proposed in [15] and is shown in Figure 15.15. For this converter, instead of just adding an additional multi-level leg, which would increase the number of converter switches, three pairs of switches are stacked on top of each other, with each pair connected to one of three bus capacitors. Each bus capacitor has a voltage that is a third of the DC bus voltage V_{in} so that the peak voltage stress across each switch is $V_{in}/3$. The converter also contains a three-phase delta-connected transformer and a three-phase output diode implementation, as was discussed above in Section 15.1.3. Blocking capacitors C_{y1} and C_{y2} are added to the primary to block any DC component from appearing at the output. Each of these capacitors has a voltage of $V_{in}/3$ across during steady-state operation. The converter gating signals are such that the maximum pulse width of the gating signals of S_1, S_3, and S_5 is $DT_s/3$, where T_s is the switching cycle and the gating signals of S_2, S_4, S_6 are complementary to the switches directly above them. PWM control is achieved by varying the gating signals of S_1, S_3, and S_5. The gating signals of S_1, S_3, and S_5 are shifted 120° with respect to each other, and the gating signals of S_2, S_4, and S_6 are shifted 120° with respect to each

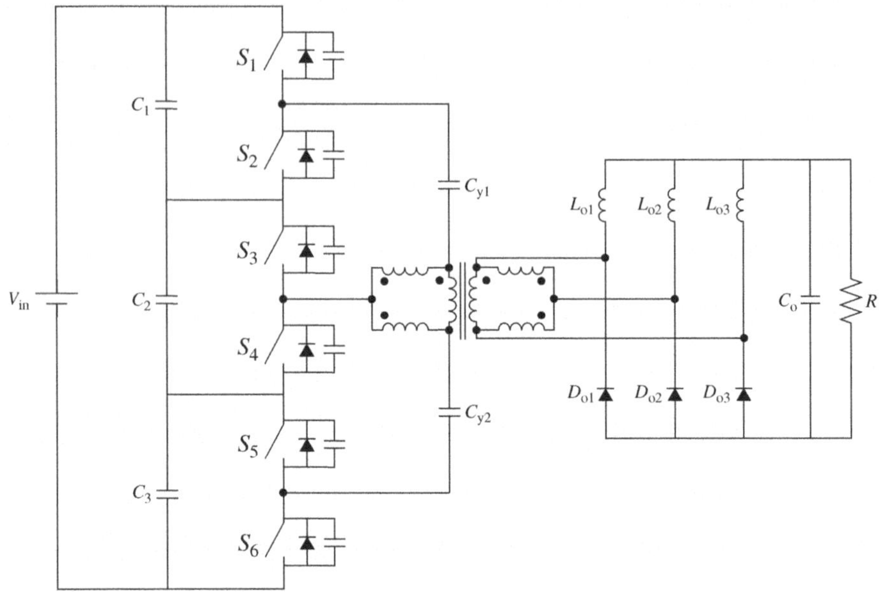

Figure 15.15 Three-phase reduced switch multi-level three-phase DC–DC converter.

other. Both switches of a switch pair are never on simultaneously so that S_1 and S_2 are never on at the same time; likewise for the S_3–S_4 and S_5–S_6 pairs. The maximum pulse width of the gating signals for S_1, S_3, and S_4 is $T_s/3$.

The main advantage of this converter is that the key benefit of multi-level converters, reduced switch stress, can be achieved in a three-phase DC–DC converter by using only six switches. The main drawback is that the gating signals are asymmetrical so that like other converters with asymmetrical gating signals, the current ratings of certain switches can be high and the overall conduction losses are higher than for converters operating with symmetrical gating signals.

15.6.2 Three-Phase DC–DC High-Gain Converter

The properties of three-phase DC–DC converters can also be used to be advantageously in high-gain converters as was proposed in [16] and shown in Figure 15.16. The converter consists of three input inductors, L_{in1}–L_{in3}, two output capacitors, C_{o1}–C_{o2}, three switches, S_1–S_3, three boost diodes that are connected to output capacitor C_{o2}, D_1–D_3, a three-phase transformer and a three-phase diode bridge. Energy can be transferred to C_{o2} through the three boost diodes and to C_{o1} through the three-phase transformer. The voltages across C_{o1} and C_{o2} are stacked on top of each other to ensure a high gain. The converter has high gain without the need for an additional clamping

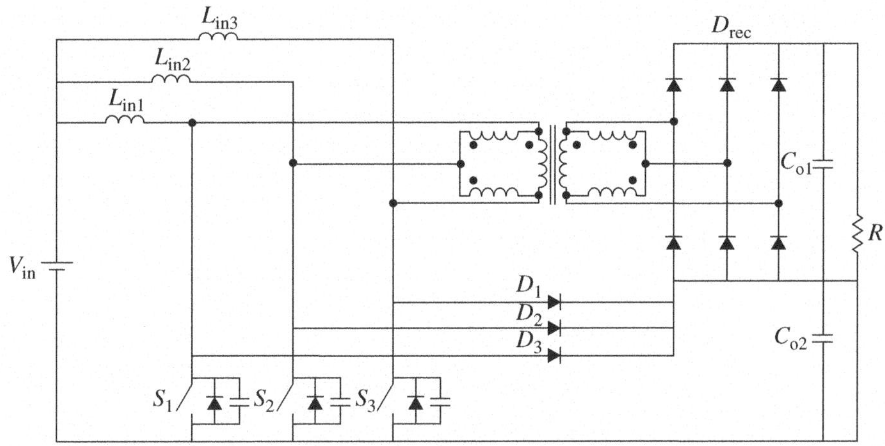

Figure 15.16 Three-phase DC–DC high-gain converter.

mechanism as C_{o2} serves as a voltage clamp, with the additional benefits of an interleaved input, which ensures a low input current ripple and a more compact magnetic structure. Moreover, a three-phase transformer with a single core can be used instead of three separate transformers or coupled inductors, thus reducing the size and weight of the converter. A full analysis of the converter shown in Figure 15.16 was presented in [17], but with a single input inductor instead of three smaller inductors.

15.7 THREE-PHASE TRANSFORMER IMPLEMENTATIONS

The three-phase transformer in the three-phase DC–DC converters that have been discussed in this chapter can be implemented in various ways and some of these have been mentioned in this chapter. The three-phase transformer can be implemented as follows:

- Three separate single-phase transformers can be used to form a three-phase transformer with their primary and secondary terminals connected in some combination of wye and/or delta connection. This is the simplest implementation as each single-phase transformer can be designed in a conventional manner or be obtained ready-made "off the shelf" and the flux contained in each transformer is independent of that of the other two transformers so that any saturation issues that may arise from any flux imbalances is constrained to single cores rather than one single overall cores. Although this is the simplest implementation, it is bulky and expensive as three individual transformers must be used.
- A single three-legged core such as an E-shaped core with each leg wound with the primary and secondary windings of each phase. Although this

implementation is less expensive and takes up less space than using three single-phase transformers, avoiding core saturation is a challenge as the flux produced in each phase must be carefully monitored to ensure that they are balanced. This is easier to achieve in current-fed topologies, where the output capacitor can appear across secondary windings, than in voltage-fed topologies, where the primary and secondary terminals are essentially floating. As a result, the design of such a transformer is not a simple matter, as discussed in publications such as [18]. The challenges involved in avoiding core saturation limit the use of three-legged cores to lower power applications unless a resonant topology is used, as is discussed later in this section.

- Cores with four legs can be used, as suggested in [19]. Three of the legs have primary and secondary windings for each phase and the fourth leg is left unwound. Ideally, the flux in the three core legs should add up to zero, but ensuring this is not a simple manner, and thus, the central fourth leg provides a path for any net flux to flow through. This makes transformer saturation more difficult and can allow for the relaxation of certain design criteria. For example, the transformer can be designed with more windings and with thicker windings as more flux can be generated in the core before saturation occurs. The main drawback with this approach is that special, nonstandard cores need to be manufactured as they are not readily available commercially. In the end, this approach takes up less space than using three separate cores but more space than a three-legged core.
- The three-phase transformer can be implemented using planar magnetics, as was proposed in [20]. With this approach, three separate planar cores are used with printed circuit board traces serving as the primary and secondary windings. The main advantage of this approach is that the benefits of using three single-phase transformers can be obtained with much smaller size and weight. The main drawback is the set of challenges that needs to be faced when designing a printed circuit board with three planar transformers, such as layout, grounding, transformer leakage, and winding arrangement.

It should be mentioned that the implementation of three-phase transformers can be simplified if they are used in resonant-type topologies, especially topologies that have a DC-blocking capacitor in their resonant tank. The DC-blocking resonant capacitor eliminates any flux imbalance that can occur in individual phases, thus making the possibility of core saturation less likely. In the case of three-legged cores, however, the possibility of flux imbalance, although reduced, is still possible so that while three-legged cores can be used for higher-power applications if they are implemented in resonant converters, care must still be taken to avoid saturation.

15.8 CONCLUSION

Although three-phase DC–DC converters with three-phase transformers in their topologies were first proposed in the early 1990s, they have not been popular due to limitations in technology. In recent years, improvements in device technology such as the introduction of new wide bandgap devices and advances in magnetics, coupled with the requirements dictated by new higher-power applications such as those related to renewable energy systems and automotive applications, have made three-phase DC–DC converters more attractive. In this chapter, a number of voltage-fed and current-fed three-phase DC–DC converters were presented, their basic operation was explained, and their strengths and weakness were stated and compared. Several options for implementing the three-phase transformer in these converters were presented as well.

In general, current-fed converters are attractive in applications where a low DC voltage, such as what is typically produced by a set of fuel cells or PV panels, needs to be stepped up. Converters with 6- or 12-switch bridges are preferred for higher-power applications, resonant converters are preferred for applications where high switching frequencies must be used to reduce converter size. Ultimately, as with all other types of converters, the choice of topology for any given application is dependent on factors such as the specifications of the application, cost, size, performance, and efficiency.

REFERENCES

1. A. R. Prasad, P. D. Ziogas, and S. Manias, "Analysis and design of a three-phase offline DC–DC converter with high-frequency isolation," *IEEE Transactions on Industry Applications*, vol. 28, no. 4, pp. 824–832, July–Aug. 1992.
2. D. de Souza Oliveira and I. Barbi, "A three-phase ZVS PWM DC/DC converter with asymmetrical duty cycle for high power applications," *IEEE Transactions on Power Electronics*, vol. 20, no. 2, pp. 370–377, Mar. 2005.
3. D. S. Oliveira and I. Barbi, "A three-phase ZVS PWM DC/DC converter with asymmetrical duty cycle associated with a three-phase version of the hybridge rectifier," *IEEE Transactions on Power Electronics*, vol. 20, no. 2, pp. 354–360, Mar. 2005.
4. A. R. Prasad, P. D. Ziogas, and S. Manias, "A three-phase resonant PWM DC–DC converter," *IEEE Power Electronics Specialists Conference (PESC)*, 1991, June 24–27, 1991, Cambridge, USA, pp. 463–473.
5. A. K. S. Bhat and R. L. Zheng, "A three-phase series-parallel resonant converter-analysis, design, simulation, and experimental results," *IEEE Transactions on Industry Applications*, vol. 32, no. 4, pp. 951–960, July–Aug. 1996.
6. H. Cha, J. Choi, and P. N. Enjeti, "A three-phase current-fed DC/DC converter with active clamp for low-DC renewable energy sources," *IEEE Transactions on Power Electronics*, vol. 23, no. 6, pp. 2784–2793, Nov. 2008.

7. K. Modepalli, A. Mohammadpour, T. Li, and L. Parsa, "Three-phase current-fed isolated DC–DC converter with zero-current switching," *IEEE Transactions on Industry Applications*, vol. 53, no. 1, pp. 242–250, Jan.–Feb. 2017.
8. C. Liu, A. Johnson, and J.-S. Lai, "A novel three-phase high-power soft-switched DC/DC converter for low-voltage fuel cell applications," *IEEE Transactions on Industry Applications*, vol. 41, no. 6, pp. 1691–1697, Nov.–Dec. 2005.
9. M. Almardy and A. K. S. Bhat, "Three-phase fixed-frequency interleaved (LC)(L)-type series-resonant converter with a capacitive output filter," *IET Journal of Engineering*, vol. 2019, no. 17, pp. 4178–4184, June 2019.
10. H. R. E. Larico and I. Barbi, "Voltage-fed three-phase push-pull DC–DC converter," *IEEE Industrial Electronics Conference (IECON)*, 2009, Nov. 3–5, 2009, Porto, Portugal, pp. 956–961.
11. H. R. E. Larico and I. Barbi, "Three-phase push–pull DC–DC converter: analysis, design, and experimentation," *IEEE Transactions on Industrial Electronics*, vol. 59, no. 12, pp. 4629–4636, Dec. 2012.
12. J. Choi, H. Cha, and B. Han, "A three-phase interleaved DC–DC converter with active clamp for fuel cells," *IEEE Transactions on Power Electronics*, vol. 25, no. 8, pp. 2115–2123, Aug. 2010.
13. K. R. Sree and A. K. Rathore, "Impulse-commutated zero-current-switching current-fed three-phase DC/DC converter," *IEEE Transactions on Industry Applications*, vol. 52, no. 2, pp. 1855–1864, Mar.–Apr. 2016.
14. A. A. Abosnina and G. Moschopoulos, "Three-phase DC–DC PWM boost converter for renewable energy applications," *IEEE Energy Conversion Congress and Exposition (ECCE)*, 2017, Oct. 1–5, 2017, Cincinnati, USApp. 447–454.
15. T. Song, H. S. Chung, and A. Ioinovici, "A high-voltage DC–DC converter with Vin/3—voltage stress on the primary switches," *IEEE Transactions on Power Electronics*, vol. 22, no. 6, pp. 2124–2137, Nov. 2007.
16. A. A. Abosnina and G. Moschopoulos, "A novel high-gain three-phase DC–DC PWM boost converter," *IEEE Applied Power Electronics Conference (APEC)*, 2018, Mar. 4–8, 2018, San Antonio, USA, pp. 1533–1539.
17. G. H. de Alcântara Bastos, L. F. Costa, F. L. Tofoli, G. V. Torrico Bascopé, and R. P. Torrico Bascopé, "Nonisolated DC–DC Converters with wide conversion range for high-power applications," *IEEE Journal of Emerging and Selected Topics in Power Electronics*, vol. 8, no. 1, pp. 749–760, Mar. 2020.
18. M. Noah, S. Endo, H. Ishibashi, K. Nanamori, J. Imaoka, K. Umetani, and M. Yamamoto, "A current sharing method utilizing single balancing transformer for a multiphase LLC resonant converter with integrated magnetics," *IEEE Journal of Emerging and Selected Topics in Power Electronics*, vol. 6, no. 2, pp. 977–992, June 2018.
19. M. Noah, S. Kimura, J. Imaoka, W. Martinez, S. Endo, M. Yamamoto, and K. Umetani, "Magnetic design and experimental evaluation of a commercially available single integrated transformer in three-phase LLC resonant converter," *IEEE Transactions on Industry Applications*, vol. 54, no. 6, pp. 6190–6204, Nov.–Dec. 2018.
20. C. Fei, R. Gadelrab, Q. Li, and F. C. Lee, "High-frequency three-phase interleaved LLC resonant converter with GaN devices and integrated planar magnetics," *IEEE Journal of Emerging and Selected Topics in Power Electronics*, vol. 7, no. 2, pp. 653–663, June 2019.

CHAPTER 16

BIDIRECTIONAL AND DUAL ACTIVE BRIDGE CONVERTERS

Bidirectional converters are used to interface two independent voltage sources so that power can be transferred from one to another in either direction. A typical application is the interface between a battery and some sort of DC bus where a battery is used to provide power to the DC bus when needed and is then charged by the DC bus sometime later. Instead of using two unidirectional power flow converters to do this, just one bidirectional converter can be used for the charging and discharging of the battery. Bidirectional converters can be found in applications where some sort of battery backup is used, such as for renewable energy systems or uninterruptable power supplies, and in applications where power flow between different voltage levels is required, such as in electric vehicles and microgrids.

In this chapter, various non-isolated and isolated DC–DC bidirectional converters are presented and their operating principles are explained. Methods for implementing ZVS operation in these converters are discussed as well. For isolated converters that are implemented with two distinct bridges, such as half-bridges or full-bridges on each side of the transformer, emphasis is placed on dual active bridge operation where both bridges are active, as opposed to single active bridge operation where just one bridge is active and the other is passive. The key points of the chapter are summarized in the conclusion.

DC–DC Converter Topologies: Basic to Advanced, First Edition. Gerry Moschopoulos.

Published 2024 by John Wiley & Sons, Inc.

16.1 BASIC NON-ISOLATED BIDIRECTIONAL CONVERTERS

The simplest DC–DC bidirectional converter is the boost/buck converter shown in Figure 16.1. If the converter is seen from left to right, then it appears to be a boost converter with its diode replaced by an active switch. If the converter is seen from right to left, then it appears to be a buck converter with its diode replaced by an active switch. This converter is typically used in low-power applications as the interface between a low-voltage battery and a high-voltage DC bus that is used to supply other converters and loads. In the circuit shown Figure 16.1, the battery would be on the left-hand side and the DC bus would be on the right-hand side. If power needs to be transferred to the DC bus, then switch S_1 is operational and switch S_2 is never turned on so that its body diode is used as the boost diode. If the battery needs to be charged, then switch S_2 is operational and switch S_1 is never turned on so that its body diode is used as the buck diode. Conduction losses can be reduced if S_1 is operated as a synchronous rectifier when acting as a buck converter diode or S_2 is operated as a synchronous rectifier when acting as a boost converter diode.

Bidirectional converters can be synthesized from the other four basic non-isolated DC–DC converters that were presented in Chapter 1 – the buck-boost, Cuk, SEPIC, and Zeta converters – simply by replacing the converter diode with an active switch. If this is done for the Cuk and the SEPIC converters, then the converters shown in Figure 16.2 are the result. Figure 16.2a is a bidirectional converter that acts as a Cuk converter regardless of the direction of power flow. Figure 16.2b is a bidirectional converter that acts as a SEPIC converter if the power flow is in one direction and as a Zeta converter if the flow of power is in the opposite direction. A bidirectional converter that is based on the buck-boost converter will act as a buck-boost converter regardless of the direction of power flow.

Non-isolated multilevel bidirectional converters that are based on the fundamental bidirectional buck-boost converter like the one shown in Figure 16.3 have been proposed for applications such as dual-voltage automotive systems to avoid the use of magnetics, as demonstrated in [1] and in high-power, high-voltage applications, as demonstrated in [2, 3]. These converters are based on the flying capacitor multilevel converters that were presented in Section 13.1.2 and are operated in a very similar manner. The capacitors are placed in a way that makes the voltage across each switch a fraction of the high-side voltage,

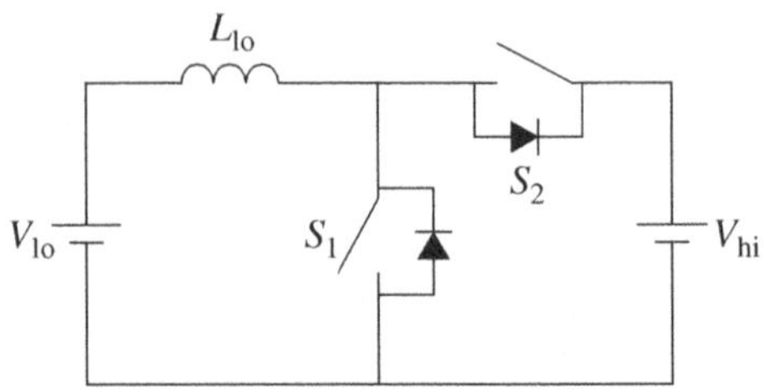

Figure 16.1 Fundamental non-isolated bidirectional DC–DC buck-boost converter.

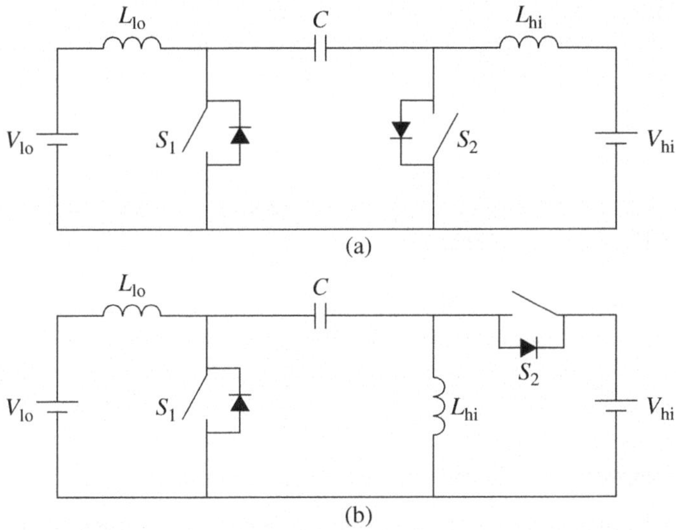

Figure 16.2 Other non-isolated bidirectional DC–DC converters based on fundamental PWM topologies: (a) Bidirectional Cuk converter. (b) Bidirectional SEPIC/Zeta converter.

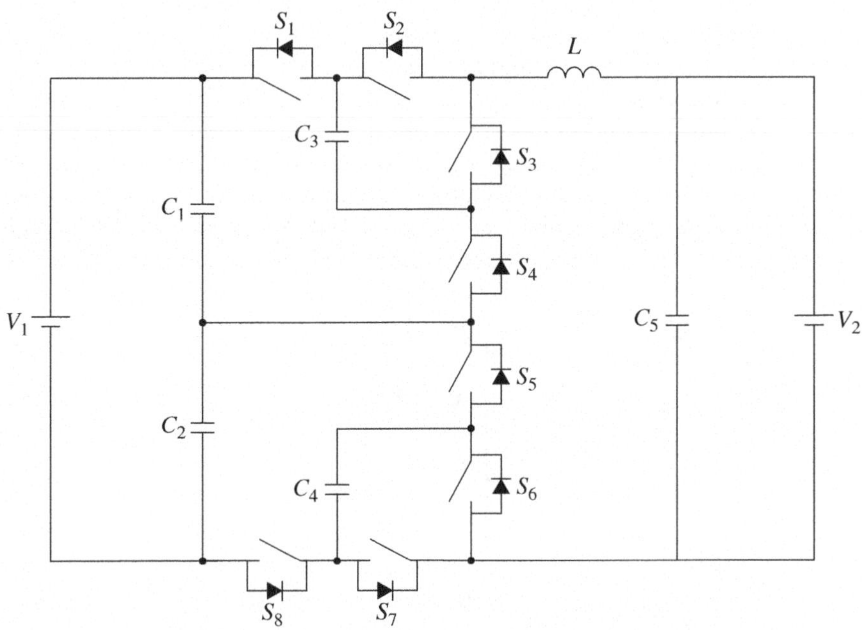

Figure 16.3 Non-isolated DC–DC bidirectional buck-boost multi-level converter.

thus allowing lower-rated and less expensive switches to be used. In some applications, it makes economic sense to use a number of cheaper, lower voltage-rated switches rather than fewer, more expensive, higher voltage-rated switches.

16.2 ZVS OPERATION OF THE FUNDAMENTAL BUCK-BOOST BIDIRECTIONAL CONVERTER

Given the popularity of the fundamental buck-boost converter shown in Figure 16.1, various methods of operating this converter with soft-switching have been proposed. Several of these methods are presented in this section.

16.2.1 Bidirectional Quasi-Square Wave Converter

The ZVS quasi-square wave converter, shown in Figure 16.4, was presented in Section 2.2 and its operation was explained [4]. This converter is a buck converter that has two active switches instead of a switch and a freewheeling diode in its topology. The second active switch, which is S_2 in Figure 16.4, is used as a way of generating a current in the reverse direction that will discharge the output capacitance of S_1 and then flow through its body diode so that the switch can be turned on with ZVS. The ZVS operation for buck mode operation was explained in Section 2.2.

If the output capacitor C_o and the load are replaced by a voltage source, then the converter shown in Figure 16.4 becomes the converter shown in Figure 16.1, but flipped over horizontally. Using operational principles that are similar to the buck ZVS quasi-square wave converter, the converter with a low-side voltage source can be made to operate as a ZVS boost converter. Referring to Figure 16.1, which has a low-side voltage source V_{lo} and a high-side voltage source V_{hi}, the converter can be made to operate as a ZVS boost

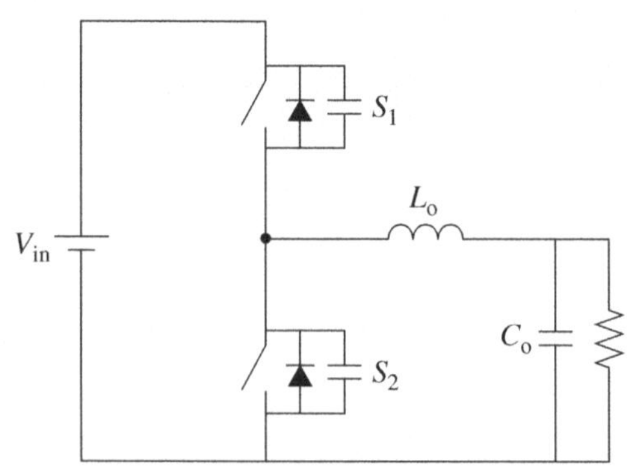

Figure 16.4 Unidirectional ZVS quasi square-wave converter.

converter with S_1 operating as a boost converter switch and S_2 operating as the boost diode and as the switch that initiates the ZVS mechanism for S_1.

Assuming that S_2 is already on, S_1 is off to avoid any short circuiting, and current flows through the switch and the inductor from right to left in Figure 16.1. When S_2 is turned off, the output capacitance of S_2, C_{s2}, begins to be charged while that of S_1, C_{s1}, begins to be discharged. Current starts to flow through the body diode of S_1 after C_{s1} has been fully discharged; S_1 can be turned on with ZVS soon afterward. If MOSFETs are used for S_1 and S_2, then the current can flow in the direction of the source to drain after S_1 has been turned on if the drop across the device is less than what it would be if it flows through the body diode. Regardless of how current flows through this device, the inductor current begins to decrease and, if the inductor is designed properly, eventually drops to zero. With S_1 on, the current then reverses direction and begins to flow from the low-voltage side to the high-voltage side, from left to right in Figure 16.4. After a sufficient current has been built up in the inductor, S_1 is then turned off and C_{s1} and C_{s2} begin to charge and discharge, respectively. After C_{s2} has been discharged, current starts to flow through the body diode of S_2, which allows S_2 to be turned on with ZVS. Eventually, the inductor current reverses direction in preparation for the ZVS turn-on of S_1. It should be noted that the current in the inductor is bidirectional but must be more positive than negative so that power can actually be transferred from the low-voltage side to the high-voltage side.

As was stated in the ZVS quasi-square discussion of Section 2.2, the main disadvantage of operating the converter in this manner is that the output inductor current flows in two directions instead of just one. This increases the peak current that the switches must handle, increases conduction losses and turn-off switching losses, and makes the ripple in the output inductor current large, which necessitates the use of a large output capacitor to filter out current harmonics.

16.2.2 Four-Switch Buck-Boost Converter

An alternative to the fundamental buck-boost converter is the four-switch buck-boost converter that was proposed in [5, 6] and shown in Figure 16.5. This converter has four switches instead of two and has the ability to step voltage down as a buck converter or step voltage up as a boost converter in either direction. For example, if V_1 is greater than V_2 and power is to be transferred from V_1 to V_2, then S_1 can be made to act as a buck converter switch, S_2 can be made to act as a buck converter diode, and S_3 and S_4 can be made inactive. The equivalent circuit then becomes a buck converter with S_1 as the switch, S_2 as the diode, and S_3 acting as a diode in series with the inductor. There is no current flow in S_4. The converter can step voltage up or down in either direction if the switches are operated as is shown in Table 16.1.

The four-switch configuration of the converter allows all its active switches to operate with ZVS. In cases where switching devices have current flowing

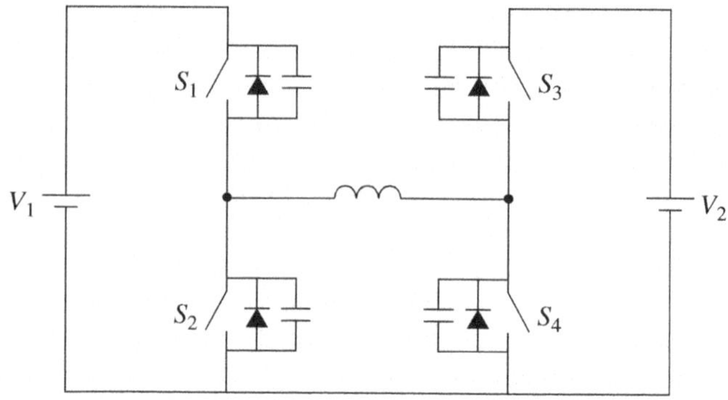

Figure 16.5 Four-switch bidirectional ZVS buck-boost DC–DC converter.

TABLE 16.1 Switch operation for the four-switch buck-boost converter.

S_1	S_2	S_3	S_4	Operation mode	Direction of current in Figure 16.5
Acts as a buck switch	Acts as a buck diode	Always OFF. Current flows through its body diode	Always OFF. No current flows through the device	Buck ($V_1 > V_2$)	From V_1 to V_2
Always ON	Acts as a boost switch	Acts as a boost diode	Always OFF. No current flows through the device	Boost ($V_1 < V_2$)	From V_1 to V_2
Always OFF. Current flows through its body diode	Always OFF. No current flows through the device	Acts as a buck switch	Acts as a buck diode	Buck ($V_2 > V_1$)	From V_2 to V_1
Acts as a boost diode	Always OFF. No current flows through the device	Always ON	Acts as a boost switch	Boost ($V_2 < V_1$)	From V_2 to V_1

through their body diode during regular step-up or step-down operation, these devices can be turned on with ZVS just before the switch that is acting as the main boost switch or the main buck switch is to be turned on. For example, if power is flowing from V_1 to V_2 and V_1 is greater than V_2, then S_1 is the main buck switch, S_2 is the main buck diode and current flows through the body diode of S_3. When S_1 is off, current flows through the body diodes of S_2 and S_3 if the inductor current is continuous. When it is time to turn on S_1, S_2, and S_3 are turned on to reverse the direction of current in the inductor. When sufficient current has been built up in the inductor, then S_2 and S_3 are turned off and current flows through the body diode of S_1 and S_4. S_1 is turned on with ZVS, but S_4 is never turned on. Eventually, the inductor current reverses direction and power flows once again from V_1 to V_2. The converter can step voltage up or down in either direction with ZVS if the switches are operated as shown in Table 16.2.

16.2.3 Active Auxiliary Circuits

If it is desired to make the bidirectional boost/buck converter shown in Figure 16.1 operate with soft-switching and fixed switching frequency PWM, then active auxiliary circuits can be added in parallel with each switch and the auxiliary circuit becomes operational when the switch that it is attached to it. Standard ZVS and ZCS methods, such as the ones presented in Chapters 2 and 3, can be used so that the converter can operate as a ZVS-PWM or a ZCS-PWM boost converter when power needs to be delivered to a DC bus or as a ZVS-PWM or a ZCS-PWM buck converter when power needs to be delivered to a battery. The main drawback to this approach is that it requires the addition of two active auxiliary circuits, which means that two switches must be added to the converter along with two sets of passive circuits; this increases the cost and size of the converter. A simpler approach that involves just a single active switch, one capacitor and a couple of inductors was proposed in [7] and shown in Figure 16.6. This converter can operate bidirectionally with ZVS for both directions of power flow with fewer components than other ZVS-PWM boost/buck converters with active auxiliary circuits. Follow-up work done by the authors in [8] demonstrated that the concept can be extended to coupled inductors as well.

The converter operates as follows: When it is operating as a boost converter and switch S_1 is turned off, the current in L_{r1} flows through the body diode of S_a and charges C_r. When it is time to turn S_1 on again, S_a is turned on first, and energy from C_r is transferred to L_{r1} through S_a and L_{r2}. When S_a is turned off, energy in L_{r1} discharges the output capacitance of S_1, C_{s1}, so that the voltage across S_1 falls to zero and the body diode of S_1 conducts. S_1 can be turned on with ZVS while its body diode conducts current. After this has been done, the current in the S_1 device reverses direction so that the converter operates as a standard PWM boost converter.

TABLE 16.2 Switch operation for the four-switch buck-boost converter operating with ZVS.

S_1	S_2	S_3	S_4	Operation mode	Direction of current in Figure 16.5
Acts as a buck switch	Acts as a buck diode and is part of the mechanism that initiates the ZVS turn-on of S_1	Always OFF during regular operation as current flows through its body diode and is part of the mechanism that initiates the ZVS turn-on of S_1	Always OFF. No current flows through the device except when current is flowing through the body diode of S_1 before it is turned on with ZVS	Buck ($V_1 > V_2$)	From V_1 to V_2
Always ON	Acts as a boost switch	Acts as a boost diode and is part of the mechanism that initiates the ZVS turn-on of S_2	Always OFF. No current flows through the device except when current is flowing through the body diode of S_2 before it is turned on with ZVS	Boost ($V_1 < V_2$)	From V_1 to V_2
Always OFF during regular operation as current flows through its body diode and is part of the mechanism that initiates the ZVS turn-on of S_2	Always OFF. No current flows through the device except when current is flowing through the body diode of S_1 before it is turned on with ZVS	Acts as a buck switch	Acts as a buck diode and is part of the mechanism that initiates the ZVS turn-on of S_3	Buck ($V_2 > V_1$)	From V_2 to V_1
Acts as a boost diode and is part of the mechanism that initiates the ZVS turn-on of S_4	Always OFF. No current flows through the device except when current is flowing through the body diode of S_4 before it is turned on with ZVS	Always ON	Acts as a boost switch	Boost ($V_2 < V_1$)	From V_2 to V_1

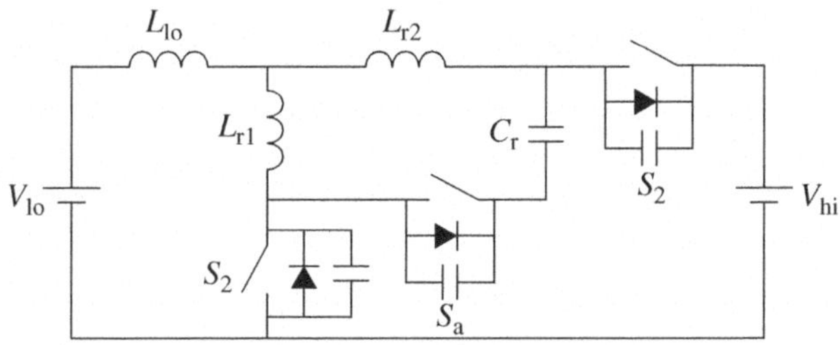

Figure 16.6 Bidirectional active-clamp ZVS buck-boost converter with single auxiliary switch.

When the converter is operating as a buck converter and switch S_2 is turned off, the current in L_{r2} flows through C_r through the body diodes of S_1 and S_a and charges C_r. When it is time to turn S_2 on again, S_a is turned on and energy in C_r is placed in L_{r2} through S_a and L_{r1}. When S_a is turned off, energy in L_{r2} discharges the output capacitance of S_2, C_{s2}, so that the voltage across S_2 falls to zero and the body diode of S_2 conducts. S_2 can be turned on with ZVS while its body diode conducts current. After this has been done, the current in the S_2 device reverses direction so that the converter operates as a standard PWM buck converter.

16.3 BIDIRECTIONAL CONVERTER TOPOLOGIES WITH TRANSFORMER ISOLATION

As with the non-isolated converters that have been discussed thus far in this chapter, bidirectional converters with a conventional single-phase two-winding transformer typically have a high voltage side and a low voltage side. A transformer is used to provide voltage gains that are more extreme than those possible with a non-isolated converter. It is common practice to have a current-fed converter on the low-voltage side and a voltage-fed converter on the high-voltage side. The current-fed converter acts as a step-up converter when power flows from the low-voltage side to the high-voltage side and the voltage-fed converter acts as a step-down converter when power flows from the high-voltage side to the low-voltage side. Several examples of isolated bidirectional converter topologies are presented in this section.

A very simple isolated bidirectional converter is the converter that was proposed in [9] and shown in Figure 16.7a. One side of this converter, the low-voltage side, is a current-fed half-bridge converter that consists of switches S_1 and S_2, bulk capacitors C_1 and C_2 and an inductor L_1 that is placed between the source and the rest of the converter. The other side of the converter is a standard voltage-fed half-bridge converter. The low-voltage side current-fed converter can also be implemented by using the two-inductor current-fed converter that was presented in Section 10.2, as was proposed in [10] and shown in Figure 16.7b.

The concept of using low-voltage current-fed/high-voltage voltage-fed converters in isolated bidirectional converters can be extended to full-bridge converters, as was proposed in [11, 12] and shown in Figure 16.7c. In this topology, the low-voltage side converter is an active-clamp current-fed converter like the one that was discussed in Section 10.6.1 and the voltage-fed converter

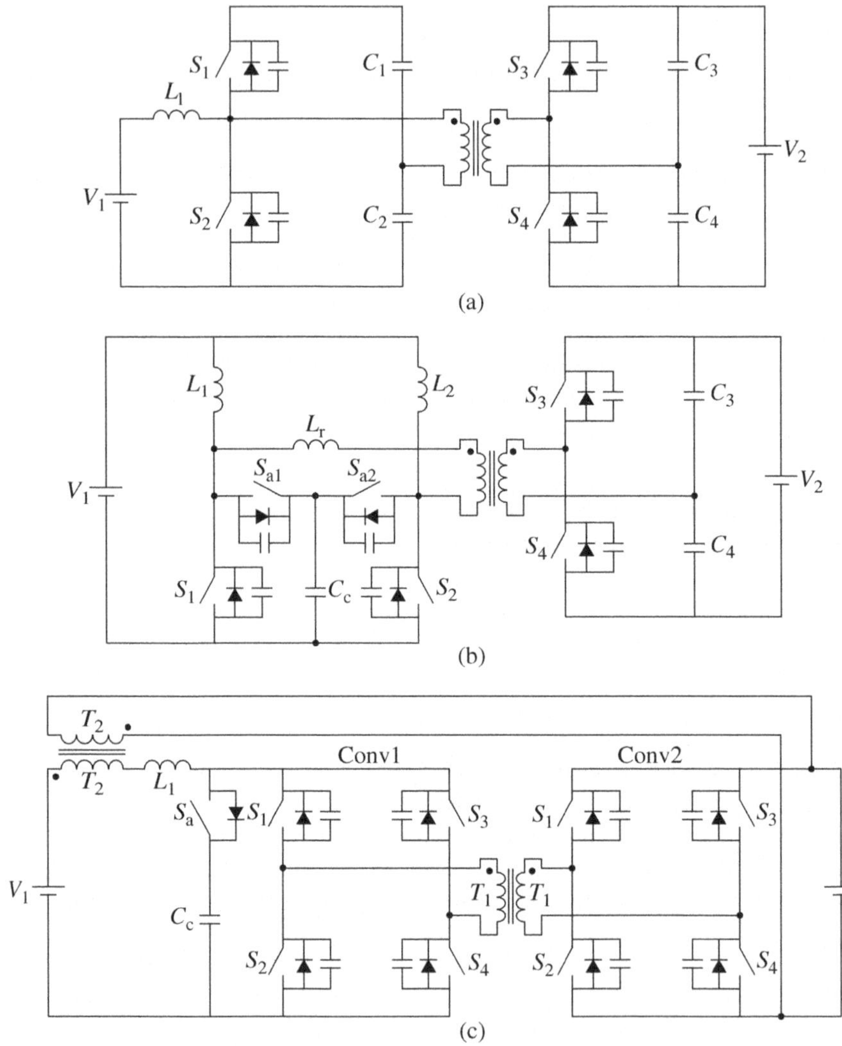

Figure 16.7 Isolated DC–DC bidirectional converter topologies: (a) Dual half-bridge converter with current-fed and voltage-fed converters. (b) Active-clamp current-fed half-bridge/voltage-fed half-bridge. (c) Dual full-bridge converter with current-fed and voltage-fed converters. (d) Dual full-bridge converter with three-level DC–DC converter. (e) Three-phase bidirectional converter. (f) Three-phase reduced switch bidirectional converter.

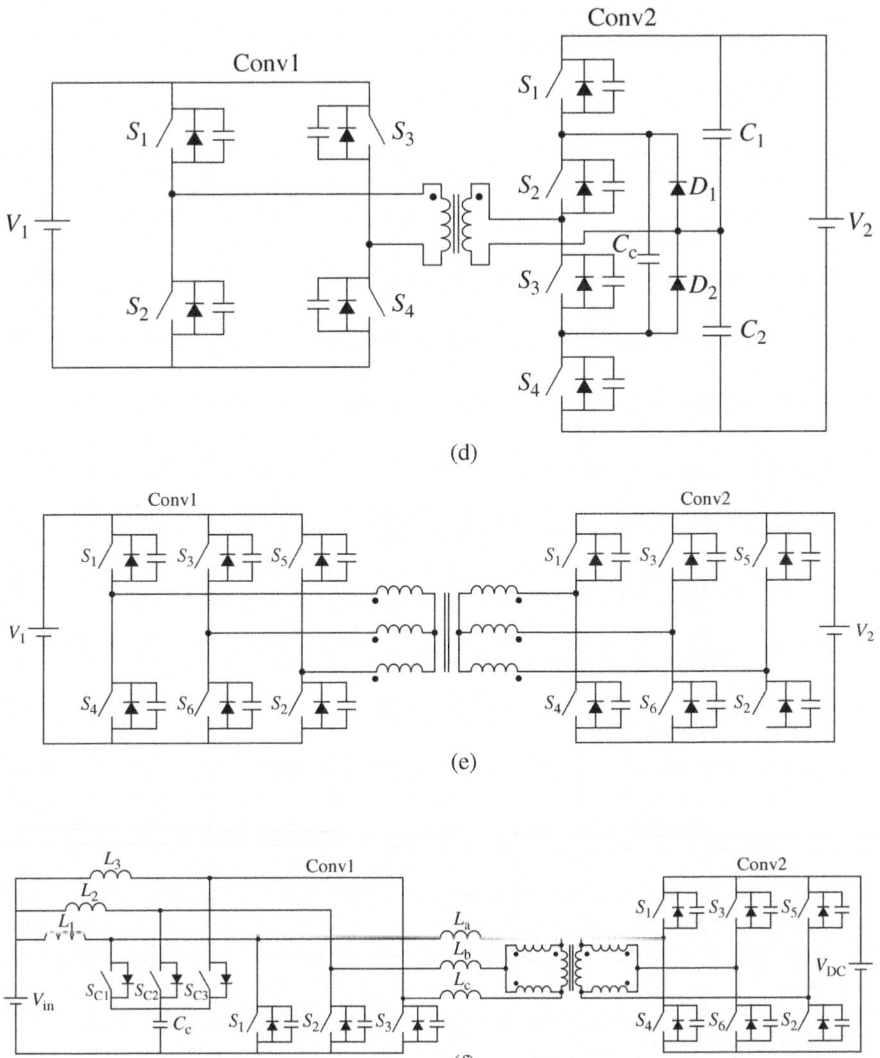

Figure 16.7 (*Continued*)

is a conventional ZVS-PWM full-bridge converter. If the voltage of the high-voltage side is especially high, then the high-voltage voltage-fed converter can be a three-level converter, as was proposed in [13] and shown in Figure 16.7d. Bidirectional converters can be implemented with three-phase topologies, as shown in Figure 16.7e, f. Figure 16.7e shows a three-phase bidirectional converter with the six-switch three-phase converter on each side of the transformer that was proposed in [14, 15] and Figure 16.7f shows a three-phase bidirectional converter that was proposed in [16] with the low-voltage current-fed side implemented with the three-phase active-clamp current-fed topology that was discussed in Section 15.5.2.

The following should be noted:

- The bidirectional converters shown in Figure 16.7 can be operated either in single active bridge mode or dual active bridge mode. When operated in single active bridge mode, only one of the two converters is active, while the other is passive. For example, when the full-bridge converter shown in Figure 16.7c is operated in single active bridge mode, the bridge that is inactive acts as a four-diode bridge rectifier. Although the single active bridge mode is simpler, the dual active bridge is more advantageous and thus is more popular. The dual active bridge mode of operation is discussed in the next section of this chapter.
- For converters that use auxiliary circuitry to assist in the soft-switching operation of the main switches, this circuitry is not used if the converter is passive in a bidirectional converter that is operating in a single active bridge mode. In the case of active clamp converters, such as the ones shown in Figure 16.7b, c, care must be taken that the active clamp capacitor is not charged to an excessive voltage level as there is no opportunity to discharge this capacitor under normal operation. The voltage across this capacitor must be monitored and this capacitor must be somehow discharged occasionally, either through some additional circuitry or by turning on the active clamp switch occasionally.
- Issues involving soft-switching with active auxiliary circuits can be avoided if bidirectional resonant converters are used like the ones shown in Figure 16.8. The resonant tank can be asymmetrical or symmetrical. The resonant converter shown in Figure 16.8a is an LLC resonant converter, which is an asymmetrical resonant tank as each of the two converters of the overall converter has different resonant tank components [17–19]. The converter shown in Figure 16.8b is a CLLC resonant converter that has a symmetrical resonant tank [20–24]. A challenge when designing bidirectional resonant converters is that the impedance transformation of the resonant elements from high-side to low-side and vice versa must be considered in the design. For example, a bidirectional resonant converter with an LLC resonant tank may operate as an LLC converter if power is transferred from left to right according to the diagram in Figure 16.8a, but may operate as a series resonant converter if power is transferred in the opposite direction. For a symmetrical converter like the CLLC converter shown in Figure 16.8b, the converter operates in the same resonant manner regardless of the direction of power flow. The difference in resonant operation may be advantageous in some applications but may not be so in others. The decision to use an asymmetrical or a symmetrical resonant tank is very much dependent on the application, specifications, and operating conditions of the bidirectional resonant converter.

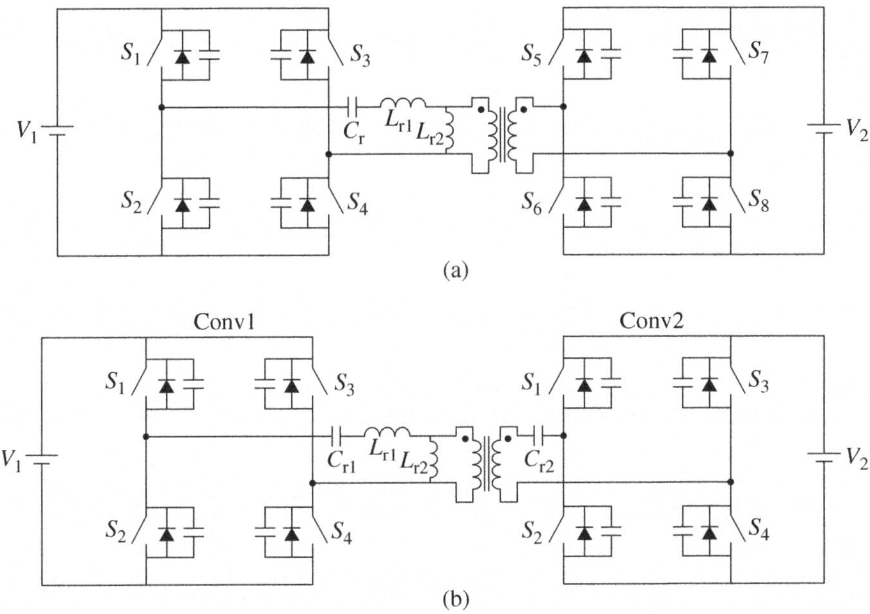

Figure 16.8 Bidirectional resonant converters: (a) LLC resonant converter (b) CLLC resonant converter.

16.4 DUAL ACTIVE BRIDGE CONVERTERS

Any isolated bidirectional converter can be considered to be a dual active bridge converter if both bridges are active during a switching cycle, as opposed to having one converter be inactive and operate as a passive rectifier, as is the case with single-active bridge control. Several examples of dual active bridge operation are presented in this section.

16.4.1 Dual Active Bridge Half-Bridge Converter

The bidirectional converter shown in Figure 16.7a can be used as an interface between two voltage busses. According to [9], it can be operated as a dual active bridge converter in the following manner: All switches are operated with a duty cycle of $D = 0.5$ and the gating signal of each switch is complementary to that of the same leg. As a result, a square-wave voltage is generated across each transformer winding as shown in Figure 16.9. If these two voltages are shifted 90° relative to each other, then the maximum amount of energy transfer can take place. In this case, a positive voltage is placed across one of the transformer windings and a negative voltage is placed across the other winding so that the current begins to be built up in the transformer. Current in

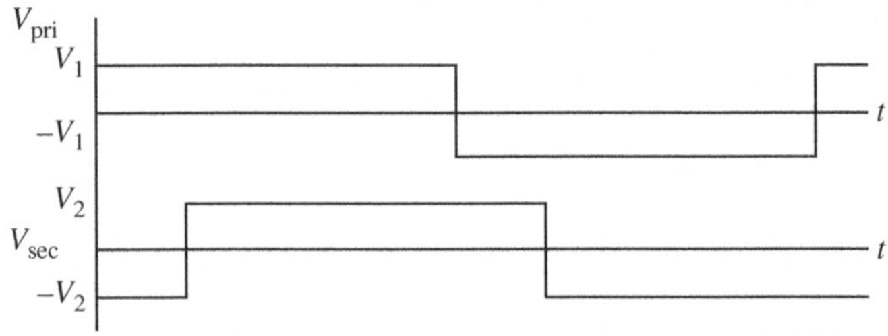

Figure 16.9 Transformer voltage waveforms for dual half-bridge converter shown in Figure 16.7a.

the transformer falls when the polarities of both voltages are reversed halfway through the switching cycle as the voltage across the transformer inductance changes polarity as well. When there is no phase shift between the two voltages, the voltage difference across the transformer inductances is minimal. Thus, little current is generated in the circuit and little energy is transferred from one side to the other. If these two voltages are shifted 180° relative to each other, then like the case when the phase shift is 0°, no energy transfer can take place.

The following should be noted about the converter:

- The duty cycles of the switches in both half-bridges are kept fixed at approximately 50% of the switching cycle. The remaining parts of the switching cycle when none of the switches of a bridge leg is on is for dead times that allow output switch capacitances to be discharged. Operating a half-bridge converter with $D = 0.5$ ensures that its switches operate with ZVS because the current is forced to flow through the body diode of a switch after its complementary switch in the same leg is turned off and that switch can be immediately turned on with ZVS. If $D < 0.5$, then the current in the body diode of a switch dies down before the switch can be turned on so that the opportunity for ZVS operation is lost.
- Operation with $D < 0.5$ should only be considered under light-load conditions. If D is fixed at $D = 0.5$, then the current-fed side will be in boost mode, putting energy in the inductor, but there will be no way for this energy to be transferred to the other side. As a result, the voltage across each bus capacitor will increase and will become excessive so that the voltage difference across the transformer inductances will become significant even though the voltage waveforms across each bridge are in phase. Reducing D when the converter is operating under light-load conditions prevents this from happening, but it is at the expense of switching losses as ZVS is lost.
- The leakage inductance of the transformer is a critical parameter that needs to be considered when designing the converter as it determines

the amount of current that is allowed to flow in the transformer and, thus, the amount of power that the converter operates with.

- The direction of power flow in the converter is determined by which bridge leads the other. For example, according to the voltage waveforms shown in Figure 16.9, if the voltage waveform of the current-fed half-bridge leads that of the voltage-fed half-bridge, the energy is transferred from the low voltage side to the high voltage side. If the order of the waveforms is changed and the voltage-fed half-bridge leads to the current-fed half-bridge, then energy is transferred from the high-voltage side to the low-voltage side.
- The converter shown in Figure 16.7a has a current-fed converter on one side and a voltage-fed converter on the other. Implementations where both sides are the same are possible and it is common to have voltage-fed converters on both sides when less voltage boosting of the low-voltage side voltage is needed.

16.4.2 Dual Active Bridge PWM Full-Bridge Converters

Dual active bridge (DAB) full-bridge (FB) converters are typically implemented with two voltage-fed converter modules, as shown in Figure 16.10. Numerous PWM methods have been proposed to operate a DAB-PWM-FB converter like the one shown in Figure 16.10. The most basic methods involve the control or variation of certain key parameters that are referred to as degrees of freedom in the literature to control the flow of power and the amount of power that is transferred. Referring to Converter #1 (Conv1) and Converter #2 (Conv2) in Figure 16.10, the basic degrees of freedom are as follows:

- The duty cycle D_1 of Converter #1
- The duty cycle D_2 of Converter #2
- The phase-shift φ between the voltage that is impressed across the transformer winding of Converter #1 and the voltage that is impressed across the transformer winding of Converter #2

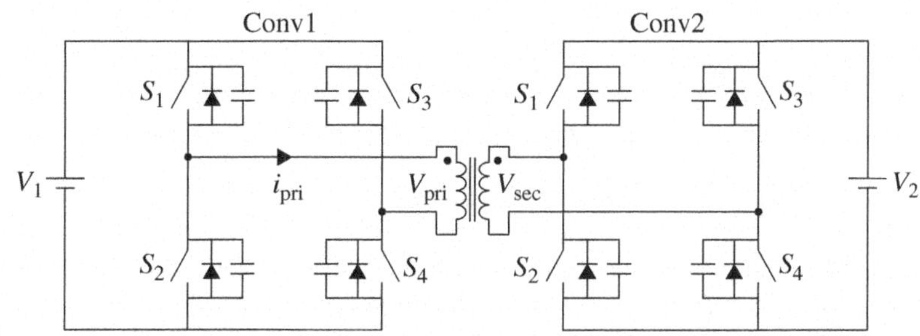

Figure 16.10 Dual active bridge converter with two voltage-fed converter modules.

- The switching frequency f_{sw} of Converter #1 and Converter #2, which is the same for both converters.

Based on these degrees of freedom, numerous PWM methods have been proposed to operate a DAB-PWM-FB converter [25–34]. The most basic of these methods are as follows:

- *Phase-Shift Modulation:* This is the PWM method that was described in Section 16.4.1 for the DAB half-bridge converter. To recap, the switching frequency of both converters is fixed and they are both operated with a fixed duty cycle of $D_1 = D_2 = 0.5$, as shown in Figure 16.11. The converter is controlled simply by shifting the gating signals of the switches so that φ is the only control variable. This is the simplest way to operate a DAB-PWM-FB converter and it allows for maximum power transfer, but it produces high RMS transformer current (thus high conduction losses) and the ZVS-operating range of the converter is limited. These drawbacks make phase-shift modulation unattractive for most situations unless $\varphi = 90°$ and maximum power transfer occurs. No power transfer occurs if $\varphi = 0°$ or $180°$.
- *Triangular Modulation:* With this method, voltage pulses across the transformer windings of the two converters are generated so that the beginning of their pulses is aligned and the secondary switches are turned off when the current drops to zero, as shown in Figure 16.12. This method gets its name from the triangular transformer current waveform it produces, as shown in Figure 16.12. The DAB-PWM-FB operates with a fixed switching frequency and with $\varphi = 0°$. The converter is controlled by varying D_1 and D_2. With triangular modulation, the secondary switches can turn on

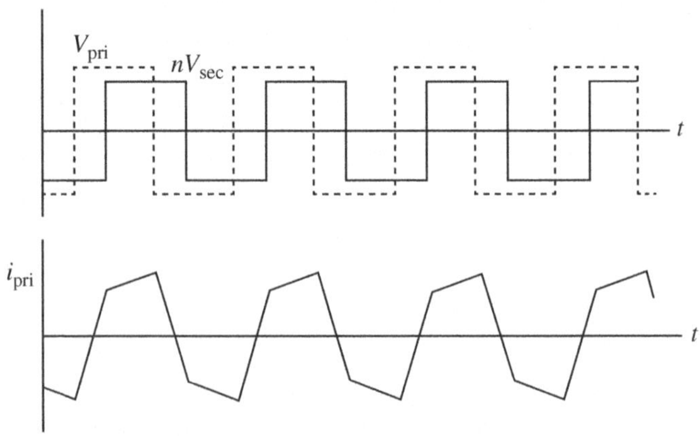

Figure 16.11 Voltage and current waveforms of a DAB-PWM-FB converter operating with phase-shift modulation: transformer primary voltage V_1, secondary voltage nV_2, and transformer primary current i_{pri}.

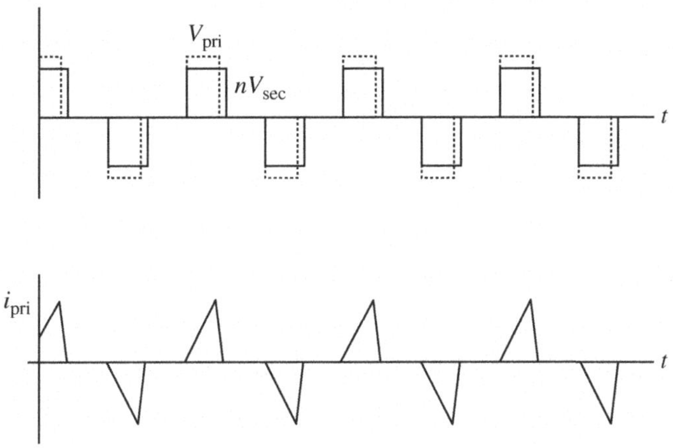

Figure 16.12 Voltage and current waveforms of a DAB-PWM-FB converter operating with triangular modulation: transformer primary voltage V_1, secondary voltage nV_2, and transformer primary current i_{pri}.

with ZCS, no matter what the operating conditions as the current through these switches is made to be discontinuous, as shown in Figure 16.12. Since current is not continuously flowing in the converter, the RMS current flowing in the transformer is reduced, and thus, so too are the conduction losses. Triangular modular, however, is more complex and difficult to implement than phase-shift modulation and cannot achieve the same amount of power flow. It should be noted that the gain of the converter is affected when converter switches are turned off at the same time instead of being turned on at the same time, as shown in Figure 16.12. According to [33], doing so increases the gain of the converter.

- *Trapezoidal Modulation:* With this method, D_1, D_2, and φ are varied, as shown in Figure 16.13. This method, in essence, combines the variable φ operation of the phase-shift modulation method with the variable duty cycle of the triangular method and derives its name from the shape of the transformer current waveform, as shown in Figure 16.13. Since this method is a combination of the phase-shift and triangular modulation methods, therefore, it allows the converter to have a combination of the features of these methods, but to a lesser degree. For example, trapezoidal modulation allows the DAB-PWM-FB to transfer more power from one converter to another than does the triangular method, but not as much as the phase-shift method can. Trapezoidal modulation allows the secondary switches to operate with ZCS, unlike phase-shift modulation, but over a narrower range than triangular modulation. Trapezoidal modulation allows the DAB-PWM-FB converter to operate its primary switches with ZVS, but over a narrower range than with phase-shift modulation and allows the DAB-PWM-FB converter to

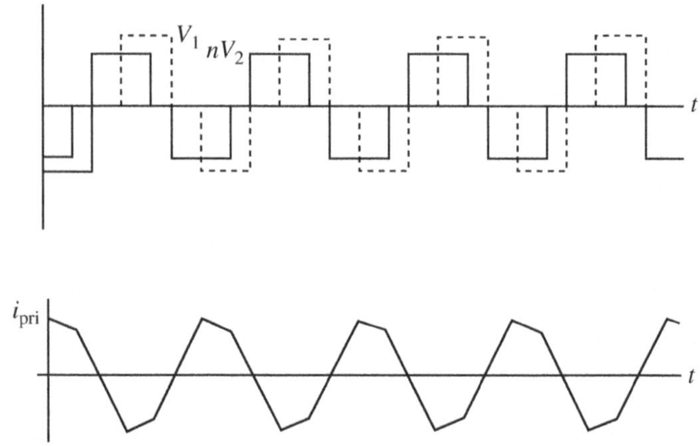

Figure 16.13 Voltage and current waveforms of a DAB-PWM-FB converter operating with trapezoidal modulation: transformer primary voltage V_1, secondary voltage nV_2, and transformer primary current i_{pri}.

operate with less RMS transformer current than phase-shift modulation for lower power but with more RMS current than triangular modulation.

- *Combined Modulation:* A combination of PWM methods can be used to operate a DAB-PWM-FB converter. For example, the converter can be operated with the triangular method for light-load, low-power applications because the discontinuous transformer current results in lower conduction losses and operated with the trapezoidal method for heavier loads and higher power applications as it allows for higher power transfer than the triangular method.

The following should be noted when operating a DAB-PWM-FB converter:

- The methods that have been described here are the most fundamental and research is ongoing to develop new PWM methods that can result in more efficient operation.
- The key challenge involved when designing a DAB-PWM-FB converter is the optimization of the various degrees of freedom to produce the best result for a specific application with specific operating conditions and specifications. This design may not be simple and researchers have applied sophisticated optimization methods in their design – methods that require expertise in optimization theory to a degree that may be too advanced for many practicing power electronics personnel.
- In multilevel DAB-PWM-FB converters, the duration of a voltage level can be controlled, which allows these converters to have even more degrees of freedom than two-level converters as the duration of a voltage level.

- Switching losses are less of a concern when wide bandgap devices are used due to their superior switching characteristics. The key emphasis in the design of these converters becomes trying to minimize circulating current in these converters while satisfying power transfer requirements without concern for ZVS operation, which relaxes constraints placed on degrees of freedom. The increasing popularity of wide bandgap devices has made DAB-PWM-FB converters more attractive than they were in the past.
- Some control of the DAB-PWM-FB converter can be achieved by using frequency modulation, as explained in publications such as [30]. With frequency modulation, the converter is controlled by changing the switching frequency f_{sw} of both converters, keeping it the same for both converters. Doing so changes the impedance of the parasitic elements of the transformer, such as the leakage inductance, which in turn changes the amount of voltage that is dropped by the transformer. This method is basically a fine-tuning of sorts and should be used with one of the other methods discussed above. It is more suitable for resonant DAB converters.
- Dual active bridge converters can be implemented as resonant converters with resonant tanks such as the ones shown in Figure 16.8. Examples of CLLC bidirectional dual active bridge converters are presented in [22–24]. Resonant DAB converters are typically operated either with frequency modulation, phase-shift modulation, or a combination of both. Triangular and trapezoidal-based modulation methods are rarely used as these methods in resonant DAB converters as they limit the converters' ability to operate with ZVS. The design of a DAB resonant converter involves designing around certain degrees of freedom, similar to what was described in the previous section for DAB-PWM converters, but with the need to consider the resonant operation and its impact on gain and ZVS operation.

16.5 CONCLUSION

Bidirectional converters are used to interface two independent voltage sources so that power can be transferred from one source to another in either direction. They are commonly used in applications such as electric vehicles, renewable energy systems, and uninterruptible power supplies where some sort of battery interface is needed. Bidirectional converters for lower power applications can be synthesized from the basic non-isolated DC–DC converters, while those for higher power applications can be synthesized by using a current-fed converter at the low-side voltage for voltage step-up and a voltage-fed converter at the high-side voltage source for voltage step-down. It is possible to synthesize a bidirectional converter with two current-fed converters or two voltage-fed converters with the latter combination being more widely used. Soft-switching methods for bidirectional converters were discussed in this chapter. Bidirectional converters can be implemented with soft-switching by using active auxiliary circuits or resonant tank components. Converters that are based on full-bridge topologies can be operated with ZVS by using standard phase-shift modulation, as is used for the ZVS-PWM full-bridge converter discussed in Chapter 7.

Bidirectional full-bridge converters can be operated in single active bridge mode or in dual active bridge mode. Operation in single active bridge mode requires that only one of the two converters in the overall bidirectional converter is active, whereas both converters are active when the converter is operated in dual active bridge mode. The main advantage of dual active bridge mode operation is that there is less current flowing in the converter, which results in fewer conduction losses. The main drawbacks are that dual active bridge operation is more complicated and the converter operates with a more limited ZVS range. The limitation in ZVS range is less critical in bidirectional converters that are implemented with wide bandgap devices and the growing use of such devices has contributed to the increasing popularity of dual active bridge converters.

REFERENCES

1. F. Z. Peng, F. Zhang, and Z. Qian, "A magnetic-less DC–DC converter for dual-voltage automotive systems," *IEEE Transactions on Industry Applications*, vol. 39, no. 2, pp. 511–518, Mar.–Apr. 2003.
2. K. Jin, X. Ruan, M. Yang, and M. Xu, "A hybrid fuel cell power system," *IEEE Transactions on Industrial Electronics*, vol. 56, no. 4, pp. 1212–1222, Apr. 2009.
3. L. F. Costa, S. A. Mussa, and I. Barbi, "Multilevel buck/boost-type DC–DC converter for high-power and high-voltage application," *IEEE Transactions on Industry Applications*, vol. 50, no. 6, pp. 3931–3942, Nov.–Dec. 2014.
4. C. P. Henze, H. C. Martin, and D. W. Parsley, "Zero-voltage switching in high frequency power converters using pulse width modulation," *IEEE Applied Power Electronics Conference and Exposition (APEC),* 1988, Feb. 1–5, 1988, New Orleans, USA, pp. 33–40.
5. F. Caricchi, F. Crescimbiri, and A. Di Napoli, "20 kW water-cooled prototype of a buck–boost bidirectional DC–DC converter topology for electrical vehicle motor drives," *IEEE Applied Power Electronics Conference (APEC)*, 1995, Mar. 5–9, 1995, Dallas, USA, pp. 887–892.
6. S. Waffler and J. W. Kolar, "A novel low-loss modulation strategy for high-power bidirectional buck + boost converters," *IEEE Transactions on Power Electronics*, vol. 24, no. 6, pp. 1589–1599, Jun. 2009.
7. P. Das, B. Laan, S. A. Mousavi, and G. Moschopoulos, "A nonisolated bidirectional ZVS-PWM active clamped DC–DC converter," *IEEE Transactions on Power Electronics*, vol. 24, no. 2, pp. 553–558, Feb. 2009.

8. P. Das, S. A. Mousavi, and G. Moschopoulos, "Analysis and design of a nonisolated bidirectional ZVS-PWM DC–DC converter with coupled inductors," *IEEE Transactions on Power Electronics*, vol. 25, no. 10, pp. 2630–2641, Oct. 2010.
9. H. Li, F. Z. Peng, and J. S. Lawler, "A natural ZVS medium-power bidirectional DC–DC converter with minimum number of devices," *IEEE Transactions on Industry Applications*, vol. 39, no. 2, pp. 525–535, Mar.–Apr. 2003.
10. H. Xiao and S. Xie, "A ZVS bidirectional DC–DC converter with phase-shift plus PWM control scheme," *IEEE Transactions on Power Electronics*, vol. 23, no. 2, pp. 813–823, Mar. 2008.
11. K. Wang, F. C. Lee, and J. Lai, "Operation principles of bi-directional full-bridge DC/DC converter with unified soft-switching scheme and soft-starting capability," *IEEE Applied Power Electronics Conference (APEC)*, 2000, Feb. 6–10, 2000, New Orleans, USA, pp. 111–118.
12. K. Wang, L. Zhu, D. Qu, H. Odendaal, J. Lai, and F. C. Lee, "Design, implementation, and experimental results of bi-directional full-bridge DC/DC converter with unified soft-switching scheme and soft-starting capability," *IEEE Power Electronics Specialists Conference (PESC)*, 2000, June 18–23, 2000, Galway, Ireland, pp. 1058–1063.
13. D. Sha, Q. Lin, F. You, X. Wang, and G. Xu, "A ZVS bidirectional three-level DC–DC converter with direct current slew rate control of leakage inductance current," *IEEE Transactions on Industry Applications*, vol. 52, no. 3, pp. 2368–2377, May–June 2016.
14. R. W. A. A. De Doncker, D. M. Divan, and M. H. Kheraluwala, "A three-phase soft-switched high-power-density DC/DC converter for high-power applications," *IEEE Transactions on Industry Applications*, vol. 27, no. 1, pp. 63–73, Jan.–Feb. 1991.
15. J. Walter and R. W. De Doncker, "High-power galvanically isolated DC/DC converter topology for future automobiles," *IEEE Power Electronics Specialists Conference (PESC)*, 2003, June 15–19, 2003, Acapulco, Mexico, pp. 27–32.
16. H. Cha, J. Choi, W. Kim, and V. Blasko, "A new bi-directional three-phase interleaved isolated converter with active clamp," *IEEE Applied Power Electronics Conference (APEC)*, 2009, Feb. 15–19, 2009, Washington DC, USA, pp. 1766–1772.
17. G. Pledl, M. Tauer, and D. Buecherl, "Theory of operation, design procedure and simulation of a bidirectional LLC resonant converter for vehicular applications," *IEEE Vehicle Power and Propulsion Conference (VPPC)*, 2010, pp. 1–5.
18. A. Hillers, D. Christen, and J. Biela, "Design of a highly efficient bidirectional isolated LLC resonant converter," *IEEE International Power Electronics and Motion Control Conference (PEMC)*, 2012, Sept. 4–6, 2012, Novi Sad, Serbia, pp.13-1–13-8.
19. T. Jiang, X. Chen, J. Zhang, and Y. Wang, "Bidirectional LLC resonant converter for energy storage applications," *IEEE Applied Power Electronics Conference (APEC)*, 2013, Mar. 17–21, 2013, Long Beach, USA, pp. 1145–1151.
20. J. Jung, H. Kim, M. Ryu, and J. Baek, "Design methodology of bidirectional CLLC resonant converter for high-frequency isolation of DC distribution systems," *IEEE Transactions on Power Electronics*, vol. 28, no. 4, pp. 1741–1755, Apr. 2013.
21. W. Chen, P. Rong, and Z. Lu, "Snubberless bidirectional DC–DC converter with new CLLC resonant tank featuring minimized switching loss," *IEEE Transactions on Industrial Electronics*, vol. 57, no. 9, pp. 3075–3086, Sept. 2010.

22. W. L. Malan, D. M. Vilathgamuwa, and G. R. Walker, "Modeling and control of a resonant dual active bridge with a tuned CLLC network," *IEEE Transactions on Power Electronics*, vol. 31, no. 10, pp. 7297–7310, Oct. 2016.
23. P. He and A. Khaligh, "Comprehensive analyses and comparison of 1 kW isolated DC–DC converters for bidirectional EV charging systems," *IEEE Transactions on Transportation Electrification*, vol. 3, no. 1, pp. 147–156, Mar. 2017.
24. F. Lin, X. Zhang, and X. Li, "Design methodology for symmetric CLLC resonant DC transformer considering voltage conversion ratio, system stability, and efficiency," *IEEE Transactions on Power Electronics*, vol. 36, no. 9, pp. 10157–10170, Sept. 2021.
25. F. Krismer, S. Round, and J. W. Kolar, "Performance optimization of a high current dual active bridge with a wide operating voltage range," *IEEE Power Electronics Specialists Conference (PESC)*, 2006, June 18–22, 2006, Jeju, South Korea, pp. 1–7.
26. G. G. Oggier, G. O. García, and A. R. Oliva, "Switching control strategy to minimize dual active bridge converter losses," *IEEE Transactions on Power Electronics*, vol. 24, no. 7, pp. 1826–1838, July 2009.
27. Y. Wang, S. W. H. de Haan, and J. A. Ferreira, "Optimal operating ranges of three modulation methods in dual active bridge converters," *IEEE International Power Electronics and Motion Control Conference (IPEMCC)*, 2009, May 17–20, Wuhan, China, pp. 1397–1401.
28. F. Krismer, "Modeling and optimization of bidirectional dual active bridge DC–DC converter topologies," *Ph.D Dissertation, Power Electronics Systems Lab, ETH*, Zurich Switzerland, 2010.
29. A. K. Jain and R. Ayyanar, "PWM control of dual active bridge: Comprehensive analysis and experimental verification," *IEEE Transactions on Power Electronics*, vol. 26, no. 4, pp. 1215–1227, Apr. 2011.
30. X. Hu, K. J. Tseng, S. Yin, R. Tjandra, and T. Wang, "Elimination of reactive power in dual active bridge by hybrid of phase shift and PWM control," *IEEE Energy Conversion Congress and Exposition (ECCE)*, 2013, Sept. 15–19, 2013, Denver, USA, pp. 4985–4990.
31. F. Jauch and J. Biela, "Generalized modeling and optimization of a bidirectional dual active bridge DC–DC converter including frequency variation," *IEEJ Journal of Industrial Applications*, vol. 4, no. 5, pp. 593–601, May 2015.
32. Y. Cui, R. Hou, P. Malysz, and A. Emadi, "Improved combined modulation strategy for dual active bridge converter in electrified vehicles," *IEEE Transportation Electrification Conference and Expo (ITEC)*, 2017, June 22–24, 2017, Chicago, USA, pp. 101–107.
33. A. Tong, L. Hang, G. Li, X. Jiang, and S. Gao, "Modeling and analysis of a dual-active-bridge-isolated bidirectional DC/DC converter to minimize RMS current with whole operating range," *IEEE Transactions on Power Electronics*, vol. 33, no. 6, pp. 5302–5316, June 2018.
34. D. Goldmann, S. Schramm, and H. Herzog, "Triangular and trapezoidal modulation for dual active bridge DC–DC converters with fast switching semiconductors," *IEEE European Conference on Power Electronics and Applications (ECCE Europe)*, 2019, Sept. 3–5, 2019, Genova, Italy, pp. P.1–P.10.

CHAPTER 17

MISCELLANEOUS DC–DC CONVERTERS

The field of power electronics encompasses numerous practical applications with each application having its own specifications and requirements. The power converters that are presented in this chapter fall outside the various categories of converters that have been presented in the previous chapters of this book, but are, nonetheless, well-suited for certain particular applications that are popular today. This final chapter on miscellaneous DC–DC converters presents the following types of converters:

- Z-source converters for renewable energy systems
- Converters with very low voltage gains for ultra-low output voltage regulator modules
- T-type converters for improved light-load efficiency
- Multi-port converters for applications where multiple DC bus voltages are required.

For each of these converter types, their use in particular industrial applications is discussed and their basic fundamental operating principles are explained.

DC–DC Converter Topologies: Basic to Advanced, First Edition. Gerry Moschopoulos.

Published 2024 by John Wiley & Sons, Inc.

17.1 Z-SOURCE CONVERTERS

Renewable energy systems are implemented with sources that can vary over a very wide voltage range. For example, in solar-based energy systems where DC voltage is produced by photovoltaic (PV) panels, the voltage produced by these panels is very much dependent on the amount of sunlight that they are exposed to throughout the course of a day, a month, a year, etc. As a result of such varying voltage, renewable energy systems need to be implemented with converters that can operate over a wide input voltage range. In some architectures, the input voltage is low and a high gain converter is needed. In other architectures, a higher input voltage is used and a converter that can step voltage up or down is needed. Although buck-type and boost-type converters can be used, these converters can only step voltage down or up and thus do not have the wide range of voltage gain that is needed. Resonant converters with series-parallel or LLC tanks such as those described in Chapter 11 can be used, but converter size in renewable energy systems is less important than efficiency so that converters in renewable energy systems are not operated with very high switching frequency. As a result, the resonant tank components of these converters make their size larger than that of pulse-width modulation (PWM) converter alternatives and thus resonant converters are not common in renewable energy systems.

Converters based on topologies such as the buck-boost, Cuk, SEPIC, and Zeta converters discussed in Chapter 1 that can step voltage down or up are preferable and can be used for lower power applications in their single-switch form, but the higher power full-bridge equivalent versions of these converters are not popular for several reasons:

- They cannot be implemented with the four-switch full-bridge intact and must have passive elements such as inductors and capacitors in series with the switches. This prevents commercially available ready-made four-switch modules that are widely used as part of higher-power full-bridge converters from being used.
- Their duty cycle variation is narrower as the gain of these converters is dependent on the ratio of $D/(1-D)$ when they are operating with continuous current. The narrower duty cycle range is a key reason why a buck-boost converter is not used in applications where voltage needs to be only stepped down or only stepped up. Ideally, what is desired is a full-bridge converter that can act like a step-down converter that can step down voltage with a duty cycle range of $0 < D < 1$ and that can act like a step-up converter that can step up voltage with a duty cycle range of $0 < D < 1$, depending on operating conditions.

Z-source converters are converters that have passive element networks in their DC bus that allows them to act as step-down or step-up converter whenever voltage needs to be stepped down or up. They were first proposed by F. Z. Peng [1] as inverters for motor drives and automotive applications, but have since been adapted by power electronic researchers for use as DC–DC converters as well. A DC–DC Z-source converter is shown in Figure 17.1a. This converter is a full-bridge converter with a passive element network in its DC bus that is referred to as a Z-source in the literature and that consists of two inductors and two capacitors. A blocking diode is placed in series with the DC source to prevent current from flowing back into the source as the voltage across the DC bus can exceed that of the input source. The converter can be transformed into a bidirectional converter if the diode is replaced with a switch, but the discussion in this section will be limited to unidirectional power flow operation. The converter has been redrawn in Figure 17.1b to offer a different perspective of how the Z-source network is placed in the converter.

When the converter is operating as a standard ZVS-PWM full-bridge converter, the equivalent circuit becomes that which is shown in Figure 17.2a. A current can flow from the source, through the blocking diode

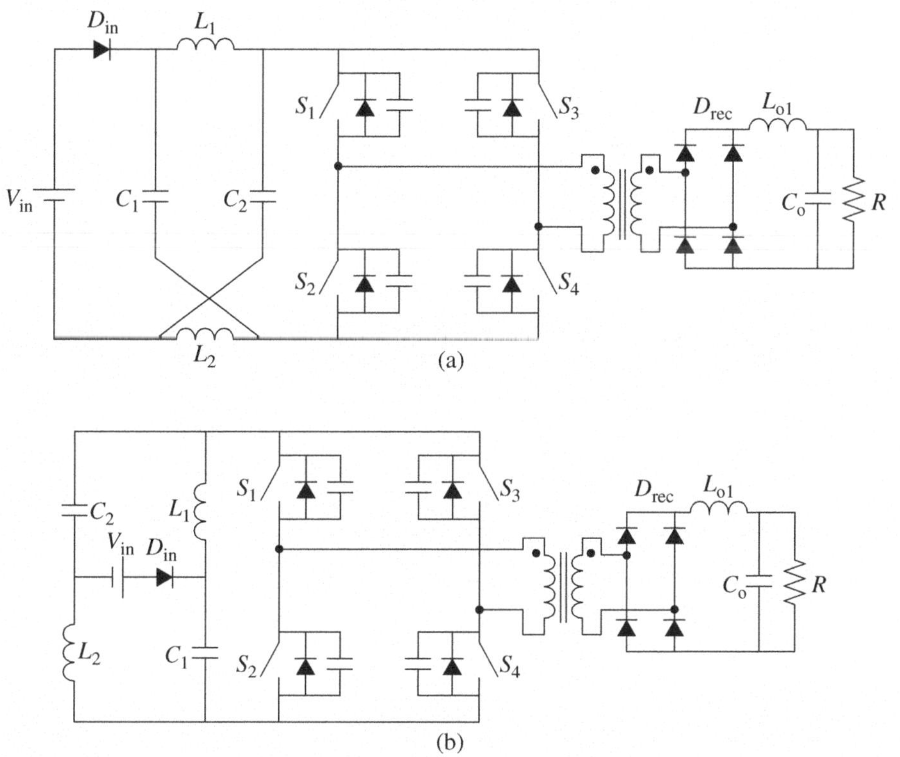

Figure 17.1 Z-source DC–DC full-bridge converter: (a) Standard circuit diagram. (b) Redrawn circuit diagram.

and L_1, through the converter, through L_2 and back to the source, with C_1 and C_2 acting as filter elements in conjunction with L_1 and L_2. If current in the full-bridge must flow through the two body diodes of two diagonal switches, which is the case whenever a transition from a freewheeling mode to a power-transfer mode is made, as explained in Section 7.2, then current can flow through capacitor C_2, through the DC source and the blocking diode, through C_1 and back into the full-bridge.

When the converter needs to step-up voltage, the positive and zero-voltage rails of the full-bridge can be periodically short-circuited by turning on all four full-bridge switches or by turning on the two switches of a full-bridge leg simultaneously. When the converter is in a shoot-through mode of operation, its equivalent circuit becomes that shown in Figure 17.2b. Capacitors C_1 and C_2 discharge through L_1 and L_2, respectively, and put energy in these inductors, just like energy is placed in the DC bus inductor of a current-fed converter like the ones discussed in Chapter 10. While this is happening, the blocking diode in series with the source becomes reverse-biased and the source is disengaged from the rest of the converter. The voltage gain of the converter depends on the duration of the shoot-through states relative to the switching cycle with longer shoot-through states resulting in higher voltage gain.

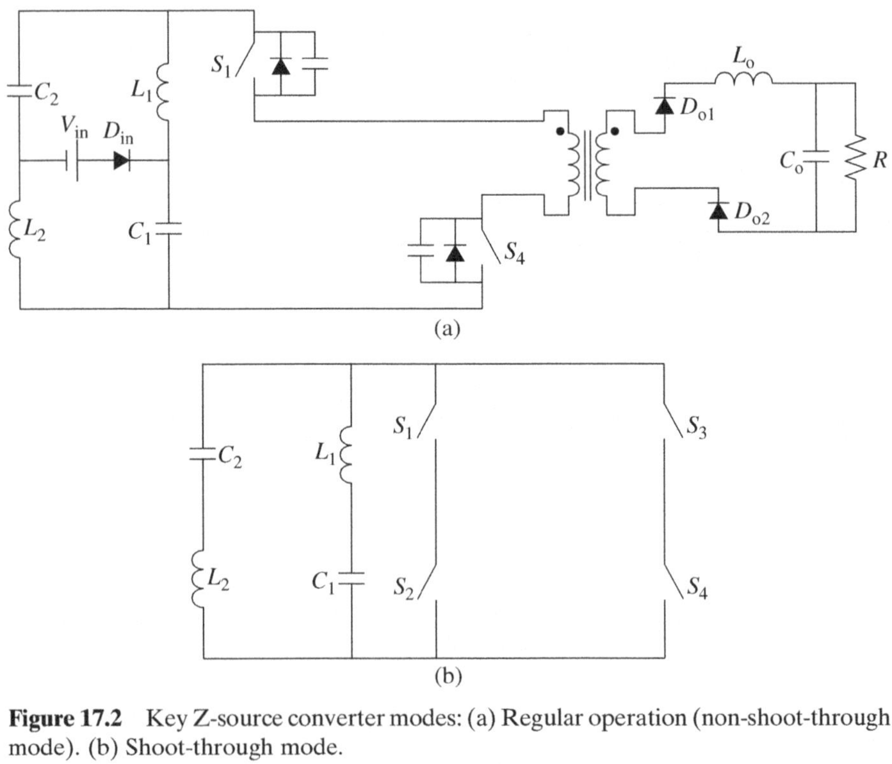

Figure 17.2 Key Z-source converter modes: (a) Regular operation (non-shoot-through mode). (b) Shoot-through mode.

Although the Z-source converter shown in Figure 17.1 can step voltage down or up, it has two significant drawbacks. The first is that the voltage across the Z-source capacitors can be high, which necessitates larger, higher voltage-rated capacitors. The second is that the input current is discontinuous whenever the converter is in a shoot-through mode of operation. The latter is a particular disadvantage for renewable energy systems as it makes the implementation of maximum power point tracking techniques (techniques that can extract the maximum amount of power that is available from PV panels) more difficult, increases peak current stresses, and reduces the lifetime of elements such as batteries. In order to overcome these drawbacks, quasi-Z source converters that were used as inverters were proposed by F. Z. Peng in several publications, including [2, 3].

A quasi Z-source DC–DC full-bridge converter that was proposed in [4] is shown in Figure 17.3. The base converter, which is the converter without the passive element network in its bus, does not have an output inductor and is similar to the zero-voltage-zero-current switching (ZVZCS) converter that was presented in Section 9.3.3. The converter works as follows: When the converter is operating with conventional phase-shift PWM and is in a power transfer mode, current flows from the input source to the four-switch full-bridge through L_1 and C_2 and flows from capacitor C_1 through L_2 as well. If current has to flow through diagonally opposed body diodes as is the case after a switch is turned off to transfer the converter out of a freewheeling mode of operation, current can flow through capacitors C_1 and C_2 and D_1. The current in L_1 flows through D_1 and into C_1 while the current in L_2 freewheels through C_2 and D_1. The converter can operate with shoot-through modes by turning on both switches of a converter leg or by turning on all the converter switches. Since there is an inductor in series with both C_1 and C_2, the converter can easily handle shoot-through modes as energy is simply placed in the two inductors

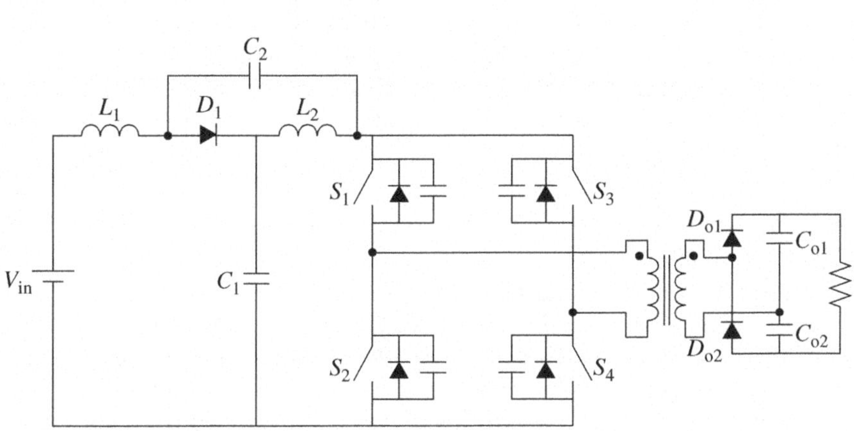

Figure 17.3 Quasi-Z-source full-bridge converter.

just like energy is placed in the boost inductor of a conventional single-switch PWM boost converter whenever the main boost switch is turned on.

Z-source and quasi-z-source converters represented the start of a trend in power electronics to experiment with passive impedance networks for DC–AC inverters and DC–DC full-bridge converters. These networks consist of various passive elements such as capacitors, regular inductors, and coupled inductors that are placed between the DC source (which can be either a voltage source or a current source) and the converter inverter or bridge. They are used to modify the characteristics of the original converter in some way, the most popular being to increase its gain. Within the past 10 years, it has been especially popular to use passive impedance networks to increase the gain of the original converter so that they can be used in applications where a high output voltage is required as an interface to a low voltage source such as a set of fuel cells or PV panels. Since these more sophisticated networks are mostly found in DC–AC inverters and rarely found in DC–DC converters, they are outside the scope of the book.

17.2 LOW VOLTAGE GAIN CONVERTERS FOR VOLTAGE REGULATOR MODULES

The term "voltage regulator module" (VRM) in power electronics is used to define a power converter that can step down voltage considerably, to produce an output with very low voltage and high current. VRMs are typically used to convert power from a 3.3 to 48 V DC bus to an output voltage of 1–1.3 V and an output current of at least 30 A and greater (even hundreds of amps) to power motherboards with electrical components such as microprocessors. They are required to be compact and must have very tight output voltage regulation, given the wide variance in load that they may encounter.

The input voltage of these converters is taken off a DC bus whose voltage depends on the system architecture of the application. For example, data centers typically have a 48 V DC bus so that VRMs can be fed off point-of-load voltages such as 5 V or 12 V that have been generated off the 48 V DC bus, or they can be fed directly off the 48 V DC bus. Typical DC bus voltages are 3.3–5 V, 12 V or 48 V and thus different approaches are required for each of these input voltages:

- If the DC bus voltage is 3.3–5 V, then some sort of non-isolated buck converter can be used.
- If the DC bus voltage is 12 V, then a modified non-isolated buck converter is used. If a standard buck converter is used, then the duty cycle of the main buck switch will be very low and thus the amount of time allowed for power to be transferred to the output will be low as well, which is an inefficient use of this switch and places considerable stress on it. What is required is some means by which the duty cycle of this

switch can be extended so that it operates with a greater duty cycle and is thus subjected to less stress. Although narrow duty cycles can be avoided using isolated converters, these converters are not without their disadvantages such as size and the spacing of components, which introduce parasitic elements that may be problems if high switching frequencies are used.

- If the DC bus voltage is 48 V, then some sort of isolated half-bridge converter is typically used. This converter can be a PWM converter or an LLC resonant converter, with the latter gaining in popularity in recent years because of its wide voltage gain [5]. Since these converters use standard topologies such as the ones that have been covered in this book and the implementation challenges are mainly concerned with the design of the magnetics, control, or printed circuit board layout to minimize parasitics as the converters are operated with very high current and high switching frequencies, VRMs with 48 V DC inputs are not discussed here.

In addition to operating VRMs with high current and high switching frequencies, another challenge that must be faced when implementing VRMs is that they must have a very fast transient response in order to comply with the needs of their loads, which are electronics that have very tight regulation requirements. In VRMs, the output capacitor is typically implemented by using a bank of numerous small, low-voltage aluminum-polymer solid capacitors with low values of equivalent series resistance (ESR) values that are paralleled to reduce the overall output capacitor ESR. Reducing ESR in this manner is important, given the large amount of output current that flows out of a low output voltage VRM.

For 3.3–5 V VRMs, an approach such as the one shown in Figure 17.4, which was suggested in papers such as [6, 7] can be used. Here two paralleled

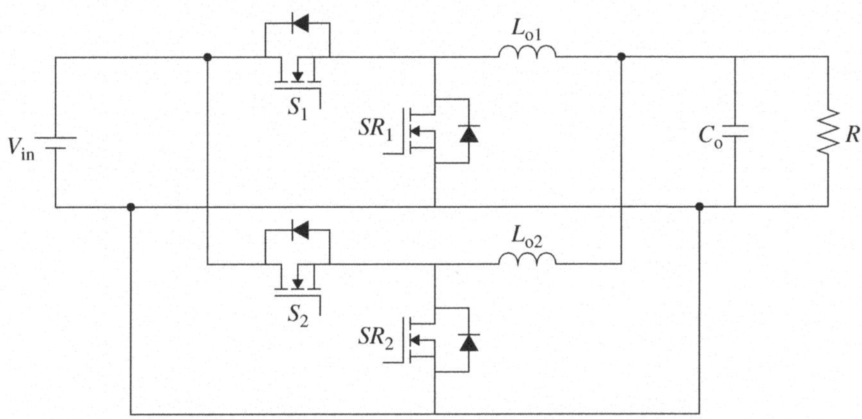

Figure 17.4 VRM with paralleled buck converter modules.

buck converter modules are used to split the current stresses in the switches and to reduce the overall output inductor size by interleaving. More paralleled modules can be used to further reduce switch current stress and output inductor size further if desired. Given that very high current flows in VRMs, the buck converter freewheeling diode is replaced with a synchronous rectifier to reduce conduction losses, as described in Section 5.1. Operating a converter with N paralleled buck converter modules with interleaving means that the gating signals of the switches of each module are shifted by 360°/N relative to those of the other modules, with the phase difference being 180° for a two-module converter, 120° for a three-module converter, 90° for a four-module converter, etc. This allows for ripple reduction at the output as the ripple of each module is reduced when added to those of the other modules. With less ripple, the size of the inductance in each module can be made small and thus the size of the overall output inductance can be less than what it would be if just a single module were to be used. An important outcome of reducing the output inductor size is that the transient response of the overall converter can be reduced, thus making it easier to comply with transient load requirements.

With higher input voltages such as 12 V, the standard buck converter topology must be modified in order to allow the converter to operate with duty cycles that are not extreme. A fundamental low voltage gain topology that has this property is the tapped inductor buck converter, of which there are numerous variations that have been proposed in the literature. An example of a tapped inductor buck converter is the one that was proposed in [8] and shown in Figure 17.5. This converter is a buck converter with two inductors, L_{in} and L_o, that are coupled together on the same core. The fundamental principle behind the converter is that the voltage across output inductor L_o, which is generated as a result of voltage appearing across L_{in} when S_1 is on, acts as a counter-voltage to reduce the amount of energy that is fed to the output. This forces the converter to operate with a greater duty cycle than it would if it was not implemented with a coupled inductor so that any problems that are caused

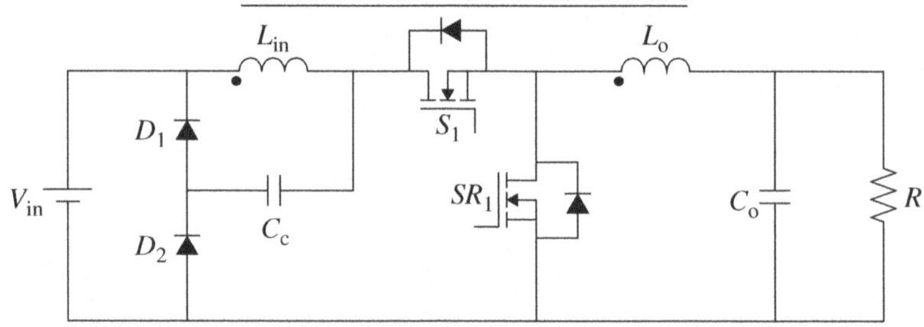

Figure 17.5 Tapped inductor buck converter with passive clamp for VRM.

by operating with an extremely narrow duty cycle in a regular duty cycle are avoided. Figure 17.5 shows the converter implemented with a clamping circuit that consists of D_1, D_2, and C_c. Such clamping circuits are a must when implementing tapped inductor converters to avoid having overvoltage spikes appear across the main buck switch.

More advanced VRMs that offer better performance than buck converters with tapped inductors have been proposed. These converters are often implemented with sophisticated coupled inductor arrangements that serve to extend the duty cycle of the converter so that the main gating pulses are not extremely narrow. Numerous converters with unique operation modes have been proposed and only a few examples are presented here. The VRM that was proposed in [9] and shown in Figure 17.6a consists of a

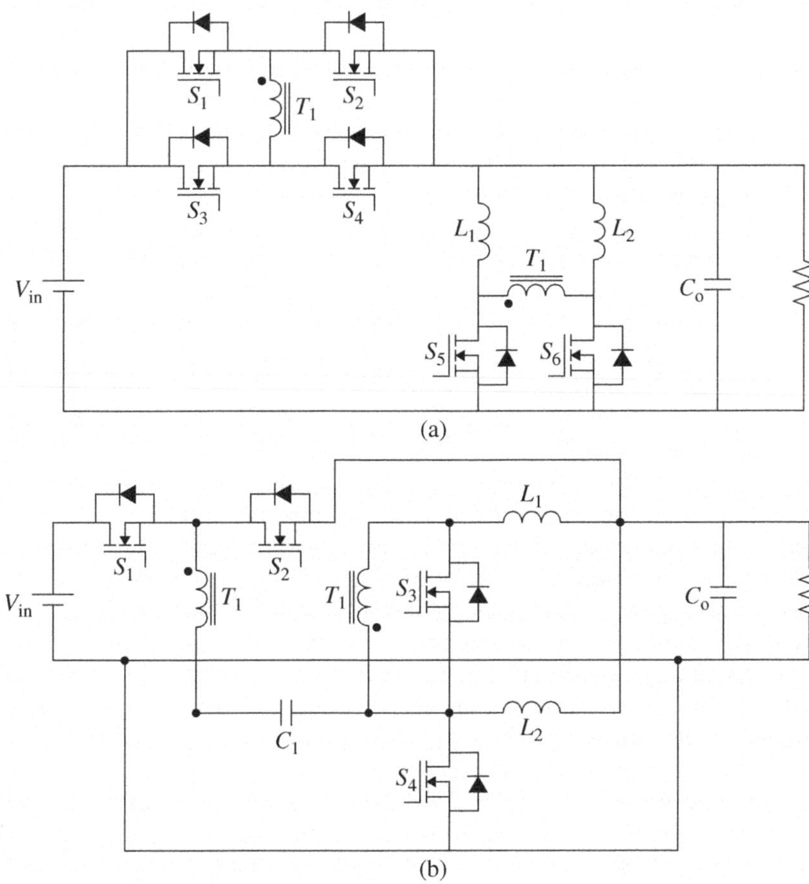

Figure 17.6 Coupled inductor VRMs with extreme voltage gains: (a) Full-bridge VRM. (b) Asymmetrical half-bridge VRM.

full-bridge converter with switches S_1–S_4 connected to a coupled inductor that is the primary winding of a transformer. Another coupled inductor that can be considered to be the secondary winding of a transformer is connected after the full-bridge in a current-doubler type rectifier arrangement with switches S_5 and S_6 that act as synchronous rectifiers (SRs), two output inductors, L_{o1} and L_{o2}, and an output capacitor C_o. Switches S_1–S_4 are operated with phase-shift PWM just like a conventional ZVS-PWM full-bridge converter with energy-transfer and freewheeling modes and switches S_5 and S_6, along with L_1 and L_2, conduct current as if they were placed at the secondary of a ZVS-PWM full-bridge converter.

The polarities of the two windings should be such that they are opposed to each other. Switches S_1–S_4 impress voltage across the primary winding whenever diagonally opposed switches are on. The secondary winding is coupled to the primary winding and the primary/secondary turns ratio of the transformer is $N : 1$ with $N > 1$. Whenever power is to be transferred to the output, current is made to flow through the primary winding and the secondary winding. Given that the polarity of the voltage across the secondary winding is opposed to that of the voltage across the primary winding, the voltage that appears at the output filter is reduced. It is this reduction in voltage that forces switch S_1–S_4 to operate with higher duty cycle in order to make up for it. Extending the duty cycle of the switches in this manner and thus avoiding narrow duty cycle operation keeps the converter switches from being exposed to the high peak current stresses that they would otherwise be exposed to and make the converter operate more efficiently. Other advantages that the converter has over tapped inductor converters such as the one shown in Figure 17.5 are lower peak voltage switch stresses due to the arrangement of the switches in the converter and the ZVS operation of switches S_1–S_4 as their operation is the same as the four switches in a ZVS-PWM FB converter.

Figure 17.6b shows a simplified variation of this converter that was proposed in [10], with the main difference being that the full-bridge section is replaced with a half-bridge. Switches S_1 and S_2 form a half-bridge along with capacitor C_1 and a winding that is the primary winding of a transformer. The half-bridge arrangement becomes clearer if another capacitor, C_2 is added across S_1, connecting the DC source to this C_2, but such a capacitor is not necessary as it would be redundant. The rest of the converter is a rectifier arrangement that is almost the same as that of the converter shown in Figure 17.6a, but rotated in the figure. Switches S_1 and S_2 are operated in an asymmetrical manner with S_1 having a duty cycle D and S_2 having a duty cycle of $(1 - D)$. S_1 and S_2 can operate with ZVS – if S_1 is turned off, energy from the primary transformer winding can be used to discharge the output capacitance of S_2 and if S_2 is turned off, then the current flowing in the primary transformer winding is made to flow through the body diode of S_1 before it is turned on. The converter shown in Figure 17.6b has the advantages of its more sophisticated counterpart and can be used in lower power applications.

17.3 T-TYPE CONVERTERS

Light-load converter efficiency has gained importance in recent years due to the proliferation of power converters in society that has been caused by an ever-increasing reliance on electrical power. Although most power converters are designed to have peak efficiency when they are operating under heavy-load conditions, the reality is that many power converters operate with light loads most of the time and are forced to operate with heavy loads occasionally. Research has been conducted into ways of improving light-load converter efficiency to try to address this issue, especially for the conventional zero-voltage switching (ZVS), pulse width modulated (PWM) full-bridge (FB) DC–DC converter, which is widely used in industry in applications requiring DC–DC power conversion for loads greater than 500 W and is shown in Figure 17.7. As explained in Chapter 7, the switches of the ZVS-PWM-FB converter can lose their ZVS operating capability when the converter operates with light loads, because there is not enough energy to discharge the output capacitances of the switches devices before they are turned on. This is especially true of the switches that are turned on to exit a freewheeling mode of operation and enter an energy-transfer mode because the only energy that is available to discharge the switch output capacitances is that stored in the primary-side leakage inductance of the converter's transformer. This energy is insufficient when the converter is operating with light loads.

A second issue with the ZVS-PWM-FB converter is that current just circulates in the converter without any energy transferred to the secondary when the converter is in a freewheeling mode of operation. This is because the voltage across the transformer primary is zero when either two top converter switches or two bottom converter switches are on so that the voltage across the transformer secondary is also zero. The circulating current flows through two converter devices and creates conduction losses. Converter switches are subjected to increased RMS current stresses as well for no beneficial reason.

One way that has been proposed to improve light load efficiency in ZVS-PWM-FB converter is to operate the converter using a hybrid-PWM method,

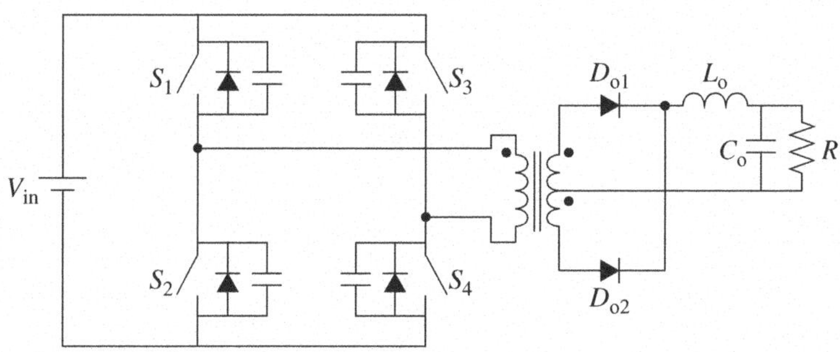

Figure 17.7 Conventional ZVS-PWM-FB DC–DC converter.

whereby the converter is made to operate with standard phase-shift PWM for heavy loads, but with some other PWM method for light loads. One such method is hybrid-PWM, just like what was described in Section 7.5. Unlike phase-shift PWM, the switch gating signals vary so that they are <50% of the switching cycle and diagonally opposed pairs of switches are turned off and on at the same time. This introduces a state where there is no voltage across the transformer and no current freewheeling in the primary as it is quickly extinguished. When it is time to turn on diagonally opposed switches, the voltage across each switch is half that of the DC bus ($V_{in}/2$) so that the switching loss of each of the two switches is reduced by 75%. It is the combination of a lack of freewheeling current plus the reduction (but not full elimination of switching losses) that helps the converter operate with a better light-load efficiency than if PS-PWM was used.

Another way to improve light-load efficiency is to use a T-type DC–DC converter such as the one shown in Figure 17.8, as was proposed in [11], instead of a ZVS-PWM FB converter. T-type converters get their name from the fact that they are synthesized by connected two back-to-back switches between the mid-point of a leg of switches and the mid-point of two capacitors in series so that this switch structure resembles the letter T rotated by 90°. The first T-type converter was the inverter proposed in [12] and T-type converters for DC–DC conversion were introduced in [13]. The DC–DC ZVS T-type converter shown in Figure 17.8 has four switches, S_1–S_4, a split-capacitor DC bus, C_1–C_2, a transformer, two secondary rectifying diodes, D_{o1}–D_{o2}, an output inductor, L_o, and an output filter capacitor, C_o. S_1 and S_2 are connected together as one converter leg, just as in a conventional ZVS-PWM-FB converter. The other two switches, S_3 and S_4, are connected in series with each other with one end of the combined series combination connected to one end of the transformer and its other end to the mid-point of the two bus capacitors in a T-type configuration. The secondary side of the converter is identical to that of a conventional ZVS-PWM-FB converter.

The converter operates as follows. Voltage is impressed across the transformer primary whenever S_1 or S_2 is on. In this case, one end of the

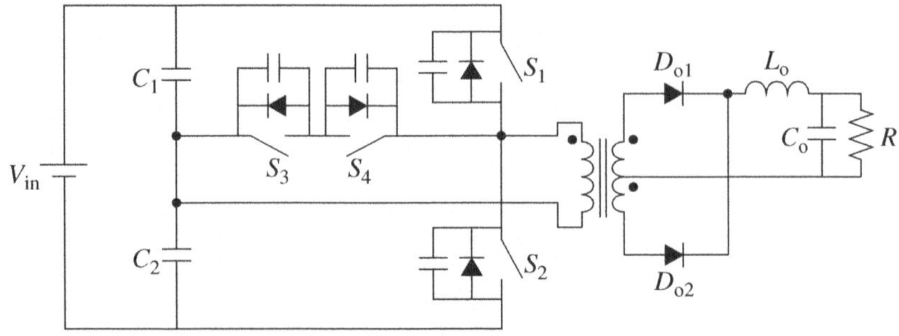

Figure 17.8 DC–DC T-type converter.

transformer primary has a voltage that is either the DC bus voltage or zero voltage and the other end of the primary has a voltage that is half the DC voltage. The voltage at this end of the primary is fixed and is always half the DC bus voltage, $V_{in}/2$. As a result, a voltage of $V_{in}/2$ is impressed across the transformer primary, either with positive polarity if S_1 is on or with negative polarity if S_2 is on. This primary voltage is stepped down by the transformer, rectified by the secondary diodes, and filtered by the output LC filter. The converter is in an energy-transfer mode of operation whenever S_1 or S_2 is on. The converter is in a freewheeling mode of operation when both S_1 and S_2 are off, and either S_3 or S_4 is on, depending on the direction of the current that circulates in the converter. In this mode, the current flows through one switch and through the body diode of the other switch, and the voltage impressed across the transformer primary is zero.

The DC–DC T-type ZVS-PWM converter has the following features:

- All its switches are exposed to a voltage of $V_{in}/2$ before they are turned on instead of V_{in}, which is not the case for the switches in the conventional ZVS-PWM-FB converter. Switches S_1 and S_2 have a voltage of $V_{in}/2$ across them before either one is turned on because the end of the transformer primary winding that is connected to the mid-point of their series connection is $V_{in}/2$, due to S_3 or S_4 being on during a freewheeling mode of operation. This means that S_1 has a voltage of $V_{in} - V_{in}/2 = V_{in}/2$ across it before it is turned on and S_2 has a voltage of $V_{in}/2 - 0 = V_{in}/2$ before it is turned on. Switches S_3 and S_4 have a voltage of $V_{in}/2$ before they are turned on because these switches are connected to the midpoint $V_{in}/2$ voltage of the DC bus capacitors. As a result, especially if the load is light and output switch capacitances have not been discharged, these switches will turn on with one-fourth of the CV^2 losses of the switches in a conventional ZVS-PWM-FB converter, which results in better light-load efficiency.
- Switches S_1 and S_2 conduct current only when the converter is in an energy-transfer mode, and switches S_3 and S_4 conduct current only when the converter is in a freewheeling mode of operation. This is in contrast to the switches in a conventional ZVS-PWM-FB converter where they all must conduct current during an energy-transfer mode and a freewheeling mode and, thus, less expensive, lower current-rated devices can be used in the converter.
- Switches S_3 and S_4 are never exposed to more than half the DC bus voltage due to their connection to the midpoint of the DC bus capacitor; therefore, cheaper, lower voltage-rated devices with lower values of on-state resistance can be used for these switches. This helps reduce conduction losses caused by circulating current in a freewheeling mode of operation. Switches S_1 and S_2, however, can be exposed to the full DC bus voltage V_{in} whenever one of these switches is on and the other is off.

- There is inherent volt–second balancing across the primary of the transformer as one end of the transformer primary winding is connected to the midpoint of the DC bus capacitors. If for some reason a net amount of voltage seconds is applied to the transformer during a switching cycle, there will be a slight adjustment in the midpoint voltage to compensate for this so that there is no more volt–second imbalance. This keeps the transformer from saturating without the need for any additional means, as is the case with the conventional ZVS-PWM-FB converter.

The T-type converter has better light-load efficiency than the conventional ZVS-PWM-FB converter for the following reasons:

- Its switches are exposed to only half the bus voltage before they are turned on and thus can turn on with fewer losses at light loads.
- As the converter is based on the standard half-bridge PWM topology, the current flowing in its primary is double that of what flows through the primary of a ZVS-PWM-FB converter, given that its transformer primary voltage is halved. With more primary current, there is more transformer leakage inductance energy available to discharge switch capacitances than what exists in the ZVS-PWM FB converter at light loads.
- Since switches S_3 and S_4 must block only half the DC bus voltage, they can be implemented with less expensive devices that have lower current and voltage ratings and less on-state resistance. If the load is very light, then the conduction losses of these switches, which are related to the product of the square of the RMS (root-mean-square) current flowing through a switch and the on-state resistance, may be lower than those found in switches in a conventional ZVS-PWM full-bridge converter. This is especially true when the converter is operating with very light loads when its duty cycle is limited and it spends most of its time in a freewheeling mode of operation during a switching cycle.

The ZVS-PWM-FB converter, however, has a better heavy-load efficiency than the T-type converter since the T-type converter does have more current flowing in its primary. What is advantageous for light-load operation becomes disadvantageous as the load is increased as the additional primary current results in the converter having more conduction losses and turn-off losses than the ZVS-PWM-FB converter under heavy-load operating conditions. In other words, there is a boundary load where the T-type converter is more efficient for loads that are lighter than this load and the ZVS-PWM-FB is more efficient for loads that are heavier than this load. This boundary load will depend on the specifications of the converter.

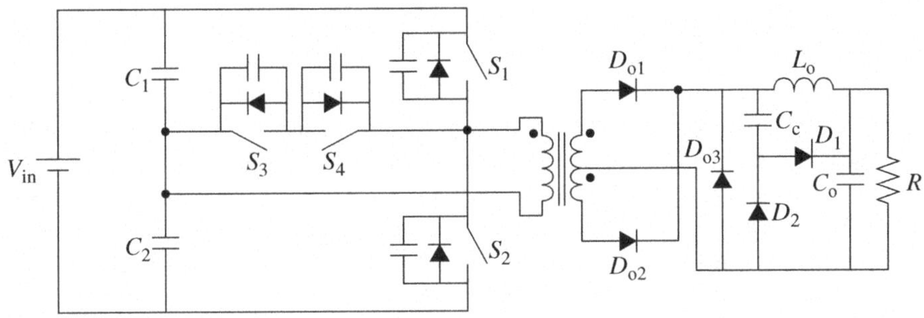

Figure 17.9 ZVZCS DC–DC T-type converter with passive secondary-side auxiliary circuit.

The light-load efficiency of the T-type DC–DC converter can be further improved if the freewheeling current is eliminated by using some sort of ZVZCS method, as discussed in Chapter 9. Such an approach was proposed in [14] and is shown in Figure 17.9. Figure 17.9 shows a T-type DC–DC converter with the secondary auxiliary circuit that was proposed in [15] and discussed in Section 9.2. As was explained in Section 9.2, the passive auxiliary circuit impresses a voltage across the transformer secondary when the converter is in the process of transferring out of an energy-transfer mode and into a freewheeling mode so that this voltage, which is reflected to the primary, acts as a counter-voltage that extinguishes the primary current. Although the elimination of freewheeling current means that the lagging switches lose their ability to turn on with ZVS, the fact that switches in a ZVS full-bridge-type converter lose their ZVS ability under light-load conditions and the fact that the lagging switches can turn on with half the DC bus voltage across them and not the full voltage means that the penalty for losing ZVS is not severe when the converter is operating under light-load conditions. Any switching that arises in this case is offset by the reduction in current-related losses.

17.4 MULTI-PORT CONVERTERS

With the exception of the multiple output converters discussed in Section 5.3, the converters that have been presented in this book have been two-port converters with a single input and a single output. There are, however, a number of applications where converters with multiple ports are required, including the following:

- Applications where a converter can be supplied by multiple input sources. A common example is a power conversion architecture where a DC voltage obtained from the output of a grid-connected AC–DC converter and a set of solar panels can be used to supply a DC load. The solar panels can be used to supply power to the load when they are exposed to sufficient sunlight and the grid can be used to supply power when there is insufficient sunlight.

- Uninterruptible power supplies (UPS), which are power supplies with battery back-up that supply power to critical loads when the main DC source (e.g. output of AC–DC converter, set of solar panels, fuel cells, etc.) is unavailable for some reason. In a UPS application, the main DC source is used to supply power to a critical load or loads and to charge a battery if needed. If the DC source cannot supply power for some reason, then power is supplied to the load(s) by the battery.
- Electric and hybrid electric vehicles with power train architectures that have multiple voltage buses of various DC voltage levels. One such architecture has a high voltage DC bus of 300–400 V that supplies power to inverter motor drives that are needed to rotate the tires of a vehicle and make it move, a lower voltage DC bus, typically 42–48 V for applications such as electrical window lifts, door locks, and side-view mirrors, and an even lower DC voltage bus, typically 12–14 V, for lower power electronics such as windshield wipers and consumer electronics.

If applications such as these were implemented with conventional two-port converters, either unidirectional or bidirectional, then the result would be expensive and bulky as several DC–DC converters would be required. Multi-port power converters allow the functions of multiple converters to be integrated into a single converter, thus reducing size, cost, and weight. Non-isolated and isolated multi-port converters are discussed in this section.

17.4.1 Non-Isolated Multi-Input Converters

The most basic multi-port converters are multi-input converters that consist of converters where the multiple input sources are either paralleled with each other or stacked one on top of the other. An example of a paralleled multi-input converter was proposed in [16] and is shown in Figure 17.10. For this type of converter, as well as for other paralleled multiple input

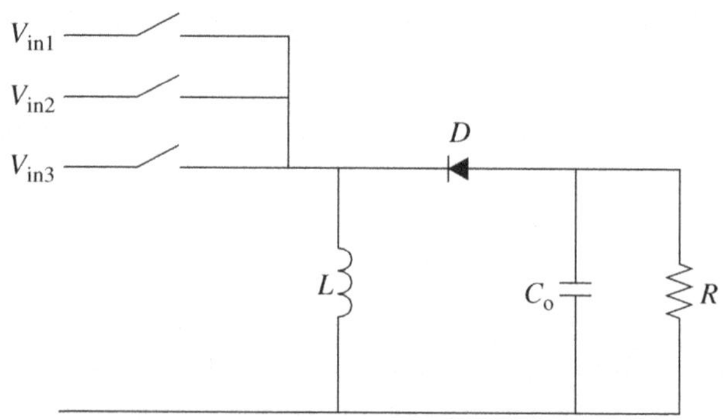

Figure 17.10 Multi-input buck-boost converter.

converters that have been proposed [17, 18], multiple DC sources are connected to a common DC–DC converter, thus eliminating the need to have individual power converters for each source. A multiple input converter such as the one shown in Figure 17.10 can be used in a variety of ways in a variety of system architectures. The DC sources can be some combination of DC voltage obtained from a grid-connected AC–DC converter, solar panels, fuel cells, batteries, etc.

Regardless of how the converter is used, only one switch is allowed to be on at any given time to avoid circulating current between input sources. The converter shown in Figure 17.10 is implemented with a buck-boost converter, but other converters can be used instead as well other types of input cells where multiple DC voltages can be tapped from an individual DC source. The concept can be extended to bidirectional converters simply by replacing unidirectional switches with bidirectional switches, as was proposed in [19]. These bidirectional switches can be simply MOSFET devices with anti-parallel body diodes that allow current to flow from source to drain, or back-to-back switches in series where current can flow through the switch of one device and the anti-parallel diode of another device when one of the switching devices is turned on.

Examples of stacked multiple input converters are shown in Figure 17.11. Figure 17.11a shows the simplest possible stacked multiple input converter,

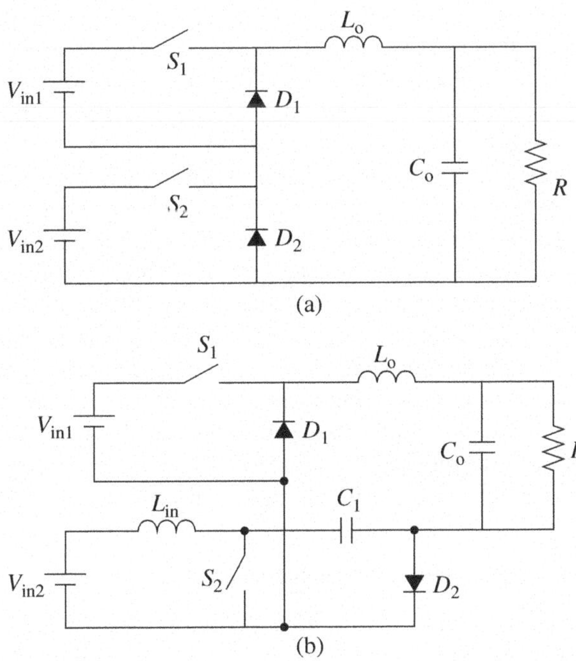

Figure 17.11 Multi-input converters with stacked inputs: (a) Dual buck converter. (b) Buck/Cuk converter.

which is a dual buck converter topology that was proposed in [20]. Unlike paralleled input topologies like the one shown in Figure 17.10, it is possible for both switches to be on at the same time. If this is the case, then current can flow through V_{in2}, S_2, V_{in1} and S_1 without any problem. If only switch is on, then current flows through that switch, an input voltage source, and a freewheeling diode. Current flows through both freewheeling diodes when none of the switches are on.

More sophisticated stacked multiple input converters can be synthesized by combining two different converters. An example of such a converter was proposed in [20] and shown in Figure 17.11b. The top converter is a Cuk converter, the bottom converter is a buck converter. As with the dual buck converter shown in Figure 17.11a, it is possible for both switches to be on simultaneously, just one switch to be on, or no switch to be on at any given time. For the buck converter part of the circuit, current flows through the V_{in1}–S_1 branch and D_2 or through D_1 and D_2 at any given time, depending on whether the buck converter switch is on or not, if there is output inductor current. No current flows in the buck converter if S_1 is off and there is no output inductor current. The flow of current through the Cuk converter follows a more complicated path. If S_2 is on, current flows through V_{in2} and L_{in}, but also flows through the loop consisting of C_1, L_o, the output capacitor and load, and D_1. Note that the direction of output inductor current in Figure 17.11b is left to right. If S_2 is off, then current flows through V_{in2}, L_{in}, C_1, and D_2 and also flows through D_1, D_2, L_o and the output capacitor and load branch. Current does not flow in the Cuk converter if the input inductor current and the output current are both zero. It should be noted that different converters cannot just be combined at random as it must be ensured that the operation of one converter does not interfere with that of the other. Generic rules for the synthesis of stacked multiple input converters are given in [20].

The decision of whether to use a paralleled multiple input converter or a stacked one will depend on the particular application and on the nature of the sources. If all the sources produce similar DC voltages and only one needs to be providing power at any given time, then paralleled multiple input converters can be used. If the sources produce vastly different DC voltages, or produce low-level DC voltages, then stacked multiple input converters are more attractive. It should be noted that it is very difficult to stack more than two converters so that paralleled multiple input converters become the preferred option if there are more than two sources. It is, however, possible for the two types of multiple input converters to be combined so that two paralleled input converters, each with multiple sources, are stacked together.

17.4.2 Isolated Multi-Port Converters

The multiple port converters that have been discussed thus far do not have transformer isolation. Numerous multi-port converters that have each port isolated from the other ports have been proposed. Some of the most popular converters of this type are so-called tri-port converters with three ports. These ports can be one for the main DC source, one for a battery, and one for a load as in the case of an UPS, or one port for high voltage bus, one for a lower DC voltage, and one for a different DC voltage, as is the case for some electric vehicle architectures. Among the first isolated triport converters were the converters that were proposed in [21, 22] in UPS systems and shown in Figure 17.12. The converter shown in Figure 17.12a is a low power UPS that can be used in telecommunication applications where critical loads must be supplied with power without interruption. The converter is essentially a flyback converter with a battery charger. When the battery needs to be charged,

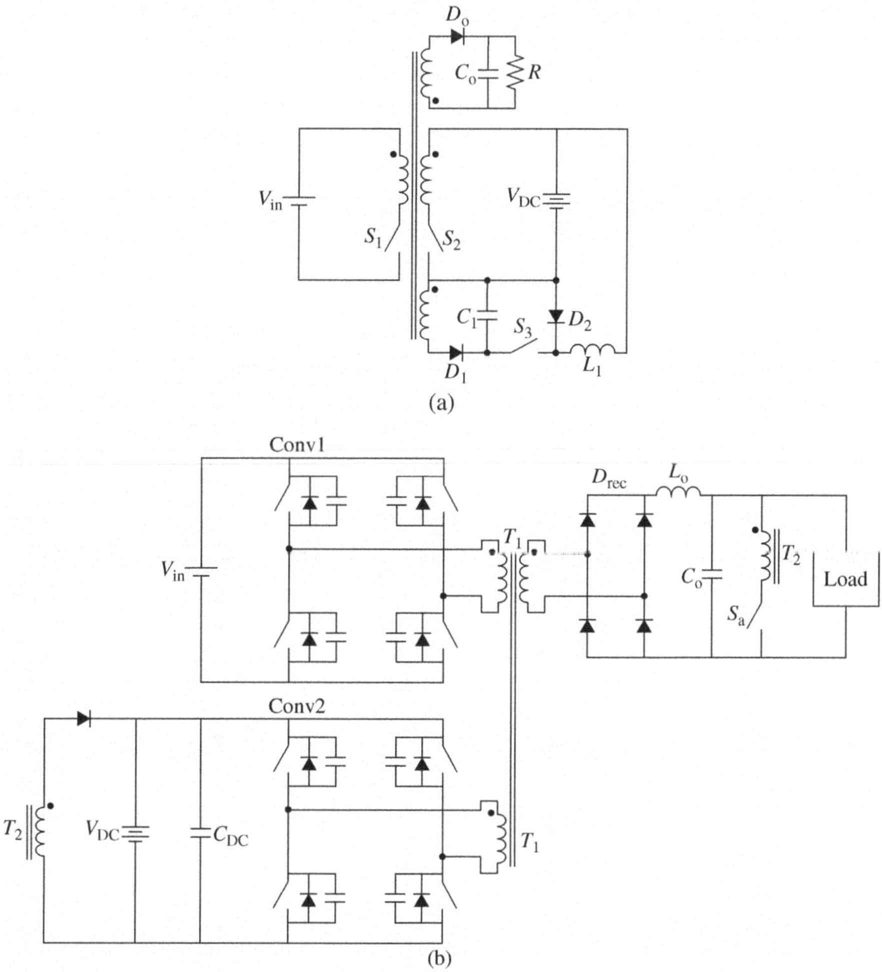

Figure 17.12 Triport converters for UPS with three isolated ports: (a) Flyback UPS converter. (b) Full-bridge UPS converter.

switch S_3 is on and power is supplied to the battery by capacitor C_1. Once the battery is charged, S_3 is opened and C_1 continues to be charged until secondary diode D_1 becomes reverse biased. D_2 is a freewheeling diode that is placed in the converter to ensure a path for L_1 current to flow through. If something should happen to the primary DC source, then power can be provided to the critical load by operating S_2 as a flyback converter with the battery being the source. The most common reason for the DC source being unavailable is that it is the output of an AC–DC converter and some interruption has occurred in the AC source.

The converter shown in Figure 17.12b is similar to the one shown in Figure 17.12a except that it has been implemented with full-bridge converters. This converter, which can also be used as a UPS, has one port for the main DC source, one port for a battery and one port for a critical load. In this converter, unlike the one shown in Figure 17.12a, the battery charging mechanism is isolated from the battery through a flyback converter that has its primary at the critical load port. This allows for a wider range of battery voltage to be used and avoids the use of a series switch such as S_3 in Figure 17.12a that can cause conduction losses.

A popular type of multi-port converter that has three ports with a reduced number of components is the triport converter that was proposed in [23, 24]. His converter has three ports – one for a high DC voltage bus (Port 1), one for a lower DC voltage bus (Port 2), and one for a critical load (Port 3). A symmetrical version with two inductors connected to Port 2 was presented in [23] and an asymmetrical version with a single inductor connected to Port 2 was presented in [24]. The converter presented in [24] is shown in Figure 17.13. The voltage for Port 2 is obtained by considering the two primary-side converter legs as DC–DC converters that are operated in an asymmetrical manner. Unlike the switches in a conventional ZVS-PWM-FB converter, the switches in this converter do not operate with a fixed pulse width that is approximately 50% of the converter's switching cycle but is

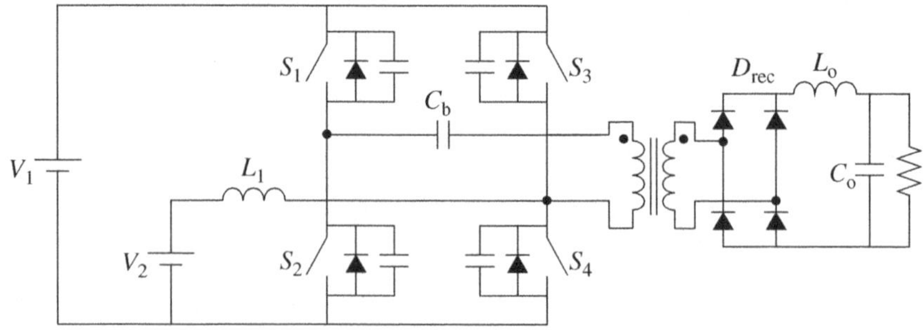

Figure 17.13 Triport converter with a non-isolated port.

allowed to change in a complimentary manner. If one switch of a leg has a duty cycle of D, then the duty cycle of the other switch in the leg is $(1 - D)$. In addition to this PWM control, the gating signals of the switches of one leg are phase-shifted relative to those of the other leg so that, in the end, the converter is controlled by a combination of PWM and phase-shift control. The direction of flow of power from Port 1 to Port 2 or vice versa is dependent on the duty cycle of the switches – if both top switches, S_1 and S_3 have a duty cycle that is $D > 0.5$, then energy is transferred from Port 1 to Port 2; if they have a duty cycle that is $D < 0.5$, then energy is transferred from Port 2 to Port 1. The duty cycle of the switches also dictates which of the two ports supplies power to the load. When Port 1 supplies power to the load, the amount of phase-shift between the gating signals of the two legs is greater than what it is when it is Port 2 that supplies power to the load as a lower effective duty cycle at the secondary is required.

The following should be noted about the converter:

- The voltage of Port 2 should be about 50% of that of Port 1 so that the duty cycle of the gating signals of all the switches approaches $D = 0.5$. This increases the load range over which phase-shift PWM can be performed to regulate the output voltage of the critical load.
- A DC blocking capacitor must be placed in series with the transformer. Without this capacitor, a DC component would be placed across the transformer due to the asymmetrical nature of the gating signals and the transformer would saturate.
- In the symmetrical version of the converter presented in [23], the converter is operated with conventional phase-shift PWM. Since switches S_1 and S_2 have gating signals that have 50% duty cycle and are 180° phase-shifted relative to each other, an additional inductor can be connected between Port 2 and the source of S_1 so that the current flowing through Port 2 is interleaved. Doing this reduces the ripple current, but does not allow for flexibility in the Port 2 voltage as it must be approximately 50% of the Port 1 voltage.

More sophisticated triport converters can be used if the third port is a voltage bus instead of a critical load. Such converters can be found in electric vehicles in power architectures with three DC buses as described above. Making the third port into a voltage bus generally requires that the diode bridge that interfaces the load with the rest of the converter is replaced with another converter, as was proposed in [25] and shown in Figure 17.14. For very high power applications such as for DC grids, three-phase bridges can be used instead of conventional bridges, as proposed in [26]. The replacement of a diode bridge with a third active bridge makes the control of triple active bridge converters a very

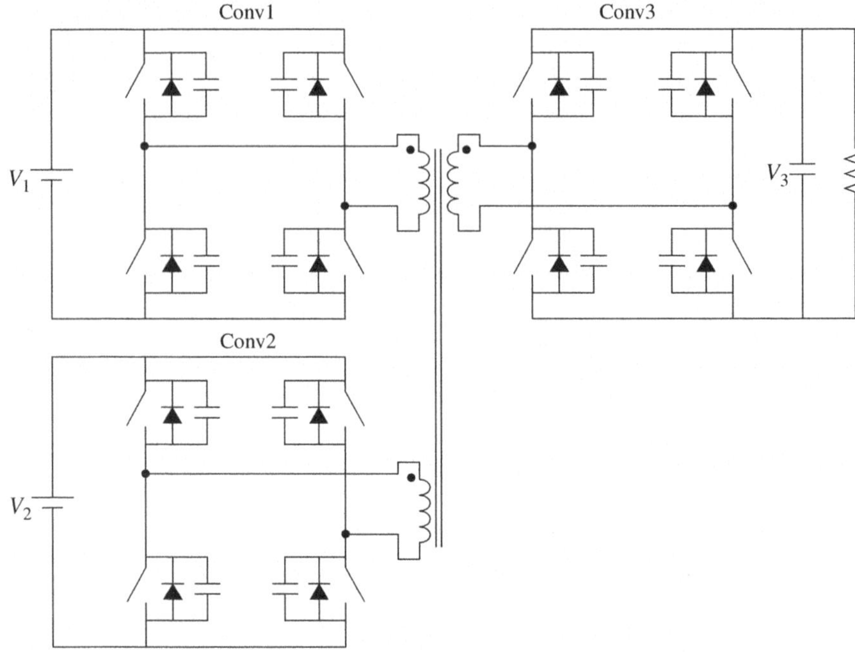

Figure 17.14 Multi-port converter with three active bridges.

complicated version of the combined PWM/duty cycle control and phase-shift control that was described for dual active bridge converters in Section 16.4. A detailed explanation of the control of triple active bridge converters is outside the scope of this book and interested readers are referred to [25] for details.

17.5 CONCLUSION

The focus of this chapter has been on several types of power converter topologies with particular characteristics that are used in certain specialized industrial applications. When all is said and done, power electronic topologies can be considered to be like tools in a toolbox that power electronics specialists have available at their disposal. When confronted with a particular application with power converter requirements that are defined by certain specifications, they can draw upon a wealth of options and tailor their designs appropriately. The topologies that have been presented in this book represent a small fraction of the vast number of DC–DC converter topologies that have been proposed in the past and that continue to be proposed. It is hoped that readers of this book can use it as a guide or a stepping stone that will enable them to understand newer and more sophisticated converters that have been and that will be presented in the literature.

REFERENCES

1. F. Z. Peng, "Z-source inverter," *IEEE Transactions on Industry Applications*, vol. 39, no. 2, pp. 504–510, Mar./Apr. 2003.
2. D. Cao and F. Z. Peng, "A family of Z-source and quasi-Z-source DC–DC converters," *IEEE Applied Power Electronics Conference (APEC)*, 2009, Feb. 15–19, 2009, Washington DC, USA, pp. 1097–1110.
3. J. Anderson and F. Z. Peng, "Four quasi-Z-source inverters," *IEEE Power Electronics Specialists Conference (PESC)*, 2008, June 15–19, 2008, Rhodes, Greece, pp. 2743–2749.
4. O. Husev, L. Liivik, F. Blaabjerg, A. Chub, D. Vinnikov, and I. Roasto, "Galvanically isolated quasi-Z-source DC–DC converter with a novel ZVS and ZCS technique," *IEEE Transactions on Industrial Electronics*, vol. 62, no. 12, pp. 7547–7556, Dec. 2015.
5. M. H. Ahmed, F. C. Lee, and Q. Li, "Two-stage 48-V VRM with intermediate bus voltage optimization for data centers," *IEEE Journal of Emerging and Selected Topics in Power Electronics*, vol. 9, no. 1, pp. 702–715, Feb. 2021.
6. X. Zhou, P.-L. Wong, P. Xu, F. C. Lee, and A. Q. Huang, "Investigation of candidate VRM topologies for future microprocessors," *IEEE Transactions on Power Electronics*, vol. 15, no. 6, pp. 1172–1182, Nov. 2000.
7. Y. Panov and M. M. Jovanovic, "Design considerations for 12-V/1.5-V, 50-A voltage regulator modules," *IEEE Transactions on Power Electronics*, vol. 16, no. 6, pp. 776–783, Nov. 2001.
8. K. Yao, M. Ye, M. Xu, and F. C. Lee, "Tapped-inductor buck converter for high-step-down DC–DC conversion," *IEEE Transactions on Power Electronics*, vol. 20, no. 4, pp. 775–780, Jul. 2005.
9. S. Ye, W. Eberle, and Y. Liu, "A novel non-isolated full bridge topology for VRM applications," *IEEE Transactions on Power Electronics*, vol. 23, no. 1, pp. 427–437, Jan. 2008.
10. Z. Zhang, W. Eberle, Y. Liu, and P. C. Sen, "A nonisolated ZVS asymmetrical buck voltage regulator module with direct energy transfer," *IEEE Transactions on Industrial Electronics*, vol. 56, no. 8, pp. 3096–3105, Aug. 2009.
11. J. Khodabakhsh and G. Moschopoulos, "A study of T-type and ZVS-PWM full-bridge converters for switch-mode power supplies," *IEEE Transactions on Power Electronics*, vol. 35, no. 7, pp. 7145–7159, Jul. 2020.
12. M. Schweizer and J. W. Kolar, "Design and implementation of a highly efficient three-level T-type converter for low-voltage applications," *IEEE Transactions on Power Electronics*, vol. 28, no. 2, pp. 899–907, Feb. 2013.
13. D. G. Bandeira, S. A. Mussa, and I. Barbi, "A ZVS-PWM T-type isolated DC–DC converter," *IEEE Brazilian Power Electronics Conference and Southern Power Electronics Conference (COBEP/SPEC)*, 2015, Nov. 29–Dec. 2, 2015, Fortaleza, Brazil, pp. 1–6.
14. J. Khodabakhsh and G. Moschopoulos, "A T-Type DC–DC Converter with reduced middle-leg switch current stress and improved light-load efficiency," *IEEE Open Journal of Power Electronics*, vol. 1, pp. 238–249, 2020.
15. J.-G. Cho, J.-W. Baek, C.-Y. Jeong, and G.-H. Rim, "Novel zero-voltage and zero-current-switching full-bridge PWM converter using a simple auxiliary

circuit," *IEEE Transactions on Industry Applications*, vol. 35, no. 1, pp. 15–20, Jan.-Feb. 1999.

16. B. G. Dobbs and P. L. Chapman, "A multiple-input DC–DC converter topology," *IEEE Power Electronics Letters*, vol. 1, no. 1, pp. 6–9, Mar. 2003.
17. A. Kwasinski and P. T. Krein, "Multiple-input DC–DC converters to enhance local availability in grids using distributed generation resources," *IEEE Applied Power Electronics Conference (APEC)*, Feb. 25–Mar. 1, 2007, Anaheim, USA, 2007, pp. 1657–1663.
18. A. Kwasinski, "Identification of feasible topologies for multiple-input DC–DC converters," *IEEE Transactions on Power Electronics*, vol. 24, no. 3, pp. 856–861, Mar. 2009.
19. A. Khaligh, J. Cao, and Y. Lee, "A multiple-input DC–DC converter topology," *IEEE Transactions on Power Electronics*, vol. 24, no. 3, pp. 862–868, Mar. 2009.
20. Y.-C. Liu and Y.-M. Chen, "A systematic approach to synthesizing multi-input DC–DC converters," *IEEE Transactions on Power Electronics*, vol. 24, no. 1, pp. 116–127, Jan. 2009.
21. P. Jain and H. Pinheiro, "High frequency triport UPS topologies for emerging fiber networks," *IEEE International Telecommunications Energy Conference (INTELEC)*, 1998, Oct. 4–8, 1998, San Francisco, USA, pp. 505–512.
22. J.-Y. Lee, Y.-M. Chang, and F.-Y. Liu, "A new UPS topology employing a PFC boost rectifier cascaded high-frequency tri-port converter," *IEEE Transactions on Industrial Electronics*, vol. 46, no. 4, pp. 803–813, Aug. 1999.
23. H. Al-Atrash, M. Pepper, and I. Batarseh, "A zero-voltage switching three-port isolated full-bridge converter," *IEEE International Telecommunications Energy Conference (INTELEC)*, 2006, Sept. 10–14, 2006, Providence, USA, pp. 1–8.
24. H. Al-Atrash and I. Batarseh, "Boost-integrated phase-shift full-bridge converter for three-port interface," *IEEE Power Electronics Specialists Conference (PESC)*, 2007, June 17–21, 2007, Orlando, USA, pp. 2313–2321.
25. C. Zhao, S. D. Round, and J. W. Kolar, "An isolated three-port bidirectional DC–DC converter with decoupled power flow management," *IEEE Transactions on Power Electronics*, vol. 23, no. 5, pp. 2443–2453, Sept. 2008.
26. M. Neubert, A. Gorodnichev, J. Gottschlich, and R. W. De Doncker, "Performance analysis of a triple-active bridge converter for interconnection of future DC-grids," *IEEE Energy Conversion Congress and Exposition (ECCE)*, 2016, Sept. 18–22, Milwaukee, USA, pp. 1–8.

APPENDIX

This appendix is divided into two sections: the first presents some steady-state analysis of the buck, boost, and buck–boost converters working in continuous conduction mode (CCM), while the second presents some analysis of the demagnetization of the transformer core of a forward converter. The purpose of the analysis presented here is to illustrate certain fundamental concepts that are related to these converters and to give examples of analysis.

A.1 EXAMPLE STEADY-STATE ANALYSIS OF BUCK, BOOST, AND BUCK–BOOST CONVERTERS OPERATING IN CONTINUOUS CONDUCTION MODE

The steady-state analysis that is performed in this section is to determine three key parameters that are related to the CCM operation of the buck, boost, and buck–boost converters: output voltage, peak converter current, and the critical inductance value that ensures that the converter operates in CCM as opposed to discontinuous conduction mode (DCM). This analysis is performed for each converter and explained in detail.

A.1.1 Buck Converter

The output voltage of a buck converter (Figure A.1a) that is working in CCM can be derived by simple integration, as was shown in Chapter 1. A more general

DC–DC Converter Topologies: Basic to Advanced, First Edition. Gerry Moschopoulos.

Published 2024 by John Wiley & Sons, Inc.

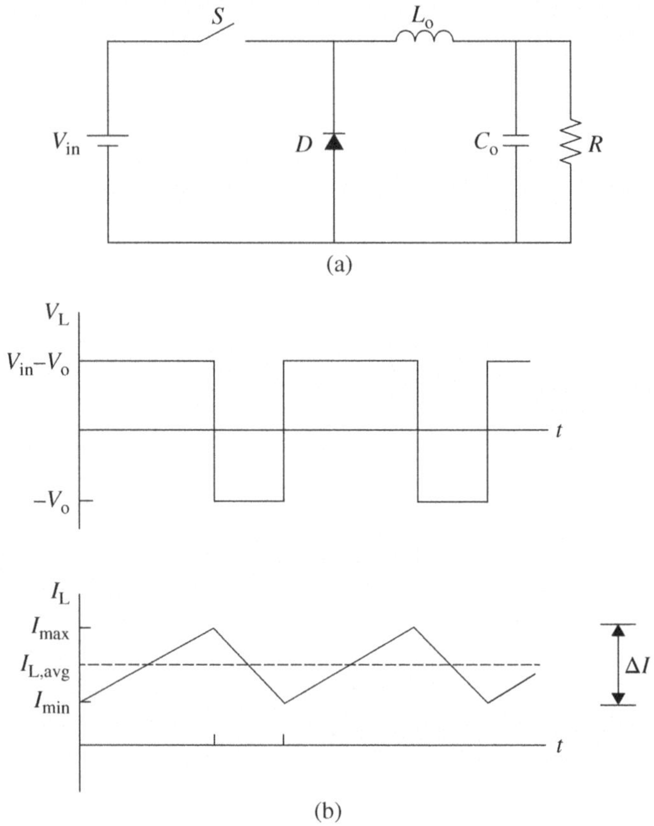

Figure A.1 Buck converter: (a) Topology. (b) Inductor voltage and current waveforms for CCM operation.

way of deriving the output voltage of any of the six fundamental single-switch PWM converters that were introduced in Chapter 1 is to use the voltage across the main inductor or the current through the main capacitor and take advantage of the fact that the average value of both must be zero over a switching period when a converter is operating under steady-state conditions. Since it is often easier to use inductor voltage rather than capacitor current, the output voltage of a buck converter operating in CCM is derived in this manner here.

The first step is to determine the average values of the positive voltage and negative voltage across the output buck inductor for a switching cycle, which is shown in Figure A.1b and equate these two values as they must be equal. This can be done by integration, but since integration is related to area and the output inductor voltage is a simple square wave, the area of the positive portion of the inductor voltage waveform can be equated to that of the negative portion as follows:

$$\left(\frac{V_{\text{in}} - V_{\text{o}}}{L}\right)(DT) - \left(\frac{V_{\text{o}}}{L}\right)(1 - D)T = 0 \tag{A.1}$$

where V_{in} is the input voltage, V_o is the output voltage, D is the duty-cycle, T is the switching period, and L is the value of the inductor. If L and T are cancelled in this equation and what is left is rearranged, then the following expression for output voltage can be obtained:

$$V_o = V_{in}D \tag{A.2}$$

which is the same as Eq. (1.3), derived in Chapter 1.

Two important parameters for a buck converter operating in CCM are the peak converter current and the critical inductance value that ensures that the converter operates in CCM instead of DCM. The first is needed to determine peak component current stress while the second is needed as part of the design of the converter. Both parameters can be determined from the output inductor current waveform that is shown in Figure A.1b.

From Figure A.1b, it can be seen that the peak value of the output inductor current waveform can be expressed as

$$I_{max} = I_{L,avg} + \frac{\Delta i_L}{2} \tag{A.3}$$

where $I_{L,avg}$ is the average value of the waveform. Given that the inductor current is split between the output capacitor and the load resistor R and that the average value of the output capacitor current over a switching period must be zero, the average value of the output inductor current must be equal to the output current I_o, which is purely DC, so that

$$I_{L,avg} = I_o = \frac{V_o}{R} \tag{A.4}$$

which takes care of half of Eq. (A.3). As for the other half, this can be determined by deriving an expression for Δi_L then dividing the result by two. Δi_L can be determined either by using the positive slope or the negative slope of the output inductor waveform. It is the negative slope of the waveform that will be used as $I_{L,avg}$ contains only V_o, as does the term for the negative slope of the waveform; a voltage of $-V_o$ is impressed across the inductor and the current falls when the switch is off. As a result, the following expression can be determined for the peak current value I_{max} by considering that period T and switching frequency f are inversely related:

$$I_{max} = \frac{V_o}{R} + \frac{1}{2}\left[\frac{V_o}{L}(1-D)T\right] = V_o\left(\frac{1}{R} + \frac{1-D}{2Lf}\right) \tag{A.5}$$

The minimum value of the output inductor current waveform can be expressed as

$$I_{min} = I_{L,avg} - \frac{\Delta i_L}{2} = \frac{V_o}{R} - \frac{1}{2}\left[\frac{V_o}{L}(1-D)T\right] = V_o\left(\frac{1}{R} - \frac{1-D}{2Lf}\right) \tag{A.6}$$

In order for a buck converter to operate in CCM, this value must be positive. If it is negative, then the converter cannot operate in CCM as this would mean that the current would have to flow in the reverse direction, back to the source, during part of a switching cycle. As this is something that is not possible for a buck converter operating in the standard manner, a negative value of I_{min} would mean that the converter is actually working in DCM, which would invalidate the CCM analysis.

The boundary condition between CCM and DCM operation occurs when I_{min} is zero, which can be expressed as follows:

$$I_{min} = 0 = V_o\left(\frac{1}{R} - \frac{1-D}{2Lf}\right) \tag{A.7}$$

Rearranging this expression results in the following equation for the critical inductance at the boundary of CCM and DCM operation:

$$L_{min} = \frac{(1-D)R}{2f} \tag{A.8}$$

There are a couple of things that should be noted about this equation:

- The critical inductance L_{min} must be increased if the switching frequency is decreased. A lower switching frequency means that the output inductor current has more time to vary and thus have a larger ripple. Increasing the switching frequency means that the value of L_{min} can be decreased and this is an argument for operating with higher switching frequencies as long as measures are taken to reduce switching losses, something that is discussed throughout this book.
- The critical inductance L_{min} must be increased if the load resistance R is increased. Increasing R will not affect the general shape of the output current waveform if the converter is operating in CCM as all AC ripple flows through the capacitor, assuming that C is infinitely large. Increasing R, however, will decrease the average value of the output inductor current, according to Eq. (A.4). A lower average value $I_{L,avg}$ for this current means that less AC ripple can be allowed if the converter is to be operated in CCM, which means that the value of L_{min} must be increased. A high value of R corresponds to light-load operation so that it is unavoidable for the converter to operate in DCM under light-load conditions unless the inductor is made to be very large, which would be impractical. Since there is little current in the converter when it is operating with light loads, any issues related to high peak currents that may occur when operating with DCM such as high peak currents for heavy-load operation can be avoided. As a result, the output inductor of a buck converter is designed so that

the converter operates in CCM with heavy loads and DCM with light loads with the boundary set anywhere between 10% and 30% of full load, depending on specifications.

A.1.2 Boost Converter

As with the buck converter, equations for output voltage, peak converter current, and the critical inductance for CCM operation can be derived for a boost converter (Figure A.2a) operating in CCM. This can be done by using the same general approach, but given that the two converters have different topologies and operate differently, there are some differences in the derivation of the equations and these are discussed here, in this section.

The output voltage of a boost converter operating in CCM can be determined by determining the average values of the positive and negative

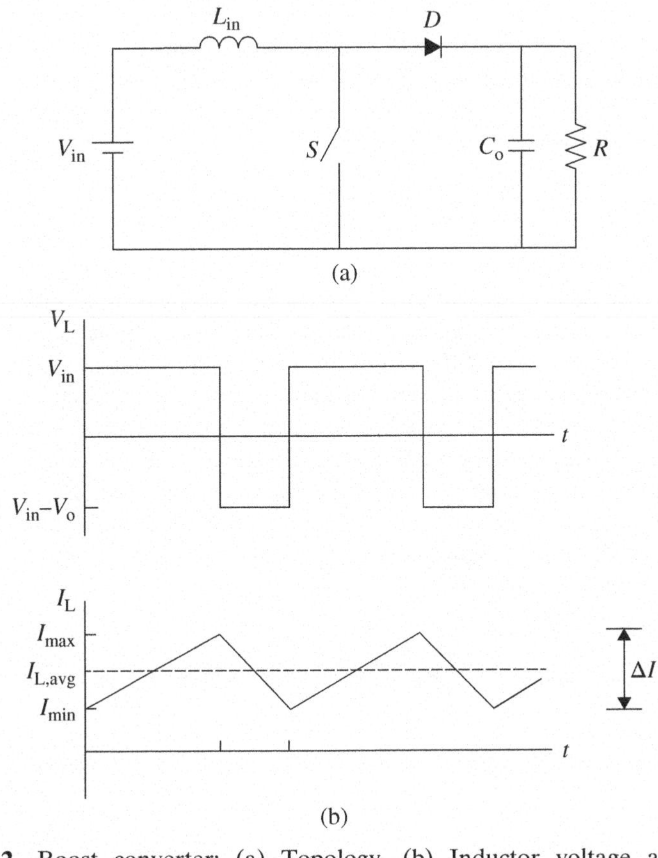

Figure A.2 Boost converter: (a) Topology. (b) Inductor voltage and current waveforms for CCM operation.

portions of the input inductor voltage, shown in Figure A.2b, and equating the two. Given that these portions can be related to the area, as was shown earlier with the buck converter, the following expression equating the areas of the positive and negative portions of the input inductor voltage waveform can be written:

$$\frac{V_{in}DT}{L} + \frac{(V_{in} - V_o)(1 - D)T}{L} = 0 \quad \text{(A.9)}$$

Cancelling L and T from this expression results in

$$V_L = V_{in}D + (V_{in} - V_o)(1 - D) = 0 \quad \text{(A.10)}$$

which can be expanded and ultimately expressed as

$$V_o = \frac{V_{in}}{1 - D} \quad \text{(A.11)}$$

The peak converter current I_{max} and critical inductance value L_{min} for CCM operation can be determined from the inductor waveform shown in Figure A.2b, as was the case for the buck converter in the previous section. Although I_{max} can be expressed as

$$I_{max} = I_{L,avg} + \frac{\Delta I}{2} \quad \text{(A.12)}$$

where ΔI is the peak-to-peak ripple, the average value of the inductor current $I_{L,avg}$ cannot be determined from the output DC load current as the inductor is located on the other side of the circuit. $I_{L,avg}$ must be determined by using the average value of the input current instead. Given that only harmonics of the same frequency contribute to real power and the input source voltage is pure DC, the only current harmonic that contributes to real power is the DC harmonic or the average component of the input current, which is $I_{L,avg}$ as the inductor is connected to the source. If the output power P_o is expressed as the square of the output voltage over the load resistance, then the following equation can be written:

$$V_{in}I_{L,avg} = \frac{V_o^2}{R} = \frac{[V_{in}/(1 - D)]^2}{R} = \frac{V_{in}^2}{(1 - D)^2}R \quad \text{(A.13)}$$

Rearranging this equation results in the following expression for $I_{L,avg}$:

$$I_{L,avg} = \frac{V_{in}}{(1 - D)^2 R} = \frac{V_o^2}{V_{in}R} = \frac{V_o I_o}{V_{in}} \quad \text{(A.14)}$$

An expression for ΔI can be determined either by using either the positive slope or the negative slope of the input inductor waveform. Since $I_{L,avg}$ contains only V_o as does the positive slope of the waveform as a voltage of V_s is impressed across the inductor and current rises when the switch is on, it is the positive slope of the waveform that will be used, unlike the case for the buck converter where it was the negative slope that was used. As a result, the following expression can be determined for the peak current value I_{max} by considering that period T and switching frequency f are inversely related:

$$I_{max} = I_{L,avg} + \frac{\Delta i_L}{2} = \frac{V_{in}}{(1-D)^2 R} + \frac{V_{in} DT}{2L} \tag{A.15}$$

The minimum value of the input inductor current waveform can be expressed as

$$I_{min} = I_L - \frac{\Delta i_L}{2} = \frac{V_o}{R} - \frac{1}{2}\left[\frac{V_o}{L}(1-D)T\right] = V_o\left(\frac{1}{R} - \frac{1-D}{2Lf}\right) \tag{A.16}$$

As with the buck converter, a negative value of I_{min} is an indication that the converter is actually operating in DCM, which would invalidate the CCM analysis. The boundary condition between CCM and DCM operation occurs when I_{min} is zero, which can be expressed as follows:

$$I_{min} = 0 = \frac{V_{in}}{(1-D)^2 R} - \frac{V_{in} DT}{2L} \tag{A.17}$$

Rearranging this expression results in the following equation for the critical inductance at the boundary of CCM and DCM operation:

$$L_{min} = \frac{D(1-D)^2 R}{2f} \tag{A.18}$$

A.1.3 Buck–Boost Converter

Equations for output voltage, peak converter current, and the critical inductance for CCM operation can be derived for a buck–boost converter operating (Figure A.3a) in CCM using the same general approach as was done for the buck and boost converters. There are, however, two key differences that need to be considered in the analysis of the buck–boost converter:

- The first is that the inductor is not at the output of the converter, as is the case with the buck converter, nor is it at the input of the converter, as is the case with the boost converter, but it is, instead in the middle of the

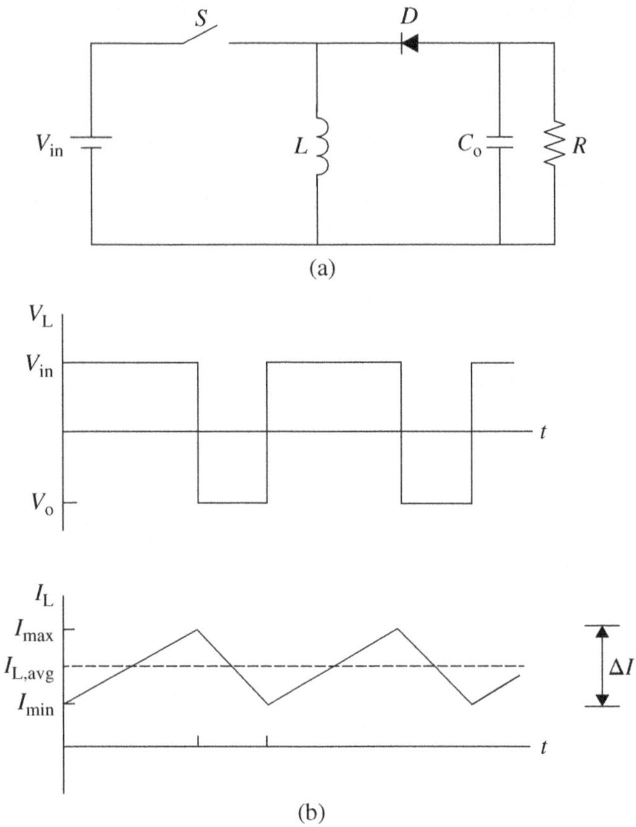

Figure A.3 Buck–Boost converter: (a) Topology. (b) Inductor voltage and current waveforms for CCM operation.

converter. This complicates the determination of the average inductor current value as there is no simple equation that can relate it to load current or input power.

- The second difference is that care must be taken in dealing with the polarity of the output voltage as it is different from that of the input voltage.

As with the other two converters, the output voltage of the converter voltage gain of the buck–boost converter can be determined from the voltage across the inductor (Figure A.3b) by equating its positive and negative areas. Doing so results in the following equation:

$$\frac{V_{in}DT}{L} + \frac{V_o(1-D)T}{L} = 0 \tag{A.19}$$

It should be noted that the two terms of the equation are added to each other instead of subtracted due to the polarity of V_o, as discussed earlier. Rearranging this equation gives

$$V_o = -V_{in}\left(\frac{D}{1-D}\right) \tag{A.20}$$

The average inductor current value can be found using the same procedure as with the boost converter if a relation between the average value of the input current and the average value of the inductor current can be established. If this can be done, then all that needs to be done is to substitute the input current in the boost converter analysis with the inductor current and proceed accordingly.

The input inductor current waveform is not a continuous waveform as current flows out of the input source only when the switch is on, which is for a duration of DT. The average value of the input current $I_{in,avg}$ is therefore related by a factor of D and can thus be expressed as follows:

$$I_{in,avg} = I_{L,avg}D \tag{A.21}$$

If this expression of $I_{in,avg}$ is substituted into the equations that were previously derived for the boost converter, then the following expression can be obtained:

$$I_{L,avg} = \frac{V_o^2}{V_{in}RD} = \frac{P_o}{V_{in}D} = \frac{V_{in}D}{R(1-D)^2} \tag{A.22}$$

$$I_{max} = I_{L,avg} + \frac{\Delta I}{2} = \frac{V_{in}D}{R(1-D)^2} + \frac{V_{in}DT}{2L} \tag{A.23}$$

$$I_{min} = I_{L,avg} - \frac{\Delta I}{2} = \frac{V_{in}D}{R(1-D)^2} - \frac{V_{in}DT}{2L} \tag{A.24}$$

$$L_{min} = \frac{(1-D)^2 R}{2f} \tag{A.25}$$

A.2 ANALYSIS OF TRANSFORMER CORE DEMAGNETIZATION OF A FORWARD CONVERTER

Ensuring that the transformer core of a fundamentally isolated converter such as a flyback converter or a forward converter (Figure A.4a) does not saturate is key to the operation of these converters. The analysis presented in this

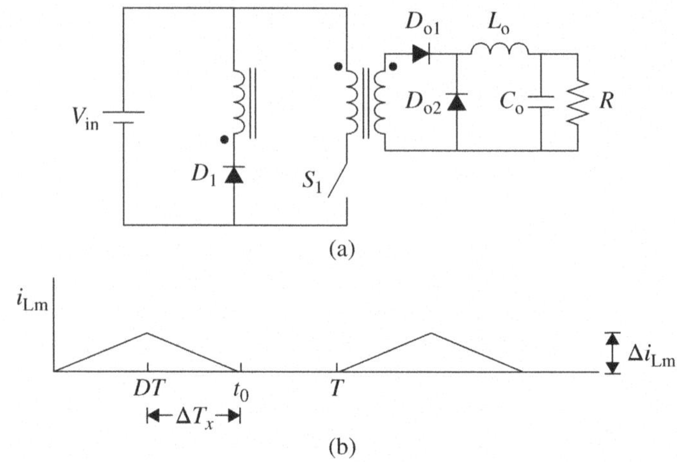

Figure A.4 Forward converter: (a) Topology. (b) Magnetizing current waveform for full core demagnetization.

section is for a forward converter and depends on the voltage that is stepped from the tertiary winding. This analysis can be adapted for a flyback converter simply by using the voltage that is stepped from the secondary winding instead.

Looking at the waveform for the magnetizing current i_{Lm} shown in Figure A.4b, it can be seen that the current rises during the first part of the switching cycle when the switch is on and falls during the next part when the switch is off. Since magnetic current is related to flux, the value of i_{Lm} at the end of a switching cycle must be the same as that at the start of the cycle, if the converter operates under steady-state conditions. If the end value is greater than the start value, this means that the forward converter transformer has a net gain of flux. If this flux gain is repeated over the next switching cycles, then the transformer will become saturated, which will degrade the performance of the converter and may result in eventual catastrophic failure.

Although a forward converter can be designed to operate with continuous magnetizing current, it is more commonly designed to operate with discontinuous magnetizing current, which means that the transformer core is fully demagnetized before the start of each switching cycle. This is done to minimize any possibility of saturation. The following procedure demonstrates how this can be done.

Assuming that the forward converter in this example is designed to operate with its output inductor current in CCM, the output voltage is simply the product of the buck converter output voltage equation and the transformer turns ratio (N_2 over N_1, the ratio of the number of secondary turns to the number of primary turns). It can be written as

$$V_o = V_{in} D \frac{N_2}{N_1} \tag{A.26}$$

When the switch is turned off, the reset voltage is applied across the tertiary winding so that the following negative voltage is applied to the magnetizing inductance:

$$v_{L_m} = v_1 = V_{in}\left(\frac{N_1}{N_3}\right) = L_m \frac{di_{L_m}}{dt} \tag{A.27}$$

where N_3 is the number of tertiary winding turns. In order for the transformer to be fully demagnetized, the magnetizing current must fall to zero at some time to, as shown in Figure A.4b, a sometime ΔT_x after the switch has been turned off. Since the positive and negative slopes of the current are

$$\Delta i_{L_m} = \frac{V_{in}DT}{L_m} = \frac{V_{in}}{L_m}\left(\frac{N_1}{N_3}\right)\Delta T_x \tag{A.28}$$

$$\Delta T_x = DT\left(\frac{N_3}{N_1}\right) \tag{A.29}$$

$$t_o = DT + \Delta T_x = DT + DT\left(\frac{N_3}{N_1}\right) = DT\left(1 + \frac{N_3}{N_1}\right) \tag{A.30}$$

This must be less than the period T, as expressed in the following manner:

$$DT\left(1 + \frac{N_3}{N_1}\right) < T \tag{A.31}$$

which can be rewritten as

$$D\left(1 + \frac{N_3}{N_1}\right) < 1 \tag{A.32}$$

If the number of turns of the tertiary winding is designed to be equal to those of the primary winding so that $N_3 = N_1$, which is a typical forward converter design, then the maximum duty cycle that the converter can operate with is $D = 0.5$. The duty cycle in practical forward converters is limited to values that are slightly below 0.5 to provide some margin. If it is desired to operate a forward converter with duty cycles that are greater than 0.5, then according to Eq. (A.32), this can be done by decreasing the value of N_3. If this is done, however, then the voltage stress of the switch, which can be expressed as

$$v_{sw} = \begin{cases} V_{in}\left(1 + \dfrac{N_1}{N_3}\right) & \text{for } DT < t < t_o \\ V_{in} & \text{for } t_o < t < T \end{cases} \tag{A.33}$$

will increase as more voltage must be applied to the transformer to demagnetize it in less time as increasing D decreases $(1 - D)$. This voltage, which is added to the source voltage, can become excessive if D is made significantly greater than 0.5. As a result, an acceptable compromise in forward converter designs is to limit D to 0.5 and keep the peak switch voltage stress at twice the input voltage. This is done by making N_3 equal to N_1.

INDEX

DC–DC Converter Topologies: Basic to Advanced, First Edition. Gerry Moschopoulos.

Published 2024 by John Wiley & Sons, Inc.

Wiley The manufacturer's authorized representative according to the EU General Product Safety Regulation is Wiley-VCH GmbH, Boschstr. 12, 69469 Weinheim, Germany, e-mail: Product_Safety@wiley.com.

Printed and bound by CPI Group (UK) Ltd, Croydon, CR0 4YY

16/02/2026

02054736-0001